Colloids and Interface Science Series
Volume 1

Colloid Stability

Edited by
Tharwat F. Tadros

1807–2007 Knowledge for Generations

Each generation has its unique needs and aspirations. When Charles Wiley first opened his small printing shop in lower Manhattan in 1807, it was a generation of boundless potential searching for an identity. And we were there, helping to define a new American literary tradition. Over half a century later, in the midst of the Second Industrial Revolution, it was a generation focused on building the future. Once again, we were there, supplying the critical scientific, technical, and engineering knowledge that helped frame the world. Throughout the 20th Century, and into the new millennium, nations began to reach out beyond their own borders and a new international community was born. Wiley was there, expanding its operations around the world to enable a global exchange of ideas, opinions, and know-how.

For 200 years, Wiley has been an integral part of each generation's journey, enabling the flow of information and understanding necessary to meet their needs and fulfill their aspirations. Today, bold new technologies are changing the way we live and learn. Wiley will be there, providing you the must-have knowledge you need to imagine new worlds, new possibilities, and new opportunities.

Generations come and go, but you can always count on Wiley to provide you the knowledge you need, when and where you need it!

William J. Pesce
President and Chief Executive Officer

Peter Booth Wiley
Chairman of the Board

Colloids and Interface Science Series
Volume 1

Colloid Stability

The Role of Surface Forces – Part I

Edited by
Tharwat F. Tadros

BICENTENNIAL 1807 WILEY 2007 BICENTENNIAL

WILEY-VCH Verlag GmbH & Co. KGaA

The Editor

Prof. Dr. Tharwat F. Tadros
89 Nash Grove Lane
Wokingham, Berkshire RG40 4HE
Great Britain

Library of Congress Card No.: applied for

British Library Cataloguing-in-Publication Data
A catalogue record for this book is available from the British Library

Bibliographic information published by the Deutsche Nationalbibliothek
The Deutsche Nationalbibliothek lists this publication in the Deutsche Nationalbibliografie; detailed bibliographic data are available in the Internet at http://dnb.d-nb.de

Printed in the Federal Republic of Germany
Printed on acid-free paper

Composition K+V Fotosatz GmbH, Beerfelden
Printing betz-druck GmbH, Darmstadt
Bookbinding Litges & Dopf Buchbinderei GmbH, Heppenheim

ISBN 978-3-527-31462-1

Contents

Colloids and Interface Science Series, Vol. 1
Colloid Stability: The Role of Surface Forces, Part I. Edited by Tharwat F. Tadros

ISBN: 978-3-527-31462-1

Preface

This is a new series of reviews that are aimed at identifying the role of colloid and interface science in various fields. The first two volumes describe some aspects of colloid stability with special reference to the role of surface forces. Several reviews with different scopes are written by leading scientists from all over the world. They cover topics such as the thermodynamic criteria of colloid stability, the role of surface forces, hydrophobic interaction, long-range forces, nanoparticles, colloid stability using polymeric surfactants, etc. A great deal of emphasis is given to foam and emulsion films, which are used fundamentally to investigate the role of surface forces in the stabilization of such films. Some other aspects covered are wetting films, both static and dynamic, and emulsion stability. The reviews are not given in any specific order and they are published on the basis of the order of receiving the manuscripts. These reviews are comprehensive, with many references, and they should be extremely useful for those engaged in fundamental studies of colloid stability and the role of surface forces both in academia and in industry. The first two volumes are dedicated to Professors Dotchi Exerowa and Dimo Platikanov (on the occasion of their 70th birthdays), both from the famous school of colloid science that was led by the late Professor A. Scheludko in Bulgaria. Photographs, biographies and lists of selected papers published by both scientists are given directly after this preface.

I would like to thank all the authors for their dedication in producing these excellent reviews, which made my editing task fairly easy. I would like also to thank the staff of Wiley-VCH for producing these two volumes quickly.

Wokingham, October 2006

Tharwat Tadros
Editor of the Series

Colloids and Interface Science Series, Vol. 1
Colloid Stability: The Role of Surface Forces, Part I. Edited by Tharwat F. Tadros

ISBN: 978-3-527-31462-1

Professor Dotchi Exerowa, DrSc, Academician

Dotchi Exerowa was born on 20 May 1935 in Varna, Bulgaria. After finishing secondary school in Sofia (1953), she studied chemistry at the University of Sofia. In 1958 she obtained the degree of Diploma-Chemist (equal to an MSc). Her thesis was in the area of colloid science, carried out in the Department of Physical Chemistry. Her scientific career began at the Institute of Physical Chemistry of the Bulgarian Academy of Sciences, which has continued to be her workplace up to the present. As a junior research associate she completed her PhD thesis in 1969, supervised by the great Bulgarian colloid scientist Professor A. Scheludko. She advanced her research work and in 1972 was awarded Habilitation, which entitled her to a senior position at the Institute. In 1983 she succeeded Professor Scheludko as Head of the Department of Colloid and Interface Science. Her DrSc thesis was successfully defended in 1987 and a year later she became Professor in Physical Chemistry. Parallel to her main duties at the Institute of Physical Chemistry she has lectured on Physicochemical Methods in Biology, being a Visiting Professor at the University of Sofia. She has also supervised many PhD and MSc students and postdoctoral fellows, thus bringing many young people into scientific research.

Dotchi Exerowa has published about 200 papers in the scientific literature. In addition, she has written a monograph, *Foam and Foam Films*, co-authored with Peter Kruglyakov (published by Elsevier, Amsterdam, 1998). Owing to her extensive publications, Dotchi Exerowa has become internationally well known in the field of colloid and interface science. She has been invited many times to lecture at international conferences and seminars in leading scientific institutions. She has been a member of many scientific committees of conferences and an editorial board member of four international scientific journals. In 1997 she was co-chairman of the 9th International Conference on Surface and Colloid Science. She was twice elected a member of the Council of the International Association of Colloid and Interface Scientists. In 2004, Professor Exerowa was elected to the Bulgarian Academy of Sciences, receiving the title Academician.

The contributions of Dotchi Exerowa are mainly in the field of thin liquid films, surfactants, foams, liquid interfaces, lung surfactant systems, etc. Many of the results obtained were on aspects that have stimulated new directions in the development of knowledge in the field of thin liquid films and also the physics and chemistry of interfacial phenomena.

In the 1960s, Dotchi Exerowa, together with her teacher Professor Scheludko, developed a unique experimental method for the study of thin liquid films based on the very useful model of a microscopic (radius ca. 100 μm) horizontal thin liquid film. This allowed the measurement of important parameters characterizing their properties: equilibrium thickness, critical thickness of rupture, disjoining pressure, contact angle film/bulk liquid, etc. The method and equipment for microscopic thin liquid film investigations are known as the Scheludko–Exerowa micro-interferometric technique. Special attention has been given to the direct measurement of the interaction forces in microscopic liquid films, the disjoining pressure/thickness isotherm and the transition from long- to short-range molecular interactions. This is now referred to as the thin liquid film pressure balance technique, and is widely used in many laboratories all over the world.

Extensive studies on surface forces in thin liquid films have been performed. A quantitative assessment of the main theory of colloid stability, the DLVO theory, was made. A new vision of the electrostatic interactions in liquid films has been developed. For the first time, the values of the diffuse electric layer potential at the liquid/air interface and the isoelectric points at this interface have been determined. These parameters are very informative for gaining an understanding of the charge nature and the electrostatic interaction, respectively. Barrier and non-barrier transitions in the disjoining pressure isotherm of foam films from liposome suspensions of phospholipids have been obtained experimentally.

DLVO and non-DLVO surface forces in liquid films from amphiphilic block copolymers (PEO–PPO–PEO type) have been determined. The transition from electrostatic to steric stabilization has been elucidated by determination of the critical electrolyte concentration, which divides the two types interactions. It was found that the electrostatic repulsion arises from the charge at the water/air interface due to preferential adsorption of OH^- ions. For the non-DLVO surface forces, brush-to-brush contact was established and the disjoining pressure isotherm followed the de Gennes scaling theory.

A new approach to amphiphile bilayers, the thinnest Newton black films, has been developed. A microscopic theory of the formation and stability of amphiphile bilayers was created. The rupture of bilayers was considered on the basis of a fluctuation mechanism of formation of nanoscopic holes in the bilayers. The hole formation was treated as a nucleation process of a new phase in a two-dimensional system with short-range intermolecular forces. Free rupture and deliberate rupture (by α-particles) of bilayers have been described. The role of bulk surfactant concentration for the formation and stability of amphiphile bilayers was demonstrated. A number of important parameters, the binding energy of an amphiphile molecule in the bilayer and the specific hole linear energy, which are important characteristics of the short-range surface forces in bilayers, were determined. Also for the first time the equilibrium surfactant concentration has been found, at which the bilayer (in contact with the bulk phase) is thermodynamically stable. In that way the ruptured, the metastable and the stable amphiphile bilayers can be clearly differentiated.

A two-dimensional chain-melting phase transition in foam bilayers was established for the first time. The binding energy of two neighboring phospholipid molecules was determined for the gel and liquid crystalline state of the bilayers from several phospholipids. It is to be expected that foam bilayers from phospholipids could be used as a model for the investigation of short-range forces in biological structures, of interactions between membranes, etc.

A new theoretical vision of polyhedric foams has been developed. It was verified experimentally for solutions of different surfactants, amphiphilic polymers and natural and technological mixtures. Methods for the differentiation of the processes connected with the syneresis (drainage) and stability of foams by creating a pressure gradient in the liquid phase have been developed. In that way the processes and factors acting in liquid channels and foam films are distinguished. On this basis, new methods and equipment have been developed, for foam stability determination at constant capillary pressure (foam pressure drop technique), rapid foam rupture, effective foam concentration and separation, water purification from surfactants, foam elimination in waste materials of nuclear fuel processing, effective foam formation at surfactants with high self-stabilizing ability during oil recovery, etc.

The lung surfactant system has been studied on the basis of theoretical and experimental investigations of amphiphile bilayers from amniotic fluid and alveolar surfactant. A new *in vitro* model for studying alveolar surface and stability, namely the microscopic foam film, has been introduced under the conditions in the lung alveolus: capillary pressure, radius, electrolyte concentration and temperature. It was shown that under these conditions, a foam bilayer stabilized by short-range interaction forces was formed and new parameters characterizing its formation and stability were introduced. On this basis, new clinical methods for the diagnosis of lung maturity and assessment of surfactant lung maturity of newborns have been created. The very good fit of the clinical results and the parameters of the *in vitro* model have allowed a new hypothesis to be created for the structure of the alveoli, namely an ordered structure in contrast to the widely accepted "monolayer" model. The most significant feature of the offered new model for the alveolar structure is that its stability is determined by the lateral short-range interactions in addition to the normal interactions between the ordered molecules. This gives the possibility of the quantitative study of the lung surfactant system and the processes related to the main physiological process – breathing.

The newly created method for fetal lung maturity assessment has a number of advantages: high precision (90%), a small quantity of liquid used (1 cm^3) and speed – the result is ready in about 20 minutes. The method for lung maturity diagnosis has also been very successfully developed for therapy control, i.e. defining the action of therapeutic surfactants, which cure the respiratory distress syndrome. This creates the possibility of looking for the most effective medicines to influence the lung surfactant system in respiratory distress.

From the above description, it is clear that Professor Exerowa has made significant original contributions in the field of colloid and interface science, for which she has been awarded the highest possible scientific position in Bulgaria, namely an Academician. A list of her most important publications is provided.

Selected Publications

1 Foam and Foam Films. Monograph in the Series *Studies in Interface Science.* D. Exerowa, P. M. Kruglyakov. Elsevier Science, Amsterdam, 1998, pp. 796.

2 Über den elektostatischen und van der Waalsschen zusätzlichen Druck in wässeringen Schaumfilmen. A. Scheludko, D. Exerowa, *Kolloid-Z.*, *168* **(1960)** 24.

3 Effect of Adsorption, Ionic Strength and pH on the Potential of the Diffuse Electric Layer. D. Exerowa, *Kolloid-Z.*, *232* **(1969)** 703.

4 Some Techniques for the Investigation of Foam Stability. D. Exerowa, Kh. Khristov, I. Penev, in: *Foams* (R. Akers, Ed.), Academic Press, London, 1976, p. 109.

5 Influence of the Pressure in the Plateau–Gibbs Borders on the Drainage and the Foam Stability. Khr. Khristov, P. M. Kruglyakov, D. Exerowa, *Colloid Polym. Sci.*, *257* **(1979)** 506.

6 Nucleation Mechanism of Rupture of Newtonian Black Films. I. Theory. D. Kashchiev, D. Exerowa, *J. Colloid Interface Sci.*, *77* **(1980)** 501.

7 Common Black and Newton Film Formation. D. Exerowa, A. Nikolov, M. Zacharieva, *J. Colloid Interface Sci.*, *81* **(1981)** 419.

8 Newtonian Black Films Rupture by Irradiation with α-Particles. I. Stochastic Model of the Phenomenon. I. Penev, D. Exerowa, *J. Colloid Interface Sci.*, *87* **(1982)** 5.

9 Influence of the Type of Foam Films and the Type of Surfactants on Foam Stability. Khr. Khristov, D. Exerowa, P. M. Kruglyakov, *Colloid Polym. Sci.*, *261* **(1983)** 265.

10 Nucleation Mechanism of Rupture of Newtonian Black Films. II. Experiment. D. Exerowa, B. Balinov, D. Kashchiev, *J. Colloid Interface Sci.*, *94* **(1983)** 45.

11 Bilayer Lipid Membrane Permeation and Rupture Due to Hole Formation. D. Kashchiev, D. Exerowa, *Biochim. Biophys. Acta*, *732* **(1983)** 133.

12 Hole-mediated Stability and Permeability of Bilayers. D. Exerowa, D. Kashchiev, *Contemp. Phys.*, *27* **(1986)** 429.

13 Method for Assessment of Fetal Lung Maturity. D. Exerowa, Zdr. Lalchev, B. Marinov, K. Ognianov, *Langmuir*, *2* **(1986)** 664.

14 Bilayer and Multilayer Foam Films – Model for Study of the Alveolar Surface and Stability. D. Exerowa, Zdr. Lalchev, *Langmuir*, *2* **(1986)** 668.

15 Direct Measurement of Disjoining Pressure in Black Foam Films. I. Films from an Ionic Surfactant. D. Exerowa, T. Kolarov, Khr. Khristov, *Colloids Surf.*, *22* **(1987)** 171.

16 Newtonian Black Films Stabilized with Insoluble Monolayers Obtained by Adsorption from the Gas Phase. D. Exerowa, R. Cohen, A. Nikolova, *Colloids Surf.*, *24* **(1987)** 43.

17 Stability and Permeability of Bilayers. D. Exerowa, D. Kashchiev, D. Platikanov, *Adv. Colloid Interface Sci.*, *40* **(1992)** 201.

18 Phase Transitions in Phospholipid Foam Bilayers. D. Exerowa, A. Nikolova, *Langmuir*, *8* **(1992)** 3102.

19 Foam Bilayer from Amniotic Fluid: Formation and Phase State. A. Nikolova, D. Exerowa, *Langmuir*, *12* **(1996)** 1846.

20 DLVO and Non-DLVO Surface Forces in Foam Films from Amphiphilic Block Copolymers.

R. Sedev, D. Exerowa, *Adv. Colloid Interface Sci., 83* (**1999**) 111.

21 Structure and Surface Energy of the Surfactant Layer on the Alveolar Surface.
D. Kashchiev, D. Exerowa, *Eur. Biophys. J., 30* (**2001**) 34.

22 Chain-melting Phase Transition and Short-range Molecular Interactions in Phospholipid Foam Bilayers.
D. Exerowa, *Adv. Colloid Interface Sci., 96* (**2002**) 75.

23 Foam Films as Instrumentation in the Study of Amphiphile Self-assembly.
E. Mileva, D. Exerowa, *Adv. Colloid Interface Sci., 100–102* (**2003**) 547.

24 Foam and Wetting Films: Electrostatic and Steric Stabilization.
D. Exerowa, N. Churaev, T. Kolarov, N. E. Esipova, N. Panchev, Z. M. Zorin, *Adv. Colloid Interface Sci., 104* (**2003**) 1.

25 Thin Liquid Films from Phospholipids: Formation, Stability and Phase Transitions.
D. Exerowa, *Prog. Colloid Polym. Sci., 128* (**2004**) 135.

26 Amphiphile Bilayers from DPPC: Bilayer Lipid Membranes (BLM) and Newton Black Films (NBF).
D. Exerowa, R. Todorov, L. Nikolov, *Colloids Surf. A, 250* (**2004**) 195.

Professor Dimo Platikanov, PhD, DrSc

Dimo Platikanov was born on 2 February 1936 in Sofia, Bulgaria. After finishing secondary school in Sofia (1953), he studied chemistry at the University of Sofia. In 1958 he obtained the degree of Diploma-Chemist (equal to an MSc). His thesis was in the area of colloid science, performed in the Department of Physical Chemistry. His scientific career began at the Department of Physical Chemistry of the University of Sofia, which has continued to be his workplace up to the present. As a Junior Assistant Professor he completed his PhD thesis in 1968, supervised by the great Bulgarian colloid scientist Professor A. Scheludko. He became Associate Professor in 1970. He spent the academic year 1973–74 at the University of Munich, Germany, as Alexander von Humboldt Foundation Research Fellow. In 1989 he succeeded Professor Scheludko as Head of the Department of Physical Chemistry at the University of Sofia. His DrSc thesis was successfully defended also in 1989 and a year later he became Professor in Physical Chemistry. During the last 30 years he had lectured in the main course of Physical Chemistry to students in the Faculty of Chemistry at the University of Sofia. He has also been supervisor of many PhD and MSc students and postdoctoral fellows, thus bringing many young people into scientific research.

The scientific results of Dimo Platikanov have been published in about 120 papers in the scientific literature. He also published two extensive chapters together with Dotchi Exerowa: "Thin Liquid Films", in Fundamentals of Interface and Colloid Science, edited by J. Lyklema (Elsevier, 2005), and "Symmetric Thin Liquid Films with Fluid Interfaces", in *Emulsions and Emulsion Stability*, edited by J. Sjoblom (Taylor and Francis, 2005). Owing to his original scientific work and publications, Dimo Platikanov has been invited many times to lecture at international conferences and seminars in leading scientific institutions. He has been a member of many scientific committees of conferences and an editorial board member of four international scientific journals. In 1997 he was co-chairman of the 9th International Conference on Surface and Colloid Science. He was elected a member of the Council and later President (2000–2003) of the International Association of Colloid and Interface Scientists (IACIS), and is currently a member of the Standing Committee and the Council of IACIS. In the past 15 years he had been member of the Standing Committee of the European Chemistry at Interfaces Conferences and since 2004 he has been member of the Physical and Biophysical Chemistry Division Committee of IUPAC. He has also been President of the Humboldt Union in Bulgaria since 2002.

The contributions of Dimo Platikanov are mainly in the field of thin liquid films, liquid interfaces, three-phase contact, wetting, etc. The many original results obtained have stimulated new directions in the development of knowledge in the field of thin liquid films and also the physical chemistry of the interfacial phenomena.

Most of the scientific results obtained by Dimo Platikanov are experimental. In most cases a unique experimental method has been developed, especially for corresponding studies. Such methods allowed the measurement of important parameters characterizing the system studied (wetting film, foam film, black film, etc.) and its properties. About 20 unique experimental techniques have been developed in his experimental work.

An original experimental cell developed for the investigation of microscopic, circular, wetting liquid films on solid surfaces proved to be very effective and it has been used by many researchers in different countries. The Scheludko–Exerowa micro-interferometric technique has been extended in conjunction with this cell. Using this method, the shape of the three-phase contact gas/liquid film + meniscus/solid surface has been studied in detail under dynamic and equilibrium conditions. The experimental data for the development of the 'dimpling' in the initial stages of the formation of a wetting film were subsequently used for elaborating the hydrodynamic theory of this phenomenon. The equilibrium profile of the transition zone between a wetting film and the bulk liquid has been experimentally obtained and data for the 'contact thickness' have been calculated using a disjoining pressure theory.

Studies on surface forces in thin wetting films have also been performed. The disjoining pressure/thickness isotherm measured for a wetting benzene film on mercury showed a complicated character of the molecular interactions in this system. Aqueous wetting films (without or with amphiphilic PEO–PPO–PEO block copolymers added) on surfaces of fused quartz and of silicon carbide have been studied in detail. The electrolyte concentration and the solid surface pretreatment strongly influenced their stability. At electrolyte concentrations where the electrostatic disjoining pressure was fully suppressed, the disjoining pressure/thickness isotherm was measured using the dynamic method. It has been interpreted by the superposition of a negative van der Waals component and a positive steric component (due to brush-to-brush contact) of the disjoining pressure, hence the electrostatic and steric stabilization of wetting films have been distinguished. Surface forces in thin, non-aqueous foam films have also been studied. The disjoining pressure/thickness isotherms for films from benzene and chlorobenzene were measured using the dynamic method. The effect of electromagnetic retardation of the dispersion molecular interactions has been experimentally established and the Hamaker constant and London wavelength calculated.

Extensive investigations on common and Newton black foam films (CBFs and NBFs) have been performed using a number of unique experimental methods. Through deformation of a black film in electric field, the reversibility of the

CBF/NBF transition and vice versa and also the electrical neutrality of the thinnest NBFs have been proved. The measured longitudinal electrical conductivity and the transport numbers of ions in black films gave information about their structure: a three-layer structure for the CBF whereas the NBF is a bilayer of amphiphilic molecules. Other new methods allowed the measurement of the film tension and the line tension of NBFs. The film tension of NBFs from sodium dodecyl sulfate was found to be constant over wide range of static and dynamic conditions; this was not the case with NBFs from phospholipids. The values of the line tension of NBFs from sodium dodecyl sulfate have been determined – positive at low and negative at high electrolyte concentrations.

The gas permeability through foam films has been determined for several cases using two newly developed methods. The gas permeability coefficient of NBFs depends strongly on the surfactant concentration. This dependence was in good agreement with the nucleation theory of fluctuation formation of nanoscopic holes responsible for the bilayer stability and permeability. A very interesting result obtained is that the gas permeability coefficient of thicker CBFs is 2–3 times larger than that for the thinnest NBFs. Another important result is that the coefficient of the CBFs increases with decreasing electrolyte concentration (increasing film thickness), passing through a maximum.

Dynamic contact angles and gas permeability coefficients of NBFs from aqueous dispersions of phospholipids have been measured by an original method. The results for two types of solutions, (1) liposome suspension and (2) ethanol + water solution of phospholipids, were found to be very different. The contact angles in case (1) vary strongly under dynamic conditions whereas in case (2) they remain almost constant. The gas permeability is larger in case (2) than in case (1). The results were discussed in connection with the thickness and structure of the NBFs from the two types of solutions, taking into account the solubility (or insolubility) and the hydration of the adsorption layers of phospholipid molecules.

Extensive investigations of black films from aqueous protein solutions showed more complicated behavior. A dynamic hysteresis of the contact angles has been established and studied. The results have been interpreted in connection with the rheological properties of the protein adsorption layers.

A combination of the Langmuir–Blodgett technique and neutron activation analysis has been used to determine the stoichiometry of the interaction between arachidic acid monolayers and cadmium or barium ions dissolved in the subsolution. The stability constants of the corresponding arachidic soaps formed in the monolayer have been calculated from the experimental data. Equations for equilibrium constants of arachidic acid monolayer–subsolution counterion ion exchange were also derived. The interaction of octadecylamine monolayers with the subsolution phosphate counterions at different pH and ionic strength have been studied by the same combination of techniques and the stability constant of octadecylammonium hydrogenphosphate has been estimated. A series of experiments on the elasticity of soluble and non-soluble monolayers on a liquid substrate have also been performed.

From the above description, it is clear that Professor Platikanov has made significant original scientific contributions in the field of colloid and interface science. His publications allowed him to become internationally known and for this reason he has been elected President of the International Association of Colloid and Interface Scientists. A list of his most important papers is provided.

Selected Publications

1. Untersuchung dünner flüssiger Schichten auf Quecksilber.
A. Scheludko, D. Platikanov, *Kolloid-Z.*, *175* (**1961**) 150.
2. Experimental Investigation on the "Dimpling" of Thin Liquid Films.
D. Platikanov, *J. Phys. Chem.*, *68* (**1964**) 3619.
3. Disjoining Pressure in Thin Liquid Films and the Electro-Magnetic Retardation Effect of the Molecular Dispersion Interactions.
A. Scheludko, D. Platikanov, E. Manev, *Discuss. Faraday Soc.*, *40* (**1965**) 253.
4. Electrical Conductivity of Black Foam Films.
D. Platikanov, N. Rangelova, in *Research in Surface Forces* (B.V. Derjaguin, Ed.), Vol. 4, Consultants Bureau, New York, 1972, p. 246.
5. Orientation of Nonionic Surfactants on Solid Surfaces: n-Alkyl Polyglycol Ethers on Montmorillonite.
D. Platikanov, A. Weiss, G. Lagaly, *Colloid Polym. Sci.*, *255* (**1977**) 907.
6. Free Black Films of Proteins.
D. Platikanov, G.P. Yampolskaya, N. Rangelova, Zh. Angarska, L.E. Bobrova, V. N. Izmailova, *Colloid J. USSR*, *42* (**1980**) 753; *43* (**1981**) 149; *52* (**1990**) 442.
7. Line Tension of Newton Black Films.
D. Platikanov, M. Nedyalkov, A. Scheludko, V. Nasteva, *J. Colloid Interface Sci.*, *75* (**1980**) 612, 620.
8. Interaction of Octadecylamine Monolayers with Phosphate Counterions.
J.G. Petrov, I. Kuleff, D. Platikanov, *J.Colloid Interface Sci.*, *109* (**1986**) 455.
9. Equilibrium Constants of Ion Exchange Reactions between Fatty Acid Monolayers and Dissolved Counterions.
J.G. Petrov, D. Platikanov, *Colloid Polym. Sci.*, *265* (**1987**) 65.
10. The Transition Region between Aqueous Wetting Films on Quartz and the Adjacent Meniscus.
Z. Zorin, D. Platikanov, T. Kolarov, *Colloids Surf.*, *22* (**1987**) 147; *51* (**1990**) 37.
11. Method for Direct Measurement of the Film Tension of Black Foam Films.
D. Platikanov, M. Nedyalkov, N. Rangelova, *Colloid Polym. Sci*, *265* (**1987**) 72; *269* (**1991**) 272.
12. On the Curvature Dependence of the Film Tension of Newton Black Films.
D. Platikanov, M. Nedyalkov, A. Scheludko, B.V. Toshev, *J.Colloid Interface Sci*, *121* (**1988**) 100.
13. Line Tension in Three-phase Equilibrium Systems.
B.V. Toshev, D. Platikanov, A. Scheludko, *Langmuir*, *4* (**1988**) 489.
14. Method for Direct Measurement of Film Tension of Newton Black Films on a Diminishing Bubble.
M. Nedyalkov, G. Schoepe, D. Platikanov, *Colloids Surf.*, *47* (**1990**) 95.
15. Disjoining Pressure, Contact Angles and Line Tension in Free Thin Liquid Films.
B.V. Toshev, D. Platikanov, *Adv. Colloid Interface Sci.*, *40* (**1992**) 157.
16. On the Mechanism of Permeation of Gas through Newton Black Films at Different Temperatures.
M. Nedyalkov, R. Krustev, A. Stankova, D. Platikanov, *Langmuir*, *8* (**1992**) 3142; *12* (**1996**) 1688.
17. Permeability of Common Black Films to Gas.
R. Krustev, D. Platikanov, M. Nedyalkov, *Colloids Surf.*, *79* (**1993**) 129; *123/124* (**1997**) 383.
18. Linear Energy with Positive and Negative Sign.
D. Exerowa, D. Kashchiev, D. Platikanov, B.V. Toshev, *Adv. Colloid Interface Sci.*, *49* (**1994**) 303.

19 Thin Liquid Films from Polyoxyethylene–Polyoxypropylene Block Copolymer on the Surface of Fused Quartz.
B. Diakova, M. Kaisheva, D. Platikanov, *Colloids Surf. A, 190* (**2001**) 61.

20 Thin Wetting Films from Aqueous Electrolyte Solutions on SiC/Si Wafer.
B. Diakova, C. Filiatre, D. Platikanov, A. Foissy, M. Kaisheva, *Adv. Colloid Interf. Sci., 96* (**2002**) 193.

21 Phospholipid Black Foam Films: Dynamic Contact Angles and Gas Permeability.
D. Platikanov, M. Nedyalkov, V. Petkova, *Adv. Colloid Interf. Sci., 101* (**2003**) 185; *104* (**2003**) 37.

22 Physico-chemical Background of the Foaming Protein Separation for Waste Minimization.
D. Platikanov, V.N. Izmailova, G P. Yampolskaya, in *Role of Interfaces in Environmental Protection* (S. Barany, Ed.), *NATO Science Series, IV/24* (**2003**) 507.

23 Thin Wetting Films from Aqueous Solutions of a Polyoxyethilene–Polyoxypropylene Block Copolymer on Silicon Carbide Surface.
B. Diakova, D. Platikanov, R. Atanassov, M. Kaisheva, *Adv. Colloid Interface Sci., 104* (**2003**) 25.

24 Thin Liquid Films.
D. Platikanov, D. Exerowa, in *Fundamentals of Interface and Colloid Science*, Vol. 5 (J. Lyklema, Ed.), Elsevier, Amsterdam, 2005, Chap. 6, p. 6.1.

25 Symmetric Thin Liquid Films with Fluid Interfaces.
D. Platikanov, D. Exerowa, in *Emulsions and Emulsion Stability*, 2nd edn. (J. Sjöblom, Ed.), CRC Press, Taylor and Francis, New York, 2006, Chap. 3, p. 127.

List of Contributors

Ludmila B. Boinovich
A. N. Frumkin Institute of Physical
Chemistry and Electrochemistry
Russian Academy of Sciences
Leninsky prospect 31
119991 Moscow
Russia

Alexandre M. Emelyanenko
A. N. Frumkin Institute of Physical
Chemistry and Electrochemistry
Russian Academy of Sciences
Leninsky prospect 31
119991 Moscow
Russia

Jan Christer Eriksson
Department of Chemistry
Surface Chemistry
Royal Institute of Technology
10044 Stockholm
Sweden

Valentin B. Fainerman
Medical Physicochemical Centre
Donetsk Medical University
16 Ilych Avenue
83003 Donetsk
Ukraine

Alexander Kamyshny
Casali Institute of Applied Chemistry
Institute of Chemistry
The Hebrew University of Jerusalem
Edmond Safra Campus
Givat-Ram
91904 Jerusalem
Israel

Volodymyr I. Kovalchuk
Institute of Biocolloid Chemistry
42 Vernadsky Avenue
03680 Kiev
Ukraine

Zdravko I. Lalchev
Faculty of Biology
Sofia University
"St. Kliment Ohridski"
8 Dragan Tsankov Str.
1164 Sofia
Bulgaria

Martin E. Leser
Nestec Ltd.
Nestlé Research Centre
Vers-chez-les-Blanc
1000 Lausanne 26
Switzerland

Colloids and Interface Science Series, Vol. 1
Colloid Stability: The Role of Surface Forces, Part I. Edited by Tharwat F. Tadros

ISBN: 978-3-527-31462-1

Shlomo Magdassi
Casali Institute of Applied Chemistry
Institute of Chemistry
The Hebrew University of Jerusalem
Edmond Safra Campus
Givat-Ram
91904 Jerusalem
Israel

Elena Mileva
Rostislaw Kaischew Institute of Physical Chemistry
Bulgarian Academy of Sciences
Acad. G. Bonchev Str., block 11
1113 Sofia
Bulgaria

Reinhard Miller
Max-Planck-Institut für Kolloid- und Grenzflächenforschung
14424 Potsdam/Golm
Germany

Mickael Nedyalkov
Department of Physical Chemistry
University of Sofia
1 James Bourchier Blvd.
1126 Sofia
Bulgaria

Hiroyuki Ohshima
Faculty of Pharmaceutical Sciences
Tokyo University of Science
2641 Yamazaki, Noda
Chiba 278-8510
Japan

Sümer Peker
Chemical Engineering Department
Ege University
Bornova
İzmir
Turkey

Alexander V. Pertsov
Chemical Faculty
Moscow State University
Vorob'evy Gory
119899 Moscow
Russia

Brita Rippner Blomqvist
Biopharmaceuticals Octapharma AB
11275 Stockholm
Sweden

Eugene D. Shchukin
Department of Geography and Environmental Engineering
34 North Charles Street
The Johns Hopkins University
Baltimore, MD 21218
USA

Cosima Stubenrauch
School of Chemical and Bioprocess Engineering
University College Dublin
Belfield, Engineering Building
Dublin
Ireland

Tharwat F. Tadros
Consultant
89 Nash Grove Lane
Wokingham, Berkshire RG40 4HE
UK

Plamen Tchoukov
Rostislaw Kaischew Institute of Physical Chemistry
Bulgarian Academy of Sciences
Acad. G. Bonchev Str., block 11
1113 Sofia
Bulgaria

Borislav V. Toshev
Department of Physical Chemistry
University of Sofia
1 James Bourchier Blvd.
1164 Sofia
Bulgaria

Liguang Wang
Center for Advanced Separation
Technologies
Virginia Polytechnic Institute and
State University
Blacksburg, VA 24061
USA

Roe-Hoan Yoon
Center for Advanced Separation
Technologies
Virginia Polytechnic Institute and
State University
Blacksburg, VA 24061
USA

1
General Principles of Colloid Stability and the Role of Surface Forces

Tharwat Tadros

Abstract

In this overview, the general principles of colloid stability are described with some emphasis on the role of surface forces. Electrostatic stabilization is the result of the presence of electrical double layers which, on approach of particles, interact, leading to repulsion that is determined by the magnitude of the surface or zeta potential and electrolyte concentration and valency. Combining this electrostatic repulsion with van der Waals attraction forms the basis of theory of colloid stability due to Deyaguin-Landau-Verwey-Overbeek (DLVO theory). This theory can explain the conditions of stability/instability of colloidal particles and it can predict the Schulze-Hardy rule. Direct confirmation of the DLVO theory came from surface force measurements using cross mica cylinders. Particles containing adsorbed or grafted nonionic surfactant or polymer layers produce another mechanism of stabilization, referred to as steric stabilization. This arises from the unfavorable mixing of the stabilizing layers when these are in good solvent conditions and the loss of entropy of the chains on significant overlap. The criteria of effective steric stabilization have been summarized. The flocculation of sterically stabilized dispersions can be weak and reversible or strong and irreversible depending on the conditions. Weak flocculation can occur when the adsorbed layer thickness is small (<5 nm), whereas strong (incipient) flocculation occurs when the solvency of the medium for the stabilizing chains become worse than that of a theta-solvent. The effect of addition of "free" non-adsorbing polymer is described in terms of the presence of a polymer-free zone (depletion zone) between the particles. This results in weak flocculation and equations are presented to describe the free energy of depletion attraction.

Colloids and Interface Science Series, Vol. 1
Colloid Stability: The Role of Surface Forces, Part I. Edited by Tharwat F. Tadros

ISBN: 978-3-527-31462-1

1.1 Introduction

The stability of colloidal systems is an important subject from both academic and industrial points of views. These systems include various types such as solid–liquid dispersions (suspensions), liquid–liquid dispersions (emulsions) and gas–liquid dispersions (foams). The colloid stability of such systems is governed by the balance of various interaction forces such as van der Waals attraction, double-layer repulsion and steric interaction [1]. These interaction forces have been described at a fundamental level such as in the well know theory due to Deryaguin and Landau [2] and Verwey and Overbeek [3] (DLVO theory). In this theory, the van der Waals attraction is combined with the double-layer repulsion and an energy–distance curve can be established to describe the conditions of stability/instability. Several tests of the theory have been carried out using model colloid systems such as polystyrene latex. The results obtained showed the exact prediction of the theory, which is now accepted as the cornerstone of colloid science. Later, the origin of stability resulting from the presence of adsorbed or grafted polymer layers was established [4]. This type of stability is usually referred to as steric stabilization and, when combined with the van der Waals attraction, energy–distance curves, could be established to describe the state of the dispersion.

This overview summarizes the principles of colloid stability with particular reference to the role of surface forces. For more details on the subject, the reader is referred to two recent texts by Lyklema [5, 6].

1.2 Electrostatic Stabilization (DLVO Theory)

As mentioned above, the DLVO theory [2, 3] combines the van der Waals attraction with the double-layer repulsion. A brief summary of these interactions is given below and this is followed by establishing the conditions of stability/instability describing the influence of the various parameters involved.

1.2.1 Van der Waals Attraction

There are generally two ways of describing the van der Waals attraction between colloids and macrobodies: the microscopic [7] and the macroscopic approach [8]. The microscopic approach is based on the assumption of additivity of London pair interaction energy. This predicts the van der Waals attraction with an accuracy of $\sim 80\%$–90%. The macroscopic approach, which gives a more accurate evaluation of the van der Waals attraction, is based on the correlation between electric fluctuations of two macroscopic phases. However, this approach requires quantification of the dielectric dispersion data, which are available for only a limited number of systems. For this reason, most results on the van der Waals attraction are based on the microscopic approach that is briefly described below.

For a one-component system, individual atoms or molecules attract each other at short distances due to van der Waals forces. The latter may be considered to consist of three contributions: dipole–dipole (Keesom), dipole–induced dipole (Debye) and London dispersion interactions. For not too large separation distances between atoms or molecules, the attractive energy G_a is short range in nature and it is inversely proportional to the sixth power of the interatomic distance r:

$$G_a = -\frac{\beta_{11}}{r^6} \tag{1}$$

where β_{11} is a constant referring to identical atoms or molecules.

Since colloidal particles are assemblies of atoms or molecules, the individual contributions have to be compounded. In this case, only the London interactions have to be considered, since large assemblies have no net dipole moment or polarization (both the Keesom and Debye forces which are vectors tend to cancel in such assemblies). The London (dispersion) interaction arises from charge fluctuations within an atom or molecule associated with the motion of its electrons. The London dispersion energy of interaction between atoms or molecules is short range as given by Eq. (1), whereby β_{11} is now the London dispersion constant (that is related to the polarizability of atoms or molecules involved). For an assembly of atoms or molecules, as is the case with colloidal particles, the van der Waals energy of attraction between two equal particles each of radius R, at a distance h in vacuum, is given by the expression

$$G_A = -\frac{A_{11}}{6}\left[\frac{2}{s^2-4} + \frac{2}{s^2} + \ln\left(\frac{s^2-4}{s^2}\right)\right] \tag{2}$$

where $s = (2R+h)/R$ and A_{11} is the Hamaker constant, which is given by

$$A_{11} = \pi^2 q^2 \beta_{11} \tag{3}$$

where q is the number of atoms or molecules per unit volume.
For very short distances of separation ($h \ll R$), Eq. (2) can be approximated by

$$G_A = -\frac{RA_{11}}{12h} \tag{4}$$

It is clear from comparison of Eqs. (1) and (4) that the magnitude of the attractive energy between macroscopic bodies (particles) is orders of magnitude larger than that between atoms or molecules. It is also long range in nature, increasing sharply at short distances of separation. G_A is also proportional to R and A_{11}.

The above expressions are for particles in vacuum and in the presence of a medium (solvent), the Hamaker constant A_{12} of particles of material 1 dispersed in a medium of Hamaker constant A_{22} is given by

$$A_{12} = A_{11} + A_{22} - 2A_{12} = \left(A_{11}^{\frac{1}{2}} - A_{22}^{\frac{1}{2}}\right)^2 \qquad (5)$$

Since with most disperse systems $A_{11} > A_{22}$ then a dispersion medium A_{12} is positive and two particles of the same material always attract each other.

The London dispersion forces exhibit the phenomenon of retardation implying that for large r the attraction decreases more rapidly with distance than at small r. This means that G_a (Eq. (1)) becomes proportional to r^{-7}. For intermediate distances there is a gradual transition from r^{-6} to r^{-7} [9]. This retardation is also reflected in the attraction between particles, whereby h^{-1} should be h^{-2} in Eq. (4). The retardation effect is automatically included in the macroscopic approach.

1.2.2
Double-layer Repulsion

Several processes can be visualized to account for charging suspended particles such as dissociation of surface groups (e.g. OH, COOH, SO_4Na) and adsorption of certain ionic species (such as surfactants). In all cases, charge separation takes place with some of the specifically adsorbed ions at the surface forming a surface charge which is compensated with unequal distribution of counter and co-ions. This forms the basis of the diffuse double layer due to Gouy and Chapman [10], which was later modified by Stern [11], who introduced the concept of the specifically adsorbed counter ions in the fixed first layer (the Stern plane). The potential at the surface ψ_0 decreases linearly to a value ψ_d (located at the center of the specifically adsorbed counter ions) and then exponentially with decrease in distance x, reaching zero in bulk solution. The Stern potential is sometimes equated with the measurable electrokinetic or zeta potential, ζ.

The extension of the double layer, referred to as double-layer thickness, depends on the electrolyte concentration and valency of the ions, as given by the reciprocal of the Debye-Hückel parameter:

$$\frac{1}{\kappa} = \left(\frac{\varepsilon_r \varepsilon_0 kT}{2n_0 Z^2 e^2}\right)^{\frac{1}{2}} \qquad (6)$$

where ε_r is the relative permittivity, ε_0 is the permittivity of free space, k is the Boltzmann constant, T is the absolute temperature, n_0 is the number of ions of each type present in the bulk phase, Z is the valency of the ions and e is the electronic charge.

The parameter $1/\kappa$ increases with decrease in electrolyte concentration and decrease in the valency of the ions. For example, for 1:1 electrolyte (e.g. KCl), the double-layer thickness is 100 nm in 10^{-5}, 10 nm in 10^{-3} and 1 nm in 10^{-1} mol dm^{-3}. As we shall see later, this reflects the double-layer repulsion, which increases with decrease in electrolyte concentration.

When two particles each with an extended double layer with thickness $1/\kappa$ approach to a distance of separation such that double-layer overlap begins to occur (i.e. $h<2/\kappa$), repulsion occurs as a result of the following effect. Before overlap, i.e. $h>2/\kappa$, the two double layers can develop completely without any restriction and in this case the surface or Stern potential decays to zero at the mid-distance between the particles. However, when $h<2/\kappa$, the double layers can no longer develop unrestrictedly, since the limited space does not allow complete potential decay. In this case the potential at the mid-distance between the particles $\psi_{h/2}$ is no longer zero and repulsion occurs. The electrostatic energy of repulsion, G_{el}, is given by the following expression which is valid for $\kappa R<3$:

$$G_{el} = \frac{4\pi\varepsilon_r\varepsilon_0 R^2 \psi_d^2 \exp(-\kappa h)}{2R+h} \tag{7}$$

Equation (7) shows that G_{el} decays exponentially with increase of h and it approaches zero at large h. The rate of decrease of G_{el} with increase in h depends on $1/\kappa$: the higher the value of $1/\kappa$, the slower is the decay. In other words, at any given h, G_{el} increases with increase in $1/\kappa$, i.e. with decrease in electrolyte concentration and valency of the ions.

1.2.3
Total Energy of Interaction (DLVO Theory)

The total energy of interaction between two particles G_T is simply given by the sum of G_{el} and G_A at every h value:

$$G_T = G_{el} + G_A \tag{8}$$

A schematic representation of the variation of G_{el}, G_A and G_T with h is shown in Fig. 1.1. As can be seen, G_{el} shows an exponential decay with increase in h, approaching zero at large h. G_A, which shows an inverse power law with h, does not decay to zero at large h. The G_T–h curve shows two minima and one maximum: a shallow minimum at large h that is referred to as the secondary minimum, G_{sec}, a deep minimum at short separation distance that is referred to as the primary minimum, $G_{primary}$, and an energy maximum at intermediate distances, G_{max} (sometimes referred to as the energy barrier). The value of G_{max} depends on the surface (Stern or zeta) potential and electrolyte concentration and valency.

The condition for colloid stability is to have an energy maximum (barrier) that is much larger than the thermal energy of the particles (which is of the order of kT). In general, one requires $G_{max}>25kT$. This is achieved by having a high zeta potential (>40 mV) and low electrolyte concentration ($<10^{-2}$ mol dm^{-3} 1:1 electrolyte). By increasing the electrolyte concentration, G_{max} gradually decreases and eventually it disappears at a critical electrolyte concentration. This is illustrated schematically in Fig. 1.2, which shows the G_T–h curves at various $1/\kappa$ values for 1:1 electrolyte. At any given electrolyte concentration, G_{max} decreases with increase in the valency of electrolyte. This explains the poor stability in the presence of multivalent ions.

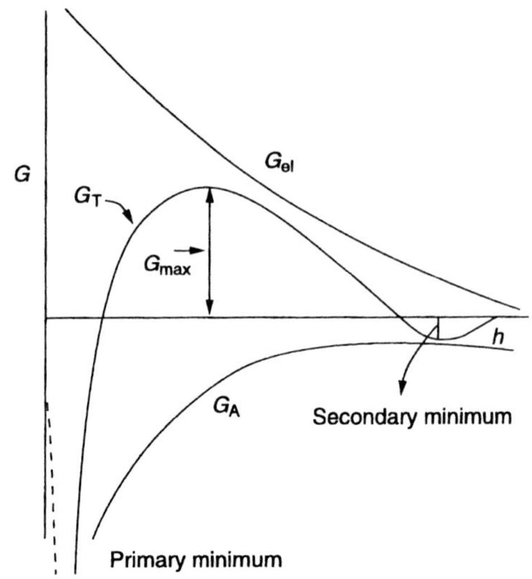

Fig. 1.1 Energy–distance curve for electrostatically stabilized systems.

The above energy–distance curve (Fig. 1.1) explains the kinetic stability of colloidal dispersions. For the particles to undergo flocculation (coagulation) into the primary minimum, they need to overcome the energy barrier. The higher the value of this barrier, the lower is the probability of flocculation, i.e. the rate of flocculation will be slow (see below). Hence one can consider the process of flocculation as a rate phenomenon and when such a rate is low enough, the systems can stay stable for months or years (depending on the magnitude of the energy barrier). This rate increases with reduction of the energy barrier and ultimately (in the absence of any barrier) it becomes very fast (see below).

An important feature of the energy–distance curve in Fig. 1.1 is the presence of a secondary minimum at long separation distances. This minimum may be-

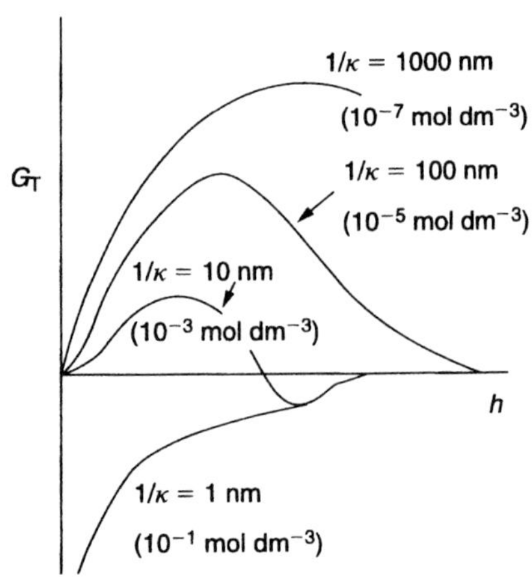

Fig. 1.2 Energy–distance curves at various 1:1 electrolyte concentrations.

come deep enough (depending on electrolyte concentration, particle size and shape and the Hamaker constant), reaching several kT units. Under these conditions, the system become weakly flocculated. The latter is reversible in nature and some deflocculation may occur, e.g. under shear conditions. This process of weak reversible flocculation may produce "gels", which on application of shear break up, forming a "sol". This process of sol↔gel transformation produces thixotropy (reversible time dependence of viscosity), which can be applied in many industrial formulations, e.g. in paints.

1.2.4 Stability Ratio

One of the useful quantitative methods to assess the stability of any dispersion is to measure the stability ratio W, which is simply the ratio between the rate of fast flocculation k_0 (in the absence of an energy barrier) to that of slow flocculation k (in the presence of an energy barrier):

$$W = \frac{k_0}{k} \tag{9}$$

The rate of fast flocculation k_0 was calculated by Von Smoluchowski [12], who considered the process to be diffusion controlled. No interaction occurs between two colliding particles occurs until they come into contact, whereby they adhere irreversibly. The number of particles per unit volume n after time t is related to the initial number n_0 by a second-order type of equation (assuming binary collisions):

$$n = \frac{n_0}{1 + k_0 n_0 t} \tag{10}$$

where k_0 is given by

$$k_0 = 8\pi DR \tag{11}$$

where D is the diffusion coefficient, given by the Stokes-Einstein equation:

$$D = \frac{kT}{6\pi\eta R} \tag{12}$$

Combining Eqs. (11) and (12):

$$k_0 = \frac{8kT}{6\eta} \tag{13}$$

For particles dispersed in an aqueous phase at 25 °C, $k_0 = 5.5 \times 10^{-18}\ m^3\ s^{-1}$.

In the presence of an energy barrier, G_{max}, slow flocculation occurs with a rate depending on the height of this barrier. In this case, a flocculation rate k

may be defined that is related to the stability ratio W as given by Eq. (9). Fuchs [13] related W to G_{max} by the following expression:

$$W = 2R \int_{2R}^{\infty} \exp\left(\frac{G_{max}}{kT}\right) h^{-2} dh \tag{14}$$

An approximate form of Eq. (14) for charge-stabilized dispersions was given by Reerink and Overbeek [14]:

$$W = \frac{1}{2} k_0 \exp\left(\frac{G_{max}}{kT}\right) \tag{15}$$

Reerink and Overbeek [14] also derived the following theoretical relationship between W, electrolyte concentration C, valency Z and surface potential ψ_0:

$$\log W = \text{constant} - 2.06 \times 10^9 \left(\frac{R\gamma^2}{Z^2}\right) \log C \tag{16}$$

$$\gamma = \frac{\exp(Ze\psi_0/2kT) - 1}{\exp(Ze\psi_0/2kT) + 1} \tag{17}$$

Equation (16) predicts that experimental plots of $\log W$ versus $\log C$ should be linear in the slow flocculation regime and $\log W = 0$ ($W = 1$) in the fast flocculation regime. This is illustrated in Fig. 1.3 for 1:1 and 2:2 electrolytes. The plots show two linear portions intersecting at a critical electrolyte concentration at which $W = 1$, i.e. the critical flocculation concentration (CFC). Note that in Fig. 1.3, W is <1 in the fast flocculation regime as a result of contribution of the van der Waals attraction.

Verwey and Overbeek [14] introduced the following criteria between stability and instability:

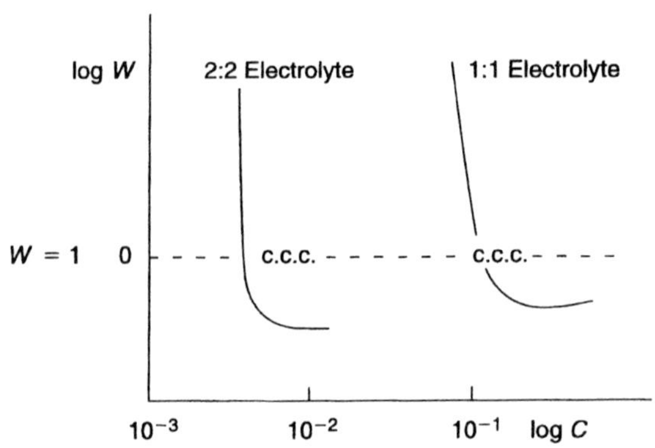

Fig. 1.3 Log$W$ versus log$C$ for 1:1 and 2:2 electrolytes.

$$G_T = G_{el} + G_A = 0 \tag{18}$$

$$\frac{dG_T}{dh} = 0 \tag{19}$$

Using Eqs. (4) and (7), one can obtain an expression for the CFC:

$$\text{CFC} = 3.6 \times 10^{-36} \frac{\gamma^2}{A^2 Z^2} \text{mol m}^{-3} \tag{20}$$

Equation (20) shows that the CFC increases with ψ_0 or ψ_d and decreases with increasing A (the Hamaker constant) or van der Waals attraction and it also decreases with increase in Z. For very high values of ψ_d, γ approaches unity and the CFC becomes inversely proportional to the sixth power of valency. However, very high values of ψ_d are not encountered in practice, in which case the CFC is proportional to Z^{-2}. This dependence of CFC on Z is the basis of the Schulze-Hardy rule.

As mentioned above, with electrostatically stabilized systems weak flocculation can occur, when a secondary minimum with sufficient depth (1–5kT) occurs in the energy–distance curve. In this case, the flocculation is reversible and in the kinetic analysis one must take into account the backward rate of flocculation (with a rate constant k_b) as well as the forward rate of flocculation (with a rate constant k_f). In this case the rate of decrease of particle number with time is given by

$$\frac{dn}{dt} = -k_f n^2 + k_b n \tag{21}$$

The rate of deflocculation k_b depends on the floc size and the exact ways in which the flocs are broken down (e.g. how many contacts are broken). This means that the second-hand term on the right hand-side of Eq. (21) should be replaced by a summation over all possible modes of breakdown, thus making the analysis of the kinetics complex. Another complication in the analysis of the kinetics of reversible flocculation is that this type of flocculation is a critical phenomenon rather than a chain (or sequential) process. Hence a critical particle number concentration, n_{crit}, has to be exceeded before flocculation occurs, i.e. flocculation becomes a thermodynamically favored process. The kinetics of weak, reversible flocculation have more in common, therefore, with nucleation kinetics, rather than with chemical (e.g. polymerization) kinetics. This is not to say that doublets, triplets, etc., will not form transiently below n_{crit}. These are thermodynamically unstable, but their effective concentrations may be calculated from a suitable kinetic analysis. This is beyond the scope of the present review.

The above flocculation process (strong or weak) is diffusion controlled and it is usually referred to as perikinetic flocculation. In other words, particle collisions arise solely from Brownian diffusion of particles and the diffusion coefficient is given by the Stokes-Einstein relation $D=kT/f$, where f is the fractional coefficient (as given by Eq. 12). If external energy is applied on the system, e.g. shear, ultrasound or centrifugal, or the system is not at thermal equilibrium (so that convection currents arise), then the rate of particle collisions is modified (usually increased) and the flocculation is referred to as orthokinetic. For example, in a shear field (with shear rate γ), the rate of flocculation is given by

$$-\frac{dn}{dt} = \frac{16}{3} \alpha^2 \gamma R^3 \tag{22}$$

where α is the collision frequency factor, i.e. the fraction of collisions which result in permanent aggregates. Although for irreversible flocculation one might expect orthokinetic flocculation conditions to lead to an increased rate of flocculation (in any given time interval), with a weakly (reversible) flocculation the opposite is the case, i.e. application of shear may lead to deflocculation.

1.2.5 Extension of the DLVO Theory

As discussed above, the basis premises of the DLVO theory are basically sound and considering van der Waals attraction and double-layer repulsion as the sole and additive contribution to the pair-wise interaction between particles in a dispersion results in a number of predictions such as the dependence of stability on surface (or zeta potential), electrolyte concentration and valency (e.g. prediction of the Schulze-Hardy rule) as well as distinction between "strong" (irreversible) flocculation and "weak" (reversible) flocculation. However, over the past five decades or so, a number of authors have attempted to extend the DLVO theory to take into account some of the unexplained results, e.g. the dependence of stability on the counter ion specificity (the so-called Hoffmeister series).

The main extension of the DLVO theory is the presence of repulsion at very short distances, which has been attributed to solvent structure-mediated forces (referred to as salvation forces). This will add an extra contribution to the pair-wise interaction, i.e.

$$G_T = G_A + G_{el} + G_{solv,str} \tag{23}$$

For full consideration of the above extensions, one should refer to the recent text by Lyklema [15].

1.2.6 The Concept of Disjoining Pressure

The concept of disjoining pressure, $\Pi(h)$, was initially introduced by Deryaguin and Obukhov [16] to account for the stability of thin liquid films at interfaces. Basically, $\Pi(h)$ is the pressure that develops when two surfaces are brought to each other from infinity to a distance h. It is the change in the Gibbs free energy (in Joules) with separation distance h (in m). Hence $\Pi(h)$ is given as force per unit area ($N\ m^{-2}$), which is the unit for pressure:

$$\Pi(h) = -\left[\frac{\delta G(h)}{\delta h}\right]_{p,T} \tag{24}$$

$\Pi(h)$ can be split into three main contributions, Π_A (the van der Waals attraction), Π_{el} (the electrostatic repulsion) and $\Pi_{solv,str}$ (arising from salvation forces):

$$\Pi(h) = \Pi_A \Pi_{el} + \Pi_{solv,str} \tag{25}$$

A schematic representation of the variation of G and Π with h is shown in Fig. 1.4. The disjoining pressure diagram has three zero points at the secondary minimum, G_{max}, and the primary minimum (at these points $dG/dh=0$). However, the zero mid-point (at the energy maximum) is not met in practice be-

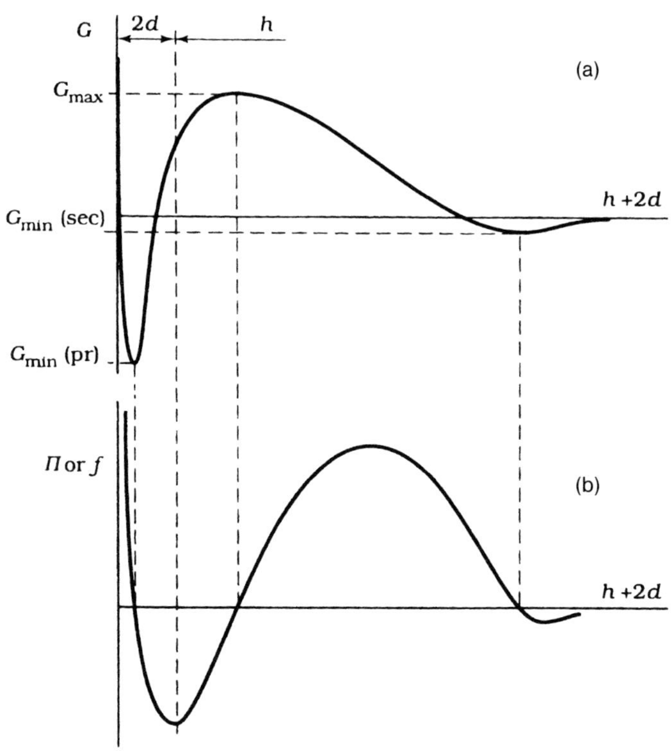

Fig. 1.4 Gibbs free energy for the DLVO type of interaction between colloidal particles (top) and the corresponding disjoining pressure (bottom).

cause it is labile: a slight displacement to the right makes Π positive (repulsive), leading to further separation of particles.

The concept of disjoining pressure has been particularly useful in describing the stability of foam and emulsion films. Model foam films produced using ionic surfactants were used to describe the stability by measuring the disjoining pressure as a function of film thickness h [17]. The results obtained could be used to describe the mechanism of stabilization and hence to test the DLVO theory. The latter considers the interaction between parallel plates (the parallel layers of two surfactant films). In this case the electrostatic contribution to the disjoining pressure is given by

$$\Pi(h)_{\mathrm{el}} = 64RTC \tanh^2\left(\frac{F\psi_0}{4RT}\right) \exp(-\kappa h) \tag{26}$$

The van der Waals contribution to the disjoining pressure is given by

$$\Pi(h)_{\mathrm{vdW}} = -\frac{A}{6\pi h^3} \tag{27}$$

Using Eqs. (26) and (27), one can calculate the total disjoining pressure $\Pi(h)$ and a direct comparison with the measured values can be obtained. This could prove the validity of the DLVO theory and any deviation could be accounted for by introducing other contributions, e.g. $\Pi(h)_{\mathrm{solv,st}}$. Such comparison is beyond the scope of the present review.

1.2.7
Direct Measurement of Interaction Forces

There are generally two main procedures for measuring the interaction forces between macroscopic bodies, both of which have some limitations. The first technique is based on measurement of interaction forces between cross cylinders of cleaved mica (a molecularly smooth surface) that was described in detail by Israelachvili and Adams [18]. Full details of the technique are beyond the scope of this review. However, as an illustration, Fig. 1.5 shows the force–distance curves for the interaction between two cross mica cylinders at various KNO_3 concentrations. The semilogarithmic $f(h)$ curves have a linear part, becoming steeper with increasing salt concentration, corresponding to the long-distance $\exp(-\kappa h)$ decay, predicted by the DLVO theory. For $h < 2.5$ nm, often a short-range repulsion is observed, which is due to the water structure-mediated solvation force.

The second technique for measuring interaction forces is based on atomic force microcopy (AFM), which will be described in detail in another chapter. In this technique, one can measure the force between a sphere and flat plate.

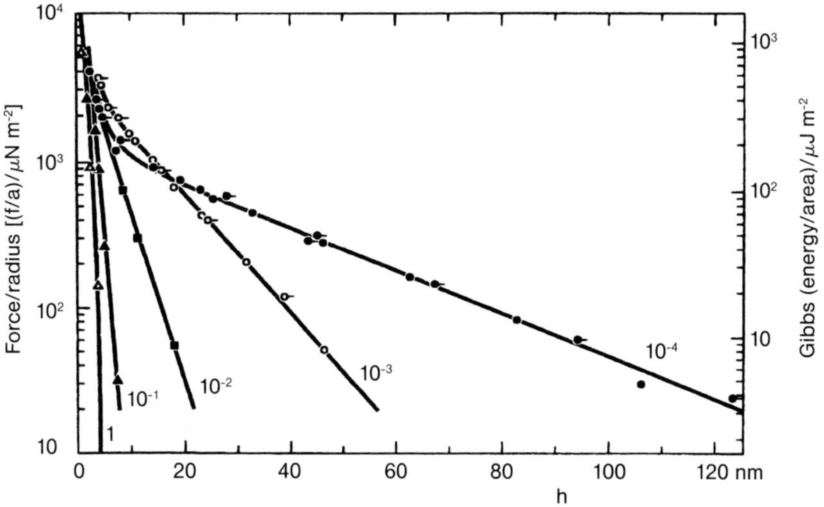

Fig. 1.5 Force–distance curves for two crossed mica cylinders at various KNO_3 concentrations.

1.3 Steric Stabilization

This arises from the presence of adsorbed or grafted surfactant or polymer layers, mostly of the nonionic type. The most effective systems are those based on A–B, A–B–A block and BA_n graft types (sometimes referred to as polymeric surfactants). Here B is the "anchor" chain that is usually insoluble in the dispersion medium and has strong affinity to the surface. A is the stabilizing chain that is soluble in the medium and strongly solvated by its molecules. To understand the role of these polymeric surfactants in the stabilization of dispersions, it is essential to consider the adsorption and conformation of the polymer at the interface. This is beyond the scope of the present review and the reader should refer to the text by Fleer et al. [19]. It is sufficient to state at this point that adsorption of polymers is irreversible and the isotherm is of the high-affinity type. The B chain produces small loops with multipoint attachment to the surface and this ensures irreversibility of adsorption. The stabilizing chains, on the other hand, extend from the surface, producing several "tails" with a hydrodynamic adsorbed layer thickness δ_h of the order of 5–10 nm depending on the molecular weight of the A chains.

When two particles with radius R and having an adsorbed layer with hydrodynamic thickness δ_h approach each other to a surface–surface separation distance h that is smaller than $2\delta_h$, the polymer layers interact resulting in two main situations [20, 21]: either the polymer chains may overlap or the polymer layer may undergo some compression. In both cases, there will be an increase in the local segment density of the polymer chains in the interaction zone. Provided that the dangling chains A are in a good solvent (see below), this local increase in segment density in the interaction zone will result in strong repulsions as a result of two main effects: (1) increase in the osmotic pressure in the overlap region as result of the unfavorable mixing of the polymer chains (when these are in good solvent conditions) [20, 21]; this is referred to as osmotic repulsion or mixing interaction and it is described by a free energy of interaction, G_{mix}; and (2) reduction of the configurational entropy of the chains in the interaction zone. This entropy reduction results from the decrease in the volume available for the chains whether these are either overlapped or compressed. This is referred to as volume restriction interaction, entropic or elastic interaction and it is described by a free energy of interaction, G_{el}.

Combination of G_{mix} and G_{el} is usually referred to as the steric free energy of interaction G_s:

$$G_s = G_{mix} + G_{el} \tag{28}$$

The sign of G_{mix} depends on the solvency of the medium for the chain. In a good solvent, i.e. the Flory-Huggins interaction parameter χ is <0.5 (see below), then G_{mix} is positive and the unfavorable mixing interaction leads to repulsion. In contrast, if $\chi > 0.5$, i.e. the chains are in poor solvent condition, then G_{mix} is negative and the interaction (which is favorable) is attractive. G_{el} is always positive regardless of the solvency and hence is some cases one can produce stable dispersions in relatively poor solvent conditions.

Several sophisticated theories are available for description of steric interaction and these has been recently reviewed by Fleer et al. in a recent book by Lyklema [22]. However, in this section, only the simple classical treatment will be described, which is certainly an oversimplification and not exact.

1.3.1
Mixing Interaction, G_{mix}

As mentioned above, this results from the unfavorable mixing of the polymer chains, when under good solvent conditions. This is represented schematically in Fig. 1.6, which shows the simple case of two spherical particles, each with radius R and each having an adsorbed layer with thickness δ. Before overlap, one can de-

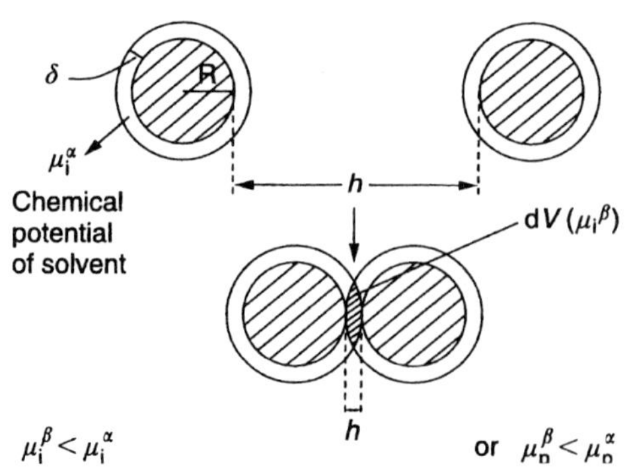

Fig. 1.6 Schematic representation of polymer layer overlap.

fine in each polymer layer a chemical potential for the solvent μ_i^{α} and a volume fraction for the polymer in the layer ϕ_2^{α}. In the overlap region (volume element dV), the chemical potential of the solvent is reduced to μ_i^{β}; this results from the increase in polymer segment concentration in the overlap region (with a volume fraction ϕ_2^{β}). This amounts to an increase in the osmotic pressure in the overlap region. As a result, solvent will diffuse from the bulk to the overlap region, thus separating the particles and, hence, a strong repulsive energy arises from this effect.

The above repulsive energy can be calculated by considering the free energy of mixing two polymer solutions, as treated for example by Flory and Krigbaum [23]. This free energy is given by two terms: an entropy term that depends on the volume fraction of polymer and solvent and an energy term that is determined by the Flory-Huggins interaction parameter χ [24]:

$$\delta(G_{\text{mix}}) = kT(n_1 \ln \phi_1 + n_2 \ln \phi_2 + \chi n_1 \phi_2) \tag{29}$$

where n_1 and n_2 are the number of moles of solvent and polymer with volume fractions ϕ_1 and ϕ_2, k is the Boltzmann constant and T is the absolute temperature.

It should be mentioned that the Flory-Huggins interaction parameter χ is a measure of the non-ideality of mixing a pure solvent with a polymer solution. This creates an osmotic pressure π that can be expressed in terms of the polymer concentration c_2 and the partial specific volume of the polymer ($v_2 = V_2/M_2$; V_2 is the molar volume and M_2 is the molecular weight):

$$\frac{\pi}{c_2} = RT\left[\frac{1}{M_2} + \left(\frac{v_2^2}{V_1}\right)\left(\frac{1}{2} - \chi\right)c_2 + \cdots\right] \tag{30}$$

where V_1 is the molar volume of the solvent, R is the gas constant and T is the absolute temperature.

The second term in Eq. (30) is the second virial coefficient B_2:

$$\frac{\pi}{c_2} = RT\left(\frac{1}{M_2} + B_2 c_2 + \cdots\right) \tag{31}$$

$$B_2 = \left(\frac{v_2^2}{V_1}\right)\left(\frac{1}{2} - \chi\right) \tag{32}$$

Note that $B_2 = 0$ when $\chi = ½$; the polymer behaves as ideal in mixing with the solvent. This is referred to as the θ-condition. In this case the polymer chains in solution have no attraction or repulsion and they adopt their unperturbed dimension. When $\chi < ½$, B_2 is positive and mixing is non-ideal, leading to positive deviation (repulsion). This occurs when the polymer chains are in good solvent conditions. When $\chi > ½$, B_2 is negative and mixing is non-ideal, leading to negative deviation (attraction). This occurs when the polymer chains are in poor solvent conditions.

Using the Flory-Krigbaum theory and definition of the χ parameter, one can derive the total change in free energy of mixing for the whole interaction zone by summing all the elements in dV:

$$G_{\text{mix}} = \frac{2kTV_2^2}{V_1} \nu_2 \left(\frac{1}{2} - \chi\right) R_{\text{mix}}(h) \tag{33}$$

where ν_2 is the number of chains per unit area and $R_{\text{mix}}(h)$ is a geometric function that depends on the form of the segment density distribution of the chain normal to the surface, $\rho(z)$.

Using the above analysis, one can derive an expression for the free energy of mixing of two polymer layers (assuming a uniform segment density distribution in each layer) surrounding two spherical particles as a function of separation distance h between the particles (21),

$$\frac{G_{\text{mix}}}{kT} = \left(\frac{V_2^2}{V_1}\right) \nu_2 \left(\frac{1}{2} - \chi\right) \left(3R - 2\delta + \frac{h}{2}\right) \left(\delta - \frac{h}{2}\right)^2 \tag{34}$$

The sign of G_{mix} depends on the Flory-Huggins interaction parameter χ: if $\chi < ½$, G_{mix} is positive and the interaction is repulsive; if $\chi > ½$, G_{mix} is negative and the interaction is attractive. The condition $\chi = ½$ and $G_{\text{mix}} = 0$ is termed the θ-condition. The latter corresponds to the case where polymer mixing is ideal, i.e. mixing of the chains does not lead to either an increase or decrease of the free energy of the system. The θ-point represents the onset of change of repulsion to attraction, i.e. the onset of flocculation (see below).

1.3.2
Elastic Interaction, G_{el}

As mentioned above, this arises from the loss of configurational entropy of the chain on the close approach of a second particle. This is represented in Fig. 1.7 for the simple case of a rod with one point attachment to the surface according to Mackor and van der Waals [25]. When the two surfaces are separated by an infinite distance (∞), the number of configurations of the rod is $\Omega(\infty)$, which is proportional to the volume of the hemisphere swept by the rod. When a second particle approach to a distance h that such that it cuts the hemisphere (losing some volume), the volume available to the chain is restricted and the number of configurations becomes $\Omega(h)$ which is less than $\Omega(\infty)$.

For two flat plates, G_{el} is given by the expression

$$\frac{G_{\text{el}}}{kT} = 2\nu_2 \ln\left[\frac{\Omega(h)}{\Omega(\infty)}\right] = 2\nu_2 R_{\text{el}}(h) \tag{35}$$

where $R_{\text{el}}(h)$ is a geometric function whose form depends on the segment density distribution $\rho(z)$.

G_{el} is always positive and could play a major role in steric stabilization. It becomes very strong when the separation between the particles becomes comparable to the adsorbed layer thickness δ. It is particularly important for the case of multipoint attachment of the polymer chain.

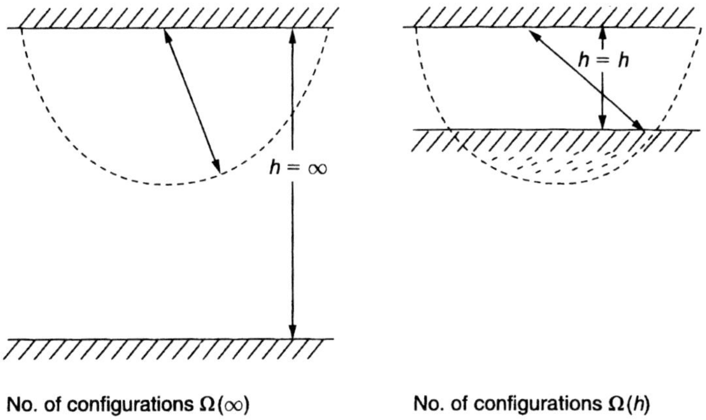

Fig. 1.7 Scheme of configurational entropy loss on the approach of a second particle [25].

1.3.3 Total Energy of Interaction

Combination of G_{mix} and G_{el} with G_A gives the total energy of interaction G_T (assuming there is no contribution from any residual electrostatic interaction):

$$G_T = G_{mix} + G_{el} + G_A \quad (36)$$

Figure 1.8 gives a schematic representation of the variation of G_{mix}, G_{el}, G_A and G_T with surface–surface separation h. G_{mix} increases very sharply with decrease in h when $h<2\delta$. G_{el} increases sharply with decrease in h when $h<\delta$. G_T versus h shows only one minimum, G_{min}, at $h \approx 2\delta$. When $h<2\delta$, G_T shows a rapid increase with further decrease in h.

Unlike the G_T–h curve predicted by the DLVO theory, which shows two minima and one maximum (see Fig. 1.4), the G_T–h curve for sterically stabilized sys-

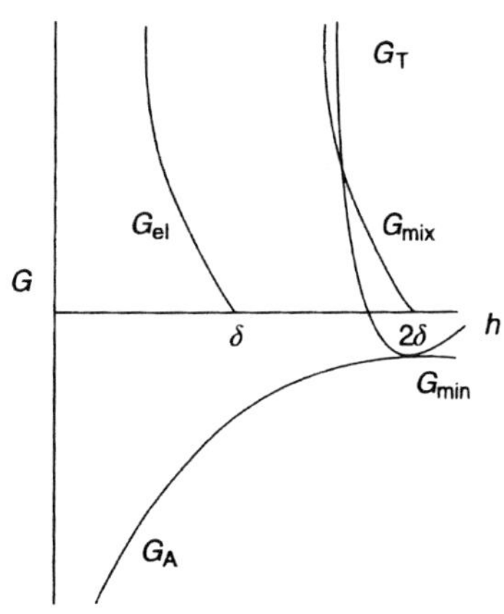

Fig. 1.8 Variation of G_{mix}, G_{el}, G_A and G_T with h for a sterically stabilized system.

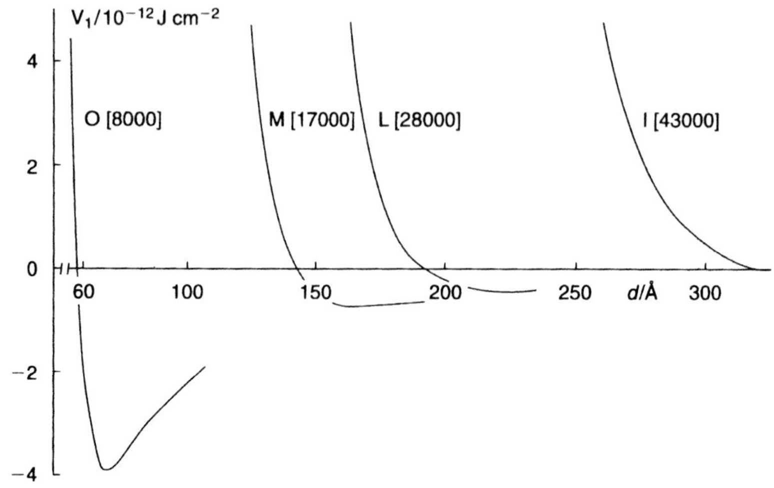

Fig. 1.9 Total interaction energy versus separation distance for particles with adsorbed PVA layers of various molecular weights (thickness).

tems show only one shallow attractive minimum followed by a rapid increase in the total energy as the surfaces approach each other closely to distances comparable to 2δ. The depth of the minimum depends on the particle radius R, the Hamaker constant A and the adsorbed layer thickness δ. For given R and A, G_{min} increases with decrease in δ, i.e. decreasing the molecular weight of the stabilizing chain. To illustrate this dependence, calculations were carried out for poly(vinyl alcohol) (PVA) polymer fractions with various molecular weights. The hydrodynamic thickness of these polymer fractions adsorbed on polystyrene latex particles was determined using dynamic light scattering [photon correlation spectroscopy (PCS)] and the results are given in Table 1.1.

Figure 1.9 shows the results of calculation for PVA of various molecular weight. As can be seen, G_{min} increases with decrease in molecular weight. When the molecular weight of PVA is >43 000, G_{min} is so small that it does not appear on the energy–distance curve. In this case, the dispersion will approach thermodynamic stability (particularly with low volume fraction dispersion).

Table 1.1 Hydrodynamic thickness of PVA with various molecular weights.

MW	δ/nm
67000	25.5
43000	19.7
28000	14.0
17000	9.8
8000	3.3

However, when the molecular weight of the polymer reaches 8000 or δ becomes 3.3 nm, G_{min} reaches sufficient depth for weak flocculation to occur. This was confirmed using freeze fracture scanning electron microscopy.

1.3.4
Criteria for Effective Steric Stabilization

1. The particles should be completely covered by the polymer, i.e. the amount of polymer should correspond to the plateau value. Any bare patches may cause flocculation either by van der Waals attraction between the bare patches or by bridging. The latter occurs when the polymer becomes simultaneously adsorbed on two or more particles.
2. The polymer should be strongly adsorbed ("anchored") to the particle surface. This is particularly the case with A–B, A–B–A block and BA_n graft copolymers where the B chain is chosen to be insoluble in the medium and has high affinity to the surface. Examples of the B chain in aqueous media are polystyrene and poly(methyl methacrylate).
3. The stabilizing chain A should be highly soluble in the medium and strongly solvated by its molecules, i.e. the Flory-Huggins interaction parameter χ should remain $<½$ under all conditions (e.g. in the presence of electrolyte and/or increase in temperature).
4. The adsorbed layer thickness δ should be sufficiently large to maintain a shallow minimum, G_{min}. This is particularly the case when a colloidally stable dispersion without any weak flocculation is required. To ensure this colloid stability, δ should be >5 nm.

1.3.5
Flocculation of Sterically Stabilized Dispersions

1.3.5.1 Weak Flocculation

This occurs when the thickness of the adsorbed layer thickness is small (usually <5 nm), particularly when the particle size and Hamaker constant are large. In this case G_{min} becomes sufficiently large (a few kT units) for flocculation to occur. This flocculation is reversible and with concentrated dispersions the system may show thixotropy (at a given shear rate the viscosity decreases with time and when the shear is removed the viscosity increases to its initial value within a time-scale that depends on the extent of flocculation). This process, sometimes referred to as sol$\leftrightarrow$gel transformation, is important in many industrial applications, e.g. in paints and cosmetics.

The depth of the minimum, G_{min}, required for flocculation depends on the volume fraction of the dispersion. This can be understood from a consideration of the free energy of flocculation, ΔG^{flocc}, which consists of two terms, an enthalpy term, ΔH^{flocc}, which is negative and determined by the magnitude of

G_{min}, and an entropy term, $T\Delta S^{flocc}$, which is also negative since any aggregation results in a decrease in entropy. According to the second law of thermodynamics

$$\Delta G^{flocc} = \Delta H^{flocc} - T\Delta S^{flocc} \tag{37}$$

Hence the second term in Eq. (37), which has a negative sign, becomes positive and therefore entropy reduction must be compensated by a high enthalpy term for flocculation to occur, i.e. for ΔG^{flocc} to become negative.

For a dilute dispersion with a low volume fraction ϕ, the entropy loss on flocculation is large, and to obtain an overall negative free energy, ΔH^{flocc} needs to be large, i.e. a large G_{min} is required. In contrast, for a concentrated dispersion with large ϕ, the entropy loss on flocculation is relatively small and a lower G_{min} is sufficient for flocculation to occur. This means that G_{min} required for flocculation decreases with increase in the volume fraction ϕ of the dispersion.

1.3.5.2 Strong (Incipient) Flocculation

This occurs when the solvency of the medium for the chains becomes worse than a θ-solvent, i.e. $\chi > ½$. This is illustrated in Fig. 1.10, which shows the situation when χ changes from a value $<½$ for a good solvent to $>½$ for a poor solvent. When $\chi < ½$, both G_{mix} and G_{el} (or G_{VR}) are positive and hence the total energy of interaction will show only a shallow minimum at a distance close to 2δ. However, when $\chi > ½$, G_{mix} becomes negative (attractive), which, when combined with the van der Waals attraction, gives a deep minimum causing "catastrophic" flocculation, usually referred to as incipient flocculation. In most cases, there is a correlation between the critical flocculation point and the θ-condition of the medium. Good correlation is found in many cases between the critical flocculation temperature (CFT) and θ-temperature of the stabilizing

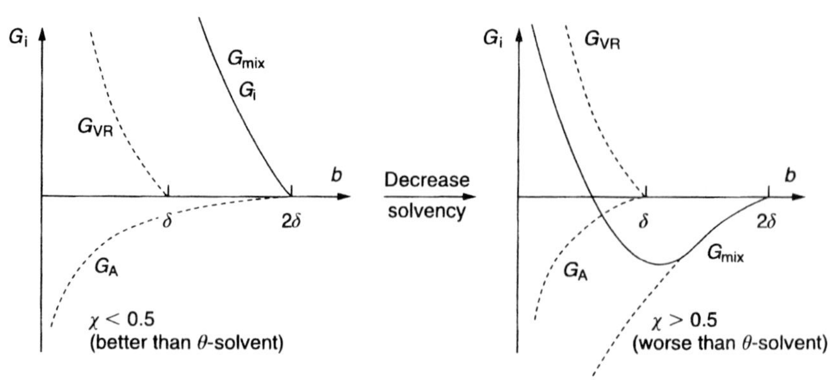

Fig. 1.10 Influence of reduction in solvency of the medium for the chains on the energy–distance curve for sterically stabilized dispersions.

chain A in solution. Good correlation is also found between the critical volume fraction (CFV) of a non-solvent for the A chain and its θ-point. However, in some cases such correlation may break down, particularly the case for polymers that adsorb with multi-point attachment. This situation has been described by Napper [20], who referred to it as "enhanced" steric stabilization.

Hence by measuring the χ parameter for the stabilizing chain as a function of the system variables such as temperature and addition of electrolytes, one can establish the limits of stability/instability of sterically stabilized dispersions. The χ parameter can be measured using various techniques, e.g. viscosity, cloud points, osmotic pressure or light scattering.

1.4 Depletion Flocculation

This is produced on the addition of "free", non-adsorbing polymer to a dispersion [26]. The free polymer cannot approach the particle surface by a distance Δ (that is approximately equal to twice the radius of gyration of the polymer, $2R_G$). This is due to the fact that when the polymer coils approach the surface, they lose entropy and this loss is not compensated by an adsorption energy. Hence the particles will be surrounded by a depletion zone (free of polymer) with thickness Δ. When the two particles approach each other to a surface–surface separation distance $<2\Delta$, the depletion zones of the two particles will overlap. At and above a critical volume fraction of the free polymer ϕ_p^+, the polymer coils become "squeezed" out from between the particles and hence the osmotic pressure outside the particle surfaces becomes higher than in between them (with a polymer-free zone) and this results in weak flocculation. The first quantitative analysis of this process was carried out by Asukara and Osawa [26], who derived the following expression for the depletion free energy of attraction:

$$G_{dep} = -\frac{3}{2}\phi_2 \beta x^2; \quad 0 < x < 1 \tag{38}$$

where ϕ_2 is the volume concentration of the free polymer:

$$\phi_2 = \frac{\frac{4}{3}\pi\Delta^3 N_2}{V} \tag{39}$$

N_2 is the total number of coils and V is the volume of solution. β is equal to R/Δ and x is given by

$$x = \frac{\Delta - (h/2)}{\Delta} \tag{40}$$

Fleer et al. [27] derived the following expression for G_{dep}:

$$G_{dep} = \left(\frac{\mu_1 - \mu_1^0}{V_1}\right)\frac{2\pi}{3}\left(\Delta - \frac{1}{2}h\right)^2\left(3R + 2\Delta + \frac{1}{2}h\right) \tag{41}$$

where μ_1^0 and μ_1 are the chemical potentials of pure solvent and polymer solution with a volume fraction ϕ_p and V_1 is the molar volume of the solvent.

References

1 Th.F. Tadros (ed.), *Solid/Liquid Dispersions*, Academic Press, London (**1987**).
2 B. Deryaguin and L.D. Landau, *Acta Physicochim. USSR*, *14*, 633 (**1941**).
3 E.W. Verwey and J.Th.G. Overbeek, *Theory of Stability of Lyophobic Colloids*, Elsevier, Amsterdam (**1948**).
4 D.H. Napper, *Polymeric Stabilization of Colloidal Dispersions*, Academic Press, London (**1983**).
5 J. Lyklema, *Fundamentals of Interface and Colloid Science*, Vol. I, Academic Press, London (**1991**).
6 J. Lyklema, *Fundamentals of Interface and Colloid Science*, Vol. II, Academic Press, London (**1995**).
7 H.C. Hamaker, *Physica (Utrecht)*, *4*, 1058 (**1937**).
8 E.M. Lifshits, *Sov. Phys. JETP*, *2*, 73 (**1956**).
9 H.B.G. Casimir and D. Polder, *Phys. Rev.*, *73*, 360 (**1948**).
10 G. Gouy, *J. Phys.*, (4) *9*, 457 (**1910**); *Ann. Phys.*, (9) *7*, 129 (**1917**); D.L. Chapman, *Philos. Mag.*, (6) *25*, 475 (**1913**).
11 O. Stern, *Z. Electrochem.*, *30*, 508 (**1924**).
12 M. Von Smoluchowski, *Phys. Z.*, *17*, 557, 585 (**1917**).
13 N. Fuchs, *Z. Phys.*, *89*, 736 (**1934**).
14 H. Reerink and J.Th.G. Overbeek, *Discuss. Faraday Soc.*, *18*, 74 (**1954**).
15 J. Lyklema, *Fundamentals of Interface and Colloid Science*, Vol. III, Academic Press, London (**2000**).
16 B.V. Deryaguin and E. Obukhov, *Zh. Fiz. Khim.*, *7*, 297 (**1936**).
17 D. Exerowa and P.M. Kruglyakov, *Foam and Foam Films*, Studies in Interface Science, Vol. 5, Elsevier, Amsterdam (**1998**).
18 J.N. Israelachvili and G.E. Adams, *J. Chem. Soc., Faraday Trans. 1*, *74*, 975 (**1978**).
19 G.J. Fleer, M.A. Cohen Stuart, J.M.H.M. Scheutjens, T. Cosgrove and B. Vincent, *Polymers at Interfaces*, Chapman and Hall, London (**1993**).
20 Th.F. Tadros and B. Vincent, in *Encyclopedia of Emulsion Technology*, P. Becher (ed.), Marcel Dekker, New York (**1983**).
21 Th. F. Tadros, in *Principles of Science and Technology in Cosmetics and Personal Care*, E.D. Goddard and J.V. Gruber (eds.), 73–113. Marcel Dekker, New York (**1999**).
22 J. Lyklema, *Fundamentals of Interface and Colloid Science*, Vol. V, Elsevier, Amsterdam (**2005**).
23 P.J. Flory and W.R. Krigbaum, *J. Chem. Phys.*, *18*, 1086 (**1950**).
24 P.J. Flory, *Principles of Polymer Chemistry*, Cornell University Press, Ithaca, NY (**1953**).
25 E.L. Mackor and J.H. van der Waals, *J. Colloid Sci.*, *7*, 535 (**1951**).
26 A. Asukara and F. Oosawa, *J. Chem. Phys.*, *22*, 1235 (**1954**); *J. Polym. Sci.*, *93*, 183 (**1958**).
27 G.J. Fleer, J.M.H.M. Scheutjens and B. Vincent, *Am. Chem. Soc. Symp. Ser.*, *240*, 245 (**1984**).

2 Thermodynamic Criterion of Spontaneous Dispersion

Eugene D. Shchukin and Alexander V. Pertsov

Abstract

The analysis is presented of the behavior of the function $\Delta F = \Delta F(r, \sigma, n)$, which describes changes in the free energy of a system due to dispersion of a macrophase at sufficiently low interfacial energy. Three different cases are considered: under a constant volume of disperse phase, constant size of particles and constant particle number. It is shown that in all three cases the condition $\Delta F = 0$ serves as a necessary thermodynamic criterion of the possibility of spontaneous dispersion and formation of a stable, lyophilic colloid system.

2.1 Introduction

The idea of the thermodynamic stability of a microheterogeneous system with very low interfacial tension between a dispersion medium and particles (droplets) participating in Brownian motion was expressed by Volmer in connection with the problem of the nature of critical emulsions [1, 2]. Following Volmer's idea, Rehbinder considered the possibility of spontaneous (under the thermal vibration effect) splitting of microblocks with size δ from the solid surface under the conditions of strong decrease in the surface energy, σ, caused by the contact of this surface with a surface-active medium, down to a critical value σ_c, defining by the work $\delta^2\sigma \lesssim kT$ [3].

Shchukin and Rehbinder proposed for the first time quantitative considerations of the condition of spontaneous dispersion of a condensed phase on the basis of the analysis of changes in the system free energy, ΔF, due to dispersion of condensed phase in a given dispersion medium [4, 5]. It was shown that the macrophase dispersion is thermodynamically favorable if the free energy change due to dispersion (isolation of n particles with radius r, at interfacial energy σ) becomes negative, i.e. $\Delta F = n \times 4\pi r^2\,\sigma - T\Delta S < 0$, where $\Delta S(C)$ is the entropy gain and C is the concentration. If there is a factor opposing dispersion to mo-

Colloids and Interface Science Series, Vol. 1
Colloid Stability: The Role of Surface Forces, Part I. Edited by Tharwat F. Tadros

ISBN: 978-3-527-31462-1

lecular dimension b, a negative minimum of ΔF at $r > b$ may occur, i.e. formation of a thermodynamically stable colloid system takes place.

These notions were then developed in several directions, in theoretical and experimental studies of our scientific school, with the participation of Kochanova and Fedoseeva [6–10], in the work of Rusanov and Kuni [11–15]. Such an approach has been used in studies of the contact interactions of disperse phase particles [16–21]. Various aspects of the problem of spontaneous dispersion and of the thermodynamic stability of microheterogeneous systems have been considered [22–36]. In most of the studies in this area, a principal question continuously discussed was what the particular characteristics of the ΔF function are that one needs to consider as corresponding to the general, universal conditions of spontaneous dispersion [26, 27, 29, 34]. For elucidation of this problem, changes in the free energy of a system, ΔF, associated with the dispersion were analyzed under three different conditions: (1) in dependence on particle size, for a given, constant disperse phase volume (or concentration), when ΔF has a virtual positive maximum; (2) in dependence on particle number, for their constant size, with a negative minimum in ΔF; and (3) in dependence on particle size, for their constant number, when ΔF is monotonic [37–40]. This analysis is presented in the following sections, in all cases, for monodisperse systems, with broad variation of σ. Examples of the real physical systems corresponding to the considered versions of the behavior of ΔF are given.

2.2
Work and Entropy of Dispersion

The involvement of n isolated particles with an average size δ in Brownian motion leads to the compensation of the required dispersion work $n\alpha\delta^2\sigma$ by the entropy gain, ΔS:

$$\Delta F = n\alpha\delta^2\sigma - T\Delta S \qquad (1)$$

where α is a numerical coefficient depending on particle shape and the state of initial phase. In the approach of the regular solution theory, the entropy increase resulting from the mixing of n particles or $N_1 = n/N_A$ mol of particles with N_2 mol of dispersion medium is given by

$$\Delta S = R\{N_1 \ln{(N_1 + N_2)/N_1]} + N_2 \ln{[(N_1 + N_2)/N_2]}\}$$

In dilute colloid dispersions, $N_1 < N_2$, and

$$\Delta S \approx R[N_1 \ln(N_2/N_1) + N_1] = nk[\ln{(N_2/N_1)} + 1]$$

Consequently,

$$\Delta F = n\alpha\delta^2\sigma - nkT[\ln(N_2/N_1) + 1] = n(\alpha\delta^2\sigma - \beta kT) \qquad (2)$$

In real dilute systems, the parameter $\beta = \ln(N_2/N_1) + 1$ can be on the order of 10–15 and more.

The work needed to isolate a spherical particle with diameter $\delta = 2r$ from a stable compact equilibrium macrophase equals $w = 4\pi r^2\sigma$. (For such initial macrophase, the factor 1/3 is not used; compare [7–11] and [17, 20].) Equations (1) and (3) then take the following form:

$$\Delta F = n \times 4\pi r^2\sigma - nkT\,[\ln(1/C) + 1] \tag{3}$$

The ratio N_2/N_1 is replaced here by a generalizing parameter, the dilution $1/C$, where C is the disperse phase concentration. One can consider the value of C either as the ratio $C = n/N$, where n is the number of disperse phase particles and N is the number of particles (molecules) of the dispersion medium, or as the ratio $C = m/M$ of numbers of moles (or of values of masses) and also as the ratio $C = v/V$ of the common volume of the disperse phase, v, and volume V of the dispersion medium, etc. Some ambiguity in the definition of this parameter in the entropy of mixing evaluation reflects known contradictions in the calculations of chemical potential in thermodynamics of small systems for the case of significantly different sizes of disperse phase particles and dispersion medium molecules (compare, e.g., [41–43] and [17a], pp. 270–273). One may say that such ambiguity corresponds to the evaluation of the disperse phase *number of moles* for two essentially different definitions of a *mole* of disperse phase: as N_A particles or as N_A molecules (such contradictions can mitigate in the case of polymer mixtures). However, this problem is not a subject of discussion in this chapter. We are confined here to two "extreme" approaches in the estimation of concentration C: first, as the ratio $C = n/N$ (in aqueous medium, $N = 3.37 \times 10^{25}$ molecules per 10^3 cm^3), and, second, as the ratio $C = v/V$ assuming $V = 10^3$ cm^3.

In accordance with such an approach, Eqs. (2) and (3) can be written for the case of $C = v/V$ as

$$\Delta F = n \times 4\pi r^2\sigma - nkT\,\{\ln[V/v] + 1\} \tag{4}$$

and

$$\Delta F/kT = (v/{}^4\!/\!_3\pi r^3)[4\pi r^2\sigma/kT - \ln(Ve/v)] \tag{5}$$

or

$$\Delta F/kT = n\{4\pi r^2\sigma/kT - \ln(Ve/{}^4\!/\!_3\pi r^3 n)] \tag{6}$$

For the case $C = n/N$ the same Eqs. (3) and (5) are as follows:

$$\Delta F = n \times 4\pi r^2\sigma - nkT\,[\ln(N/n) + 1] \tag{7}$$

and

$$\Delta F/kT = (v/\tfrac{4}{3}\pi r^3)\{4\pi r^2\sigma/kT - \ln[(V/v_m)e/(v/\tfrac{4}{3}\pi r^3)]\} \tag{8}$$

or

$$\Delta F/kT = n[4\pi r^2\sigma/kT - \ln(Ne/n)] \tag{9}$$

where v_m is the volume of a dispersion medium molecule.

These expressions will be used in the following sections where ΔF is analyzed as a function of various parameters and under various conditions, in all cases for systems of spherical particles monodisperse at every given stage. The dispersion medium volume is assumed to be constant and equal to, e.g., $V = 1000\ \text{cm}^3$. At 300 K, the kT value is 4.14×10^{-21} J.

Such an approach to the quantitative evaluation of changes in the free energy of a system allows one to formulate the general thermodynamic conditions of the transition to spontaneous dispersion and to the possibility of formation of a stable colloid–disperse system. The ΔF value is positive for coarse dispersions (with respect to the initial compact phase) and high values of σ and falls rapidly for low values of σ and sufficiently high dispersion. The transition from positive to negative values of the free energy changes at $\Delta F = 0$ gives us a criterion of the thermodynamically favorable spontaneous dispersion of a macrophase. In accordance with Eqs. (2) and (3) and the following equations, the possibility of such a transition of ΔF to negative values needs, for a given particle size, $\delta = 2r$, very low values of the interfacial energy:

$$\sigma \lesssim \sigma_c = kT[\ln(1/C) + 1]/\alpha\delta^2 = \beta kT/4\pi r^2$$

Correspondingly, for a given value of σ, such a transition needs sufficiently small values of $r \lesssim (\beta kT/4\pi\sigma)^{1/2}$.

For some smaller particle size and in the presence of definite additional conditions (e.g. increase in σ when approaching molecular dimensions), a negative minimum of the free energy is possible, corresponding to the formation of a thermodynamically stable, lyophilic colloid-disperse system.

If the initial system is not a compact phase, but an aggregate of insoluble or partially insoluble particles (coagulation structure or a structure with phase contacts [21, 44–46]), the term $4\pi r^2\sigma$ is changed by the product of cohesion energy in a contact between particles and the average coordination number.

Special attention in the following sections will be paid to overcoming the traditional, formal approach in the analysis of this problem. A number of examples of real physical processes and states that can correspond to the described types of ΔF behavior will be given. These include peptization/coagulation transition in colloid sols, particle bridging in the course of hydration hardening of mineral binders, etc.

One has to stress that the presented analysis concerns *thermodynamic* conditions of spontaneous dispersion and formation of a stable colloid system. In this chapter, only minor attention is paid to kinetic factors, which need independent consideration, particularly with respect to solid phases [4, 25, 35, 47].

2.3 Behavior of $\Delta F(r)$ When v=Constant

In order to evaluate the level of thermodynamic advantage (disadvantage) of various states of a disperse system, let us consider first the $\Delta F(r)$ function in the case when v=constant and the system remains monodisperse at every stage of such virtual grinding [4–12, 17]. In this case the number of particles, n, is inversely proportional to the third power of particle size, i.e. $n = v/(\frac{4}{3}\pi r^3)$ and Eqs. (4) and (5) acquire a simplified, illustrative form:

$$\Delta F/kT = c_1(1/r) - c_2(1/r^3)\ .$$

Equations (7) and (8) are then expressed as

$$\Delta F/kT = c_1(1/r) - (c_2 + c_3 + c_4 \ln r)(1/r^3)$$

where c_1, c_2, c_3 and c_4 are constants.

In both cases, as the grinding of compact condensed phase into smaller and smaller spherical particles of equal sizes proceeds, the ΔF value first grows relatively slowly from zero up to some maximum value ΔF_{max} at $r^{(max)}$ and then falls rapidly, passes 0 at $r=r^{(0)}$ and becomes negative. For the critical value of $r=r^{(0)}$ one can write

$$4\pi r^{(0)2}\sigma = kT\,[\ln(1/C) + 1]$$

A positive maximum of the ΔF function is obtained from the equation $d\Delta F/dr=0$. This maximum corresponds to the particle radius $r^{(max)} = 3^{1/2}r^{(0)}$. This is the case with both approaches: for $C = v/V$ and $C = n/N$. Such a maximum has some kind of a virtual, "figurative" character [17a, pp. 463–464] and [40]. A *continuous* ("quasi-continuous") approach to this maximum from a macrophase, while maintaining constant volume, means first splitting of one molecule from every particle and second using these molecules to form a number of new particles of the same size as the residual ones. Such a process, however, cannot be realized kinetically. This process is in principle impossible when the disperse phase is insoluble in the dispersion medium as is assumed in [34]. In reality, such a consideration of changesin $\Delta F(r)$ does not correspond to physical conversion of a given system into more and more finely dispersed particles, but only reflects one's (mental) approach to a comparison of series of systems [4–6, 9, 10, 17]. This approach represents a correct evaluation of critical parameters, $r^{(0)}$ and σ_c.

In [34], the described $\Delta F(r)$ maximum is viewed as a real potential barrier. That is, if during process of dispersion–mechanical grinding the particle size of $r^{(max)}$ is reached, the system disperses spontaneously. However, a process of dispersion in the form of such *discrete* transition from one monodisperse system to another (for a spherical particle, such a process corresponds to dividing into two equal ones with a 1.26-fold decrease in radius) cannot be physically realized either. These estimates, however, do not constitute a paradox if one is not limited to the consideration of spontaneous dispersion only for the case of the $\Delta F(r)$ dependence for ν=constant.

It was stressed in [4–6] that a transition to a particle size less than $r^{(max)}$ or $r^{(0)}$ does not by itself constitute a sufficient condition for the formation of a stable colloid–disperse system to take place. Obviously, a combination of only two so far considered factors, $c_1/r – c_2/r^3$, does not provide for any restrictions on dispersion down molecular size. One needs to consider obligatory a third factor ("factor 3"), which prevents the system being dividing down to individual molecules. Formally, this factor can be expressed by the $\sigma(r)$ dependence (e.g. [48–51]): increase in σ for small r. This can be associated with an asymmetric structure of the disperse phase molecules, particularly of the diphilic molecules of surface-active substances in micelles. It can also be some given domain structure of a solid phase or a globular structure with open porosity, with coagulation or phase contacts between particles, and, of course, a polymer system in which particles are represented by macromolecules.

Depending on the particular character of this third factor, changes in ΔF in the small particle size range can proceed in various ways [5, 6, 9, 10, 40]. The two extreme cases are, on the one hand, the dispersion down to individual molecules or ions, dissolution (the third factor is very weak or completely absent) and, on the other, a monotonic increase in ΔF until molecular (ionic) dimensions are reached ("very strong factor 3", high σ values, i.e. the most common case of a lyophobic system, of a strong solid in the absence of surface-active medium). Between these two extreme situations, a minimum in ΔF is possible for particles of colloidal dimensions. Such a minimum can be positive; the system is then metastable and tends to return to the compact state. If, however, $\Delta F = 0$ is achieved at some $r = r^{(0)}$, then, for a "moderately strong factor 3", a negative minimum in the ΔF function can exist at a definite $r^{(eq)} < r^{(0)}$, corresponding to a stable equilibrium colloid system. It is important that the critical size, $r^{(0)}$, exceeds the molecular size b. Since $r^{(0)} = (\beta kT/4\pi\sigma)^{½}$, this condition can be formulated, after Rehbinder and Shchukin, as a criterion

$$RS = 2r^{(0)}/b = (\beta kT/ab^2\sigma)^{\frac{1}{2}} \gg 1$$

where $a = \pi$ for spherical particles with diameter $\delta = 2r$ [9, 10, 17a (p. 464)].

Along with this, one has to keep in mind, however, that such critical values of $r^{(0)}$, σ_c, $w_c = 4\pi r^{(0)2}\sigma$ (and, generally, the work w_c) do not appear by themselves (*per se*) as absolute, single-meaning characteristics, even within a frame of monodisperse systems. In the case of equilibrium system, concentration also represents a critical parameter, which for a dilute system is given by the factor $\beta = \ln(1/C)+1$. In the case of polydisperse systems, the critical condition parameters are represented by distribution histograms.

Some typical quantitative data of $\Delta F(r)$ calculations for ν=constant are given in Table 2.1 and Fig. 2.1. Table 2.1 illustrates the dependence of the critical $r^{(0)}$ value on concentration C and interfacial energy σ in the $C=\nu/V$ scheme. In a very large region of these two parameters, the $r^{(0)}$ values do not in practice exceed limits of one order of magnitude.

In Fig. 2.1, the function $\Delta F(r)/kT$ is presented for the case of ν=constant= 10^{-5} cm^3 for both calculation schemes $C=\nu/V$ and $C=n/N$. For illustrative comparison, the σ values are selected in such a way that the particle critical radius is the same: $r^{(0)}=10^{-6}$ cm. Both schemes yield qualitatively similar results [similar in their shape are also the graphs $\Delta F(r)/kT$ for $\nu=10$ cm^3, when the quantitative difference in their values is of six orders of magnitude]. In Fig. 2.1, the dependence $\Delta F(r)/nkT$ is presented also with respect to one particle; it is monotonic and grows rapidly in the area of positive values.

The sharp dependence $1/r^3<0$ predetermines the leap-like transition from positive values of the free energy changes, ΔF, to negative values as r decreases, i.e. the well-defined critical position of $r^{(0)}$ [and, correspondingly, of $r^{(\max)} = 3^{1/2} r^{(0)}$]. Under the conditions of transition to a spontaneous dispersion (to *lyophilicity*), the range of possible values of interfacial energy σ at the boundary between disperse phase and dispersion medium is predominantly within the limits between 10^{-3} and 10^{-1} erg cm^{-2} (mJ m^{-2}).

Interfacial tension measurements have been carried out [8–10] in a two-component, two-phase mixtures of hydrocarbons with a polar organic substance, 8-hydroxyquinoline, in the vicinity of the critical mixing temperature T_c and compared with the independently provided stability observations (time t_τ during which the turbidity maintained constant) of corresponding microemulsions formed in the vicinity below T_c with a width of several degrees. In these investigations, the formation of stable, direct and inverse microemulsions, and, close to T_c, bicontinual microemulsions were observed. In this case the values of σ are on the order of $\sim 10^{-2}$ mN m^{-1} or lower.

Table 2.1 Critical size of a disperse phase particle radius, $r^{(0)}$ (10^{-6} cm), corresponding to the equation $\Delta F=0$, as a function of the free interfacial energy value, σ (mJ m^{-2}), and disperse phase concentration C, in the $C=\nu/V$ scheme.

$C=\nu/V$	σ			
	0.001	**0.01**	**0.1**	**1.0**
10^{-2}	4.3	1.36	0.43	0.14
10^{-4}	5.8	1.84	0.58	0.18
10^{-6}	7.0	2.2	0.70	0.22
10^{-8}	8.0	2.5	0.80	0.25

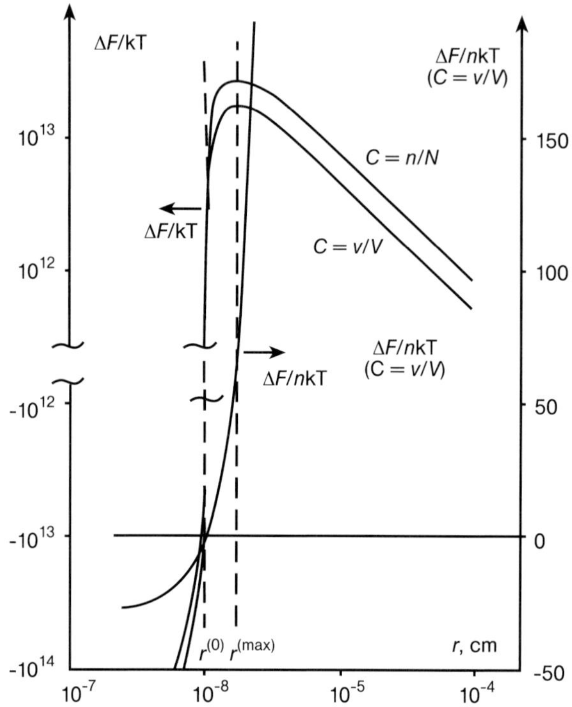

Fig. 2.1 Changes in the free energy, $\Delta F/kT$, as a function of particle radius, r (cm), in the case of the dispersion of constant volume, $\nu = 10^{-5}$ cm^3, of compact phase in $V = 1000$ cm^3 of dispersion medium: in the $C = \nu/V$ calculation scheme, for $\sigma = 0.064$ mJ m^{-2}, and in the $C = n/N$ scheme, $N = 3.37 \times 10^{25}$ molecules, for $\sigma = 0.103$ mJ m^{-2}. The $r^{(0)}$ and $r^{(max)}$ values are 10^{-6} and 1.73×10^{-6} cm, respectively, for both schemes. The change in the ΔF value for a considered monodisperse system is leap-like, of many orders of magnitude in the vicinity of a transition from positive to negative values at $r = r^{(0)}$.

Dilution values, $1/C$, can fall into a very broad range, e.g. 10^1–10^8 or more (it should be remembered, however, that a logarithm of this parameter is considered). Simply speaking, the critical values of r are inversely proportional to $\sigma^{1/2}$. As a result, at room temperature, the $r^{(0)}$ values are more or less close in their order of magnitude to 10^{-6} cm.

Such consideration of $\Delta F(r)$ for ν=constant leads one to the correct evaluations of critical relation between parameters: dispersity and σ, which govern the possibility of spontaneous dispersion. However, the condition ν=constant intrudes the value of concentration and does not correspond to a real physical process of an *equilibrium* system formation. One can say that in some sense such an approach is opposite in its logic to conditions of a real dispersion process which can result in the formation of a stable (saturated) colloid–disperse system.

Using any schemes $C=v/V$ and $C=n/N$ for evaluating of entropic factor gives in both cases qualitatively similar results, with an abrupt transition from positive to negative $\Delta F(r)$ values, i.e. with the clearly expressed correlation between $r^{(0)}$ and σ or the work $w_c \sim r^{(0)2}\sigma$. However, the quantitative difference may be significant. For instance, in the $C=n/N$ scheme, for the same values of σ, the $r^{(0)}$ is positioned more to the right {or the critical value of $w[r^{(0)}]=w_c$ becomes greater}. Indeed, the ratio $N/n> V/v$ and is higher the coarser the particles are. Of course, if one compares *concentrations* corresponding to a given value of the logarithmic term $\ln(1/C)+1$ in the entropy factor, this difference is very large; e.g. $r^{(0)}=10^{-6}$ cm, $\sigma=0.064$ mJ m^{-2} and, consequently, $w_c/kT=19.4$ represent the condition $19.4=\ln(1/C)+1$, i.e. dilution $1/=10^8$. In the $C=v/V$ scheme, for $V=10^3$ cm^3 this gives $v=10^{-5}$ cm^3 and a number of particles with given size $n^{(0)}=2.39\times10^{12}$, whereas in the $C=n/N$ scheme, for $N=3.37\times10^{25}$ (water molecules in 1 L), the same dilution $1/C=10^8$ gives particle number $n^{(0)}=N/10^8=3.3\times10^{17}$, i.e. 1.4×10^5 times greater. However, one need not to consider such discrepancies as a sign of any internal contradictions in the scheme of calculations presented here. Conversely, it would be interesting to turn to experiments with highly precise quantitative data for $r^{(0)}$, σ and C for the elucidation: which particular scheme for taking concentration C into account corresponds to the correct evaluation of the entropy factor ΔS, and, correspondingly, to the calculations of chemical potential in colloidal systems.

2.4 Behavior of $\Delta F(n)$ When r=Constant

In this section we turn to the analysis of the $\Delta F(n)$ dependence for the case of a constant degree of dispersity r=constant, which presented for the first time by our group [37–40]. A situation where unlimited fracturing is forbidden is here rather radical (formally, σ increases in a leap-like fashion up to high values). This situation occurs in a number of real systems.

Such a consideration corresponds to the generally accepted notion of the nature of the stability of micellar systems of the colloid surface-active substances as an entropic factor [14, 15, 52–58]; these also can be vesicles. This approach is close to principles of the statistical analysis of the behavior of polymer molecules [59–64]. As an extreme manifestation of the sharp $\sigma(r)$ dependence, the peptization of insoluble (partially soluble) particles of a coagulate [44–46, 65, 66] or of a disperse structure with phase contacts [46, 66–68] can be considered. Swelling and dispersion of highly hydrophilic clays, and also disaggregation of solids with definite microstructure [18, 69–71], can be also included here.

Equations (2–4 and 7) then appear in the $C=v/V$ scheme in the form of a single equation, Eq. (6), and in the $C=n/N$ scheme in the form of Eq. (9). Let us discuss this second situation. In this situation, the essence of the competition between the work of dispersion necessary for isolation of a spherical particle from a macrophase and the entropic gain due to the involvement of an isolated particle in Brownian motion can be clearly demonstrated by the following simple relation:

$$\Delta F/kT = n(c' - c'' \ln n) \tag{10}$$

where c′ and c″ are constants depending on the values of σ and r ($r>b$). As n *decreases* (from large values which are outside the range considered), the function given by Eq. (10) passes through zero at some $n^{(0)}$ and then [at some lesser $n^{(eq)}$] reveals a negative minimum corresponding to the thermodynamically stable, equilibrium colloid–disperse system, referred to as *lyophilic* following Rehbinder. In both this scheme of concentration accounting and in the $C=\nu/V$ scheme, the values of $n^{(eq)}$ and $n^{(0)}$ differ from each other by the same constant factor e: $n^{(eq)}=n^{(0)}/e$.

In the $C=n/N$ scheme, the point $\Delta F=0$ is associated with particle number, $n^{(0)}=Ne\exp[-(4\pi r^2\sigma/kT)]$. The equilibrium number of particles, i.e. the *colloid solubility* of the same size particles corresponding to the ΔF minimum equals [17a, pp. 268, 269] $n^{(eq)}=N\exp[-(4\pi r^2\sigma/kT)]$.

This minimum is fairly shallow, $\Delta F[r^{(eq)}]/kT=n^{(eq)}$, i.e. only one kT per particle (by absolute value). Such a system exists only as a very mobile dynamic equilibrium between colloidal solution of particles and their macrophase.

Some examples are given in Figs. 2.2 to 2.4. For realistic concentrations of the disperse phase, correlation between the dispersity, r, and interfacial energy, σ, or the work, w, is very critical, i.e. a transition from macroscopic state to colloid solution (*monodisperse* system) is abruptly manifested. Such a transition becomes diffuse for polydisperse systems.

In Fig. 2.2, the $\Delta F(n)/kT$ function is presented for the $C=n/N$ scheme, with $r=\text{constant}=10^{-6}$ cm and $\sigma=0.064$ mJ m^{-2}. The function has a characteristic negative minimum at $n^{(eq)}=n^{(0)}/e$. In this case, the leap in ΔF, associated with the transition from positive to negative values at $n=n^{(0)}$, reaches 30 orders of magnitude. In the $C=\nu/V$ scheme, the $\Delta F(n)/kT$ graph has a similar appearance but with substantially lower absolute values (compare Fig. 2.6). The $\Delta F/nkT$ function is monotonic; for the equilibrium (saturated) system at $n^{(eq)}$, this function equals –1.

Typical examples of real physical systems to which the approach of considering ΔF as a function of n at constant r can be applied are represented, on the one hand, by colloidal solutions of micelle-forming surfactants and, on the other, by dispersion (peptization) of a globular structure with a given strength of contacts between particles, p_1, and the work of contact rupture, u_1.

The studies of the structure and properties of micellar systems, thoroughly covered in the literature (e.g. [14, 15, 54, 72–78]), include also estimations of the surface energy at the boundary between a micelle and a medium. The next simple evaluation can illustrate a possible, semi-quantitative estimation based on the approach presented here. Let us use the value of $r=0.25\times10^{-6}$ cm as a typical radius of a micelle. According to Eqs. (9) and (6), the realistic concentrations of surfactant in micellar state correspond in this case to relatively high σ values

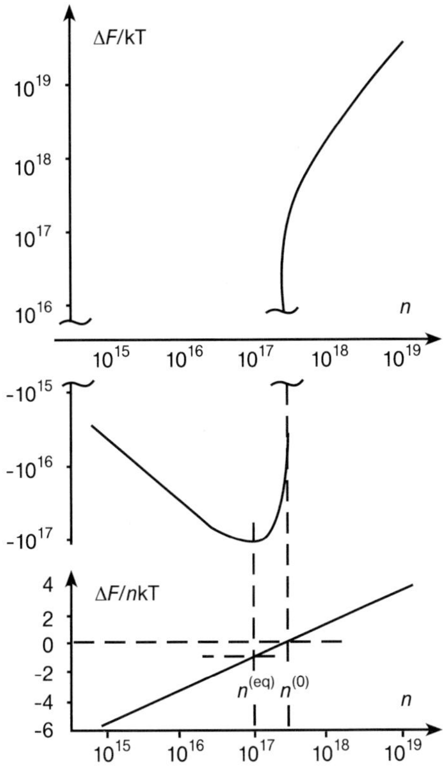

Fig. 2.2 Changes in the free energy, $\Delta F/kT$, as a function of particle number, n, in a monodisperse system consisting of spherical particles with radius $r=10^{-6}$ cm, for $\sigma=0.064$ mJ m^{-2} in the $C=n/N$ scheme, $N=3.37\times10^{25}$ molecules. The corresponding values of $\Delta F/nkT$ per one particle are also shown.

at some (imaginary) boundary between a micelle and a medium: these σ values are on the order of parts of an mJ m^{-2} or even higher (Fig. 2.3). This agrees with estimations which have recently been considered [14, 55, 57, 58, 74–78].

In the case of a dispersion structure, considered first by Pertsov [7] and independently by Martynov and Muller [79, 80], the work of particle isolation from initial phase can be formulated in a different way, namely as $w=\frac{1}{2}zu_1$, where u_1 is the free energy of cohesion in an individual contact (by the absolute value) and z is average coordination number. In this example, let us restrict ourselves to the coagulation–mechanically reversible contacts (cohesion in such contacts is defined mainly by the surface dispersion interactions [81–86]; see also [17] and [44–46]). In this case the cohesion force in an individual contact, p_1, equals $p_1=\pi r\Delta\sigma_f$, where $\Delta\sigma_f$, is the free energy of interaction of these surfaces in a given medium [17, 81]. Let us utilize the approximation $\Delta\sigma_f=2\sigma_{12}$. Then one can write that $w\approx(z/2)p_1h_0\approx(z/2)\pi r^2\sigma/h_0$, where h_0 is the gap in an equilibrium contact, which is on the order of molecular dimensions. Assuming that in a relatively loose coagulation structure $z=4$, one obtains an estimate $w=4\pi r\sigma h_0=4\pi r^2\sigma h_0/r$.

This relatively coarse estimation is nevertheless very characteristic: it shows the essential difference in the energies of dispersion of a compact phase and of an ag-

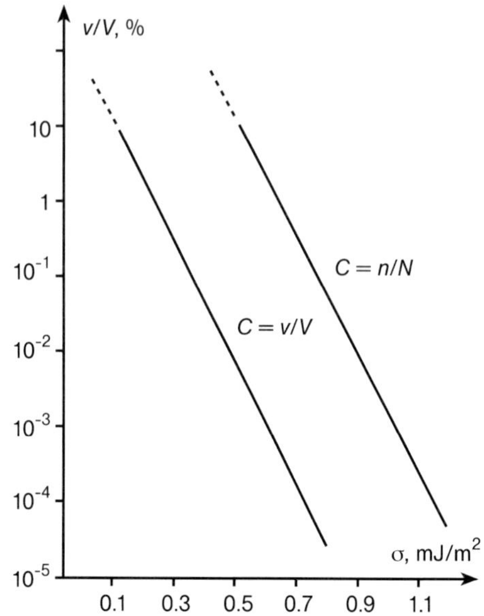

Fig. 2.3 Dependence of the concentration (% by volume) of disperse phase consisting of spherical particles with radius r=constant $=25\times10^{-8}$ cm, in equilibrium with compact macrophase, corresponding to the negative minimum of the $\Delta F/kT$ function, on the interfacial energy, σ, mJ/m^2: in the $C=\nu/V$ scheme, $V=10^3$ cm^3, and in the $C=n/N$ scheme, $N=3.37\times10^{25}$ molecules (water).

gregate of particles. In the latter case, the order of magnitude of the dispersion work is r/h_0 times lower than in the former case. This difference is the greater the larger the particles are. Consequently, for a particle aggregate the conditions of spontaneous dispersion are strongly or even very strongly (qualitatively) different from such conditions for a compact phase (Fig. 2.4). For $r\approx10^{-6}$ cm, the disperse phase concentration can reach here parts of 1% or more with interfacial energy σ values as high as several mJ m^{-2}. These are the conditions under which the peptization/coagulation transition takes place in experiments with hydrophobized Aerosil silica in alcohol (ethyl, propyl) mixtures with water [21, 46, 65, 66]. Alongside studies of coagulation (as an increase in turbidity once a particular water concentration, i.e. medium polarity, is reached), the free energy of interaction between similar macroscopic surfaces in the same media was independently measured. The critical ½$\Delta\sigma_f$ value (approximately equal to σ_c) reaches here several mJ m^{-2}. Similar observations were also performed with the "reverse system" (with respect to polarities of phases), namely hydrophilic, unmodified silica in aqueous cetylpyridinium bromide solutions [45, 46, 66].

Similarly to the $\Delta F(r)$ behavior at ν=constant, the $\Delta F(n)$ analysis at r=constant may yield a significant difference in $C=\nu/V$ and $C=n/N$ schemes. However, such a difference is only of a quantitative character: the positions of $r^{(0)}$ and, correspondingly, of $n^{(0)}$ [i.e. $\ln n^{(0)}$] are shifted. At the same time, full similarity of the ΔF dependences on the relation between the work of dispersion in an individual event of free particle isolation, w, and entropy factor is preserved. One must stress that it is particularly this potential barrier, w, which serves here as a real and universal physical characteristic of systems described [but not a virtual macroscopic maximum in the $\Delta F(r)$ graph; see above].

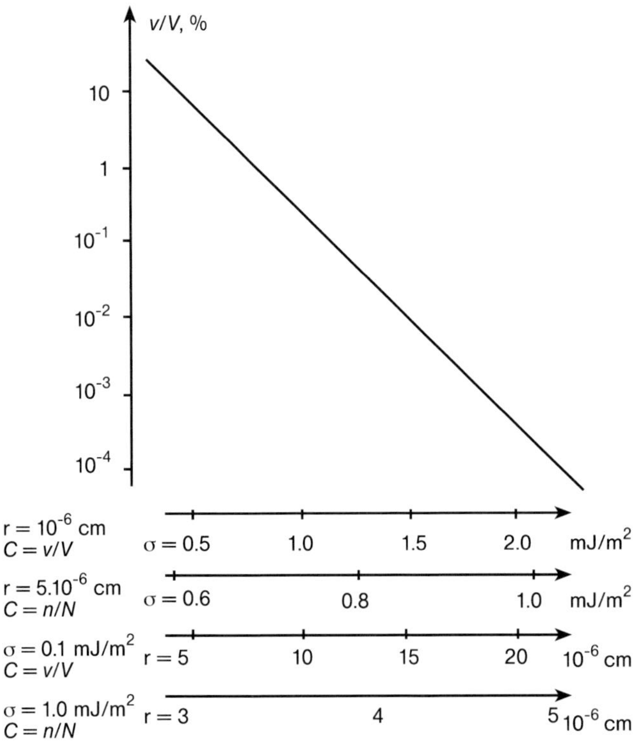

Fig. 2.4 Dependence of the concentration (% by volume) of disperse phase consisting of spherical particles with radius r (insoluble or poorly soluble in a given dispersion medium), in equilibrium with connected globular structure at the negative minimum of $\Delta F/kT$ function, on the particle radius, r (cm), for σ=constant, and on the interfacial energy, σ (mJ/m^2), for r= constant, in the $C=v/V$ scheme, $V=10^3$ cm^3, and in the $C=n/N$ scheme, $N=3.37\times10^{25}$.

2.5
Behavior of $\Delta F(r)$ When n=Constant

Let us briefly consider now one more interpretation of $\Delta F(r)$ changes which leads to the possibility of transition to a spontaneous dispersion and formation of a thermodynamically stable disperse system. Let us turn to the $\Delta F(r)$ behavior when n=constant [37–40]. It is obvious that in this case (with a given dilution, constant in the $C=n/N$ scheme), in the vicinity of the critical value $4\pi r^2\sigma = kT\ln[(1/C)+1]$, the value of w changes rapidly, as r^2. A quantitative description of the entropy factor is similar here to that used in the case of $\Delta F(r)$ for v=constant in the $C=v/V$ scheme. In contradiction to this, a formulation of the $\Delta F(r)$ dependence for n=constant in the $C=v/V$ scheme needs taking into account changes in the disperse phase volume, $v = n \times {}^4\!/_3\pi r^3$, which also contributes to the conditions of dispersion process for decreasing r.

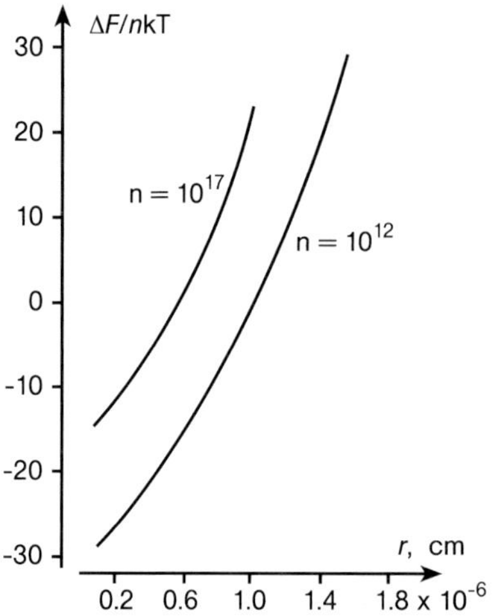

Fig. 2.5 The free energy value $\Delta F/nkT$ per particle as a function of the spherical particle radius, r (cm), in monodisperse systems for σ=0.064 mJ m^{-2}, for n=10^{12} and n=10^{17}, in the C=ν/V scheme, V=10^3 cm^3. The $\Delta F/kT$ function is also *monotonic* and undergoes a rapid change for the considered monodisperse systems with a leap of many orders of magnitude in the vicinity of $r^{(0)}$.

The $\Delta F(r)$ curve falls *monotonically* with decreasing r and passes through zero, of course, at the same ratio of critical characteristics as in the cases considered above. Obviously, only the condition $\Delta F=0$ serves here as a criterion for a transition to spontaneous dispersion and possibility of formation of a stable colloid system. In Fig. 2.5, the function $\Delta F/nkT$ in dependence upon r, also monotonic, is presented for two different concentrations.

It may appear that such an interpretation of ΔF for monodisperse (at each given stage) systems is of somewhat speculative nature, as in the case of consideration of $\Delta F(r)$ maximum for ν=constant. However, this is not the case at all. This interpretation is reflected in real physical systems. These include, on the one hand, gradual dissolution (or evaporation) of particles, e.g. when changes in the medium content occur, with transition from coagulation to peptization, and, on the other, opposing the growth of disperse phase nuclei particles, with a possible loss of stability of a free disperse system and transition (whether desirable or not) to a connected disperse structure. Such transitions take place in industrial processes based on sol–gel transitions, e.g. in aluminosilicate systems and in the hydration hardening process of mineral binders in construction [44, 46, 66, 67]. Of course, many of the well-known methods for the preparation of colloidal (particularly monodisperse) systems belong to this circle.

2.6
Effect of σ, Effect of T

The role of the free interfacial energy, σ, as a principal parameter of ΔF is reflected in all schemes of calculations considered above. There is no need for additional tables and graphs with quantitative data after a detailed description of the behavior of ΔF as a function of various parameters, including a generalizing one, namely the work of particle isolation, w, which is proportional to σ in the frame of the approach used. It is worth mentioning, however, some ways by which this parameter, a defining physical–chemical factor, may change.

First, these are broad possibilities for changing the polarity (or non-polarity) of a medium by changing its contents. The stability–coagulation transition caused by the addition of water to hydrophobic Aerosil dispersions in propyl or ethyl alcohol was mentioned above [21, 46, 65]. In turn, small additives of various surface-active substances, both diphilic organic surfactants and inorganic substances, can have a radical effect on the structure, stability and other properties of disperse systems and conditions of the discussed transitions [17, 87–92].

One should stress here that the discussed problem of establishing the quantitative criteria of spontaneous dispersion is by no means an isolated area of research. It is closely connected with the entire complex of colloid and surface phenomena studies, with their common interest in low interfacial energy. This refers primarily to studies of the formation and structure of stable micellar systems, solubilizing micellar systems and microemulsions in dependence upon variations in the composition and temperature of multicomponent water–hydrocarbon–surfactant–alcohol systems (e.g. [93–104]). These studies, providing a detailed analysis of phase equilibria and covering all kinds of systems – dilute and concentrated, various multiphase compositions, etc. – devote a special attention to the role of low and "ultra-low" values of surface energy at interfaces. The authors' first observations of the areas of stable microheterogeneity in two-, three- and four-component systems [105–108] can serve as an example.

Alongside this, the presented area remains a rather specific one, with its own characteristic features, addressing the solid phases, both compact ones and structures of particles, etc. Numerous investigations of the influence of media and small additives of surfactants on the cohesion of disperse phase particles by the authors' group have been reported [21, 44–46, 65, 66, 87].

Rehbinder's doctrine on the stability of emulsions and other disperse systems also includes the notion of the lyophilic *structure–mechanical barrier*, introduced as a factor of strong stabilization of disperse systems. This notion has been broadly used by the authors' group [56, 109–112]. Such an interfacial adsorption layer is created by diphilic, predominantly high molecular weight surface-active substances (so-called *protective colloids*), which are capable of forming a mechanically strong structure which is resistant to rupture and prevents the coalescence of droplets. At the same time, this layer has a very high affinity towards the dispersion medium, it is *lyophilic*, and has a low value of σ on its outer side, exposed to the medium (this means that the layer is characterized by small values of the complex Hamaker constant and free energy of interaction). This factor prevents particle coagulation. Under such conditions, a decrease in σ to hundredth and thousandth parts of mJ m^{-2} can provide stabilization even to dispersions with micron-sized particles, e.g. milk, natural latex and crude oil in the state of tertiary recovery, which represents an inverse emulsion of salty water in hydrocarbon medium.

It is worth now turning our attention to the *Rehbinder effect*, the universal physical–chemical phenomenon constituting the influence of lowering of the surface energy of a solid due to adsorption from an ambient medium on the mechanical properties of that solid [5, 68, 113–115]. This effect can be observed, under proper conditions and in various manifestations, with all types of solids, due both to a reversible physical–chemical decrease in σ and to chemisorption and catalytic interactions [116–118], etc. An active medium can strongly affect the stability and damageability of a solid surface and can facilitate and accelerate processes of *mechanical* dispersion, fracture and wear [118–121]. All of this relates to both compact solids and various dispersion structures. An actual example of the Rehbinder effect encountered in everyday practice is the accelerated wear and attrition of catalyst granules under the influence of an active medium in heterogeneous catalysis [122–125]. With respect to hydrophilic materials, water plays the role of such a strongly surface-active medium [126]; in suspensions of bentonite clays, equilibrium is established between a swollen solid phase and a very highly disperse (very dilute) colloid solution, etc. The influence of the medium on the contact interaction of particles in disperse structures to a great extent defines the rheological properties of these structures and provides ways of controlling them [17, 18, 68, 127–130].

The extreme case of the Rehbinder effect is a many-fold decrease in σ and in the strength caused by the contact of a solid body with a liquid phase that is strongly surface-active with respect to that solid. A particular example of such behavior is the influence that a metal melt has on the solid metal in the case of high similarity of the physical–chemical characteristics of the two metals. This is a known effect of liquid metal embrittlement (LME), comprising a dramatic decrease in the strength and plasticity of metals and alloys, to the point of the appearance of a tendency towards spontaneous dispersion [4–6, 69]. Similar phenomena also take place when rocks and minerals come in contact with some low-melting-point salts [69–71, 131]. A typical example of approaching the conditions of spontaneous dispersion is swelling of highly hydrophilic montmorillonite clays [68, 132, 133].

This short deviation from the discussion of our main topic of spontaneous dispersion to the area of *physical–chemical mechanics* of disperse systems and materials [5, 18, 68, 114, 115] is stipulated by the need to mention here the close relation of all considered phenomena of cohesive forces overcoming in a condensed phase, which represent a continuous spectrum: from pure mechanical fracture to the spontaneous dispersion caused by the medium, at the opposite ends of this spectrum, and with gradual transition between them under various combinations of these two factors and thermal fluctuations [114].

Finely, let us address the effect of temperature. It is directly included in equations for conditions of spontaneous dispersion as a factor "opposing" the values of σ and elementary work w. In this respect, it would be interesting to apply the presented approach also to dispersion at substantially high temperatures, say, of graphite (and diamond?) in liquid iron, of magma in the Earth's crust, etc.

Alongside this, significant changes in the state of a system can also be caused by relatively moderate deviations in T, e.g. by the changes in the mutual solubility of both the disperse phase (droplets) and the dispersion medium components, leading to a radical decrease in σ. As a typical example, one can use the above-mentioned experiments with approach to the critical point, T_c (lower than T_c), in binary mixtures of alkanes with moderately polar organic substances [8–10, 134, 135].

2.7 Conclusion

Within the restriction to the case of the discussed dilute monodisperse systems, the analysis of the ratio between elementary work of dispersion, w, designed as the work of particle isolation from a compact phase or disperse structure and the entropy factor, $kT[\ln(1/C)+1]$, represents a general, universal approach to the evaluation of the possibility of the spontaneous dispersion of a macrophase into colloid-sized particles (Fig. 2.6). It is the equality of these two factors, corresponding to the transition from positive values of the changes in the free energy of a system, ΔF, characteristic of the dispersion (interface growth), to negative values, determines the *possibility* of spontaneous dispersion. This takes place for all versions of the ΔF formulation as a function of different variables: r, σ, C, T, ν, n, etc. In the case of the description of the behavior of the $\Delta F(r)$ function for a constant volume of the disperse phase, ν=constant, the ΔF function shows a maximum. When $\Delta F(n)$ is analyzed as a function of particle number, n, for the case of constant dispersity, r, $\Delta F(n)$ reveals a negative minimum corresponding to a stable system. For n=constant, the $\Delta F(r)$ function is monotonic.

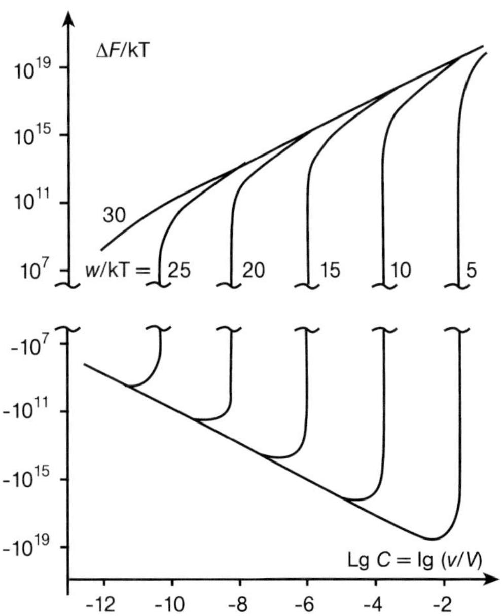

Fig. 2.6 The dependence between changes in the free energy, $\Delta F/kT$, of a monodisperse system and the concentration, C, of disperse phase, in the $C=\nu/V$ scheme, for various values of work w (in kT units) for the isolation of an individual particle from a macrophase: compact phase ($w = a\delta^2\sigma$) or globular structure of particles of the same size ($w = \frac{1}{2}zu_1$). The clearly defined, leap-like transitions from positive to negative values of ΔF are typical, alongside the negative minima at higher dilutions.

Within any approach, the equation $\Delta F=0$ corresponds to the critical ratio of system parameters. If the values of these parameters are physically realistic ($r>b$, etc.), the $\Delta F=0$ equation represents the most general formulation of the criterion describing the possibility of spontaneous dispersion and formation of thermodynamically stable (*lyophilic* according to Rehbinder) colloid–disperse system. This criterion essentially formulates the *necessary* conditions of the latter. The complex of *sufficient* conditions includes also a set of definite restrictions for dispersion down to molecular dimensions.

Under normal temperatures and low concentrations, these systems correspond to small $r \approx 10^{-6}$ cm and low $\sigma \approx 10^{-2}$–10^{-1} mJ m^{-2}. These conditions become "easier" for dispersion of an aggregate (e.g. σ on the order of units) and "more difficult" for highly concentrated systems (with σ down to 10^{-3} mJ m^{-2}).

There are no principal complications in the transition to an approximate description of similar spontaneous dispersion conditions for concentrated and highly concentrated systems. One needs in such a case to take into account both terms in the initial equation for the entropy of mixing, ΔS. The minimal gain in entropy (with respect to a particle) occurs at $N_1=N_2$. Such an approach can be used for a description of conditions of spontaneous dispersion of polymer phases (compare with the notion of a *good solvent* [59, 63]). Within the frame of the approximation used, the microheterogeneous system containing equal amounts of two phases and having a particle size of $r=0.25\times10^{-6}$ cm would be stable at $\sigma \approx 0.04$ mJ m^{-2}.

Among other factors that are worthy of quantitative examination in the future, one can suggest the internal stresses in a solid phase that play a role of a factor of initial metastability, facilitating dispersion. This is directly related to the subject of physical–chemical mechanics. Of special interest are biological objects, particularly because of the role that the thermodynamic stability of disperse systems plays in the vital activity and of the possible dependence of such disperse systems upon changes (especially small ones) in temperature or other conditions. Of principal importance is a thorough quantitative consideration of polydisperse systems [17, p. 269]. Although such a consideration may cause substantial complications in particular calculations, it could eliminate a number of contradictions within the existing approaches. Of significant theoretical interest is providing a basis for using a correct scheme for accounting for the concentration in the calculations of the entropy factor. Generally, one can employ the above-mentioned "reverse" *a posteriori* scheme. That is, this factor can be evaluated by using highly precise experimental data on equilibrium concentrations of the disperse phase in lyophilic systems. It would also be of interest to apply this approach to disperse systems that are lyophobic under normal conditions but stable at very high dilutions, such as aerosols of fragrant substances or even dispersions in cosmic vacuum (tail of comets? [16]).

The development of the spontaneous dispersion and formation of lyophilic disperse systems doctrine provides for a general scientific basis for practical applications. This gives an important and useful contribution to the means of optimization of various technological processes and controlling natural systems in order to reach the goal of either obtaining and maintaining high stability of disperse systems or overcoming it. One of the most important ecological problems, the elimination of contamination from stable colloids in an ambient medium, clearly illustrates this. The actual problem of further improvement of the sol–gel technology for manufacturing composite materials is another example of the importance of understanding conditions of stability and destabilization of highly disperse lyophilic colloid systems.

Acknowledgments

With great gratitude, the authors acknowledge the memory of their teacher Peter A. Rehbinder, who established the founding principles of this doctrine. The authors are also indebted to L. A. Kochanova, the main participant in developing these ideas.

References

1 Volmer M. Zur Theorie der lyophilen Kolloiden. *Z. Phys. Chem.* **1927**, *125*, 151–157.
2 Volmer M. Die kolloidale Natur von Fluessigkeitsgemischen in der Umgebung des kritischen Zustandes. *Z. Phys. Chem.* **1957**, *207*, 307–320.
3 Rehbinder, P. A. *New Problems in Physical–Chemical Mechanics.* Report to the Permanent Colloquium on Solid Phases of Variable Composition Together with the Moscow Colloid Colloquium, 26 January 1956. MKhTI: Moscow, **1956**.
4 Shchukin, E. D.; Rehbinder, P. A. The formation of new surfaces during the deformation and rupture of a solid in a surface active medium. *Colloid J.* **1958**, *20*, 645–654.
5 Likhtman, V. I.; Shchukin, E. D.; Rehbinder, P. A. *Physicochemical Mechanics of Metals.* Izd. Akad. Nauk SSSR: Moscow, **1962**; English translation: Israel Program for Scientific Translations: Jerusalem, **1964**.
6 Pertsov, A. V.; Mirkin, L. I.; Pertsov, N. V.; Shchukin, E. D. On the spontaneous dispersion under conditions of the strong decrease in the free interfacial energy. *Dokl. Akad. Nauk SSSR* **1964**, *158*, 1166–1168.
7 Pertsov, A. V. Studies of dispersion processes under conditions of the strong decrease in the free interfacial energy. Cand. Sci. Thesis, Moscow State University, **1967**.
8 Shchukin, E. D.; Fedoseeva, N. P.; Kochanova, L. A.; Rehbinder, P. A. On the conditions of the formation of stable emulsions in the system paraffin–oxyquinoline in vicinity of critical mixing area. *Dokl. Akad. Nauk SSSR* **1969**, *189*, 123–126.
9 Shchukin, E. D.; Kochanova, L. A.; Pertsov, A. V. On the nature of stability of lyophilic colloid emulsions. In *Physical-Chemical Mechanics and Lyophilicity of Disperse Systems*, Vol. 11. Naukova Dumka: Kiev, **1979**; 15–31.
10 Shchukin, E. D.; Kochanova, L. A. Physico-chemical fundamentals of the production of microemulsions. *Colloid J.* **1983**, *45*, 726–736.
11 Rusanov, A. I.; Shchukin, E. D.; Rehbinder, P. A. On the dispersion theory. 1. Thermodynamics of monodisperse systems. *Colloid J.* **1968**, *30*, 573–580.
12 Rusanov, A. I.; Kuni, F. M.; Shchukin, E. D.; Rehbinder, P. A. On the dispersion theory. 2. Dispersion in vacuo. *Colloid J.* **1968**, *30*, 735–743; 3. Dispersion in liq-

uid medium. *Colloid J.* **1968**, *30*, 744–753.

13 Kligman, F. I.; Rusanov, A. I. Thermodynamic equilibrium states of disperse systems with disperse particles. *Colloid J.* **1977**, *39*, 33–36.

14 Rusanov, A. I. Thermodynamics of micellar systems. 1. Equilibrium distribution and surface tension of micelles. *Colloid J.* **1981**, *43*, 890–902.

15 Rusanov, A. I. Micellization in surfactant solutions. In *Chemistry Reviews*, Vol. 22, Part 1, Volpin, M. E. (ed.). Harwood Academic: Amsterdam, **1997**.

16. Shchukin, E. D.; Amelina, E. A.; Yaminsky, V. V. Thermodynamic equilibrium between coagulation and peptization. *J. Colloid Interface Sci.* **1982**, *90*, 137–142.

17 (a) Shchukin, E. D.; Pertsov, A. V.; Amelina, E. A.; Zelenev, A. S. *Colloid and Surface Chemistry.* Elsevier: Amsterdam, **2001**; (b) Shchukin, E. D.; Pertsov, A. V.; Amelina, E. A. *Colloid Chemistry.* Vysshaya Shkola: Moscow, **2004**.

18 Shchukin, E. D.; Pertsov, N. V.; Osipov, V. I.; Zlochevskaya, R. I. (eds.). *Physical–Chemical Mechanics of Natural Disperse Systems.* Izd. MGU (Moscow State University): Moscow, **1985**.

19 Shchukin, E. D. (ed.). *Advances in Colloid Chemistry and Physical–Chemical Mechanics.* Nauka: Moscow, **1992**.

20 Pertsov, A. V. Spontaneous and mechanical dispersion and the stability of obtained disperse systems. Dr. Sci. Thesis, Moscow State University, **1992**.

21 Shchukin, E. D. Surface modification and contact interaction of particles. *J. Dispers. Sci. Technol.* **2003**, *24*, 377–395.

22 Mayer, J.; Harrison, S. Statistical mechanical of condensing systems. 3, 4. *J. Chem. Phys.* **1938**, *6*, 87–104.

23 Band, W. J. Dissociation treatment of condensing systems. *J. Chem. Phys.* **1939**, *7*, 324–326; 927–931.

24 Rice, O. Introduction to the Symposium on Critical Phenomena: general consideration of critical phenomena. *J. Phys. Colloid Chem.* **1950**, *54*, 1293–1305.

25 Frenkel, Ya. I. *Kinetic Theory of Liquids*, Vol. 3. Izd. Akad. Nauk SSSR: Leningrad, **1959**.

26 Glazman, Yu. M. Two-phase disperse systems in thermodynamic equilibrium. *Colloid J.* **1967**, *29*, 478–480.

27 Barboi, V. M.; Glazman, Yu. M.; Fuks, G. I. Nature of the aggregation stability of colloidal solutions. Conditions for existence of two-phase disperse systems in thermodynamic equilibrium. *Colloid J.* **1970**, *32*, 321–326.

28 Reiss, H. Entropy-induced dispersion of bulk liquids. *J. Colloid Interface Sci.* **1975**, *53*, 61–70.

29 Ruckenstein, E. J. On the thermodynamic stability of microemulsions. *J. Colloid Interface Sci.* **1978**, *66*, 369–371.

30 Lee, G. W.; Tadros, Th. F. Formation and stability of emulsions prodused by dilution of emulsifiable concentrates. *Colloids Surf.* **1982**, *5*, 105–115.

31 Skuse, D. R.; Tadros, Th. F.; Vincent, B. Controlled heteroflocculation of nonaqueous silica dispersions. *Colloids Surf.* **1986**, *17*, 343–360.

32 Evans, D. F.; Wennerstroem, H. *The Colloidal Domain. Where Physics, Chemistry, Biology and Technology Meet*; Marcel Decker: New York, **1994**.

33 Shahidzadeh, N.; Bonn, D.; Aguerre-Chariol, O.; Meunier J. Spontaneous emulsification: relation to microemulsion phase behavior. *Colloids Surf. A* **1999**, *147*, 375–380.

34 Rusanov, A. I. Thermodynamics of dispergation: development of Rehbinder's ideas. *Colloids Surf. A* **1999**, *160*, 79–87.

35 Babak, V. G.; Stebe, M. J. A review on higly concentrated emulsions: physicochemical principles of formulation. *J. Dispers. Sci. Technol.* **2002**, *23*, 1–22.

36 Lopez-Montilla, J. C.; Herrera-Morales, P. E.; Pandey, S.; Shah, D. O. Spontaneous emulsification. Mechanisms, physicochemical aspects, modeling and applications. *J. Dispers. Sci. Technol.* **2002**, *23*, 219–268; Lopez-Montilla, J.; Herrera-Morales, P.; Shah, D. O. New method to quantitatively determine the spontaneity of emulsification process. *Langmuir* **2002**, *18*, 4258–4262.

37 Shchukin, E. D.; Pertsov, A. V. The criterion of spontaneous dispersion. In *Abstracts of the II International Conference*

"Colloid 2003", Belorussian State University: Minsk, **2003**; 14.
38 Shchukin, E. D. Conditions of spontaneous dispersion and formation of thermodynamically stable colloid system. In *Abstracts of the "Particles 2004" Conference on Particle Synthesis, Characterization and Particle-Based Advanced Materials*, Orlando, Florida, **2004**; 56.
39 Shchukin, E. D. Criterion of spontaneous dispersion and formation of thermodynamically stable colloid system. In *Abstracts of the 228th ACS National Meeting, Division of Environmental Chemistry*, Philadelphia, PA, **2004**; Abstr. No. 763546.
40 Shchukin, E. D. Conditions of spontaneous dispersion and formation of thermodynamically stable colloid systems. *J. Dispers. Sci. Technol.* **2004**, *25*, 875–893.
41 Kirkwood, J.; Buff, F. The statistical mechanical theory of surface tension. *J. Chem. Phys.* **1949**, *17*, 338–343.
42 Hill, T. *Thermodynamics of Small Systems*, Vols. 1 and 2. Benjamin: New York, **1963–64**.
43 Davis, H. T. *Statistical Mechanics of Phases, Interfaces and Thin Films*. Wiley: New York, **1996**.
44 Shchukin, E. D.; Amelina, E. A. Contact interactions in disperse systems. *Adv. Colloid Interface Sci.* **1979**, *11*, 235–287.
45 Yaminsky, V. V.; Pchelin, V. A.; Amelina, E. A.; Shchukin, E. D. *Coagulation Contacts in Disperse Systems*. Khimiya: Moscow, **1982**.
46 Shchukin, E. D. Some colloid-chemical aspects of the small particles contact interactions. In *Fine Particles Science and Technology*, Pelizzetti, E. (ed.). Kluwer: Dordrecht, **1996**; 239–253.
47 Martynov, G. A. Stability of colloids and the Deryagin-Landau-Verwey-Overbeek theory of surface forces. *J. Phys. Chem., Moscow* **1999**, *73*, 1567–1576.
48 Tolman, R. The effect of droplet size on surface tension: the superficial density of matter at a liquid–vapor boundary. *J. Chem. Phys.* **1949**, *17*, 118–127; 333–337.
49 Koenig, F. On the thermodynamic relation between surface tension and curvature. *J. Chem. Phys.* **1950**, *18*, 449–459.
50 Rusanov, A. I. *Phase Equilibria and Surface Phenomena*. Khimiya; Leningrad, **1967**.
51 Shcherbakov, L. M. The evaluation of excess in the free energy of small objects. In *Studies in the Area of Surface Forces*. Nauka: Moscow, **1967**; 17–25.
52 Hartley, G. S. *Aqueous Solutions of Paraffin Cain Salts*. Hermann: Paris, **1936**.
53 McBain, J. W.; Woo, T. Spontaneous emulsification and reactions overshooting equilibrium. *Proc. R. Soc. London, Ser. A* **1937**, *163*, 182–188.
54 Shinoda, K.; Nakagawa, T.; Tamamushi, B.; Isemura, T. *Colloidal Surfactants*. Academic Press: New York, **1963**.
55 Mittal, K. L. (ed.). *Micellization, Solubilization and Microemulsions*. Plenum Press: New York, **1977**.
56 Rehbinder, P. A. *Selected Works: Surface Phenomena in Disperse Systems. Colloid Chemistry*. Nauka: Moscow, **1978**.
57 Jonsson, B.; Lindman, B.; Kronberg, B., Holmberg, K. *Surfactants and Polymers in Aqueous Solutions*. Wiley: Chichester, **1998**.
58 Mittal, K. H.; Shah, D. O. (eds.). *Adsorption and Aggregation of Surfactants in Solutions*. Proceedings of the 13th International Symposium held 11–16 June 2000 in Gainesville, Florida. Marcel Decker: New York, **2003**.
59 Flory, P. J. *Principles of Polymer Chemistry*. Cornell University Press: Ithaca, NY, **1953**.
60 Vol'kenshtein, M. V. *Configuration Statistics of Polymer Chains*. Izd. Akad. Nauk SSSR: Moscow, **1959**.
61 Kuleznev, V. N.; Krokhina, L. S.; Dogadkin, B. A. Mutual solubilizing and spontaneous emulsification of the components in the polymer–polymer–solvent system. *Colloid J.* **1969**, *31*, 853–859.
62 Abaturova, N. A.; Vlodavets, I. N.; Rehbinder, P. A. Features of separation of new disperse phase from metastable solution of acethylcellulose in diethylene glycol ethyl ether. *Colloid J.* **1972**, *34*, 315–320.
63 Tager, A. A. *Physico-Chemistry of Polymers*. Khimiya: Moscow, **1978**.
64 Kwak, J. C. T. (ed.). *Polymer–Surfactant Systems*. Marcel Decker: New York, **1998**.

65 Shchukin, E. D.; Yaminsky, V. V. Thermodynamic factors in the sol–gel transition: 1. Reversible coagulation of organophilic and hydrophilic colloids. *Colloids Surf.* **1988**, *32*, 19–32; 2. The role of short- and long-range forces in thermodynamics of colloid stability. *Colloids Surf.* **1988**, *32*, 33–55.
66 Shchukin, E. D. Surfactants effects on the cohesive strength of particle contacts: measurements by the cohesive force apparatus. *J. Colloid Interface Sci.* **2002**, *256*, 159–167.
67 Shchukin, E. D.; Amelina, E. A.; Kontorovich, S. I. Formation of contacts between particles and development of internal stresses during hydration processes. In *Materials Science of Concrete III*, Skalny, J. (ed.). American Ceramic Society; Westerville, OH, **1992**; 1–35.
68 Rehbinder, P. A. *Selected Works: Surface Phenomena in Disperse Systems. Physico-chemical Mechanics.* Nauka: Moscow, **1979**.
69 Rehbinder, P. A.; Shchukin, E. D. Surface phenomena in solids during deformation and fracture processes. In *Progress in Surface Science*, Vol. 3, Part 2, Davison, S. G. (ed.). Pergamon Press: Oxford, **1972**; 97–188.
70 Pertsov, N. V. The influence of adsorption-active liquid media on the mechanical properties of solids. Dr. Sci. Thesis, Moscow State University, **1971**.
71 Pertsov, N. V. The physico-chemical influence of liquid phases on the rock fracture. In *Physical–Chemical Mechanics of Natural Disperse Systems*, Shchukin, E. D.; Pertsov, N. V.; Osipov, V. I.; Zlochevskaya, R. I. (eds.). Izd. MGU (Moscow State University): Moscow, **1985**; 107–117.
72 Markina, Z. N.; Bovkun, O. P.; Rehbinder, P. A. Thermodynamics of micelle formation in surfactants in aqueous medium. *Colloid J.* **1973**, *35*, 833–837.
73 Berezin, I. V.; Martinek, K.; Yatsimirsky, A. K. Physical–chemical principles of micellar catalysis. *Usp. Khim.* **1973**, *42*, 1729–1756.
74 Friberg, S., Jansson, O. Surfactant association structure and emulsion stability. *J. Colloid Interface Sci.* **1976**, *55*, 614–623.
75 Mittal, K. L. (ed.). *Solution Chemistry of Surfactants.* Plenum Press: New York, **1979**.
76 Hubbard, A. T. (ed.). *Encyclopedia of Surface and Colloid Science.* Marcel Decker: New York, **1996**; Shah, D. O. (ed.). *Micelles, Microemulsions and Monolayers.* Marcel Decker: New York, **1998**.
77 Patist, A.; Kanicky, J. R.; Shukla, P. K.; Shah, D. O. Importance of micellar kinetics in relation to technological processes. *J. Colloid Interface Sci.* **2002**, *245*, 1–15.
78 Aguiar, J.; Molina-Bolivar, J. A.; Peula, G. J. M.; Carnero, R. C. Thermodynamics and micellar properties of tetradecyltrimethylammonium bromide in formamide–water mixtures. *J. Colloid Interface Sci.* **2002**, *255*, 382–390; Aguiar, J.; Carpena, P.; Molina-Bolivar, J. A.; Carnero R. C. On the determination of the critical micelle concentration by the pyrene 1:3 ratio method. *J. Colloid Interface Sci.* **2003**, *258*, 116–122.
79 Martynov, G. A.; Muller, V. M. On the role of disintegration in the mechanism of aggregation stability of colloid systems. *Dokl. Akad. Nauk SSSR* **1972**, *207*, 370–373; 1161–1164.
80 Muller, V. M. Theory of aggregation transformations and stability of hydrophobic colloids. Dr. Sci. Thesis, Leningrad State University, **1983**.
81 Derjaguin, B. V. (ed.). *Studies in the Area of Surface Forces.* Nauka: Moscow, **1967**.
82 Derjaguin, B. V. (ed.). *Surface Forces in Thin Films and Stability of Colloids.* Nauka: Moscow, **1974**.
83 Derjaguin, B. V.; Churaev, N. V.; Muller, V. M. *Surface Forces.* Nauka: Moscow, **1985**.
84 Israelachvili, J. N. *Intermolecular and Surface Forces*; Academic Press: Orlando, FL, **1985**.
85 Claesson, P. M.; Ederth, T.; Bergeron, V.; Rutland, M. W. Techniques for measuring surface forces. *Adv. Colloid Interface Sci.* **1996**, *67*, 119–183.
86 Christensen, H. K.; Claesson, P. M. Direct measurements of the force between hydrophobic surfaces in water. *Adv. Colloid Interface Sci.* **2001**, *91*, 391–436.

87 Shchukin, E. D. Surfactants and their influence on the properties of disperse systems. *Vestn. Akad. Nauk SSSR* **1978**, (5), 78–95.

88 Friberg, S. E. Emulsion stability. *Food Sci. Technol. (New York)* **1997**, *81*, 1–55.

89 Mittal, K.L.; Kumar, P. (eds.). *Emulsions, Foams and Thin Films*. Marcel Decker: New York, **2000**.

90 Palla, B. J.; Shah, D. O. Stabilization of high ionic strength slurries using surfactant mixings: molecular factors that determine optimal stability. *J. Colloid Interface Sci.* **2002**, *256*, 143–152.

91 Holmberg, K.; Shah, D. O.; Schwuger, M. J. (eds.). *Handbook of Applied Surface and Colloid Chemistry*, Vol. 2. Wiley: Chichester, **2002**.

92 Holmberg, K. (ed.). *Novel Surfactants*. Marcel Decker: New York, **2003**.

93 Wilson, P. M.; Brandner, C. F. Aqueous surfactant solutions which exhibit ultralow tension at the oil/water interface. *J. Colloid Interface Sci.* **1977**, *60*, 473–479.

94 Grasia, A.; Lachaise, J.; Bourrel, M.; Schechter, R. S.; Wade, W. H. Differentiation of continuous and discontinuous phases in water-in-oil microemulsions using dialysis. *J. Colloid Interface Sci.* **1988**, *122*, 83–91.

95 Chan, S. Y.; Rosano, H. L. A theory of the formation of microemulsions. *J. Dispers. Sci. Technol.* **1989**, *9*, 523–535.

96 Fontell, K.; Mandell, L. Phase equilibria and phase structure in the ternary systems sodium or potassium octanoate–octanoic acid–water. *Colloid Polym. Sci.* **1993**, *271*, 974–991.

97 Fukuda, K.; Soederman, O.; Shinoda, K.; Lindman, B. Microemulsions formed by alkyl polyglycosides and an alkyl glycerol ether. *Langmuir* **1993**, *9*, 2921–2925.

98 Kahlweit, M.; Faulhaber, B.; Busse, G. Microemulsions with mixing of nonionic and ionic amphiphiles. *Langmuir* **1994**, *10*, 2528–2532.

99 Heenan, D. M.; Friberg, S. E. Phase equilibria of methoxydimethyloctylsilane, tetramethyldioctyldisiloxane in organic solvents. *J. Dispers. Sci. Technol.* **1996**, *17*, 607–629.

100 Friberg, S. E.; Zhang, Z.; Patel, R.; Campbell, G.; Aikens, P. A, Kinetics of formation of structures in a three-phase system water/lamellar liquid crystal/water in oil microemulsion after shear. *Prog. Colloid Polym. Sci.* **1998**, *108*, 9–16.

101 Kumar P., Mittal, K. L. (eds.). *Handbook of Microemulsion Science and Technology*. Marcel Decker: New York, **1999**.

102 Aramaki, K.; Kabir, M. H.; Nakamura, N.; Kunieda, H.; Ishitobi, M. Formation of cubic-phase microemulsions in sucrose alkanoate systems. *Colloids Surf.* **2001**, *183–185*, 371–379.

103 Sjoblom, J. (ed.). *Encyclopedic Handbook of Emulsion Technology*. Marcel Decker: New York, **2001**.

104 Mishchuk, N. A.; Miller, R.; Steinchen, A.; Sanfeld, A. Conditions of coagulation and flocculation in dilute mini-emulsions. *J. Colloid Interface Sci.* **2002**, *256*, 435–450.

105 Kochanova, L. A; Kuchumova, V. M.; Fedoseeva, N. P.; Pertsov, A. V.; Rehbinder, P. A. The critical emulsions as lyophilic colloid systems: 1. On the nature of stability of critical emulsion systems. *Colloid J.* **1973**, *35*, 838–842; 2. Experimental study of the conditions and of formation and stability of critical emulsion systems. *Colloid J.* **1973**, *35*, 843–847.

106 Fedoseeva, N. P.; Kuchumova, V. M.; Kochanova, L. A.; Shchukin, E. D. Studies of surface phenomena and phase equilibria in the three-component systems with restricted solubility. *J. Phys. Chem.* **1976**, *50*, 1900–1906.

107 Fedoseeva, N. P.; Kuchumova, V. M.; Kochanova, L. A.; Shchukin, E. D. Stability of emulsions formed in the area of restricted solubility of components in the water–heptane–*tert*-butanol system at 20°C. *Colloid J.* **1977**, *39*, 1199–1202.

108 Fedoseeva, N. P.; Kuchumova, V. M.; Kochanova, L. A.; Shchukin, E. D. The sodium dodecylsulfate effect on behavior of the three-component system: water–heptane–*tert*-butanol in vicinity of the critical point. *Colloid J.* **1978**, *40*, 578–580.

109 Izmailova, V. N.; Alekseeva, I. N.; Shchukin, E. D.; Rehbinder, P. A. Rheological studies of the interfacial layers of proteins and surface-active polymers. *Dokl. Akad. Nauk SSSR* **1972**, *206*, 1150–1153.

110 Izmailova, V. N.; Rehbinder, P. A. *Structure Formation in Protein Systems.* Nauka: Moscow, **1974**.

111 Shchukin, E. D. Development of Rehbinder's doctrine on strong stabilization factors in disperse systems. *Colloid J.* **1997**, *59*, 270–284.

112 Izmailova, V. N.; Yampolskaya, G. P.; Tulovskaya, Z. D. Development of Rehbinder's concept on structure–mechanical barrier in stability of dispersions stabilized by proteins. *Colloids Surf. A* **1999**, *160*, 89–106.

113 Rehbinder, P. A. On the effect of changes in the surface energy upon cleavage, hardness and other crystal properties. In *The VIth Congress of Russian Physicists.* OGIZ: Moscow, **1928**; 29; Rehbinder, P. A. *Selected Works: Surface Phenomena in Disperse Systems. Physicochemical Mechanics.* Nauka: Moscow, **1979**, 142.

114 Shchukin, E. D. The physical–chemical theory of the strength of disperse structures and materials. In *Physical–Chemical Mechanics of Natural Disperse Systems,* Shchukin, E. D.; Pertsov, N. V.; Osipov, V. I.; Zlochevskaya, R. I. (eds.). Izd. MGU (Moscow State University): Moscow, **1985**, 72–90.

115 Shchukin, E. D. Physical-chemical mechanics in the studies of Peter A. Rehbinder and his school. *Colloids Surf. A* **1999**, *149*, 529–537.

116 Rehbinder, P. A.; Shchukin, E. D. Surface phenomena in solids in deformation and fracture processes. *Usp. Fiz. Nauk* **1972**, *108*, 3–39.

117 Shchukin, E. D. Environmentally-induced lowering of surface energy and the mechanical behavior of solids. In *Surface Effects in Crystal Plasticity,* Latanision, R. M.; Fourie J. F. (eds.). Noordhoff: Leyden, **1977**; 701–736.

118 Westwood, A. R. C.; Ahearn, J. S.; Mills, J. J. Developments in the theory and applications of chemomechanical effects. *Colloids Surf.* **1981**, *2*, 1–35.

119 Shchukin, E. D. Physical–chemical principles of the new methods for intensifying treatment of solids. *Vestn. Akad. Nauk SSSR* **1973**, (11), 30–40.

120 Shchukin, E. D.; Kochanova, L. A.; Savenko, V. I. Electric surface effects in solid plasticity and strength. In *Modern Aspects of Electrochemistry,* Vol. 24, White, R. E.; Conway, B. E.; Bockris, J. O'M. (eds.). Plenum Press: New York, **1993**; 245–298.

121 Savenko, V. I.; Shchukin, E. D. New applications of the Rehbinder effect in tribology. A review. *Wear* **1996**, *194*, 86–94.

122 Shchukin, E. D; Kontorovich, S. I.; Romanovsky, B. V. Porous materials sintering under conditions of catalytic reaction. *J. Mater. Sci.* **1993**, *28*, 1937–1940.

123 Shchukin, E. D.; Margolis, L. Ya.; Kontorovich, S. I.; Polukarova, Z. M. The influence of the medium on the mechanical properties of catalysts. *Usp. Khim.* **1996**, *65*, 881–891.

124 Shchukin, E. D. The reciprocal influence of the solid phase and medium in processes of the heterogeneous catalysis. *Khim. Prom.* **1997** (6), 28–35 (412–419).

125 Abukais, A.; Burenkova, L. N.; Zhilinskaya, E. A.; Lamonie, J.-F.; Murav'eva, G. P.; Romanovsky, B. V.; Sokolova, L. N.; Shchukin, E. D. The influence of alcohol catalytic conversion on the strength of porous materials of ZrO_2 and ZrO_2 + Y_2O_3. *Inorg. Mater.* **2003**, *39*, 602–608.

126 Rehbinder, P. A. Water as a surface-active substance. Surface activity and adsorption forces. In Rehbinder, P. A. *Selected Works: Surface Phenomena in Disperse Systems. Colloid Chemistry.* Nauka: Moscow, **1978**; 140–157.

127 Korolev, V. A.; Nikolaeva, S. K.; Osipov, V.I. The rheology of clayey grounds. In *Physical–Chemical Mechanics of Natural Disperse Systems,* Shchukin, E. D.; Pertsov, N. V.; Osipov, V. I.; Zlochevskaya, R. I. (eds.). Izd. MGU (Moscow State University): Moscow, **1985**; 222–233.

128 Ur'ev, N. B.; Potanin, A. A. *Fluidity of Suspensions and Powders.* Khimiya: Moscow, **1992**.

129 Macosko, C. W. *Rheology. Principles, Measurements and Applications.* Wiley: New York, **1994**.

130 Shchukin, E. D. The role of contact interactions in the rheological behavior of a fibrous suspension. *Colloid J.* **2001**, *63*, 855–858.

131 Pertsov, N. V.; Traskin, V. Yu. The Rehbinder effect in Nature. In *Advances in Colloid Chemistry and Physical–Chemical Mechanics*, Shchukin, E. D. (ed.). Nauka: Moscow, **1992**; 155–165.

132 Nikolaeva, S. K.; Korolev, V. A.; Osipov, V. I.; Sokolov, V. N. The thixotropy of clayey grounds. In *Physical–Chemical Mechanics of Natural Disperse Systems*, Shchukin, E. D.; Pertsov, N. V.; Osipov, V. I.; Zlochevskaya, R. I. (eds.). Izd. MGU (Moscow State University): Moscow, **1985**; 158–167.

133 Giese, R. F.; van Oss, C. J. *Colloid and Surface Properties of Clays and Related Materials.* Marcel Decker: New York, **2001**.

134 Fedoseeva, N. P.; Kochanova, L. A.; Shchukin, E. D. On surface phenomena in a binary system in vicinity of the critical temperature of mixing. In *Physical Chemistry of the Surface Phenomena in Melts*, Naidich, Yu. V. (ed.). Naukova Dumka: Kiev, **1971**; 25–33.

135 Fedoseeva, N. P.; Kuchumova, V. M.; Kochanova, L. A.; Shchukin, E. D. The conditions of formation and stability of emulsions arising in vicinity of the critical temperature of mixing in a system tricosane/8-hydroxyquinoline. In *Surface Phenomena in Liquids and Liquid Solutions*, Vol. 1, Rusanov, A. J. (ed.). Izd. LGU (Leningrad State University): Leningrad, **1972**; 157–165.

3
Electrostatic Interactions Between Colloidal Particles – Analytic Approximation

Hiroyuki Ohshima

Abstract

The electrostatic interaction between colloidal particles dispersed in an electrolyte solution plays an essential role in determining the electric behavior of the dispersion of the particles. In this chapter we give approximate analytic expressions for the force and potential energy of the electrostatic interaction between two colloidal particles for various cases.

3.1
Introduction

According to the theory of Derjaguin-Landau and Verwey-Overbeek (DLVO theory), the stability of a suspension of charged colloidal particles in a liquid containing electrolytes can be explained by the balance of the electrostatic interaction and the van der Waals interaction between two particles [1–7]. This theory is based on the Poisson-Boltzmann equation for the electric potential around the particles. In this chapter, we treat the electrostatic interaction of two parallel plates and that of two spherical particles. For the case of two parallel plates, the Poisson-Boltzmann equation can be solved exactly but the results are complicated, involving elliptic integrals [8–11]. For spherical particles, no exact solution has been derived except where the potential is low enough to allow linearization of the Poisson-Boltzmann equation. Hence for practical purposes it is convenient to use simple approximate analytic expressions for the force and potential energy of the electrostatic interaction between two colloidal particles. In this chapter, on the basis of several approximation methods, we derive various analytic equations.

Colloids and Interface Science Series, Vol. 1
Colloid Stability: The Role of Surface Forces, Part I. Edited by Tharwat F. Tadros

ISBN: 978-3-527-31462-1

3.2 An Electrical Double Layer Around a Colloidal Particle: the Poisson–Boltzmann Equation

When a charged colloidal particle is immersed in an electrolyte solution, the electrolyte ions together with the particle surface charge form an electrical double layer [1–7]. Let the electrolyte be of symmetrical valence z and bulk concentration (number density) n. For simplicity we treat a plate and take an x-axis perpendicular to the plate with its origin at the plate surface so that the region $x>0$ corresponds to the solution phase. The potential distribution across the electrical double layer is described by the Poisson-Boltzmann equation as shown below. We denote by $\rho(x)$ the density of electrolyte ions at an arbitrary position x in the solution phase. The electric potential $\psi(x)$ at position x, measured relative to the bulk solution phase, where ψ is set equal to zero, is related to the charge density $\rho(x)$ at the same point by the Poisson equation, viz.,

$$\frac{d^2\psi(x)}{dx^2} = -\frac{\rho(x)}{\varepsilon_r\varepsilon_0} \tag{1}$$

where ε_r is the relative permittivity of the electrolyte solution and ε_0 is the permittivity of a vacuum. We assume that the distribution of the electrolyte ions obeys Boltzmann's law, viz., the concentration of cations $n_+(x)$ and that of anions $n_-(x)$ are given by

$$n_\pm(x) = ne^{\mp y(x)} \tag{2}$$

with

$$y(x) = \frac{ze\psi(x)}{kT} \tag{3}$$

where e is the elementary electric charge, k is Boltzmann's constant, T is the absolute temperature and y is the scaled potential. The charge density $\rho(x)$ is thus given by

$$\rho(x) = ze[n_+(x) - n_-(x)] = -2zen\sinh[y(x)] \tag{4}$$

Combining Eqs. (1) and (4) gives the Poisson–Boltzmann equation:

$$\frac{d^2y}{dx^2} = \kappa^2\sinh y \tag{5}$$

with

$$\kappa = \left(\frac{2z^2e^2n}{\varepsilon_r\varepsilon_o kT}\right)^{\frac{1}{2}} \tag{6}$$

where κ is the Debye-Hückel parameter. Let the surface potential of the plate be ψ_0, then the solution to Eq. (5) is

$$y(x) = 2\ln\left(\frac{1+\gamma e^{-\kappa x}}{1-\gamma e^{-\kappa x}}\right) \tag{7}$$

with

$$\gamma = \tanh\left(y_0/4\right) \tag{8}$$

where

$$y_0 = \frac{ze\psi_0}{kT} \tag{9}$$

is the scaled surface potential.

The charge density σ on the plate surface is related to the potential derivative normal to the particle surface as

$$\varepsilon_{\mathrm{rp}}\frac{\partial\psi}{\partial x} - \varepsilon_{\mathrm{r}}\frac{\partial\psi}{\partial x} = \frac{\sigma}{\varepsilon_0} \tag{10}$$

where $\varepsilon_{\mathrm{rp}}$ is the relative permittivity of the plate. If the internal fields inside the particle can be neglected, then the boundary condition (10) reduces to

$$\frac{\partial\psi}{\partial x} = -\frac{\sigma}{\varepsilon_{\mathrm{r}}\varepsilon_0} \tag{11}$$

Substitution of Eq. (7) into Eq. (10) gives the following ψ_0/σ relationship:

$$\sigma = \frac{\varepsilon_{\mathrm{r}}\varepsilon_0\kappa kT}{e}\sinh\left(\frac{y_0}{2}\right) \tag{12}$$

For the case of general electrolytes composed of N ionic mobile species of valence z_{i} and bulk concentration (number density) $n_{\mathrm{i}}^{\infty}(i=1, 2, \ldots, N)$, the Poisson-Boltzmann equation becomes

$$\frac{\mathrm{d}^2 y}{\mathrm{d}x^2} = -\frac{\kappa^2\sum_{i=1}^{N} z_{\mathrm{i}}n_{\mathrm{i}}\exp(-z_{\mathrm{i}}y)}{\sum_{i=1}^{N} z_{\mathrm{i}}^2 n_{\mathrm{i}}} \tag{13}$$

with

$$\kappa = \left(\frac{1}{\varepsilon_{\mathrm{r}}\varepsilon_0 kT}\sum_{i=1}^{N} z_{\mathrm{i}}^2 e^2 n_{\mathrm{i}}^{\infty}\right)^{\frac{1}{2}} \tag{14}$$

If the potential ψ is low, viz.

$$\frac{ze|\psi(x)|}{kT} \ll 1 \tag{15}$$

then Eqs. (5) and (13) reduce to the Debye-Hückel equation:

$$\frac{d^2\psi}{dx^2} = \kappa^2\psi\,, \tag{16}$$

with the solution

$$\psi(x) = \psi_0 e^{-\kappa x} \tag{17}$$

and Eq. (12) reduces to

$$\psi_0 = \frac{\sigma}{\varepsilon_r\varepsilon_0\kappa} \tag{18}$$

3.3
Double-layer Interactions at Constant Surface Potential and at Constant Surface Charge Density

Hydrostatic pressure and osmotic pressure of electrolyte ions act on a single uncharged colloidal particle in an electrolyte solution. When the particle is charged, the particle charge and electrolyte ions (mainly counter ions) form an electrical double layer around the particle. The electrolyte ions in the electrical double layer exert an excess osmotic pressure on the particle. At the same time, the coulombic attraction acts between the charges on the particle surface and the counter ions within the electrical double layer, which is called Maxwell's stress. Thus the electrical double layer exerts two additional forces on the particle surface: the excess osmotic pressure and the Maxwell stress. When two charged colloidal particles approach each other, their electrical double layers overlap so that electric forces are acting between the particles.

There are two methods for calculating the energy of interaction between two colloidal particles [1, 2].

In the first method, one calculates the interaction force $\boldsymbol{f}$ (the excess osmotic pressure $\Delta\Pi$ and the Maxwell stress) acting on an arbitrary closed surface Σ enclosing either one of the two interacting particles, which is written as

$$\boldsymbol{f} = \int_{\Sigma} \left[\left(\Delta\Pi + \frac{1}{2}\varepsilon_r\varepsilon_0 E^2 \right) \boldsymbol{n} - \varepsilon_r\varepsilon_0 (\boldsymbol{iE}\cdot\boldsymbol{n})\boldsymbol{E} \right] \mathrm{d}S \tag{19}$$

where $\boldsymbol{E}$ is the electrostatic field vector with magnitude E, $\Delta\Pi$ is the excess osmotic pressure, i.e. the difference between the osmotic pressure on the surface Σ and the bulk solution phase and $\frac{1}{2}\varepsilon_r\varepsilon_0 E^2$ is the Maxwell stress. The potential energy V of the double-layer interaction is then obtained by integrating the force $\boldsymbol{f}$ with respect to the particle separation.

In the second method, the interaction energy is directly obtained from the difference in free energy of the system of two interacting particles:

$$V = F - F(\infty) \tag{20}$$

The form of the free energy F depends on the type of the origin of surface charges on the interacting particles. Usually two types, (1) interaction at constant surface charge density and (2) interaction at constants surface potential, are considered.

(A) Interaction at constant surface charge density [12–20]:

The first case corresponds to the situation in which the particle surface charge (with a density σ) is caused by dissociation of ionizable groups. If, further, the dissociation is complete, then the free energy per F unit surface area per particle (relative to the free energy for a neutral surface) is given by

$$F = \int_0^{\sigma} \psi(\sigma')\mathrm{d}\sigma' \tag{21}$$

which is the electric work of charging the particle surface and $\psi(\sigma')$ is the surface potential at a stage at which the surface charge density is σ' during the charging process.

(B) Interaction at constants surface potential [1, 2]:

In the second case, the particle surface charge is due to adsorption of ions. If N ions of valence Z adsorb on the surface, then the surface charge density σ is $\sigma = ZeN$ and the free energy increase associated by ion adsorption is given by

$$F_s = \int_0^{\sigma} \psi(\sigma')\mathrm{d}\sigma' + \mu_0 N - TS \tag{22}$$

The first term is the electric work and the other two terms are chemical components. The electrochemical potential μ of adsorbed ions is

$$\mu = \frac{\partial F_s}{\partial N} = Ze\psi_0 + \mu_0 - T\frac{\partial S}{\partial N} \tag{23}$$

If S is independent of N, then the term TS can be dropped. The total free energy change is the sum of F_s and the free energy change in the bulk solution phase $-\mu N$, viz.

$$F = F_s - \mu N = \int_0^{\sigma} \psi(\sigma')\mathrm{d}\sigma' - (Ze\psi_0)N = \int_0^{\sigma} \psi(\sigma')\mathrm{d}\sigma' - \sigma\psi_0 = -\int_0^{\psi_0} \sigma(\psi')\mathrm{d}\psi' \tag{24}$$

For low potentials, ψ and σ are proportional to each other (Eq. 18) so that Eqs. (21) and (24) become, respectively,

$$F = -\frac{1}{2}\psi_0\sigma \text{ (constant surface charge density)} \tag{25}$$

$$F = \frac{1}{2}\psi_0\sigma \text{ (constant surface potential)} \tag{26}$$

If the dissociation of the ionizable groups on the particle surface is not complete or S depends on N, then neither ψ_0 nor σ remains constant during interaction. These cases can be treated by the charge regulation model [9, 21–24]. Another type of interaction model has been proposed [25–28] for the interaction between ion-penetrable particles (or spherical polyelectrolytes), which we term "soft particles". If the particle size is much larger than the Debye length $1/\kappa$, then the potential deep inside the interacting particles is always equal to the Donnan potential. This type of interaction is termed "Donnan potential regulated" interaction.

3.4 Interaction Between Two Parallel Plates

For the case of two parallel plates, the nonlinear Poisson-Boltzmann equation, Eq. (5), can be solved exactly and Honig and Mul [10] tabulated the numerical values of the interaction energy and force between two parallel similar plates. Devereux and de Bruyn tabulated the exact numerical values of the interaction energy of two parallel dissimilar plates [8]. As stated in the Introduction, the above results are complicated. We give below approximate analytic expressions for the interaction force and energy.

3.4.1 Low Potentials

For low potentials, the Poisson-Boltzmann equation, Eq. (5), can be linearized to Eq. (16) so that we can obtain simple analytic approximate expressions for the force and potential energy of the electrostatic interaction between two parallel plates 1

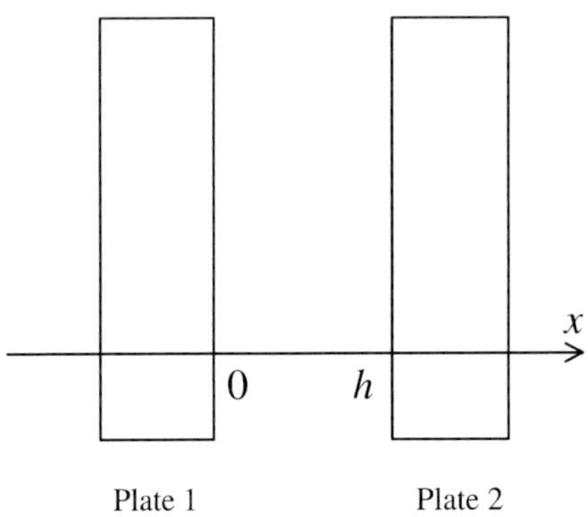

Fig. 3.1 Interaction between two parallel plates 1 and 2 at separation h.

and 2, as shown below. Consider two parallel similar plates places at $x=0$ and $x=h$, carrying surface potential ψ_0 at separation h in a solution of symmetrical electrolyte of valence z and bulk concentration n. Here we have taken an x-axis perpendicular to the plates with its origin 0 at the surface of the left plate (Fig. 3.1). We choose planes $x=h/2$ and $x=-\infty$ as a closed surface enclosing the left plate. The electric field ($-d\psi/dx$) is zero on the plane $x=x_1=h/2$ (as a result of symmetry) and on the plane $x=x_2=-\infty$ [the electric potential $\psi(x)$ itself is zero]. Hence it follows from Eq. (19) that the interaction force $P(h)$ per unit area is given by

$$P(h) = [n_+(h/2) + n_-(h/2)]kT - 2nkT = 4nkT \sinh^2\left[\frac{ze\psi(h/2)}{2kT}\right] \quad (27)$$

which, for low potentials, reduces to

$$P(h) = \frac{1}{2}\varepsilon_r\varepsilon_0\kappa^2\psi^2(h/2) \quad (28)$$

By solving the linearized Poisson-Boltzmann equation, Eq. (16), subject to the boundary conditions

$$\psi(0) = \psi(h) = \psi_0 \quad (29)$$

we find that Eq. (28) becomes

$$P(h) = \frac{1}{2}\varepsilon_r\varepsilon_0\kappa^2\psi_0^2 \operatorname{sech}^2(h/2) \quad (30)$$

The corresponding potential energy per unit area is obtained by integrating Eq. (30) with respect to h:

$$V(h) = \varepsilon_r\varepsilon_0\kappa^2\psi_0^2[1 - \operatorname{thanh}(h/2)] \quad (31)$$

For the case of constant surface charge density, if the internal fields can be neglected, the boundary conditions are given by Eq. (11), viz.

$$\left.\frac{\partial\psi}{\partial x}\right|_{x=0^+} = -\frac{\sigma}{\varepsilon_r\varepsilon_0}, \quad \left.\frac{\partial\psi}{\partial x}\right|_{x=h^-} = \frac{\sigma}{\varepsilon_r\varepsilon_0} \quad (32)$$

The final results for the interaction force $P(h)$ and energy $V(h)$ per unit area of two parallel plates 1 and 2 carrying surface charge densities σ_1 and σ_2 are

$$P(h) = \frac{1}{2}\varepsilon_r\varepsilon_0\kappa^2\psi_0^2 \operatorname{cosech}^2(h/2) \quad (33)$$

$$V(h) = \varepsilon_r\varepsilon_0\kappa^2\psi_0^2 [\coth(h/2) - 1] \quad (34)$$

where

$$\psi_0 = \frac{\sigma}{\varepsilon_r\varepsilon_0\kappa} \quad (35)$$

is the unperturbed surface potential of the two plates. Equations (33) and (34), however, do not take into account the effect of the induced field within the interacting plates, which is the characteristic of the interaction at constant surface charge density [15–20].

One can extend the above treatment to the case of two parallel dissimilar plates. Expressions for the interaction energy $V(h)$ per unit area for the constant surface potential case and for the constant surface charge density case have been derived by Hogg et al. [29] and by Wiese and Healy [30], respectively. The results are

$$V_1 = \frac{2nkT}{\kappa}\left\{\left(\frac{y_{01}+y_{02}}{2}\right)^2\left[1-\tanh\left(\frac{\kappa h}{2}\right)\right]-\left(\frac{y_{01}-y_{02}}{2}\right)^2\left[\coth\left(\frac{\kappa h}{2}\right)-1\right]\right\} \quad (36)$$

(constant surface potential case)

$$V_1 = \frac{2nkT}{\kappa}\left\{\left(\frac{y_{01}+y_{02}}{2}\right)^2\left[\coth\left(\frac{\kappa h}{2}\right)-1\right]-\left(\frac{y_{01}-y_{02}}{2}\right)^2\left[1-\tanh\left(\frac{\kappa h}{2}\right)\right]\right\} \quad (37)$$

(constant surface charge density case).

where y_{0i} is the scaled surface potential of plate i (i = 1, 2).

3.4.2 Moderate Potentials

Equations (36) and (37) are correct to the second power of surface potentials and can be applied for low potentials. To obtain analytic expressions applicable for higher potentials, one needs make corrections to higher powers of surface potentials. Levine and Suddaby [31] obtained the interaction energy between two parallel similar plates at constant surface potential correct to the sixth power in surface potentials. Ohshima et al. [32] derived the corresponding expression for the interaction energy per unit area of dissimilar plates having constant surface potentials ψ_{01} and ψ_{02} at separation h in a symmetrical electrolyte solution of bulk concentration n and valence z. The result is

$$\begin{aligned} V = {} & \frac{2nkT}{\kappa}[Y_+^2\{1-\tanh(\kappa h/2)\} - Y_-^2\{\coth(\kappa h/2)-1\}] \\ & + \frac{2nkT}{\kappa}\left[\frac{1}{48}(Y_+^4+3Y_+^2Y_-^2)\{1-\tanh(\kappa h/2)\}\right. \\ & - \frac{1}{48}(Y_-^4+3Y_+^2y_-^2)\{\coth(\kappa h/2)-1\} - \frac{Y_+^4}{32}\frac{\tanh(\kappa h/2)}{\cosh^2(\kappa h/2)} \\ & \left. + \frac{Y_-^4}{32}\frac{\coth(\kappa h/2)}{\sinh^2(\kappa h/2)} - \frac{(\kappa h/2)}{32}\left\{\frac{Y_+^2}{\cosh^2(\kappa h/2)} - \frac{Y_-^2}{\sinh^2(\kappa h/2)}\right\}^2\right] \end{aligned}$$

$$+\frac{2nkT}{\kappa}\left[\frac{Y_+^2}{5760}\left\{Y_+^4+\frac{15}{8}Y_-^2(7Y_+^2+Y_-^2)\right\}\{1-\tanh(\kappa h/2)\}\right.$$

$$-\frac{Y_-^2}{5760}\left\{Y_-^4+\frac{15}{8}Y_+^2(7Y_-^2+Y_+^2)\right\}\{\coth(\kappa h/2)-1\}$$

$$+\frac{Y_+^4}{1536}\left(Y_+^2+\frac{3}{2}Y_-^2\right)\frac{\tanh(\kappa h/2)}{\cosh^2(\kappa h/2)}-\frac{Y_-^4}{1536}\left(Y_-^2+\frac{3}{2}Y_+^2\right)\frac{\coth(\kappa h/2)}{\sinh^2(\kappa h/2)}$$

$$-\frac{Y_+^6}{1024}\frac{\tanh(\kappa h/2)}{\cosh^4(\kappa h/2)}-\frac{Y_-^6}{1024}\frac{\coth(\kappa h/2)}{\sinh^4(\kappa h/2)}$$

$$-\frac{7(\kappa h/2)}{3072}\left\{\frac{Y_+^2}{\cosh^2(\kappa h/2)}-\frac{Y_-^2}{\sinh^2(\kappa h/2)}\right\}^3$$

$$+\frac{(\kappa h/2)}{384}\left\{\frac{Y_+^4}{\cosh^4(\kappa h/2)}-\frac{Y_-^4}{\sinh^4(\kappa h/2)}\right\}[Y_+^2\tanh^2(\kappa h/2)-Y_-^2\coth 2(\kappa h/2)]$$

$$+\frac{(\kappa h/2)}{256}\left\{\frac{Y_+^2}{\cosh^2(\kappa h/2)}-\frac{Y_-^2}{\sinh^2(\kappa h/2)}\right\}^2$$

$$\left.\times\left\{Y_+^2\frac{\tanh(\kappa h/2)}{\cosh^2(\kappa h/2)}-Y_-^2\frac{\coth(\kappa h/2)}{\sinh^2(\kappa h/2)}\right\}\right] \tag{38}$$

with

$$Y_+=\frac{y_{01}+y_{02}}{2},\quad Y_-=\frac{y_{01}-y_{02}}{2} \tag{39}$$

where y_{01} and y_{02} are the scaled surface potentials.

Honig and Mul [10] suggested that better approximations can be obtained if the interaction energy is expressed as a series of $\tanh(ze\psi_0/4kT)$ instead of ψ_0 and derived the interaction energy correct to $\tanh^4(ze\psi_0/4kT)$. The interaction energy correct to $\tanh^6(ze\psi_0/4kT)$ has also been derived [33].

3.4.3
Linear Superposition Approximation

Expressions for the interaction force and energy obtained using the linear superposition approximation (LSA) are applicable for arbitrary surface potentials at large separations. In this approximation, the potential is approximated by the sum of the asymptotic values of the two unperturbed potentials which is produced by the respective plates in the absence of interaction [2]. For two similar plates:

$$\psi(h/2) \approx 2\psi_s(h/2) \tag{40}$$

and Eq. (27) asymptotes to Eq. (28). This approximation is correct in the limit of large κh. The asymptotic value of $\psi_s(h/2)$ is obtained from Eq. (7), viz.

$$\psi_s(h/2) = 4\left(\frac{kT}{ze}\right)\gamma \exp(-\kappa h/2) \tag{41}$$

Hence

$$\psi(h/2) = 8\left(\frac{kT}{ze}\right)\gamma \exp(-\kappa h/2) \tag{42}$$

and Eq. (22) becomes

$$P(h) = 64\gamma^2 nkT \exp(-\kappa h) \tag{43}$$

Equation (43) is the required approximate expression for the double-layer interaction force per unit area of two parallel plates at separation h. This approximation holds irrespective of the types of the boundary condition on the particle surface: constant surface potential or constant surface charge density. The factor nkT in Eq. (43) implies that the interaction force arises essentially from the osmotic pressure of electrolyte ions. The factor $\exp(-\kappa h)$ means that the interaction is caused by overlapping the double layers around the respective plates. The corresponding interaction energy per unit area is obtained by integrating Eq. (43):

$$V(h) = \frac{64\gamma^2 nkT}{\kappa} \exp(-\kappa h) \tag{44}$$

The LSA method can be extended to the case of dissimilar plates on the basis of the concept of effective surface potential introduced by Bell et al. [34] and the method of Brenner and Parsegian [35]. Bell et al. [34] introduced the effective surface potential of a single colloidal particle. Ohshima et al. [36, 37] applied the method of Brenner and Parsegian [35], which was applied originally to the case of two interacting cylinders, to obtain the LSA expression for the case of two parallel plates. Consider two interacting parallel plates 1 and 2 having surface potentials ψ_{01} and ψ_{02}, respectively, at separation h between their surfaces in a solution of general electrolytes. According to the method of Brenner and Parsegian [35], the asymptotic expression for the interaction energy $V(h)$ per unit area is given by

$$V(h) = \sigma_{\text{eff1}}\psi_2(h) = \sigma_{\text{eff2}}\psi_1(h) \tag{45}$$

where $\sigma_{\text{eff}j}$ ($j=1, 2$) is the hypothetical charge per unit area of an infinitely thin plate that would produce the same asymptotic potential as produced by plate j

and $\psi_j(h)$ is the asymptotic form of the unperturbed potential at a large distance h from the surface of plate j, when each plate exists separately.

Consider the electric field produced by the above hypothetical infinitely thin plate having a surface charge density σ_{effj}. The electric field produced by this hypothetical plate at a distance h from its surface is thus given by

$$E_j(h) = \frac{\sigma_{\text{effj}}}{2\varepsilon_r\varepsilon_0} e^{-\kappa h} \tag{46}$$

The corresponding electric potential $\psi_j(h)$ is given by

$$\psi_j(h) = \frac{\sigma_{\text{effj}}}{2\varepsilon_r\varepsilon_0\kappa} e^{-\kappa h} \tag{47}$$

Let us now express the asymptotic expression for the unperturbed potential of plate j (j=1, 2) at a large distance h from the surface of plate j as

$$\psi_j(h) = \left(\frac{kT}{e}\right) Y_j e^{-\kappa h} \tag{48}$$

where Y_j is the asymptotic constant. If we introduce the scaled potential of plate j, viz.

$$y = \frac{e\psi}{kT} \tag{49}$$

then Eq. (48) reduces to

$$y_j(h) = Y_j e^{-\kappa h} \tag{50}$$

Note that in the limit of low potentials, the asymptotic constant Y_j tends to the scaled surface potential y_{0j}, which is related to the unperturbed surface potentials ψ_{0j} of plate j by

$$y_{0j} = \frac{e\psi}{kT} (j = 1, 2) \tag{51}$$

Thus Y_j in Eq. (48) can be interpreted as a scaled effective surface potential of the plate.

Comparison of Eqs. (47) and (48) leads to an expression for the hypothetical charge σ_{effj} for plate j, viz.

$$\sigma_{\text{effj}} = 2\varepsilon_r\varepsilon_0\kappa \left(\frac{kT}{e}\right) y_j \tag{52}$$

By combining Eqs. (48), (52) and (45), we obtain

$$V(h) = 2\varepsilon_r\varepsilon_0\kappa \left(\frac{kT}{e}\right)^2 Y_1 Y_2 e^{-\kappa h} \tag{53}$$

which is the required result for the asymptotic expression for the interaction energy per unit area of two parallel plates 1 and 2 at large separations.

We obtain the scaled effective surface potential Y for a plate having a surface potential y_0 immersed in a solution of general electrolytes. We take an x-axis perpendicular to the plate surface with its origin at the plate surface so that the region $x>0$ corresponds to the electrolyte solution. The Poisson–Boltzmann equation for the electric potential $\psi(x)$ is given by Eq. (13). Integration of Eq. (13) yields

$$\kappa x = \int_{y}^{y_0} \frac{dy}{f(y)} \tag{54}$$

where

$$f(y) = (1 - e^{-y}) \left\{ \frac{2 \sum_{i=1}^{N} n_i [\exp(-z_i y) - 1]}{(1 - e^{-y})^2 \sum_{i=1}^{N} z_i^2 n_i} \right\}^{\frac{1}{2}} \tag{55}$$

It can be shown that the asymptotic form of $y(x)$ satisfies

$$y(x) = Y e^{-\kappa x} \tag{56}$$

with

$$Y = y_0 \exp\left\{ \int_0^{y_0} \left[\frac{1}{f(y)} - \frac{1}{y} \right] dy \right\} \tag{57}$$

Equation (57) is the required expression for the scaled effective surface potential (or the asymptotic constant) Y. Wilemski [38] derived an expression for Y (Eq. 15 with Eq. 12) [38], which can be shown to be equivalent to Eq. (57).

For a plate having a surface potential y_0 in a 1:1 electrolyte solution of bulk concentration n, we obtain

$$Y = 4 \sinh(y_0/2) \tag{58}$$

so that Eq. (53) gives

$$V(h) = \frac{64 \gamma_1 \gamma_2 n k T}{\kappa} e^{-\kappa h} \tag{59}$$

where γ_j is given by Eq. (8) (with $z=1$), viz.

$$\gamma_j = Y_j/4 = \tanh(y_{0j}/4) (j = 1, 2] \tag{60}$$

Note that Eq. (59) is an extension of Eq. (44) to the case of two dissimilar plates.

Similarly, for the case of a plate having a scaled surface potential y_0 in a 2:1 electrolyte solution of concentration n, Y can be expressed analytically by

$$Y = 6\frac{(2e^{y_0}/3+1/3)^{\frac{1}{2}}-1}{(2e^{y_0}/3+1/3)^{\frac{1}{2}}+1} \tag{61}$$

which yields

$$V(h) = \frac{192\gamma_1'\gamma_2' kT}{\kappa}e^{-\kappa h} \tag{62}$$

with

$$\gamma_j' = \frac{Y_j}{4} = \frac{3}{2}\frac{(2e^{y_0}/3+1/3)^{\frac{1}{2}}-1}{(2e^{y_0}/3+1/3)^{\frac{1}{2}}+1}\,(j=1,2) \tag{63}$$

3.4.4
Alternative Method of Linearization of the Poisson–Boltzmann Equation

A novel linearization method has been proposed for simplifying the nonlinear Poisson-Boltzmann equation to derive an accurate analytic expression for the interaction energy between two parallel similar plates in a symmetrical electrolyte solution [39, 40]. This method is different from the usual linearization method (i.e. the Debye-Hückel linearization approximation) in that in this method the Poisson–Boltzmann equation is linearized with respect to the deviation of the electric potential from the surface potential, whereas in the usual method linearization is made with respect to the potential itself.

Consider first two interacting parallel similar plates 1 and 2 with constant surface potential ψ_0 at separation h in a symmetrical electrolyte solution of valence z and bulk concentration n. We assume that the surface potential ψ_0 remains unchanged independent of h during interaction. We take an x-axis perpendicular to the plates with its origin 0 at the surface of plate 1. We linearize Eq. (5) with respect to the deviation $\Delta y = y - y_0$ of the potential y from the surface potential y_0. Then, we obtain

$$\frac{d^2\Delta y}{dx^2} = \kappa^2(\sinh y_0 + \cosh y_0\cdot\Delta y) = \kappa^2\cosh y_0(\tanh y_0 + \Delta y) \tag{64}$$

By solving Eq. (64) under appropriate boundary conditions, we obtain

$$y(x) = y_0 - \tanh y_0\left\{1 - \frac{\cosh\left[\sqrt{\cosh y_0}\kappa(h/2-x)\right]}{\cosh\left[\sqrt{\cosh y_0}\kappa h/2\right]}\right\},\ 0 \le \kappa\frac{h}{2} \tag{65}$$

which gives the following expression for the interaction energy per unit area between two parallel similar plates with constant surface potential:

$$V(h) = \frac{8nkT}{\kappa}\left\{\frac{2}{\kappa h}\ln\left[\frac{\cosh(\kappa h/2)}{\cosh\left(\sqrt{\cosh y_0}\kappa h/2\right)}\right] + 2\left[\cosh\left(\frac{y_0}{2}\right) - 1\right]\right\} \tag{66}$$

The same linearization method can be applied to the interaction at constant surface charge density with the result that

$$V(h) = \frac{8nkT}{\kappa}\left\{\frac{2}{\kappa h}\ln\left[\frac{\cosh(\kappa h/2)}{\cosh\left(\sqrt{\cosh y(0)}\kappa h/2\right)}\right] + 2\left[\cosh\left(\frac{y_0}{2}\right) - 1\right] + [y(0) - y_0]\sinh\left(\frac{y_0}{2}\right)\right\} \tag{67}$$

with

$$y(0) = \text{arcsin}\,h\left[\frac{2\sinh(y_0/2)}{\tanh(\kappa h/2)}\right] = \ln\left\{\frac{2\sinh(y_0/2)}{\tanh(\kappa h/2)} + \sqrt{\left[\frac{2\sinh(y_0/2)}{\tanh(\kappa h/2)}\right]^2 + 1}\right\} \tag{68}$$

where $y(0)$ is the surface potential at separation h and y_0 is the scaled unperturbed surface potential ($h \to \infty$). The present linearization approximation works fairly well for small separations for all values of the reduced surface potential y_0 but becomes less accurate at larger separations.

3.5 Interaction Between Two Spheres

The spherical Poisson–Boltzmann equation for the two-sphere system has not been solved analytically and there are several approximation methods. We begin with Derjaguin's approximation [41].

3.5.1 Derjaguin's Approximation

In Derjaguin's approximation [41], the potential energy of the electrostatic double-layer interaction between two similar spheres of radius a at separation H between their surfaces (Fig. 3.2) is calculated from the corresponding energy of interaction between two parallel plates $V_{pl}(h)$ by using the following equation [41]:

$$V(H) = \pi a \int_H^\infty V_{pl}(h)dh \tag{69}$$

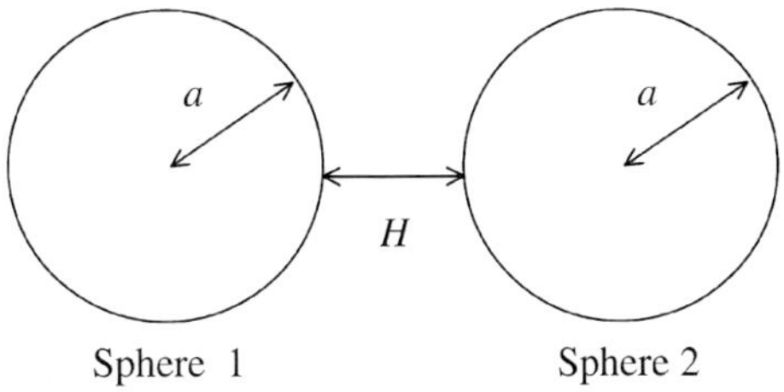

Fig. 3.2 Interaction between two similar charged spheres 1 and 2 of radius a at a separation R between their centers. H (=R–2a) is the closest distance between their surfaces.

Equation (69) is applicable for large particle radii and small particle separations, viz.

$$\kappa a \gg 1 \text{ and } \kappa H \ll a \tag{70}$$

The following two equations, which are obtained via Derjaguin's approximation, are usually used:

(1) for the low potential case:

$$V(H) = 2\pi\varepsilon_r\varepsilon_0 a\psi_0^2 \ln(1 + e^{-\kappa h}) \tag{71}$$

(constant surface potential)

$$V(H) = 2\pi\varepsilon_r\varepsilon_0 a\psi_0^2 \ln\left(\frac{1}{1 - e^{-\kappa h}}\right) \tag{72}$$

(constant surface charge density);

(2) for high potentials:

$$V(H) = \frac{64\pi a\gamma^2 nkT}{\kappa^2} e^{-\kappa H} \tag{73}$$

Equations (71) to (73) are obtained by substituting Eqs. (31), (34) and (44), respectively, into Eq. (69).

This method can be extended to the case of two dissimilar spheres 1 and 2 of radii a_1 and a_2, in which case Eq. (69) becomes

$$V(H) = \frac{2\pi a_1 a_2}{a_1 + a_2} \int_h^\infty V_{pl}(h)\mathrm{d}h \tag{74}$$

For two spheres with constant surface potentials ψ_{01} and ψ_{02}, Hogg, Healy and Fuerstenau (HHF) [29] found that, for the low-potential case, the interaction energy $V(h)$ is given by

$$V(H) = \pi\varepsilon_r\varepsilon_0 \frac{a_1 a_2}{a_1 + a_2}\left[(\psi_{01} + \psi_{02})^2 \ln\left(1 + e^{-\kappa H}\right) + (\psi_{01} - \psi_{02})^2 \ln\left(1 - e^{-\kappa H}\right)\right] \tag{75}$$

For the case of constant surface charge density case, Wiess and Healy [26] derived

$$V(H) = \pi\varepsilon_r\varepsilon_0 \frac{a_1 a_2}{a_1 + a_2}\left[-(\psi_{01}+\psi_{02})^2 \ln\left(1-e^{-\kappa H}\right) - (\psi_{01}-\psi_{02})^2 \ln\left(1+e^{-\kappa H}\right)\right] \tag{76}$$

where ψ_{01} and ψ_{02} are the unperturbed surface potentials of the respective spheres. Equations (75) and (76) are obtained by substituting Eqs. (36) and (37) into Eq. (74). For dissimilar spheres, we obtain by substituting Eq. (53) into Eq. (74)

$$V(H) = \frac{64\pi a\gamma_1\gamma_2 nkT}{\kappa^2} e^{-\kappa H} \tag{77}$$

An expression for the interaction in the mixed case where one particle has a constant surface potential and the other has a constant surface charge density has also been derived by Kar et al. [42].

3.5.2
Curvature Correction to Derjaguin's Formula and HHF Formula

The physical basis of Derjaguin's method [41] when the surface potential is low was elucidated by Ohshima et al. [43], who also derived next-order correction terms of order $1/\kappa a_i$. By the method of successive approximations, Ohshima et al. [43] obtained

$$\begin{aligned} V(H) =& \pi\varepsilon_r\varepsilon_0 \frac{a_1 a_2}{a_1 + a_2}\left[(\psi_{01}+\psi_{02})^2 \ln\left(1+e^{-\kappa H}\right) + (\psi_{01}-\psi_{02})^2 \ln\left(1-e^{-\kappa H}\right)\right] \\ & -\frac{1}{2}\left(\frac{1}{\kappa a_1}+\frac{1}{\kappa a_2}\right) V_1(H) \end{aligned} \tag{78}$$

with

$$\begin{aligned} V_1(H) =& \pi\varepsilon_r\varepsilon_0 \frac{a_1 a_2}{a_1 + a_2}\left[(\psi_{01}+\psi_{02})^2\left\{\frac{1}{2}\left(\frac{1}{3}-\left[\frac{a_1-a_2}{a_1+a_2}\right]^2\right)\kappa H \ln(1+e^{-\kappa H})\right.\right. \\ & \left. + \frac{\kappa H e^{-\kappa H}}{3(1+e^{-\kappa H})^2} - \frac{e^{-\kappa H}}{3(1+e^{-\kappa H})} + \frac{1}{3} I(\kappa H)\right\} \\ & + (\psi_{01}-\psi_{02})^2\left\{\frac{1}{2}\left(\frac{1}{3}-\left[\frac{a_1-a_2}{a_1+a_2}\right]^2\right)\kappa H \ln(1-e^{-\kappa H})\right. \\ & \left. - \frac{\kappa H e^{-\kappa H}}{3(1-e^{-\kappa H})^2} + \frac{e^{-\kappa H}}{3(1-e^{-\kappa H})} + \frac{1}{3} J(\kappa H)\right\} \\ & \left. - (\psi_{02}^2 - \psi_{02}^2)\left(\frac{a_1-a_2}{a_1+a_2}\right)\{\kappa H \ln(1-e^{-2\kappa H}) + K(\kappa H)\}\right] \end{aligned} \tag{79}$$

with

$$I(\kappa H) = \int_{\kappa H}^{\infty} \ln(1 + e^{-t}\mathrm{d})t = \sum_{n=1}^{\infty} \frac{(-1)^{n-1}}{n^2} e^{-n\kappa H} \tag{80}$$

$$J(\kappa H) = \int_{\kappa H}^{\infty} \ln(1 - e^{-t})\mathrm{d}t = -\sum_{n=1}^{\infty} \frac{1}{n^2} e^{-n\kappa H} \tag{81}$$

$$K(\kappa H) = I(\kappa H) + J(\kappa H) = \int_{\kappa H}^{\infty} \ln(1 - e^{-2t})\mathrm{d}t = -\sum_{n=1}^{\infty} \frac{1}{2n^2} e^{-2n\kappa H} \tag{82}$$

The first term in the brackets on the right-hand side of Eq. (78) agrees with the HHF Eq. (75), which is correct to order $1/(\kappa a_i)^0$, and the second term is the next-order correction of order $1/\kappa a_i$.

3.5.3
Correction to the Sixth Power of Surface Potentials in HHF Formula

Interaction equations correct to the sixth power in the surface potentials [and correct to order $1/(\kappa a)^0$] are given in Ref. [43].

$$V = \frac{8\pi a_1 a_2 n k T}{\kappa^2 (a_1 + a_2)} \left[Y_+^2 \ln(1 + e^{-\kappa H}) + Y_-^2 \ln(1 - e^{-\kappa H}) \right]$$

$$+ \frac{8\pi a_1 a_2 n k T}{\kappa^2 (a_1 + a_2)} \left[-\frac{1}{48}(Y_+^4 + 3Y_+^2 Y_-^2) \left(\frac{\kappa H}{2}\right) \left\{ 1 - \tanh\left(\frac{\kappa H}{2}\right) \right\} \right.$$

$$+ \frac{1}{48}(Y_-^4 + 3Y_+^2 Y_-^2) \left(\frac{\kappa H}{2}\right) \left\{ \coth\left(\frac{\kappa H}{2}\right) - 1 \right\}$$

$$\left. - \frac{Y_+^4}{96} \frac{1 - (\kappa H/2)\tanh(\kappa H/2)}{\cosh^2(\kappa H/2)} - \frac{Y_-^4}{96} \frac{(\kappa H/2)\coth(\kappa H/2) - 1}{\sinh^2(\kappa H/2)} \right]$$

$$+ \frac{8\pi a_1 a_2 n k T}{\kappa^2 (a_1 + a_2)} \left[-\frac{Y_+^2}{5760} \left\{ Y_+^4 + \frac{15}{8} Y_-^2 (7Y_+^2 + Y_-^2) \right\} \left(\frac{\kappa H}{2}\right) \left\{ 1 - \tanh\left(\frac{\kappa H}{2}\right) \right\} \right.$$

$$+ \frac{Y_-^2}{5760} \left\{ Y_-^4 + \frac{15}{8} Y_+^2 (7Y_-^2 + Y_+^2) \right\} \left(\frac{\kappa H}{2}\right) \left\{ \coth\left(\frac{\kappa H}{2}\right) - 1 \right\}$$

$$+ Y_+^6 \frac{17 + 4(\kappa H/2)\tanh(\kappa H/2)}{46080 \cosh^2(\kappa H/2)} - Y_-^6 \frac{4(\kappa H/2)\coth(\kappa H/2) + 17}{46080 \sinh^2(\kappa H/2)}$$

$$+ Y_+^4 Y_-^2 \frac{1+(\kappa H/2)\tanh(\kappa H/2)}{1024\cosh^2(\kappa H/2)} - Y_+^2 Y_-^4 \frac{(\kappa H/2)\coth(\kappa H/2)+1}{1024\sinh^2(\kappa H/2)}$$

$$- Y_+^6 \frac{1-11(\kappa H/2)\tanh(\kappa H/2)}{15360\cosh^4(\kappa H/2)} + Y_-^6 \frac{11(\kappa H/2)\coth(\kappa H/2)-1}{15360\sinh^4(\kappa H/2)}$$

$$\left. + \frac{(\kappa H/2)^2}{1536}\left\{\frac{Y_+^2}{\cosh^2(\kappa H/2)} - \frac{Y_-^2}{\sinh^2(\kappa H/2)}\right\}^3\right] \tag{83}$$

3.5.4
Linear Superposition Approximation for Sphere–Sphere Interaction

The linear superposition approximation (LSA) descried in Section 3.3.3 can be applied also to the interaction between two spheres [2, 36–40]. Consider next two interacting spherical colloidal particles 1 and 2 of radii a_1 and a_2 having surface potentials ψ_{01} and ψ_{02}, respectively, with a separation R between their centers in a solution of general electrolytes. The result is

$$V(R) = 4\pi\varepsilon_r\varepsilon_0 a_1 a_2 e^{\kappa a_j}\left(\frac{kT}{e}\right)^2 Y_1 Y_2 \frac{1}{R}\exp\left[-\kappa(R-a_1-a_2)\right] \tag{84}$$

with

$$Y_j = y_{0j}\exp\left[\int_0^{y_{0j}}\left\{\frac{1}{F(y)} - \frac{1}{y}\right\}dy\right] \quad (j=1,2) \tag{85}$$

$$F(y) = \frac{\kappa a}{\kappa a+1} f(y)\left[1 + \frac{2(2\kappa a+1)}{(\kappa a)^2}\frac{1}{f^2(y)}\int_0^y f(u)du\right] \tag{86}$$

where Y_j is the effective surface potential of sphere j and $f(y)$ is given by Eq. (55).

For a sphere of radius a having a scaled surface potential y_0 in a 1 : 1 electrolyte solution of concentration n, we find that Y is given by

$$Y = \frac{8\tanh(y_0/4)}{1 + \left[1 - \dfrac{2\kappa a+1}{(\kappa a+1)^2}\tanh^2(y_0/4)\right]^{\frac{1}{2}}} \tag{87}$$

The exact numerical values of the scaled effective surface potential of a sphere can be found in a book by Loeb et al. [44]. Equation (87) is found to give excellent approximations to the exact numerical values.

3.5.5 Exact Solution for Sphere–Sphere Interaction

An analytic exact solution to the linearized spherical Poisson-Boltzmann equation for the system of two interacting charged spherical (or cylindrical) particles immersed in an electrolyte solution was obtained by Ohshima et al. [45–56] on the basis of the method of images. In this section we show how the linearized spherical Poisson–Boltzmann equations for the system of two interacting spheres can be solved exactly and explicit analytic expressions for the interaction energy for this system can be obtained without recourse to Derjaguin's approximation [33] or numerical methods [57–67].

Consider now two interacting spheres 1 and 2 of radii a_1 and a_2, respectively, separated by a distance $R=H+a_1+a_2$ between their surfaces between their centers O_1 and O_2 (H being the closest distance), immersed in an electrolyte solution. We first treat the case where the surface potentials of spheres 1 and 2 both remain constant at ψ_{01} and ψ_{02}, respectively, during interaction independent of R. We write the solution to the spherical Poisson-Boltzmann equation in the following form:

$$\begin{aligned}\psi =&\psi_1^{(0)} + \psi_2^{(0)} + \left[\psi_1^{(1)} + \psi_1^{(2)} + \psi_1^{(3)} + \ldots + \psi_1^{(2v)} + \psi_1^{(2v+1)} + \ldots\right] \\ &+ \left[\psi_2^{(01)} + \psi_2^{(2)} + \psi_2^{(3)} + \ldots + \psi_2^{2v} + \psi_2^{(2v+1)} + \ldots\right]\end{aligned} \tag{88}$$

As the zeroth-order approximate solutions $\psi_1^{(0)}$ and $\psi_2^{(0)}$ we choose the unperturbed potentials produced by spheres 1 and 2. We construct the functions $\psi_1^{(i)}$ and $\psi_2^{(i)}$ ($i=2, 3, \ldots$) as follows. The unperturbed potential $\psi_1^{(0)}$ satisfies the boundary condition on sphere 1. The boundary condition on sphere 2, on the other hand, which has been satisfied by $\psi_2^{(0)}$, is now violated, since $\psi_1^{(0)}$ gives rise to a non-zero value on sphere 2. We therefore construct the first-order correction term $\psi_1^{(1)}$ so as to cancel $\psi_1^{(0)}$ on sphere 2.

The function $\psi_1^{(1)}$ can be interpreted as the "image" potential of $\psi_1^{(0)}$ with respect to sphere 2 by analogy with "the method of images" in electrostatics. By repeating this procedure, one can construct $\psi_1^{(k)}$ from $\psi_1^{(k-1)}$ ($k=1, 2, \ldots$). In a similar way, by starting from $\psi_2^{(0)}$ instead of $\psi_1^{(0)}$, one can obtain $\psi_2^{(k)}$ from $\psi_2^{(k-1)}$ ($k=1, 2, \ldots$).

On the basis of the obtained results for the potential distribution (Eq. 88), we finally obtain the required result for the interaction energy between spheres 1 and 2:

$$\begin{aligned}V(R) =&4\pi\varepsilon_r\varepsilon_0\psi_{01}\psi_{02}a_1a_2\frac{\exp\left[-\kappa(R-a_1-a_2)\right]}{R} \\ &+ 2\pi\varepsilon_r\varepsilon_0\psi_{01}^2a_1^2\frac{\exp(2\kappa a_1)}{R}\sum_{n=0}^{\infty}(2n+1)G_n(2)K_{n+1/2}^2(\kappa R) \\ &+ 2\pi\varepsilon_r\varepsilon_0\psi_{02}^2a_2^2\frac{\exp(2\kappa a_2)}{R}\sum_{n=0}^{\infty}(2n+1)G_n(1)K_{n+1/2}^2(\kappa R)\end{aligned}$$

$$
\begin{aligned}
&+ 4\pi\varepsilon\varepsilon_r\varepsilon_0\psi_{01}\psi_{02}a_1a_2\frac{\exp[\kappa(a_1+a_2)]}{R}\sum_{n=0}^{\infty}\sum_{m=0}^{\infty}\\
&(2n+1)(2m+1)B_{\mathrm{nm}}G_{\mathrm{n}}(2)G_{\mathrm{m}}(1)K_{\mathrm{n}+1/2}(\kappa R)K_{\mathrm{m}+1/2}(\kappa R)+\ldots\\
&+ 2\pi\varepsilon\varepsilon_r\varepsilon_0\psi_{01}\psi_{02}a_1a_2\frac{\exp[\kappa(a_1+a_2)]}{R}\sum_{n_1=0}^{\infty}\sum_{n_2=0}^{\infty}\cdots\sum_{n_{2v}=0}^{\infty}\\
&[L_{12}(n_1,n_2,\ldots,n_{2v})+L_{21}(n_1,n_2,\ldots,n_{2v})]\\
&\times K_{\mathrm{n}_1+1/2}(\kappa R)K_{\mathrm{n}_{2v}+1/2}(\kappa R)+2\pi\varepsilon_r\varepsilon_0\sum_{n_1=0}^{\infty}\sum_{n_2=0}^{\infty}\cdots\sum_{n_{2v-1}=0}^{\infty}\\
&(2n_{2v-1}+1)B_{\mathrm{n}_{2v-2}\mathrm{n}_{2v-1}}\\
&\times\left[\psi_{01}^2a_1^2\frac{\exp(2\kappa a_1)}{R}L_{21}(n_1,n_2,\ldots,n_{2v-2})G_{\mathrm{n}_{2v-1}}(2)\right.\\
&\left.+\psi_{02}^2a_2^2\frac{\exp(2\kappa a_2)}{R}L_{12}(n_1,n_2,\ldots,n_{2v-2})G_{\mathrm{n}_{2v-1}}(1)\right]\\
&\times K_{\mathrm{n}_1+1/2}(\kappa R)K_{\mathrm{n}_{2v-1}+1/2}(\kappa R)+\ldots
\end{aligned}
\tag{89}
$$

with

$$
\begin{aligned}
L_{21}(n_1,n_2,\ldots,n_{2v-2}) &= (2n_1+1)(2n_2+1)\ldots(2n_{2v}+1)\\
&\times B_{\mathrm{n_1n_2}}B_{\mathrm{n_2n_3}}\ldots B_{\mathrm{n}_{2v-1}\mathrm{n}_{2v}}G_{\mathrm{n}_1}(2)G_{\mathrm{n}_2}(1)\times\ldots\times G_{\mathrm{n}_{2v-1}}(2)G_{\mathrm{n}_{2v}}(1),\ v=1,2,\ldots
\end{aligned}
\tag{90}
$$

$$
\begin{aligned}
L_{12}(n_1,n_2,\ldots,n_{2v}) &= (2n_1+1)(2N_2+1)\ldots(2n_{2v}+1)\\
&\times B_{\mathrm{n_1n_2}}B_{\mathrm{n_2n_3}}\ldots B_{\mathrm{n}_{2v-1}\mathrm{n}_{2v}}G_{\mathrm{n}_1}(1)G_{\mathrm{n}_2}(2)\times\ldots\times G_{\mathrm{n}_{2v-1}}(1)G_{\mathrm{n}_{2v}}(2)\ \ v=1,2,\ldots
\end{aligned}
\tag{91}
$$

$$
G_{\mathrm{n}}(i) = -\frac{I_{\mathrm{n}+1/2}(\kappa a_{\mathrm{i}})}{K_{\mathrm{n}+1/2}(\kappa a_{\mathrm{i}})}
\tag{92}
$$

$$
B_{\mathrm{nm}} = \sum_{r=0}^{\min\{\mathrm{n,m}\}} A_{\mathrm{n\,mr}}\left(\frac{\pi}{2\kappa R}\right)^{1/2}K_{\mathrm{n+m-2r+1/2}}(\kappa R)
\tag{93}
$$

$$
A_{\mathrm{nmr}} = \frac{\Gamma(n-r+1/2)\Gamma(m-r+1/2)\Gamma(r+1/2)(n+m-r)!(n+m-2r+1/2)}{\pi\Gamma(m+n-r+3/2)(n-r)!(m-r)!r!}
\tag{94}
$$

where $\Gamma(z)$ is the gamma function.

The above method can be applied also to the case where the charge density on the particle surface (instead of the surface potential) remains constant during interaction. Consider two spheres 1 and 2 having unperturbed surface potentials ψ_{01} and ψ_{02} and relative permittivities ε_{rp1} and ε_{rp2}, respectively [51]. It can be shown that the interaction energy expression is obtained by replacing $G_n(i)$ in Eq. (89) with $H_n(i)$, defined by

$$H_n(i) = \frac{I_{n-1/2}(\kappa a_i) - (n+1+n\varepsilon_{rpi}/\varepsilon_r)I_{n+1/2}(\kappa a_i)/\kappa a_i}{K_{n-1/2}(\kappa a_i) + (n+1+n\varepsilon_{rpi}/\varepsilon_r)K_{n+1/2}(\kappa a_i)/\kappa a_i} \tag{95}$$

For the case where $\kappa a_1 \gg 1$ and $\kappa a_2 \gg 1$, $H \ll a_1$ and $H \ll a_2$, the interaction energy expression becomes

$$\begin{aligned} V(R) =& 4\pi\varepsilon_r\varepsilon_0 \frac{a_1 a_2}{a_1+a_2} \psi_{01}\psi_{02} e^{-\kappa H} + \frac{1}{4}\psi_{02}^2 e^{-2\kappa H}\left\{1 - \sqrt{\pi}\frac{\varepsilon_{rp1}}{\varepsilon_r}\sqrt{\frac{1}{\kappa a_1}+\frac{1}{\kappa a_2}}\right\} \\ &+ \frac{1}{4}\psi_{01}^2 e^{-2\kappa H}\left\{1 - \sqrt{\pi}\frac{\varepsilon_{rp2}}{\varepsilon_r}\sqrt{\frac{1}{\kappa a_1}+\frac{1}{\kappa a_2}}\right\} + O(e^{-3\kappa H}) \end{aligned} \tag{96}$$

If the terms of the order of $1/\sqrt{\kappa a_i}$ ($i=1, 2$) are neglected, Eq. (96) agrees with Wiese and Healy's equation, Eq. (75). Also, we see from Eq. (96) that the next-order curvature correction to Derjaguin's approximation is of the order of $1/\sqrt{\kappa a_i}$ ($i=1, 2$). This was first suggested by Dukhin and Lyklema [68] and discussed by Kijlstra [69]. In the case of the interaction between particles with constant surface potential, on the other hand, the next-order curvature correction to Derjaguin's approximation is of the order of $1/\kappa a_i$ ($i=1, 2$), since in this case no electric fields are induced within the interacting particles.

3.5.6
Interaction at Small Separations

In Section 3.4.4, we derived approximate expressions (Eqs. 66 and 67) for the interaction energy between two parallel plates, which are in excellent agreement with the exact numerical results, especially for small separations. By substituting these results into Derjaguin's equation, Eq. (69), one can obtain the corresponding energy for the interaction between two spheres [70]. In the present case, however, Eq. (69) is not appropriate because Eqs. (66) and (67) become less accurate for large separations except for low potentials. Therefore, instead, we employ an alternative expression for $V(H)$, viz.

$$V(H) = V(0) - \pi a \int_0^H V_{pl}(h)\mathrm{d}h \tag{97}$$

where

$$V(0) = \pi a \int_0^\infty V_{pl}(h)\mathrm{d}h \tag{98}$$

is the interaction energy between two spheres at contact ($H=0$). One can obtain an expression for $V(0)$ by combining the limiting form of $V(0)$ for low potentials and the numerical data provided by Honig and Mul [10]. For the case of constant surface potential, we obtain

$$V(0) = \frac{64\pi a n k T}{\kappa^2} \tanh^2\left(\frac{y_0}{4}\right) \ln\left[2\cosh\left(\frac{y_0}{4}\right)\right]\left[1 - \left\{a_1 \tanh^2\left(\frac{y_0}{4}\right) + a_2 \tanh^4\left(\frac{y_0}{4}\right) + a_3 \tanh^6\left(\frac{y_0}{4}\right) + a_4 \tanh^8\left(\frac{y_0}{4}\right)\right\}\operatorname{sech}(a_5 y_0)\right] \tag{99}$$

with $a_1=0.2912$, $a_2=0.06448$, $a_3=-0.03613$, $a_4=0.1660$ and $a_5=0.2786$, and for the case of constant surface charge density case, we have

$$V(0) = \frac{64\pi a n k T}{\kappa^2} \tanh^2\left(\frac{y_0}{4}\right) \left\{\frac{3|y_0|}{4\tanh(|y_0|/2)} - \ln\left[2\sinh\left(\frac{|y_0|}{2}\right)\right] + a_1 \tanh^4\left(\frac{y_0}{4}\right)\operatorname{sech}(a_2 y_0)\right\} \times \left[1 - \left\{a_3 \tanh^2\left(\frac{y_0}{4}\right) + a_4 \tanh^2\left(\frac{y_0}{4}\right) + a_5 \tanh^6\left(\frac{y_0}{4}\right) + a_6 \tanh^8\left(\frac{y_0}{4}\right)\right\}\operatorname{sech}(a_7 y_0)\right] \tag{100}$$

with $a_1=3.712$, $a_2=0.4035$, $a_3=-1.448$, $a_4=5.191$, $a_5=0.8604$, $a_6=-1.829$ and $a_7=-0.5037$.

References

1 B.V. Derjaguin, L.D. Landau, *Acta Physicochim.* **1941**, *14*, 633.
2 E.J.W. Verwey, J.Th.G. Overbeek, *Theory of the Stability of Lyophobic Colloids*, Elsevier, Amsterdam, **1948**.
3 B.V. Derjaguin, N.V. Churaev, V.M. Muller, *Surface Forces*, Consultants Bureau, New York, **1987**.
4 R.J. Hunter, *Foundations of Colloid Science*, Vols. 1 and 2, Clarendon Press, Oxford, **1989**.
5 J.N. Israelachvili, *Intermolecular and Surface Forces*, 2nd edn., Academic Press, London, **1991**.
6 J. Lyklema, *Fundamentals of Interface and Colloid Science, Vol. I, Solid–Liquid Interfaces*, Academic Press, London, **1995**.
7 H. Ohshima, K. Furusawa (eds.), *Electrical Phenomena at Interfaces, Fundamentals, Measurements and Applications*, 2nd edn., Marcel Dekker, New York, **1998**.
8 O.F. Devereux, P.L. de Bruyn, *Interaction of Plane-Parallel Double Layers*, MIT Press, Cambridge, MA, **1963**.
9 B.W. Ninham, V.A. Parsegian, *J. Theor. Biol.* **1970**, *31*, 405.
10 E.P. Honig, P.M. Mul, *J. Colloid Interface Sci.* **1973**, *36*, 258.
11 D. McCormack, S.L. Carnie, D.Y.C. Chan, *J. Colloid Interface Sci.* **1995**, *169*, 177.
12 G. Frens, J.Th.G. Overbeek, *J. Colloid Interface Sci.* **1982**, *38*, 376.
13 D. Gingell, *J. Theor. Biol.* **1967**, *17*, 451; **1968**, *19*, 340.
14 V.A. Parsegian, D. Gingell, *Biophys. J.* **1972**, *12*, 1192.
15 V.N. Shilov, *Kolloid Zh.* **1972**, *34*, 147.

16 H. Ohshima, *Colloid Polym. Sci.* **1974**, *252*, 158.
17 H. Ohshima, *Colloid Polym. Sci.* **1974**, *252*, 257.
18 H. Ohshima, *Colloid Polym. Sci.* **1974**, *253*, 150.
19 H. Ohshima, *Colloid Polym. Sci.* **1975**, *254*, 484.
20 H. Ohshima, *Colloid Polym. Sci.* **1979**, *257*, 630.
21 D. Y. C. Chan, T. W. Healy, L. R. White, *J. Chem. Soc., Faraday Trans. 1*, **1975**, *71*, 1046.
22 D. Y. C. Chan, T. W. Healy, L. R. White, *J. Chem. Soc., Faraday Trans. 1*, **1976**, *72*, 2844.
23 H. Ohshima, T. Mitsui, *J. Colloid Interface Sci.* **1978**, *63*, 525.
24 H. Ohshima, Y. Inoko, T. Mitsui, *J. Colloid Interface Sci.* **1982**, *86*, 57.
25 H. Ohshima, T. Kondo, *J. Theor. Biol.* **1987**, *128*, 187.
26 H. Ohshima, T. Kondo, *J. Colloid Interface Sci.* **1988**, *123*, 136.
27 H. Ohshima, T. Kondo, *Biophys. Chem.* **1988**, *32*, 161.
28 H. Ohshima, T. Kondo, *J. Colloid Interface Sci.* **1989**, *133*, 523.
29 R. Hogg, T. W. Healy, D. W. Fuerstenau, *Trans. Faraday Soc.* **1966**, *62*, 1638.
30 G. R. Wiese, T. W. Healy, *Trans. Faraday Soc.* **1970**, *66*, 490.
31 S. Levine, A. Suddaby, *Proc. Phys. Soc. A* **1951**, *64*, 287.
32 H. Ohshima, T. W. Healy, L. R. White, *J. Colloid Interface Sci.* **1982**, *89*, 484.
33 H. Ohshima, T. Kondo, *J. Colloid Interface Sci.* **1988**, *122*, 591.
34 G. M. Bell, S. Levine, L. N. McCartney, *J. Colloid Interface Sci.* **1970**, *33*, 335.
35 S. L. Brenner, V. A. Parsegian, *Biophys. J.* **1974**, *14*, 327.
36 H. Ohshima, T. W. Healy, L. R. White, *J. Colloid Interface Sci.* **1982**, *90*, 17.
37 H. Ohshima, *J. Colloid Interface Sci.* **1995**, *174*, 45.
38 G. Wilemski, *J. Colloid Interface Sci.* **1982**, *88*, 111.
39 H. Ohshima, *Colloids Surf. A* **1999**, *146*, 213.
40 H. Ohshima, *J. Colloid Interface Sci.* **1999**, *212*, 130.
41 B. V. Derjaguin, *Kolloid Z.* **1934**, *69*, 155.
42 G. Kar, S. Chander, T. S. Mika, *J. Colloid Interface Sci.* **1973**, *44*, 347.
43 H. Ohshima, D. Y. C. Chan, T. W. Healy, L. R. White, *J. Colloid Interface Sci.* **1983**, *92*, 232.
44 A. L. Loeb, J. Th. G. Overbeek, P. H. Wiersema, *The Electrical Double Layer Around a Spherical Colloid Particle*, MIT Press, Cambridge, MA, **1961**.
45 H. Ohshima, T. Kondo, *J. Colloid Interface Sci.* **1993**, *155*, 499.
46 H. Ohshima, T. Kondo, *J. Colloid Interface Sci.* **1993**, *157*, 504.
47 H. Ohshima, T. Kondo, *Colloid Polym. Sci.* **1993**, *271*, 1191.
48 H. Ohshima, *J. Colloid Interface Sci.* **1994**, *162*, 487.
49 H. Ohshima, *J. Colloid Interface Sci.* **1995**, *174*, 432.
50 H. Ohshima, *Adv. Colloid Interface Sci.* **1994**, *53*, 77.
51 H. Ohshima, *J. Colloid Interface Sci.* **1994**, *168*, 255.
52 H. Ohshima, *J. Colloid Interface Sci.* **1995**, *170*, 432.
53 H. Ohshima, *J. Colloid Interface Sci.* **1995**, *176*, 7.
54 H. Ohshima, E. Mishonova, E. Alexov, *Biophys. Chem.* **1996**, *57*, 189.
55 H. Ohshima, *Colloid Polym. Sci.* **1996**, *274*, 1176.
56 H. Ohshima, *J. Colloid Interface Sci.* **1998**, *198*, 42
57 S. Marcelja, D. J. Mitchell, B. W. Ninham, M. J. Sculley, *J. Chem. Soc., Faraday Trans. 2* **1977**, *73*, 630.
58 A. B. Glendinning, W. B. Russel, *J. Colloid Interface Sci.* **1983**, *93*, 95.
59 D. W. Weaver, D. L. Feke, *J. Colloid Interface Sci.* **1985**, *103*, 267.
60 J. W. Krozel, D. A. Saville, *J. Colloid Interface Sci.* **1992**, *150*, 365.
61 S. L. Carnie, D. Y. C. Chan, *J. Colloid Interface Sci.* **1993**, *155*, 297.
62 S. L. Carnie, D. Y. C. Chan, J. S. Gunning, *Langmuir* **1994**, *10*, 2993.
63 M. L. Grant, D. A. Saville, *J. Colloid Interface Sci.* **1995**, *171*, 35.
64 K. Ståhlberg, U. Applegren, B. Jönsson, *J. Colloid Interface Sci.* **1995**, *176*, 397.

65 J. Stankovich, S.L. Carnie, *Langmuir* **1996**, *12*, 1453.

66 P. Warszynski, Z. Adamczyk, *J. Colloid Interface Sci.* **1997**, *187*, 283.

67 S. Bhattacharjee, M. Elimelech, *J. Colloid Interface Sci.* **1997**, *193*, 273.

68 S.S. Dukhin, J. Lyklema, *Colloid J. USSR Engl. Transl.* **1989**, *51*, 244.

69 J. Kijlstra, *J. Colloid Interface Sci.* 1992, *153*, 30.

70 H. Ohshima, *J. Colloid Interface Sci.* **2000**, *225*, 204.

4
Role of Surface Forces on the Formation and Stability of Fractal Structures

Sümer Peker

4.1
Introduction

Particulate materials are used in various fields of chemical production, such as catalysts, medicines, cosmetics, cement and ceramics, pigments and electronics. Particulate matter is also present naturally in the environment, as in the case of clays, bacteria and blood cells, to give just a few examples from diverse fields. In chemical production, particle size distribution and morphology are of prime importance and have to be kept within controlled limits, which become narrower with the advancement of new production processes and materials [1–5]. Aggregation in a magnetic field [6] and light-induced cluster formation by ferromagnetic fluids have also been reported [7]. There are cases when aggregation is unavoidable, such as in the aggregation of bacteria and phytoplankton sedimenting in seas and aquatic media [8–12], which constitute part of the carbon cycle. Aggregation of colloidal suspensions may even be desirable if separation of solids from the carrier liquid is intended, as in the case of activated sludge processes in waste water treatment [13–16]; further, controlling the aggregation is necessary when dealing with soil and clay minerals [17] in areas such as environmental pollution and agriculture [18, 19].

Control of morphology is very difficult in colloidal solutions. However, emerging technologies involving production of nanoparticles in micellar liquids require strict control of particle size and shape to make full use of their vast surface area. Size control constitutes a critical step for their large-scale production. These require a deeper understanding of the aggregation processes.

4.2
Fractals as a Special Case of Aggregation

As these clusters are formed by the attractive forces between discrete particles, a porous structure is expected to be formed even under compact packing conditions. Pattern formation in the aggregation of particles and steric effects in the

Colloids and Interface Science Series, Vol. 1
Colloid Stability: The Role of Surface Forces, Part I. Edited by Tharwat F. Tadros

ISBN: 978-3-527-31462-1

case of clusters cause an increase in porosity to the point that the shape of the clusters can no longer be described by any of the forms of Euclidean geometry. Fractal theory developed by Mandelbrot [20] defines the volume, v, of the particles with a non-digit power of the characteristic dimension, L:

$$v \approx L^{d_f} \tag{1}$$

where d_f, the fractal dimension, is a non-digit number, a fraction, whence the name fractal originates. The fractal dimension d_f is always less than the Euclidean dimensions of the geometric form, encompassing the fractal object. For example, the fractal dimension of a cluster has a value of $d_f < 3$, which is less than the length dependence of the volume of an enveloping sphere in three dimensional space, $d = 3$, where

$$v \approx L^3 \tag{2}$$

The relation between the mass, M, and the characteristic length, such as the average radius, R, of a fractal aggregate, arises from the self-similar nature of fractal aggregates:

$$M(R, r) \approx C_0 m(R/r)^{\mathrm{D}} \tag{3}$$

where r is the radius of the component particles of mass m in a homogeneous suspension and C_0 is a geometric constant, equivalent to the constant terms in the mass balance equation. For homogeneous clusters made up of the same type of particles, the relation between the mass and the characteristic length of an aggregate is given by a simpler equation [21]:

$$M(R) \approx (R)^{d_f} \tag{4}$$

This equation also describes the scale invariance or self similarity of the fractal aggregates in that the geometric shape remains the same under isotropic rescaling of lengths:

$$d_f = \log(M_2/M_1)/\log(R_2/R_1) \tag{5}$$

Scale invariance in all directions or self-similarity is not valid for all types clusters, such as in the case of gels described by percolation theory [21, 22]. These types of clusters are called statistically self-similar fractals, as the statistical quantities used in characterization remain invariant to a change in length scales, provided that the length scale remains in the range between the characteristic length of the particles and the cluster [21].

Fractal clusters are characterized by the radius of the cluster, fractal dimension and, related to it, the density of the aggregates. In homogeneous fractal aggregates, an increase in the characteristic radius R at a ratio of λ is reflected in the mass of the aggregate as λ^{d_f}:

$$M(\lambda R) \approx C_1 \lambda^{d_f} (R)^{d_f} \tag{6}$$

A direct consequence of this equation is a decrease in the density of the aggregates with an increase in their size [21, 22]. Fractal cluster $Zn(OH)_2$ precursor particles in the production of ZnO [24] are given in Fig. 4.1. The close-up view exemplifies the ramifications which cause the density to decrease as the size of the aggregate increases.

The characteristic radius of the aggregate R in these equations can be defined by different properties of the aggregate. The three definitions commonly used in the literature are the radius of gyration R_g, the average distance of each particle from the center of gravity of the cluster; the hydrodynamic radius R_h, the radius of the sphere experiencing the same drag as the porous aggregate in the Stokes regime; and the collision radius R_c, the radius of the smallest sphere that will envelope the aggregate [25]. Inherent difficulties with the experimental techniques used to evaluate these radii led to the development of computational techniques for the prediction of hydrodynamic radius [26]. Torres et al. [27] related the hydrodynamic radius with the radius of gyration through the fractal dimension:

$$R_h^2 = \frac{d_f + 2}{d_f} R_g^2 \tag{7}$$

Fractal dimension d_f is a measure of the degree of ramification of the aggregate structure and is directly related to the degree of interaction between the constituent primary particles. Imaging techniques such as confocal laser microscopy and polarized light microscopy [28]; measurement of rheological properties such as yield stress [1] and viscoelastic moduli [29, 30]; scattering techniques, such as static and dynamic light scattering [31, 32], small angle (SAXS) and ultra small angle X-ray scattering [22, 30, 33], are some of the most frequently used methods for the experimental determination of the fractal dimension. The limitations

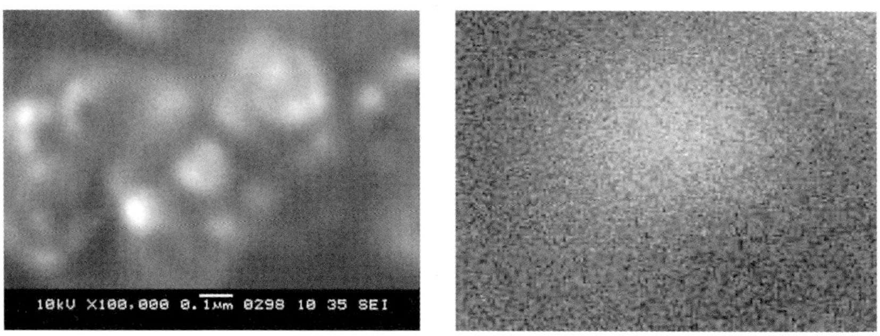

(a) (b)

Fig. 4.1 (a) $Zn(OH)_2$ precursor particles in the form of fractal clusters in the production of ZnO. (b) Close-up view of a cluster.

of these techniques have been reviewed for clusters of micron size range particles [25].

Fractal dimension is a macro-scale representation of the mechanisms effective in the aggregation process; namely, the effect of surface forces binding the components of the cluster and the hydrodynamics of the system, which may be effective in bringing together or tearing apart the components of the cluster.

Two types of cluster growth mechanisms are envisaged in modeling of aggregation processes [34]. In the first type, the growth is local and depends only on the immediate environment of the position to which a new particle or cluster is to be added. Percolation clusters and gels belong to this group. In the non-local type of cluster formation, the structure of the whole cluster can affect the probability of addition to a site at a given position. Clusters formed under diffusion-limited or reaction-limited conditions and clusters undergoing simultaneous aggregation-fragmentation processes during the growth process belong to this group. The composition of the medium between the particles also affects the mechanism of interaction between the particles. There is an enormous amount of literature on the surface forces in relation to aggregation and their effect on the fractal dimension. As a detailed account of all the aspects of this phenomena cannot be made in a limited review, only some of the recent work on the role of surface forces effective in the structure formation of aggregates will be addressed in this review.

4.3 Kinetics of Cluster Formation

The crucial step in the destabilization of colloids is the formation of a dimer at the initiation of cluster formation. If formed, these dimers then aggregate through further collisions. Once a nucleus is formed, the initial growth process proceeds by the addition of individual particles. Depending on the prevailing intensity of shear in the aggregation medium, cluster–cluster unions may also take place.

Formation of a dimer is determined by the balance of forces acting on the particles on collision. In a recent study, Dukhin et al. [35] analyzed the conditions for the stability of nano-sized particles with respect to aggregation within a mini-emulsion at low Reynolds and Stanton numbers through a force balance where the attractive forces between the two particles were also taken into consideration. They state that colloidal stability with respect to aggregation may be possible if the primary potential pit is shallow [36] while the particles are large, causing an enhancement of the hydrodynamic detaching force. Attractive forces can be decreased through surface roughnesses [37] or coating through surfactant or polymer adsorption, which causes the distance between the particles to increase.

The aggregation process involves two basic steps: The first step is the transportation of the particles to the aggregation site, with a sufficient difference in the magnitude and direction of their velocities for collision to be possible. Transportation can be brought about by random thermal Brownian motion (perikinetic aggregation), by the effect of a shear field inducing different velocities to the particles (orthokinetic aggregation) or through a difference in the terminal velocities in differential sedimentation. A transportation step results with an inevitable collision between the particles. The frequency with which the particles collide with each other per unit time and unit volume is called the collision frequency, which depends on the hydrodynamics of the medium and the volume occupied by the ramified fractal clusters [34, 38] or the number concentration of the particles, which, in turn, is a complex function of the strength of the bonds between the particles and the shear stress to which the clusters are subjected in the process.

The second step is the formation of a bond between the two particles. The success of the collision in causing such a bond is termed the collision efficiency, which depends on the shape of the interaction potential between the particles, such as given in Fig. 4.2.

A deep minimum in the potential with a negligible repulsive barrier leads to a diffusion-limited aggregation (DLA) regime, whereas an increase in the repulsive energy barrier causes the reaction step to be rate limiting, leading on to colloidal stability in the limiting case of an insurmountable energy barrier under the prevalent conditions.

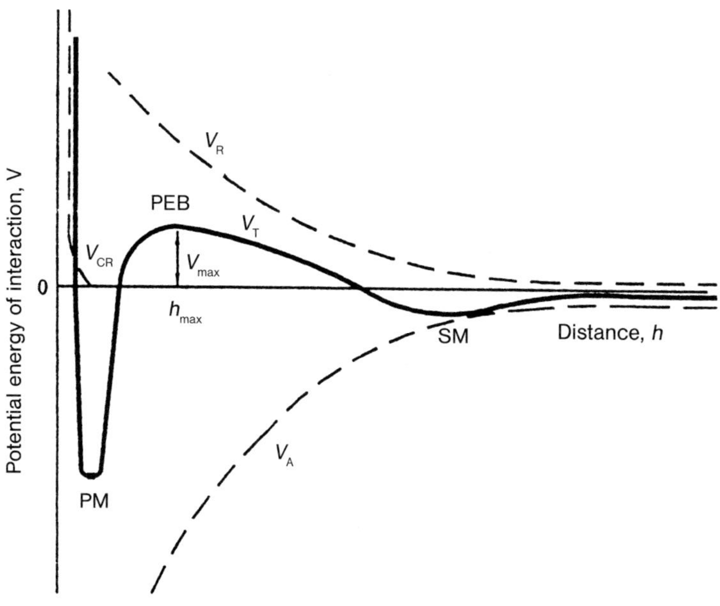

Fig. 4.2 Variation of interaction potential with distance. PM, primary minimum; SM, secondary minimum; PEB, primary energy barrier; V_{CR}, core repulsion.

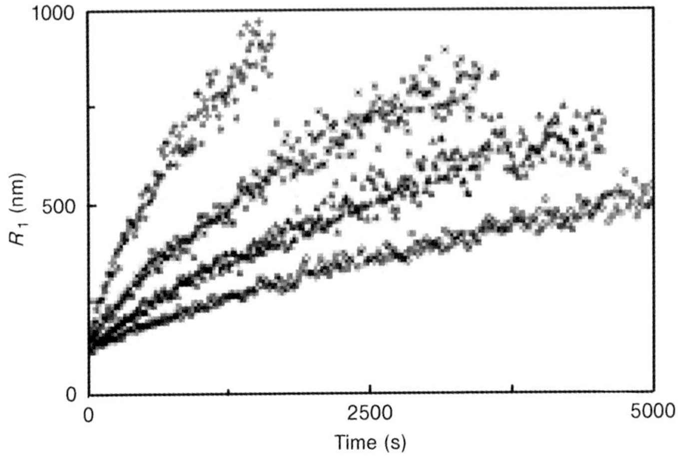

Fig. 4.3 Growth in DLA regime. (Reprinted with permission from Berka and Rice [17], copyright 2005 American Chemical Society).

In the case of DLA, the rapid growth rate of the aggregates can be described with a power law relation. In reaction-limited aggregation (RLA), the growth is much slower, following an exponential trend. Typical growth curves of the aggregates of kaolinite particles in the DLA regime obtained by Berka and Rice [17] and that of γ-alumina in the RLA regime obtained by Waite et al. [32] are given in Figs. 4.3 and 4.4.

The presence of a high repulsive barrier is not enough to prevent destabilization if there is a second minimum observed in the potential energy profile. The presence of a second attractive minimum at distances larger than the repulsive barrier in the potential energy curve signifies reversibility, ending up with simultaneous aggregation and fragmentation processes during growth, since the escape of a captured particle cannot be prevented if the depth of the minimum is not large enough. If the captured particle is in the primary minimum that is not deep enough with a relatively surmountable energy barrier, then external forces applied can only bring about restructuring.

In view of the issues stated above, the rate process for aggregation can be given by the expression

$$r = \alpha(i,j)\beta(i,j)n_i n_j \tag{8}$$

where r is the rate of aggregation in ($m^{-3}\ t^{-1}$), α is the dimensionless collision efficiency factor, β is the collision frequency factor ($m^3\ t^{-1}$) and n_i and n_j are the number concentrations of components i and j, respectively.

Smoluchowski [39] expressed the aggregation rate within the framework of a population balance for the first time in 1917 and it is still used today with some modifications and revisions:

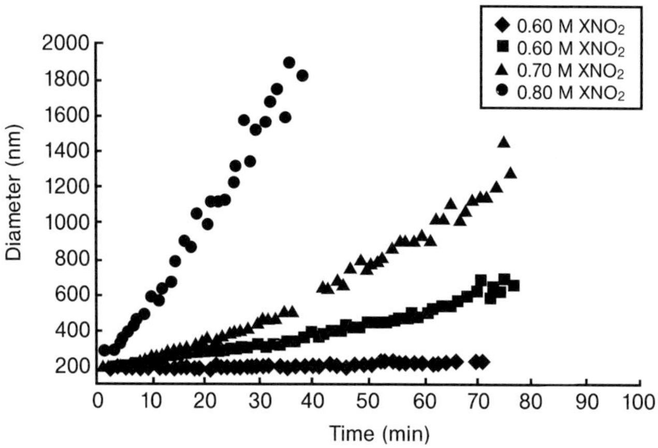

Fig. 4.4 Growth in RLA regime. (Reprinted from Waite et al. [32], with permission from Elsevier).

$$\frac{dn_k}{dt} = \frac{1}{2}\sum_{i+j=k} \beta(i,j)n_i n_j - \sum_{i=1}^{\infty} \beta(i,k)n_i n_k \tag{9}$$

This balance equation written for the conservation of particles of size k states that rate of accumulation of particles of size k is equal to that formed by the collision of particles of size i and j (the first term on the right-hand side) minus the loss of particles of size k, due to enlargement through collision and aggregation with another particle of size i. This equation is written for all sizes of particles within the range of high probability of occurrence. As this leads to an almost infinite number of equations, the sizes for which a balance is made are spaced apart by equal increments (arithmetic series) or by multiples at a constant ratio (geometric series). As these equations are non-linear and their solutions not easy, Smoluchowski made a number of simplifying assumptions: (1) the collision efficiency factor, α, is unity for all collisions; (2) fluid motion undergoes laminar shear; (3) the particles are monodispersed, that is, they are all of the same size; (4) no breakage of flocs occurs; (5) all particles are spherical in shape and remain so after collision; and (6) collisions involve only two particles.

These assumptions are critically reviewed and the recent developments discussed by Thomas et al. [40]. Except for the first assumption, these assumptions pertain to the collision frequency factor or, as more generally stated, the collision kernel, the system of equations describing the aggregation process. The collision efficiency factor, α, is generally taken into account by the stability ratio W, proposed by Fuchs in 1934 [41], defined as the inverse of the collision efficiency factor α or as the ratio of the rate constant of the fast reaction (as in DLA) to the rate constant of the slow reaction, that is, the RLA:

$$W = \frac{k_{\text{fast}}}{k_{\text{slow}}} \tag{10}$$

The stability ratio can also be defined in terms of the potential energy of interaction:

$$W = (r_{\text{i}} + r_{\text{j}}) \int_{r_{\text{i}}+r_{\text{j}}}^{\infty} \frac{\exp(V_{\text{T}}/kT)}{s^2} \, \mathrm{d}s \tag{11}$$

where r_{i} and r_{j} are the radii of the colliding particles, V_{T} the total energy including the repulsive and attractive forces, k the Boltzmann constant, T the absolute temperature and s the distance between the centers of the particles.

Kusters et al. [5] modeled clusters as an impermeable core surrounded by a porous shell. As the primary particles inside the cluster would have no effect on the aggregation mechanism, they suggested the collision efficiency factor α be calculated based on primary particles at the surface only. V_{T} is generally calculated with the DLVO theory. If other forces are also effective, the necessary terms should be added to the V_{T} expression. This is not an easy task, however, in two respects: The affecting non-DLVO force may not be easily identifiable in some cases and that the fundamental theoretical foundations are not developed for all types of forces. For this reason, the DLVO theory and the criticisms directed toward its capabilities will be treated first, followed by only a brief summary of the developments related to these issues.

4.4 Surface Forces Effective on the Collision Efficiency Factor

After the recognition of the resistance to aggregation in some systems, the potential barrier to aggregation was taken into account by the DLVO theory [42, 43]. As initially formulated by the authors, the theory assumes that the total potential energy of two interacting particles can be expressed as the sum of the electrical double layer repulsion and Van der Waals attraction:

$$V_{\text{T}}(h) = V_{\text{att}}(h) + V_{\text{rep}}(h) \tag{12}$$

For two spherical particles of radius R,

$$V_{\text{T}}(h) = \pi R \left[-\frac{H_{121}}{12\pi} \frac{1}{h} + \frac{64 c_{\text{i0}}^{*} kT \Gamma_0^2 \exp(-\kappa h)}{\kappa^2} \right] \tag{13}$$

where h is the separation between the surfaces of two particles (m), H_{121} is the Hamaker constant (m), for two particles (phase 1) in a liquid medium (phase 2), c_{i0}^{*} is the concentration of ions species in the solution (mol m^{-3}) when the potential $\Phi=0$, k is the Boltzmann constant (1.38×10^{-23} J K^{-1}), T is the absolute temperature (K) and κ (m^{-1}) is the Debye constant, the inverse of which ($1/\kappa$) gives the Debye screening length of the potential. Γ_0 is a term including the potential in the form

$$\Gamma_0 = \frac{\exp(ze\Phi_0/2kT) - 1}{\exp(ze\Phi_0/2kT) + 1} \tag{14}$$

where z is the valence, e is the electrical charge (C) and Φ_0 is the surface potential (V).

These two forces are present in a wide range of colloidal systems under moderate conditions. DLVO theory has been criticized by many authors on the grounds that the potential and charge density may not always remain constant and that it is not comprehensive enough to deal with forces that can be effective in the aggregation processes. The situations which can not be handled by the DLVO theory can be summarized as (1) high ionic strengths, (2) heteroaggregation, (3) discrete surface charges, (4) strong electrostatic interactions as in the case of multivalent counterions or low dielectric constant of the solvent, (5) specific ion effects and (6) non-DLVO forces (such as steric, hydrophobic, depletion) and very short-range hydration forces. The list increases as new types of forces are measured. Perhaps all of these shortcomings can be handled with a single unified theory. Alternative procedures based on Gibbs equation have been suggested [44] and promising results were obtained by Lyklema and Duval [45], but as the theoretical foundations are not yet fully developed, DLVO theory is still used with extensions and revisions.

It is acknowledged by many authors that the DLVO theory is valid at low to moderate ionic strengths of electrolyte solutions, where the distances between surfaces are large in comparison with molecular dimensions. This has been confirmed also by atomic force microscopy measurements in many studies. In other cases, the experimental results could not be described by the DLVO theory. A brief account of the recent developments in these widely criticized issues is given below, with the reservation that the review is by no means comprehensive:

For the case of high ionic strengths, Behrens et al. [46] noted that the position of the energy maximum shifted toward shorter distances, of the order of a few Ångstroms, with an increase in the ionic strength of the solution. As molecular dimensions are approached in these length scales, the interfacial structure of both the colloid and the solvent becomes important, as also confirmed by the appearance of non-DLVO forces in the experimental measurements of Lin et al. [47]. Behrens et al. [46] suggest that the position of the maximum in the energy barrier should also be taken into account, along with its magnitude, in the analysis of interactions. Another interesting result of the work is the demonstration of the gradual transition from fast aggregation to stabilization (through the RLA regime) with the equistability lines (Fig. 6 in their paper). This gradual transition also has a bearing on the fractal dimension, as will be discussed in the next section.

Heteroaggregation is another process which cannot be handled by the DLVO theory. Heteroaggregation involves dissimilar components in terms of shape, size, surface and other physical properties. In a review by Islam et al. [48], the contradiction between experimental results and theory arising from the assumptions of constant surface potential and constant surface charge was pointed out. This issue was approached theoretically in a recent article on regulation capacity by Lyklema and Duval [45]. Regulation is defined as the adjustment of the charge and potential when the distance between two surfaces is reduced. The regulation concept is generalized, so that it covers both charges and potentials, and can be applied both to the surface and to any part of the double layer. The spatial part of the regulation is treated as a series of electrical capacitances. The surface regulation is defined by the Gibbs energy of adsorption, number of adsorption sites available and the magnitude of the surface charge. Through the numerical solution of the model developed, an explanation was provided of phenomena such as heterointeraction and charge reversal, which could not be understood with the DLVO theory. The charge regulation concept was successfully used in modeling experimental data obtained with concentrated suspensions, where the charge on a single particle depends also on its proximity to other particles [49].

Another criticism directed toward the DLVO theory was the inability to differentiate between ions, other than with respect to their valence. Electrolytes screen the double-layer repulsions but not the van der Waals interactions [44]. Specific ion (Hofmeister) effects manifested through the propensity in bringing about aggregation might be related to these interactions. Two articles appearing at about the same time addressed this issue [50, 51].

Discrete surface charges are another issue that cannot be handled with the DLVO theory. In the mean field approximation of the Poisson-Boltzmann distribution, the surfaces are assumed to have uniform charge, denoted by the surface charge density. Discretization of the surface charge creates centers of attraction for counter ions, which then determine the interaction between surfaces. Khan et al. [52] calculated the force between two parallel charged flat surfaces, with discrete surface charges for different values of electrostatic coupling. They were able to show the type of interaction (repulsion/attraction) expected as a function of distance and coupling strength between the surfaces.

Large repulsive forces at very short distances cannot be accounted for by the DLVO theory. These are generally attributed to structuring of water molecules at the surface. Direct measurement of surface forces of silica observed at distances less than 10 nm show the existence of such repulsive forces. Various hypotheses have been proposed to explain the origins of these repulsive forces, which were reviewed and tested against experimental findings by Adler et al. [53] and Valle-Delgado et al. [54]. The hypotheses supported by experimental work include structuring of water at the surface (structural force) [55, 56] and gel formation at the surface resulting in steric repulsion of silicic acid chains [57]. These and various other recent theoretical models were compared with experimental data [53, 54]. All models approached the experimental findings but the validity of none could be corroborated.

Forces at very short distances between phospholipid bilayers were attributed to induced polarization in the presence of a large difference between the dielectric constants of the lipid and aqueous media by Jönsson and Wennerström [58]. Recently, Lin et al. [47] measured the attractive forces in water between double-chained surfactant monolayers physisorbed on mica. In this remarkable measurement, they could come down to distances below 100 Å, the lowest distance ever reached so far. The presence of very strong forces not present in the DLVO theory were recorded, yet the foundations on which a theory could be based still remained obscure. Cohen et al. [59] studied Newton black film (NBF) formation in thin films of aqueous rhamnolipid solutions, another double-chained surfactant. They proposed the continuous transition from common black film (CBF) to Newton black film (NBF) as evidence of the action of repulsive forces that are not described by the DLVO theory of colloid stability.

Recent developments in the area of nanofluids directly applicable to nanoparticle technologies were reported by Wasan and coworkers [60–62]. At extremely short distances where liquids can no longer be regarded as a continuum, oscillating forces arise. Structure changes upon confinement, depletion forces and their relation to rheology [63] are all fascinating subjects in the advancement of surface forces.

The effect of polymers on colloidal stability depends on the bonds formed with the particles, and also on the chemical or electrolytic nature of the polymer. Hence a classification in terms of non-adsorbed and adsorbed polymers and, in the case of the latter, neutral or polyelectrolyte types of polymers would be appropriate in terms of their effect on colloid stability. This classification should be preferred over a classification in terms of surface forces, since several forces can be simultaneously effective in particle–polymer interactions. By adjusting the amount and molecular weight of the polymer and conditions such as pH and temperature of the continuous medium, different effects can be obtained and control of particle sizes maintained. An interesting case was reported recently by Ditsch et al. [64], where magnetic particles of nanometer size were subjected to partial coating by random copolymers of acrylic acid–styrenesulfonic acid–vinylsulfonic acid (AA–SSA–VSA). By applying the polymers in amounts below that required for full coating and adjusting the hydrophobicity by SSA and attachment density by AA, the cluster size could be kept under control. Polyelectrolyte-mediated surface interactions were recently reviewed by Claesson et al. [65] and the areas which need further research were pointed out. The work of Biggs and coworkers [66–68] and Huang and Berg [69] will be considered in the next section, in relation to the effect of surface forces on fractal dimensions.

In summary, there have been immense developments, in theory and experiment, in the realm of surface forces. The extent to which this knowledge been used in the analysis and modeling of the formation mechanism and stability of fractal aggregates will be the topic of the next section.

4.5
Effect of Surface Forces on the Fractal Dimension

The role of surface forces in the structure of resulting aggregates is represented on the macroscale by a single parameter, the fractal dimension, d_f, for perikinetic aggregation. When only DLVO forces are effective in the process, the magnitude of d_f gives an indication of the kinetics of aggregation during the formation and the extent of fragmentation and restructuring in later stages. Non-DLVO forces operative in the aggregation process affect either the magnitude or the variation of the magnitude with the rate of aggregation or the size of the fractal cluster. These issues are discussed below for clusters formed in three-dimensional space, where the volume of the cluster is given by Eq. (2).

In solutions of indifferent electrolytes, the colloids are destabilized under the action of DLVO forces. The value of d_f is then determined only by the mechanism controlling the rate of aggregation. When the transport to the aggregation site is the only rate-limiting factor, the value of d_f was shown to vary in the range 1.7–1.8 by theoretical considerations [70] and confirmed by experimental results of the aggregation of various colloids in solutions of different electrolytes [71]. When adhesion to the cluster is the rate-limiting step, d_f was found to be in the range 2.1–2.2 through simulations and experiments with electrolytes indifferent to the surfaces of the particles. The repetition of these values in different cases gave an impression of "universality" in fractal dimensions for aggregation in DLA and RLA conditions. These values are still taken as criteria for testing the effect of a variable on the structure of the aggregates formed.

4.5.1
Effect of Stability Ratio on Fractal Dimension

The rate of aggregation proceeding slowly owing to potential energy barriers or steric hindrance is related to the rate of diffusion-controlled aggregation through the stability ratio W [41], as discussed above. Aggregation may well proceed at a rate intermediate between the two limiting cases of DLA and RLA, representing zero and maximum potential energy barriers for dimer formation. Results of light-scattering experiments with dilute suspensions of natural kaolinite of spherical equivalent hydraulic radius $R_h \approx 100 \pm 5$ nm showed a linear dependence of the fractal dimension on the logarithm of the stability ratio [17], as depicted in Fig. 4.5.

The equation obtained by linear regression for the intermediate regime in the range $5 < W < 100$:

$$d_f = 0.17(\log W) + 1.67 \tag{15}$$

is in agreement in form with those in the literature [72, 73]. The limiting values of d_f estimated with this equation, 1.78 and 2.1 for the DLA and RLA regimes, respectively, are in agreement with the universality range given above. These

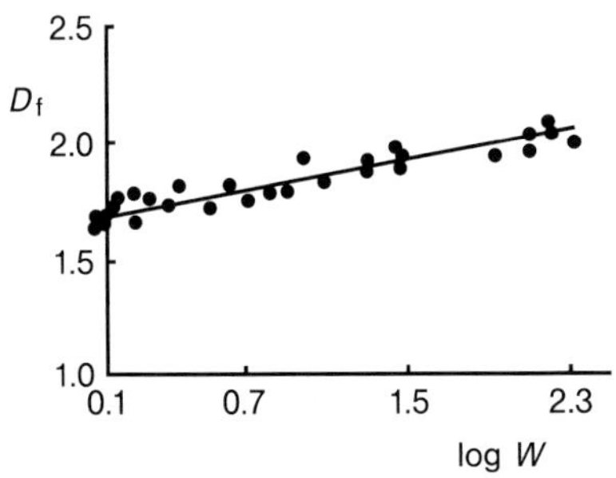

Fig. 4.5 Variation of fractal dimension with the stability ratio. y=0 17x+1.67; r^2=0.90. (Reprinted with permission from Berka and Rice [17], copyright 2005 American Chemical Society).

values were obtained with early time kinetic studies at times not greater than a few τ, the "half-time of rapid coagulation", defined as the time required for half of the original dispersed particles to coagulate, so agreement with Smoluchowski's rate equation is not surprising. As the authors acknowledge, this agreement is valid for very low volume fractions of the colloid. In the more concentrated suspensions employed in industrial applications, the compaction of the clusters and, hence, the fractal dimension may change during aggregation.

Regimes intermediate between DLA and RLA were investigated by Tirado-Miranda et al. in salt-induced aggregation of polystyrene particles, together with the effect of coating the particles with a polyelectrolyte, monomeric bovine serum albumin molecules [74]. The fractal dimension of uncoated particles varied linearly from 1.75 to 2.1 with increase in the electrolyte concentration at pH 4.8, the isoelectric point of bovine serum albumin (BSA). Partial coating of the bare particles with BSA caused an overall increase in d_f of 0.3 units irrespective of the pH: At pH 4.8, the isoelectric point, the polyelectrolyte is at its most compact configuration. At pH 9, both the polystyrene particles and the polyelectrolyte are negatively charged. In this case, compactness of the aggregate is directly proportional to the degree of shielding of the charges by the electrolyte ions, as confirmed experimentally. At very low concentrations of the electrolyte, the polystyrene particles were found to be stable. Polyelectrolyte-coated particles were stable at pH 9 but aggregated with a close-packed structure of $d_f \approx 2.2$ at pH 4.8. The authors attributed the observed instability to the partial shielding of the particle charges by the neutral polyelectrolyte at its isoelectric point, causing the particles to come as close as 3 nm, a distance comparable to the length of polymer segments which "bridge" them within a compact aggregate.

The variation of the structure factor with the stability ratio for sterically stabilized colloids has also been reported [69] for silica particles stabilized in cyclohexane by grafted polystyrene chains. Destabilization was brought about by reducing the temperature below the coil-to-globule transition of the grafted polymer. The stability ratio W was found to decrease with decrease of temperature below the transition point of the polymer. The fractal dimension d_f was found to be a function of both time and temperature: it increased from 1.75 at the transition temperature to its equilibrium value of 2.13 with time while this equilibrium value increased to 2.84 as the temperature was decreased towards the end of the transition interval. Transformation to a more compact structure as a function of time was taken to indicate spontaneous restructuring. A two-step mechanism of reduced rate aggregation followed by particle movement through lubrication of the grafted polymer into a denser configuration was proposed to explain this spontaneous restructuring.

Fractal aggregates are also observed in gels of polymeric [30, 69] or micro-emulsion origin [4, 75, 76], in fat–crystal networks [28] and in sol–gel processes [22, 33]. The fractal dimension becomes an important parameter in determining the compactness of the aggregate structure and the rheological behavior of concentrated suspensions [77], migration of charges resulting in an increase in conductivity [75, 76]. Relaxation phenomena in critical microemulsion systems is attributed to the formation of transient polydispersed fractal aggregates of $d_f \approx 2.5$. A fractal dimension of ~ 2.5 was also obtained in different gel systems in experimental observations [22, 28, 33, 41]. The effect of the aggregation regime, represented by the Fuchs stability ratio W, on the fractal dimension d_f has been investigated in model studies only [78, 79].

Mellema et al. [78] incorporated intermediate-range repulsive pair interactions and short-range irreversible bonding in their simulation model through Brownian dynamics, based on the Langevin equation. As the gel matrix exhibits fractal properties over a limited range of length scales, the upper cut-off limit of the fractal configuration, ξ, was taken to be the average radius of the gel pores and assumed to be proportional to the radius of gyration, R_g. The radius of gyration can be described by power law kinetics:

$$R_g \propto t^a \tag{16}$$

where $a = 1/d_f$, in diffusion-limited growth and with an exponential relation in reaction-limited growth:

$$R_g \propto e^\beta \tag{17}$$

The results of the simulations showed that aggregation was not controlled by the rate of the reaction and d_f was independent of the shape of the repulsive barrier in DLA and leveled off at high values of W. The authors attributed the difference between the simulations and experimental results of practical systems to rearrangements taking place within the aggregates leading to short-range compaction which could not be accounted for in the simulations.

4.5.2
Effect of Polyelectrolytes on the Fractal Dimension

Various forces, such as depletion forces in the case of non-adsorbing polymers and bridging forces among adsorbed polymers, come to be effective, in addition to the DLVO forces in the aggregation of particles in the presence of polymers. The effects of ionic strength, polyelectrolyte bulk concentration, charge density of the surface and the polymer on the interactions of charged and uncharged surfaces coated with polyelectrolyte-type polymers were systematically classified in a review by Claesson et al. [65]. For only a few of the different parameters shown to be effective in the interaction has an investigation of the effect on the fractal dimension been made: The effect of polyacrylic acid concentration on the fractal structure of hematite flocculates formed under RLA and DLA regimes was investigated by Zhang and coworkers [31, 80, 81]. Three different regimes of aggregation, denoted pre-DLA, DLA and post-DLA by the authors, were identified, in terms of increasing, constant and decreasing collision efficiency factors as a function of the polyelectrolyte/hematite ratio. At very low ratios, the RLA regime was observed with a typical d_f value of 2.08 ± 0.08 based on the use of an exponential cut-off in the correlation function, in the evaluation of light-scattering experiments [31]. The aggregates formed were found to be more tenuous, with a fractal dimension of 1.84 ± 0.02 in the DLA regime encountered at higher electrolyte/hematite ratios with a collision efficiency factor $\alpha \approx 1.0$. The fractal dimensions found for these two regimes were in accordance with those expected by universality for aggregation with charge neutralization. At still higher values of the electrolyte/hematite ratio, the collision efficiency factor decreased rapidly with increase in the value of the ratio, even though the aggregate structure retained its tenuity and the fractal dimension of 1.88 ± 0.02 as in the DLA regime. This controversial situation was explained by a self-affine type of growth model by the authors. Steric and electrostatic forces would be expected to be stronger at the contact zones of particles in the presence of polymers with a very high molecular weight. Since the penetration and adhesion of an approaching particle would be very difficult in these regions, the tips of the particle would be preferred, leading to a more tenuous structure.

Atomic force microscopy studies showed that a secondary minimum in the potential energy profile of the particles may result because of the presence of non-adsorbing polymers [82, 83]. Further support for this finding came from work on the aggregation of polystyrene latex particles in the presence of polyacrylic acid by Burns et al. [84]. The results of atomic force microscopy, over the same concentration range of polymer and pH, as was used in the aggregation experiments, is redrawn from the original work in Fig. 4.6.

A secondary minimum appears at concentrations of polymer $>1\ \mathrm{g\ L^{-1}}$. The depth of the minimum increases and its position changes to lower separation distances as the concentration of polymer increases. The deepening of the secondary minimum was attributed, by the authors, to the reduction in electrical double layer repulsion in the presence of the charged polymer. This effect of the polymer was deduced to be counteracted by a decrease in the depletion forces due to the reduction in the size of the polymer coils at increased concentrations, resulting in a leveling off of the total force as shown in the figure. The combined effect of the reduction in the double layer and depletion layer thickness coupled with an increase in osmotic pressure shifts the position of the secondary minimum to shorter interparticle distances.

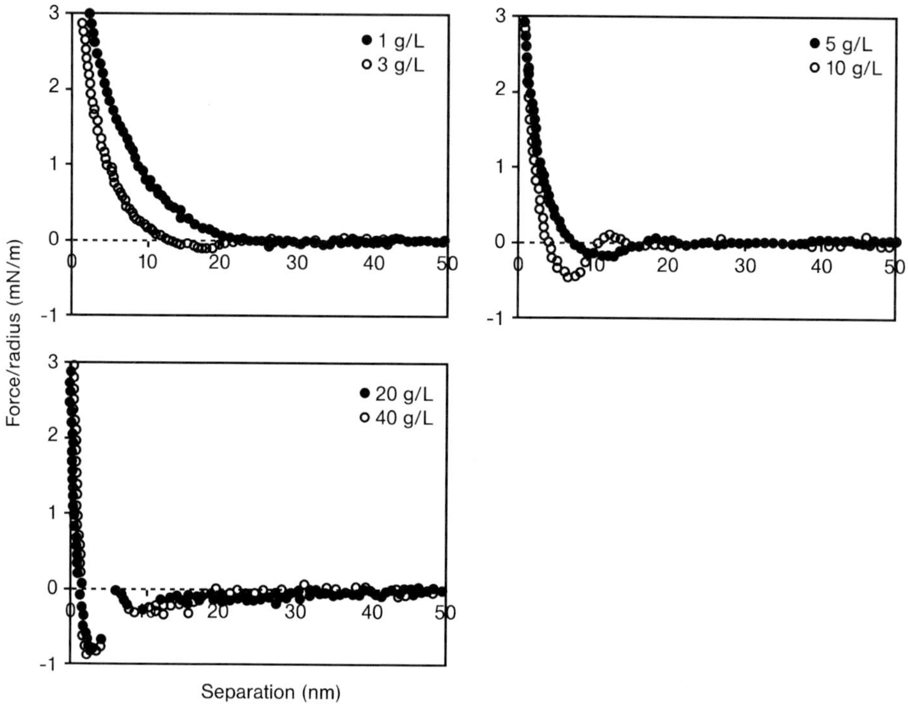

Fig. 4.6 Effect of polyacrylic acid concentration on the variation of interactive forces with separation. (Reprinted from Burns et al. [84], with permission from Elsevier).

The finite depth of the secondary minimum could allow the flocs to escape from their initial location and move around until they find an energetically more favorable place without steric effects. This wandering was physically observed by the authors under a microscope [84]. The ease with which a floc can detach is determined by the depth of the secondary minimum. A deep well located at a very short distance to the surface of the particle, observed at high polymer concentrations, causes the aggregation process to proceed in the RLA regime. When the minimum is some distance away from the surface and not very deep, the approaching flocs are free to find a place in which they can nest. The scattering exponent (*SE*), equivalent to the fractal dimension d_f, found by the authors confirmed this physical model: At very high concentrations of the polymer *SE* was found in the range 2.02–2.2, inversely proportional to the particle concentration. At polymer concentrations $<5\ \mathrm{g\ L^{-1}}$, *SE* increased sharply with decreasing polymer concentration, reaching a limit of 2.99 at $1\ \mathrm{g\ L^{-1}}$.

4.5.3
Effect of Fragmentation and Restructuring on the Fractal Dimension

The mechanical strengths of the aggregates towards impact in collision with other aggregates and towards shear stresses in orthokinetic aggregation depend directly or indirectly on the surface forces. The direct effect of surface forces on deformation, compression or extension of the aggregate under the effect of external forces is an expected sequel. The surface forces have an indirect effect due to the fractal nature of the aggregates, such as through the compaction of fractal structure, contacts per particle, porosity due to irregular structure brought about by the fast kinetics of the reaction and the composition of interparticle media leading to non-DLVO forces such as bridging, depletion and osmotic forces.

Quantitative expressions for tensile strength were obtained in terms of the shape and packing factors and the bonding forces between the primary particles of the aggregates [85]. Fractal dimensions, experimentally determined through light scattering measurements, were used in these quantitative expressions:

$$\sigma_T = 1.1 \frac{1-\varepsilon}{\varepsilon} \frac{F}{d_0^2} \tag{18}$$

where σ_T is the tensile strength (Pa), d_0 is the diameter of the primary particles (m) and ε is the porosity of the aggregates, given by the equation

$$\varepsilon = 1 - \theta^{\frac{d_f}{3}} \left(\frac{\delta}{\delta_0}\right)^{\frac{d_f}{3}-1} \left(\frac{R}{r_0}\right)^{d_f-3} \tag{19}$$

where θ is the packing factor of the aggregates and δ_0 and r_0 are the shape factor and radius (m) of the primary particles, respectively. Assuming the binding forces could be given by the DLVO theory, the tensile strengths of the aggregates formed by latex particles under diffusion- and rate-limited kinetics were calculated using the results of light-scattering measurements to evaluate the packing and shape factors. The tensile stress of the aggregates was found to decrease and level off to an equilibrium value with increase in the cross-sectional area of the aggregates. Whereas the calculated tensile stress approximates the experimentally found compressional stress values in the RLA regime, tensile stress was found to be at least an order of magnitude less than the compressional strength of the aggregates obtained under the DLA regime. As a reliable method of measuring the tensile stress has not yet been developed, experimental verification of these relations could not be made [85]. From the results obtained, the tensile stress of the aggregates is expected to decrease with increase in the size of the aggregate and of the primary particles making up the aggregate and a decrease in the bonding forces. This expectation was verified by another study on the aggregation mechanism of latex particles in a controlled shear environment [86, 87].

An aggregate subjected to stress may undergo structural revision in two ways and possibly with a combination of the two mechanisms. If there is a weak bond in the structure, it may fracture and break apart, recombining with the same or another aggregate under more favorable conditions, such as nesting in and forming bonds at more than a few sites. If the weak points can withstand the stress, then parts of the aggregate can rotate and form more bonds within itself, becoming more compact in the process. The former revision is known as a fragmentation–aggregation process and the latter as restructuring. Both processes result in an increase in fractal dimension due to compaction of the cluster. Under low to moderate shear rates ($\dot{\gamma} \leq 100\ \mathrm{s}^{-1}$), restructuring was found to predominate over fragmentation in aggregates composed of particles in the range of a few hundred nanometers. Restructuring is depicted as a peak in the dimensionless diameter equivalent to R/r_0 versus dimensionless coagulation time τ plots. The latter denotes the ratio of coagulation time t to the characteristic aggregation time for fractal aggregates, t_c, as the time required for growth from primary particle size, r_0, to the aggregate size at which orthokinetic aggregation starts to dominate, which is a function of the shear rate, viscosity, primary particle size, thermal energy and the fractal dimension of the aggregate. The diameter of the aggregates increases almost linearly up to the maximum diameter and then decreases and levels off to the equilibrium size range. The time when the peak value is observed and the equilibrium diameter attained by the aggregate decrease with increase in the diameter of the primary particles. As the decrease in diameter to the equilibrium value is not accompanied by a parallel decrease in mass, restructuring is deemed to take place.

Restructuring is best expressed with the fractal dimension. A complication which arises with the light-scattering measurement of aggregates undergoing restructuring is a change in the slope of scattered light intensity I versus the scattering vector $\boldsymbol{q}$ plots, which violates the Rayleigh-Gans-Debye approximation. The slope of these plots were designated the scattering exponent (SE), rather than the fractal dimension d_f, in view of these uncertainties [84, 86]. Nevertheless, a strong correlation between these two indices were accepted by both parties.

The scattering index of aggregates formed by small diameter particles of the order of a few hundred nanometers was found by Selomulya et al. [86] to approach the 2.6–3.0 range with time, much above the RLA limit of 2.1 for electrolyte-induced aggregation. On the other hand, when the diameter of the primary particles approaches the micrometer range of the scale, the dimensionless size ratio R/r_0 increases smoothly to an equilibrium value and remains constant with time. The lack of a peak value was attributed to the predominance of the fragmentation–reaggregation process over restructuring with relatively large primary particles. The scattering index corresponding to the fractal dimension asymptotically approaches an equilibrium constant value of 2.6–2.8 with time. For all particle sizes, the rate of approach to the equilibrium scattering index (SE) was directly proportional with the shear rate.

4.5.4
Effect of Specific Ions on the Fractal Dimension

If aggregation is brought about by the reaction of specific ions with sites on the particle surface or if the electrolyte has a special affinity for the surface of the particle, then d_f deviates from universality. In the aggregation of hematite induced by phosphate adsorption, d_f was consistently found to be <1.8 in both the DLA and RLA regimes by Chorover et al. [88]. The decrease in the fractal dimension in the RLA regime was attributed to non-random distribution of surface charge, which facilitates growth in unconstrained directions resulting in a more ramified structure. In another case [89], the existence of a short-range repulsive force found by atomic force microscopy measurements below the domain of DLVO forces was found to be effective on the fractal dimension in two respects: in the case of indifferent ions, the fractal dimension was found to remain constant within the range 2.10–2.25 under all rates of agglomeration, increasing slightly as the rate of formation increased, contrary to the expectations from universality. Specifically, binding of sulfate ions at concentrations much below the concentrations of other indifferent ions was found to reduce the d_f values to the 1.85–1.91 range for all aggregation rates, increasing again with increase in the rate of flocculation [32].

4.5.5
Effect of the Shape Factor of Primary Particles on the Fractal Dimension

In all of the cases above, the primary particles were assumed to be spherical. With the advent of nanotechnologies, hexagonal micelles, rod aggregates and nanotubes came under consideration. Even though stability and thorough dispersion of the nanoparticles are of primary concern in the development of the technologies, there is a scant amount of information on the stability of non-spherical colloids. Recently, Mohraz et al. [90] studied the aggregation behavior of rod type boehmite colloids with aspect ratios $\zeta=3.9$, 8.6 and 30.1, the results of which they compared with spheres of $\zeta=1$. Destabilization was brought about by the addition of Na_2SO_4 or NaOH, through DLA and RLA kinetics, manifested by the rate of growth of the aggregates as a function of time. The fractal dimension of the clusters formed under DLA conditions was found to be an increasing function of the monomer aspect ratio, with values of $d_f=1.81$, 1.94, 2.14 and 2.21 for $\zeta=1$, 3.9, 8.6 and 30.1, respectively. The Monte Carlo technique was used with Smoluchowski's initial assumptions of rectilinear motion, fast reaction and no internal rearrangement due to Brownian motion to simulate the aggregation mechanisms. The simulated fractal dimensions varied in the range $1.80<d_f<2.16$ for $1<\zeta<11$, in good agreement with experimental results. The authors reported that rod aggregation led to a loss of distinction between DLA and RLA terminal structures as ζ was increased, with a more branched and compact structure in comparison with those formed by spherical primary particles. Above an aspect ratio of ~ 8.6, the d_f values of the aggregate structures were approximately the same for both the DLA and RLA regimes.

This was explained by the excluded volume effect of the rods, which prevented them from penetrating into the pores of the aggregate structure, which were already compact owing to the orientation of the rods.

Another study by Chen et al. [2] on the aggregation behavior of nanotubes supported the results obtained by Mohraz et al. [90]. The inevitable entanglement among the nanotubes, due to strong attractive dispersion forces, their flexibility and large aspect ratios, can only be resolved with sonication, steric stabilization by surfactants or acid treatment for hydrophilization of the end groups. Determination of the fractal dimension by light scattering gave $d_f = 2.27$ for sterically stabilized and $d_f = 2.5$ for acid-treated nanotubes, both corresponding to tight aggregations formed under RLA conditions. Over a period of time, aggregates of acid-treated longer particles reverted into looser structures with $d_f \approx 1.7$, and shorter particles into more compact structures with $d_f \approx 2.8$. Nanotubes, sterically stabilized with Triton X-100, were found to be more stable than acid-treated nanotubes: surfactant addition to acid-treated nanotubes was found to stabilize them with a fractal dimension of 2.05.

4.5.6
Effect of Multilevel Fractal Structures on the Fractal Dimension

The primary particles making up the aggregates considered up to this point were homogeneous, the only difference between them being the diameter and/or the aspect ratio equivalent to the ratio of the length of the particle to the diameter. Naturally occurring aggregates may be made up of different materials with different physical and interfacial properties in addition to having different shape factors and sizes. In such cases, multilevel structures are formed, calling for delineation with different fractal dimensions. Wu et al. [15] determined the fractal dimensions of sludge flocs by light scattering to elucidate the structure of the micro flocs of kaolin and activated sludge, presumably mixtures of kaolin and bacteria, captivated within macro clusters formed by a cationic polyelectrolyte. The fractal dimension found by free settling methods elucidated the structure of the macro clusters, since the terminal velocity of settling is controlled by the counter-flow of fluid through the pores of the cluster. In the absence of the polyelectrolyte, the fractal dimension of kaolin aggregates were ~ 2.0 and those of activated sludges ~ 2.12, indicating close-packed structures formed by electrostatic attractions. Addition of a cationic polyelectrolyte to the kaolin flocs caused a reduction in d_f, determined by light scattering down to approximately DLA levels. Fractal dimensions determined by free settling tests were greater than those found by light scattering and passed through a minimum at the charge neutralization concentration of the polyelectrolyte, increasing again with increase in polymer concentration. The compaction brought about by polyelectrolyte addition was attributed by the authors to heterogeneities in surface charge, leading to interaction of oppositely charged sites. Similar trends were found for activated sludge flocs, although the magnitude of d_f measured by light scattering diminished to a lesser extent in proportion to the reduced amount of polyelectrolyte required for charge neutralization. The fractal dimensions of activated sludge clusters determined by free settling tests were much less, of the order of 1.33–1.48, denoting a loose, porous structure which controls the settling rate. As much smaller amounts of polymer are required for charge neutralization, the authors proposed that the flocculation mechanism of activated sludge could be through a bridging mechanism rather than a charge neutralization mechanism of kaolinite.

4.6 Modeling of Fractal Aggregates

Analysis of the formation and stability of fractal aggregates is based on interdisciplinary knowledge of fractal theory, surface forces and interfacial chemistry, aggregation kinetics and particle dynamics, hydrodynamics and rheology. As there is no established comprehensive theory covering all aspects of aggregation as yet, simulation and modeling studies are made with the motivation of finding a theoretical basis with which to compare experimental results. The endeavor toward this goal is hampered by two obstacles: Understanding of the surface forces is not yet complete and the solution of the mathematical equations developed is very difficult, requiring drastic simplifications to be made.

Somasundaran and Runkana [91] reviewed progress in mathematical modeling, pointing out the areas in which development has been achieved and areas where further work is necessary. In summary, developments have been made in incorporating both the aggregation and fragmentation into population balance equations and taking into account the effects and the consequences of the porous structure of the aggregates. Further research is suggested on the effects of surface and colloid chemistry, especially in the presence of polymers; coupling of population balances with computational fluid dynamics and effects of perturbations on the dynamics of flocculation.

After Somasundaran and Runkana's paper was published, further developments in modeling were observed in the subsequent publications. Runkana et al. [92] applied the population balance equations to polymer-induced flocculation by charge neutralization. As the polymer was a polyelectrolyte, only electrostatic forces were effective in the aggregation. The mass fractal dimension determined by dynamic scaling was also incorporated into the model. The results were validated by experimental results. In a recent study [93], they incorporated the short-range interaction energy, along with the van der Waals attraction and double-layer repulsion, in the model equations for RLA. Excellent agreement was obtained and the discrepancies were interpreted as restructuring and transition to DLA kinetics at high salt concentrations.

In a recent article, Kim and Kramer [94] summarized fractal dimensions found experimentally for different materials under varying hydrodynamic conditions (Table 1 in their paper), critically evaluated techniques used in the population balances and proposed algorithms which were computationally efficient.

In a series of papers Morbidelli and coworkers [79, 95–98] applied population balances to various aggregation cases. Population balance equations with selected aggregation kernels were used to find the cluster mass distributions. The cluster properties of gels and suspensions were used to determine cluster properties which were validated experimentally by light-scattering methods.

Selomulya et al. [87] incorporated permeability, in agreement with the core-shell model, into the collision efficiency term, the fractal dimension through the collision radius into the collision efficiency equation, variation of fractal dimension with time and fragmentation terms into the population balance equations. The results of the calculations showed that the final structure was determined by the relative extent of restructuring in comparison with fragmentation, as discussed above under restructuring.

4.7
Conclusion

The effects of surface forces on fractal properties have been reviewed. Emerging technologies and development in the measurement techniques have diverted the attention of surface scientists to fractal aggregates, bringing in radical changes in approach. Surface science itself is going through a stage of metamorphosis, where fundamental issues are being challenged. At present, modeling is a step behind in the analysis of fractal structures, which in turn is a step behind the underpinning surface sciences. Time will show how this accumulation will be reflected in the analysis of fractal systems.

References

1 W. J. Tseng, L. Kuang-Chih, *Materials Science and Engineering*, **2003**, *A355*, 186–192.
2 Q. Chen, C. Saltiel, S. Manickavasagam, L. S. Schadler, R. W. Siegel, H. Yang, *Journal of Colloid and Interface Science*, **2004**, *280*, 91–97.
3 T. Mason, M. Y. Lin, *Journal of Chemical Physics*, **2003**, *119*, 565–571.
4 L. Letamendia, E. Louisor, E. Pru-Lestret, J. Rouch, F. Sciortino, P. Tartaglia, H. Ushiki, *Colloids and Surfaces A: Physicochemical and Engineering Aspects*, **1998**, *140*, 289–293.
5 K. A. Kusters, J. G. Wijers, D. Thoenes, *Chemical Engineering Science*, **1997**, *52*, 107–121.
6 Z. Wang, C. Holm, H. Müller, *Physical Review E*, **2002**, *66*, 021405.
7 R. R. Kellner, W. Köhler, *Journal of Applied Physics*, **2005**, *97*, 034910.
8 X. Li, Y. Yuan, *Water Research*, **2002**, *36*, 3110–3120.
9 M. Han, H. Lee, *Colloids and Surfaces A: Physicochemical and Engineering Aspects*, **2002**, *202*, 23–31.
10 T. Serra, B. E. Logan, *Environmental Science and Technology*, **1999**, *33*, 2247–2251.
11 A. J. H. Pieterse, A. Cloot, *Water Science Technology* **1997**, *36*, 111–118.
12 G. Chen, K. A. Strevett, *Journal of Environmental Engineering*, **2002**, *128*, 408–415.
13 V. Chaignon, B. S. Lartiges, A. El Samrani, C. Mustin, *Water Research*, **2002**, *36*, 676–684.

14 C. P. Chu, D. J. Lee, *Chemical Engineering Science*, **2004**, *59*, 1875–1883.
15 R. M. Wu, D. J. Lee, T. D. Waite, J. Guan, *Journal of Colloid and Interface Science*, **2002**, *252*, 383–393.
16 B. M. Wilen, B. Jin, P. Lant, *Water Research*, **2003**, *37*, 3632–3645.
17 M. Berka, J. A. Rice, *Langmuir*, **2005**, *21*, 1223–1229.
18 Y. Adachi, S. Koga, M. Kobayashi, M. Inada, *Colloids and Surfaces A: Physicochemical and Engineering Aspects*, **2005**, *265*, 149–154.
19 F. Lang, H. Egger, M. Kaupenjohann, *Colloids and Surfaces A: Physicochemical and Engineering Aspects*, **2005**, *265*, 95–103.
20 B. B. Mandelbrot, *The Fractal Geometry of Nature*, W. H. Freeman, New York, **1983**.
21 P. Meakin,, *Fractals, Scaling and Growth Far From Equilibrium*, Cambridge University Press, Cambridge, **1998**.
22 S. Berthon, O. Barbieri, F. Ehrburger-Dolle, E. Geissler, P. Achard, F. Bley, A. Hecht, F. Livet, G. M. Pajonk, N. Pinto, A. Rigacci, C. Rochas, *Journal of Non-Crystalline Solids*, **2001**, *285*, 154–161.
23 J. Gregory, *Water Science Technology*, **1997**, *36*, 1–13.
24 B. İkizler, MS Thesis, Chemical Engineering Department, Ege University, Izmir, **2005**.
25 G. C. Bushell, Y. D. Yan, D. Woodfield, J. Raper, R. Amal, *Advances in Colloid and Interface Science*, **2002**, *95*, 1–50.
26 M. Lattuada, H. Wu, M. Morbidelli, *Journal of Colloid and Interface Science*, **2003**, *268*, 106–120.
27 F. E. Torres, W. B. Ressel, W. R. Schowalter, *Journal of Colloid and Interface Science*, **1991**, 142, 554–574.
28 S. S. Narine, A. G. Marangoni, *Journal of Crystal Growth*, **1999**, *198/199*, 1315–1319.
29 J. Lopes da Silva, J. A. P. Coutinho, *Rheologica Acta*, **2004**, *5*, 433–441.
30 M. Takenaka, T. Kobayashi, K. Saijo, H. Tanaka, N. Iwase, T. Hashimoto, M. Takahashi, *Journal of Chemical Physics*, **2004**, *121*, 3323–3328.
31 R. Ferreti, J. Zhang, J. Buffle, *Journal of Colloid and Interface Science*, **1998**, *208*, 509–517.
32 T. D. Waite, J. K. Cleaver, J. K. Beattie, *Journal of Colloid and Interface Science*, **2001**, *241*, 333–339.
33 L. A. Chiavacci, C. V. Santilli, S. H. Pulcinelli, C. Bourgaux, V. Briois, *Chemicals and Materials*, **2004**, *16*, 3995–4004.
34 T. Vicsek, *Fractal Growth Phenomena*, 2nd ed., World Scientific, Singapore, **1999**.
35 S. Dukhin, C. Zhu, R. N. Dave, Q. Yu, *Advances in Colloid and Interface Science*, **2005**, *114/115*, 119–131.
36 S. S. Dukhin, N. A. Mishchuk, G. Loglio, L. Liggieri, R. Miller, *Advances in Colloid and Interface Science*, **2003**, *100/102*, 47–81.
37 J. Y. Walz, *Advances in Colloid and Interface Science*, **1998**, *74*, 119–168.
38 K. S. Birdi, *Fractals in Chemistry, Geochemistry and Biophysics*, Plenum Press, New York, **1993**.
39 M. Smoluchowski, *Z. Phys. Chem.*, **1917**, *92*, 129–168.
40 D. N. Thomas, S. J. Judd, N. Fawcett, *Water Research*, **1999**, *33*, 1579–1592.
41 N. Fuchs, *Zeitschrift für Physik*, **1934**, *89*, 736–743.
42 B. V. Derjaguin, L. D. Landau, *Acta Physicochimica URSS*, **1941**, *14*, 633.
43 E. J. W. Verwey, J. T. G. Overbeek, *Theory of Stability of Lyophobic Colloids*, Elsevier, Amsterdam, **1948**.
44 B. W. Ninham, *Advances in Colloid and Interface Science*, **1999**, *83*, 1–17.
45 J. Lyklema, J. F. L. Duval, *Advances in Colloid and Interface Science*, **2005**, *114/115*, 27–45.
46 S. H. Behrens, D. I. Christl, R. Emmerzael, P. Schurtenberger, M. Borkovec, *Langmuir*, **2000**, *16*, 2566–2575.
47 Q. Lin, E. E. Meyer, M. Tadmor, J. N. Israelachvili, T. L. Kuhl, *Langmuir*, **2005**, *21*, 251–255.
48 A. M. Islam, B. Z. Chowdhry, M. J. Snowden, *Advances in Colloid and Interface Science*, **1995**, *62*, 109–136.
49 P. M. Biesheuvel, *Journal of Physics: Condensation of Matter*, **2004**, *16*, L499–L504.

50 S.A. Edwards, D.R.M. Williams, *Current Opinion in Colloid and Interface Science*, **2004**, *9*, 139–144.
51 P. Koelsch, H. Motschmann, *Current Opinion in Colloid and Interface Science*, **2004**, *9*, 87–91.
52 M.O. Khan, S. Petris, D.Y.C. Chan, *Journal of Chemical Physics*, **2005**, *122*, 104705.
53 J.J. Adler, Y.I. Rabinovic, B.M. Moudgil, *Journal of Colloid and Interface Science*, **2001**, *237*, 249–258.
54 J.J. Valle-Delgado, J.A. Molina-Bolivar, F. Galisteo-Gonzalez, M.J. Galvez-Ruiz, A. Feiler, M.W. Rutland, *Journal of Chemical Physics*, **2005**, *123*, 034708.
55 B.V. Derjaguin, N.V. Churaev, *Journal of Colloid and Interface Science*, **1974**, *49*, 249.
56 S. Marcelja, N. Radic, *Chemical Physics Letters*, **1976**, *42*, 129.
57 T.F. Tadros, J. Lyklema, *Journal of Electroanalytical Chemistry*, **1968**, *17*, 267.
58 B. Jönsson, H. Wennerström, *Journal of the Chemical Society Faraday Transactions*, **1983**, *79*, 2.
59 R. Cohen, D. Exerowa, I. Pigov, R. Heckmann, S. Lang, *Journal of Adhesion*, **2004**, *80*, 875–894.
60 D. Henderson, D.T. Wasan, A. Trokhymchuck, *Journal of Chemical Physics*, **2003**, *119*, 11989–11997.
61 D.T. Wasan, A.D. Nikolov, F. Aimetti, *Advances in Colloid and Interface Science*, **2004**, *108/109*, 187–195.
62 D. Wasan, A. Nikolov, B. Moudgil, *Powder Technology*, **2005**, *153*, 135–141.
63 L. Krishnamurthy, N.J. Wagner, *Journal of Rheology*, **2005**, *49*, 475–499.
64 A. Ditsch, P.E. Laibinis, D.I.C. Wang, T.A. Hatton, *Langmuir*, **2005**, *21*, 6006–6018.
65 P.M. Claesson, E. Poptoshev, E. Blomberg, A. Dedinaite, *Advances in Colloid and Interface Science*, **2005**, *114/115*, 173–187.
66 Y. Yan, J.L. Burns, G.J. Jameson, S. Biggs, *Chemical Engineering Journal*, **2000**, *80*, 23–30.
67 S. Biggs, J.L. Burns, Y. Yan, G.J. Jameson, P. Jenkins, *Langmuir*, **2000**, *16*, 9242–9248.
68 J.L. Burns, Y. Yan, G.J. Jameson, S. Biggs, *Journal of Colloid and Interface Science*, **2002**, *247*, 24–32.
69 A.Y. Huang, J.C. Berg, *Journal of Colloid and Interface Science*, **2004**, *279*, 440–446.
70 R. Jullien, R. Botet, *Aggregation and Fractal Aggregates*, World Scientific, Singapore, **1987**.
71 M.Y. Lin, R. Klein, H.M. Lindsay, D.A. Weitz, R.C. Ball, P. Meakin, *Journal of Colloid and Interface Science*, **1990**, *137*, 263–280.
72 A.Y. Kim, J.C. Berg, *Langmuir*, **2000**, *16*, 2101–2104.
73 M. Norgren, H. Edlung, L. Wagberg, *Langmuir*, **2002**, *18*, 2859–2865.
74 M. Tirado-Miranda, A. Schmitt, J. Callejas-Fernandez, A. Fernandez-Barbero, *Journal of. Chemical Physics*, **2003**, *119*, 9251–9259.
75 F. Bordi, C. Cametti, J. Rouch, F. Sciortino, P. Tartaglia, *Journal of Physics: Condensed Matter*, **1996**, *8*, A19–A37.
76 B. Antalek, A.J. Williams, J. Texter, Y. Feldman, N. Garti, *Colloids and Surfaces A: Physicochemical and Engineering Aspects*, **1997**, *128*, 1–11.
77 R. Lapasin, M. Grassi, S. Pricl, *The Chemical Engineering Journal and the Biochemical Engineering Journal*, **1996**, *64*, 99–106.
78 M. Mellema, J.H.J. van Opheusden, T. Van Vliet, *Journal of Chemical Physics*, **1999**, *111*, 6129–6135.
79 E. Bonanomi, P. Sandkühler, J. Sefcik, M. Morari, M. Morbidelli, *Industrial and Engineering Chemistry Research*, **2004**, *43*, 4740–4752.
80 J. Zhang, J. Buffle, *Journal of Colloid and Interface Science*, **1995**, *174*, 500.
81 R. Ferretti, J. Zhang, J. Buffle, *Colloids and Surfaces A: Physicochemical and Engineering Aspects*, **1997**, *121*, 203–215.
82 A.J. Milling, *Journal of Physical Chemistry*, **1996**, *100*, 8986–8993.
83 A.J. Milling, B. Vincent, *Journal of the Chemical Society Faraday Transactions*, **1997**, *93*, 3179.
84 J.L. Burns, Y. Yan, G.J. Jameson, S. Biggs, *Colloids and Surfaces A: Physicochemical and Engineering Aspects*, **1999**, *162*, 265–277.

85 S. Tang, Y. Ma, C. Shiu, *Colloids and Surfaces A: Physicochemical and Engineering Aspects*, **2001**, *180*, 7–16.
86 C. Selomulya, G. Bushell, R. Amal, T. D. Waite, *Langmuir*, **2002**, *18*, 1974–1984.
87 C. Selomulya, G. Bushell, R. Amal, T. D. Waite, *Chemical Engineering Science*, **2003**, *58*, 327–338.
88 J. Chorover, J. Zhang, M. K. Amistadi, J. Buffle, *Clays and Clay Minerals* **1997**, *45*, 690–708.
89 M. E. Karaman, R. M. Pashley, T. D. Waite, S. J. Hatch, H. Bustamante, *Colloids and Surfaces*, **1997**, *129/130*, 239–255.
90 A. Mohraz, D. B. Moler, R. M. Ziff, M. J. Solomon, *Physical Review Letters*, **2004**, *92*, 155503.
91 P. Somasundaran, V. Runkana, *International Journal of Mineral Processing*, **2003**, *72*, 33–55.
92 V. Runkana, P. Somasundaran, P. C. Kapur, *Journal of Colloid and Interface Science*, **2004**, *270*, 347–358.
93 V. Runkana, P. Somasundaran, P. C. Kapur, *AIChE Journal*, **2005**, *51*, 1233–1245.
94 J. W. Kim, T. A. Kramer, *Colloids and Surfaces A: Physicochemical and Engineering Aspects*, **2005**, *253*, 33–49.
95 M. Lattuada, H. Wu, M. Morbidelli, *Journal of Colloid and Interface Science*, **2003**, *268*, 96–105.
96 M. Lattuada, H. Wu, M. Morbidelli, *Journal of Colloid and Interface Science*, **2003**, *268*, 106–120.
97 M. Lattuada, H. Wu, P. Sandkühler, J. Sefcik, M. Morbidelli, *Chemical Engineering Science*, **2004**, *59*, 1783–1798.
98 P. Sandkühler, M. Lattuada, H. Wu, J. Sefcik, M. Morbidelli, *Advances in Colloid and Interface Science*, **2005**, *113*, 65–83.

5
Hydrophobic Attraction in the Light of Thin-Film Thermodynamics

Jan Christer Eriksson and Roe-Hoan Yoon

5.1
Introduction

To gain an improved understanding of the molecular basis of hydrophobicity and hydrophobic attraction forces is of crucial importance for several scientific and technological fields such as surface and colloid science, biochemistry and mineral processing. The formation of surfactant micelles and biomembranes, for instance, can be traced back to the tendency of hydrocarbon chains to associate when present in a water medium at a high enough (although still very low) concentration. Further, a large number of enzymatic reactions take place while the substrate is locked in a hydrophobic pocket. In the field of flotation, on the other hand, long-range hydrophobic attraction forces contribute to establishing the mineral particle/air bubble attachment needed to achieve selective ore enrichment.

Early experiments indicating the existence of long-range hydrophobic attraction forces were carried out by Blake and Kitchener [1] on thin water films sandwiched between hydrophobic surfaces. The instability observed was considered to be due to attraction forces operating across the film, which arise because of the contact of the water with the hydrophobic surfaces. Early on, Derjaguin et al. [2], referring to a large body of experimental results, discussed the same effect in terms of "the structural component of the disjoining pressure".

Most of the many investigations of hydrophobic attraction during recent decades have been based on employing the surface force apparatus (SFA), developed by Israelachvili and Tabor [3] (Fig. 5.1) or atomic force microscopy (AFM) or some similar direct surface force measurement device, such as the MASIF (measurement and analysis of surface interaction forces) due to Parker [4] and Yaminsky's interfacial gauge [5].

This particular research area, which is focused on thin aqueous films between hydrophobic surfaces, was inaugurated in 1982 by Israelachvili and Pashley using mica surfaces and successively adding the cationic surfactant hexadecylammonium bromide (CTAB) to the water medium [6]. They established that in

Colloids and Interface Science Series, Vol. 1
Colloid Stability: The Role of Surface Forces, Part I. Edited by Tharwat F. Tadros

ISBN: 978-3-527-31462-1

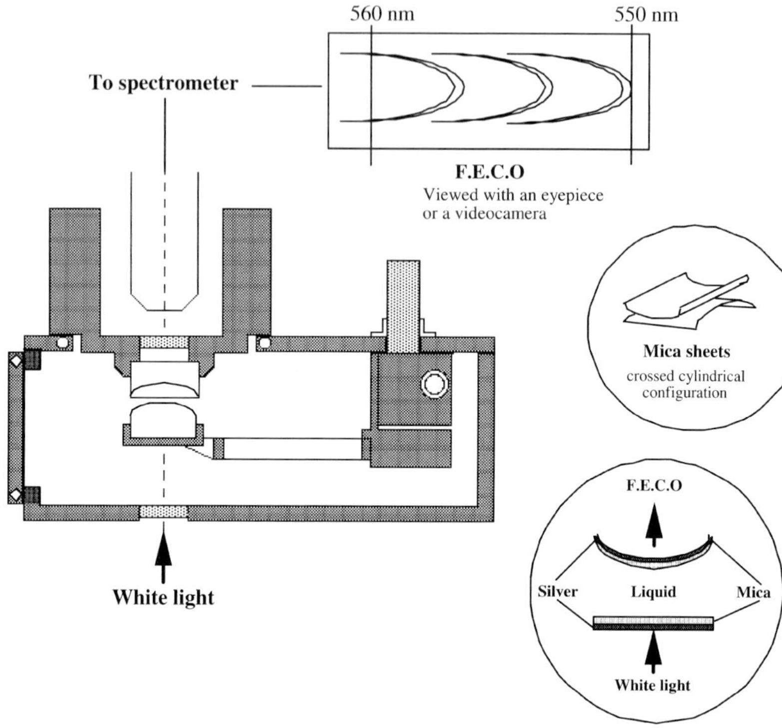

Fig. 5.1 Surface force apparatus (SFA) based on the original design by Tabor and Winterton (1969) and Israelachvili and Tabor (1972) [3]. The interacting (mica) surfaces have cylindrical shape and their separation is determined optically by means of an interferometric technique. The lower surface is attached to a cantilever spring, the deflection of which serves to determine the interaction force.

addition to the normal DLVO (Derjaguin-Landau-Verwey-Overbeek) forces due to dispersion and electrostatic interactions, a long-range attractive interaction force operates in vicinity of the surface charge neutralization concentration, presumably arising as a consequence of the hydrocarbon–water contact. Hence they identified this surface force as the "hydrophobic force", a term originally coined by Blake and Kitchener [1].

A few years later, Claesson and Christenson [7, 8] experimented with mica surfaces which had been modified by Langmuir-Blodgett (LB) deposition of dioctadecyldimethylammonium (DODA) bromide or, alternatively, a double-chain cationic fluorocarbon surfactant. For these cases, the hydrophobic attraction was found to be considerably stronger and of longer range than observed by Israelachvili and Pashley (about 100 times stronger than the van der Waals attraction; decay length about 15 nm), being measurable up to about 80 nm. Independently, measurements were conducted in Moscow by Rabinovich and Derjaguin [9] using silanized silica filaments which yielded largely equivalents results (Fig. 5.2).

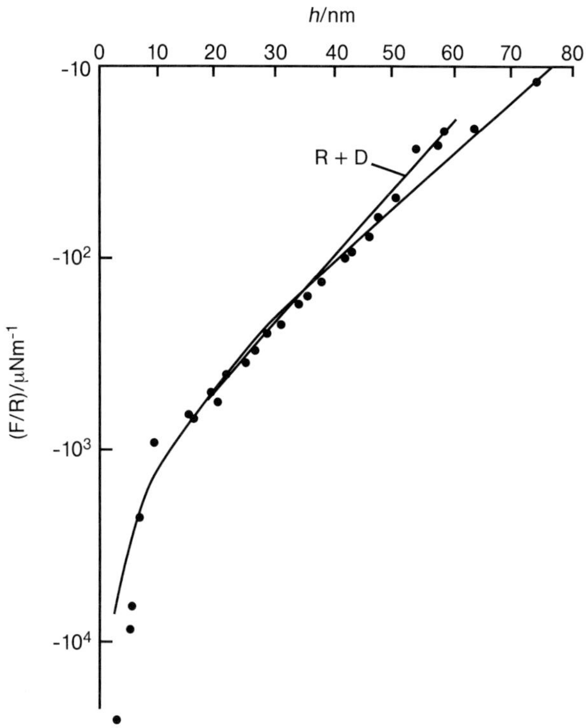

Fig. 5.2 The attractive surface force, F/R, plotted against the surface separation, h, for the case of dioctadecyldimethylammonium (DODA)-covered mica surfaces in contact with pure water. The points demark experimental values measured by Claesson and Christenson [8]. The full line represents Eq. (25) obtained by inserting $B = 0.600$ mJ m^{-2} and $b^{-1} = 15.8$ nm in Eq. (11). For comparison is shown the surface force function determined by Rabinovich and Derjaguin [9] (line marked R+D) for silica filaments immersed in 0.1 mM KCl solution. The corresponding experimental points are scattered around the R+D line within a factor of about 3.

So far so good, but subsequent control experiments employing a variety of differently prepared hydrophobic surfaces have only partially confirmed the early findings and notions. It is even difficult to summarize the current research situation, with a host of bewildering results presented in more or less disparate manners by a large number of investigators. Nevertheless, a few years ago a rather heroic effort to this end was made by Christenson and Claesson, resulting in a valuable review paper in which they present a detailed account of the scientific state-of-the-art from an experimental perspective [10]. They concluded by stating that the concept of a hydrophobic interaction force by and large seems to stand for:

1. A fairly short-range but strongly attractive force, much stronger than the van der Waals force, between stable hydrophobic surfaces.
2. An attraction of variable strength and range caused by the presence of small bubbles sporadically adhering to hydrophobic surfaces.

3. A very long-range attractive force with exponential decay operating between a variety of hydrophobic surfaces, in particular those hydrophobized by means of LB deposition on mica or silica, or adsorption from solution of an ionic amphiphile.

The reader who wants a detailed survey of the experimental field up to 2000 is recommended to consult the paper by Christenson and Claesson [10]. A complementary paper mostly dealing with the theoretical understanding of the hydrophobic force was presented some years ago by Spalla [11]. Furthermore, papers by Tsao et al. [12] and Craig et al. [13] incorporate illuminating discussions about the possible origin of the hydrophobic force. More recently, Israelachvili and coworkers began a critical re-examination of the whole issue and extensive reference lists are included in their papers [14, 15].

In the following, we concentrate on some typical results obtained for the main hydrophobic surfaces investigated so far, in particular for hydrophobic surfaces which have been documented to be of superior quality. The ultimate purpose is to scrutinize whether or not there is enough evidence today in support of the existence of what might be called a "true" hydrophobic force, i.e. a surface force of long range which is related to the structural response of a thin water film between hydrophobic surfaces.

5.2 The Molecular Organization of Water at Interfaces

Thanks to the advent of novel spectroscopic and computational methods, in recent years considerable progress has been made in the probing and modeling of the molecular state of bulk and surface water. In particular, these novel methods entail sum frequency generation (SFG) spectroscopy, X-ray absorption spectroscopy (XAS) and X-ray Raman scattering (XRS) [16, 17]. Hence it has become firmly established that about 20–25% of the water molecules in the top monolayer of the water–air interface have a non-hydrogen-bonded OH group as a unique feature.

The spectral peak of the dangling OH group remains essentially the same for a hydrocarbon–water interface and for the free water surface towards air and can be rationalized on the notion of oriented oxygen double layers similar to those present in the ice Ih lattice. Evidently, a large part of the comparatively high surface energy of water is connected with the excess of broken hydrogen bonds in the surface, compared with the situation in bulk water.

Surprisingly, for bulk water it has recently been found that there is also a tendency to form *linear* aggregates of water molecules (based on *two* strong hydrogen bonds rather than *four* of medium strength), resulting in rings and chains of water molecules [18]. It remains to be assessed, however, to what extent the novel findings and ideas about the structural aspects of liquid water will necessitate revising the conventional molecular picture based on four-coordinated water molecules in small clusters having external surfaces with less well-bonded molecules [19].

Thus, when considering hydrophobicity and hydrophobic surface forces, we have to bear in mind the on-going endeavors to probe and to simulate interfacial and bulk water by means of molecular dynamics, which may result in a substantially revised basis for a comprehensive theoretical treatment of these phenomena [18]. The classical simulations of Lee et al. [20], which are often referred to in this context, may turn out to have been based on a far too simplistic water potential (of ST2 type). This is indicated already by the circumstance that the predictions of the melting-point of ice using water models of this kind are for the most part grossly in error.

In broad terms, we may expect an atomically smooth, chemically inert solid surface that is subject to thermal motions of small amplitudes:

1. to force the water molecules to pile up against the surface, as would actually be the case for any liquid; and
2. to force the hydrogen bond network to rearrange in its vicinity so as to limit the number of broken hydrogen bonds, thus forming a clathrate-resembling contact monolayer of water molecules.

On average, this would result in a roughly tangential alignment of the water dipoles. In a cooperative fashion, such a contact monolayer in turn imposes bonding constraints on the successive layers of water molecules beneath it so as to generate a surface-attached network. A crucial question is, of course, how deep towards the bulk such a network may prevail, a much debated question we shall come back to repeatedly in the following. Interestingly, the SFG spectra recorded by Shen and coworkers (16) indicate that a well-ordered, ice-like water structure predominates at a *solid* hydrophobic surface whereas a more disordered water structure prevails for the water–air and water–hexane interfaces.

The above-mentioned simulations by Lee et al. and similar ones by other research groups [21] indicate that a significant surface effect is noticeable only for short distances in liquid water, of the order of 1 nm. By contrast many surface force measurements (detection limit ~10 $\mu N\ m^{-1}$) are indicative of non-DLVO attractions extending as far as 50 nm and, occasionally, even longer. In the end, of course, the reasons behind these widely contrasting results will have to be unfolded before a convincing theoretical description can be achieved of why and how the hydrophobic attraction arises.

5.3 Thermodynamic Aspects of Surface Force Measurements

Let us consider an idealized thin film/surface force experiment using two plane-parallel, atomically smooth, laterally homogeneous hydrophobic surfaces with a thin water solution film between them (Fig. 5.3). The thermodynamic variables of prime interest as determined in the adjacent bulk water solution are the temperature T and the solute chemical potential μ_s (or the concentration c_s of solute).

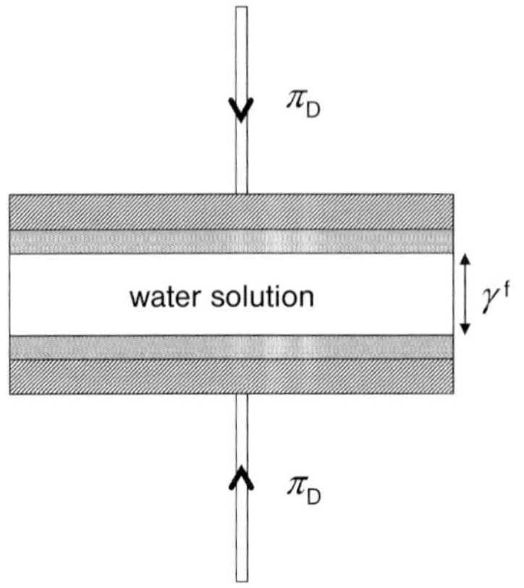

Fig. 5.3 Sketch of a thin liquid film between two planar solid surfaces. The excess (interaction) pressure operating perpendicular to the film is, using Derjaguin's terminology, the disjoining pressure, π_D. In lateral directions a film tension, γ^f, is acting. The surface force is defined as 2π times the difference in film tension between a thin film with interacting film faces and an infinitely thick film.

To establish equality of the chemical potentials everywhere, a lateral tension γ^f develops in the film which at large thicknesses h will be equal to twice the interfacial tension of the hydrophobic surface–water solution interface, i.e. about 100 mN m^{-1}. In order to keep the film at a certain thickness h, an extra pressure, positive (repulsive) or negative (attractive), the so-called (Derjaguin) disjoining pressure has to be applied [2]. This pressure has alternatively been called surface "interaction pressure".

At constant temperature and pressure, the thin-film system obeys the following thermodynamic fundamental equation to a very good approximation:

$$-d\gamma^f = \Gamma_s^{f,ex} d\mu_s + \pi_D dh \tag{1}$$

which constitutes an extension of the classical Gibbs surface tension equation to cover the case of interacting solid surfaces separated by the film thickness h (cf. Appendix).

The film thickness may most conveniently be defined as the separation between the hydrophobic surfaces devoid of any loosely adsorbed solute. This means that we are actually employing the Gibbs equimolecular dividing surfaces of the (hydrocarbon-covered) hydrophobic surfaces to delimit the water film in the direction perpendicular to the film faces.

Furthermore, $n_s^{f,ex} \equiv A\Gamma_s^{f,ex}$ (where A denotes the area of the film) is an *excess* quantity in the following sense:

$$n_s^{f,ex} = n_s^f - n_w^f c_s / c_w \tag{2}$$

where n_s^f and n_w^f denote the actual numbers of moles of surfactant and water in the film, respectively, and c_s and c_w are the corresponding bulk phase concentrations. In other words, $n_s^{f,ex}$ represents the film content of the solute component *in excess* of what we would have for a corresponding slab of bulk solution containing the same amount of water as is present in the film.

From Eq. (1), it is seen that the disjoining pressure π_D is generally related to the film tension by the derivative:

$$\left(\frac{\partial \gamma^f}{\partial h}\right)_{T,p,\mu_s} = -\pi_D \tag{3}$$

Hence an *attractive* surface force, characterized by $\pi_D<0$, will result when the film tension γ^f decreases as the thickness decreases, whereas $\pi_D>0$ means repulsion and a film tension that diminishes with increasing h.

In addition, Eq. (1) includes an interesting Maxwell relation:

$$\left(\frac{\partial \Gamma_s^{f,ex}}{\partial h}\right)_{T,p,\mu_s} = \left(\frac{\partial \pi_D}{\partial \mu_s}\right)_{T,p,h} \tag{4}$$

stating that the amount of (surfactant) solute in the film will become less on decreasing its thickness if the disjoining pressure tends to increase on raising the chemical potential of the solute.

It is also worth noting that the film tension γ^f is formally an Ω-potential per unit area:

$$\gamma^f = \Omega/A \equiv G^{f,ex}/A = G^f/A - \Gamma_w^f \mu_w - \Gamma_s^f \mu_s \tag{5}$$

Thus, when equilibrium prevails at constant T, p, μ_s, the film tension γ^f is necessarily at a minimum.

To derive the film tension change, $\Delta\gamma^f$, arising as a result of diminishing the thickness from $h=\infty$ where $\pi_D=0$ (that is, beyond the range of the surface forces), we can integrate Eq. (3) as follows while assuming that the bulk phase state remains the same:

$$\Delta\gamma^f \equiv \gamma^f(h) - \gamma^f(\infty) = -\int_{h=\infty}^{h} \pi_D \mathrm{d}h \tag{6}$$

Now, in practice, it is not infeasible to employ a plane-parallel experimental set-up such as that indicated in Fig. 5.3. Instead, one has to resort to sphere–sphere, sphere–plane or crossed cylinder–cylinder configurations. At this point, the powerful *Derjaguin approximation* (22) comes into the picture, which in effect constitutes a geometric integration of Eq. (3), yielding an additional geometry-dependent factor such that the measured surface force F becomes equal to

$$F = 2\pi R \Delta\gamma^f$$

for sphere–plane and crossed cylinder–cylinder cases and

$$F = \pi R \Delta\gamma^f$$

for the sphere–sphere case. These simple relations effect the transformation from a curved to a planar geometry. They are valid for all kinds of surface interaction forces insofar as the ranges of these forces are much less than the inverse curvature of the surfaces involved. Although they actually are based on the very circumstance that (equilibrium) surface forces represent little else but *film tension changes* that arise when the interacting surfaces are brought close to contact, in the literature it has become common practice to report surface forces "normalized" in the form F/R, corresponding to $2\pi\Delta\gamma^f$ and $\pi\Delta\gamma^f$ for the above two standard cases.

5.4
The Ideal Hydrophobic Surface Versus Real Hydrophobic Surfaces

We tend to take it for granted that for experimental purposes we are able to realize a near to perfect hydrophobic surface and, hence, that it is meaningful to consider an *ideal* hydrophobic surface in the Platonic philosophical sense. The question immediately arises about the most desirable properties of such a surface regarding smoothness, hydrophobicity, molecular constitution, motional state, fluidity, stability towards water solutions and other media, etc.

Concerning *smoothness*, the answer seems simple: we wish to have minimal roughness at the atomic level, on average approaching 0.1 nm peak-to-valley distance. First of all, this requires a sufficiently smooth substrate: mica, silica or glass is usually preferred. In addition, however, the hydrocarbon chains used as modifying agents have to be attached in such a manner that the smoothness is maintained. Chemical reaction and polymerization schemes for the bonding of alkyl or silyl chains can be risky in this regard as they may involve strong mechanical pressures/tensions operating in the surface during the reactions.

Smoothness is likely to be a crucial factor for the occurrence of a long-range hydrophobic attraction force and for this reason it should be fully verified. The main experimental difficulty encountered in this context has to do with obtaining a sufficient degree of stability for the hydrophobic layer attached, while at the same time preserving surface smoothness at the atomic level.

Concerning *hydrophobicity*, the answer may seem almost self-evident at first: maximal hydrophobicity calling for fluorocarbon rather than hydrocarbon chains. By making such an extreme choice, however, it has turned out that one may easily end up with studying the effects of vapor/air cavities and bridging bubbles rather than the hydrophobic attraction *per se*. For this reason, one might well prefer surfaces exhibiting a contact angle against water slightly less 90°, for which capillarity phenomena of this extraneous nature, in principle at least, should not occur.

Concerning the *molecular constitution*, one would preferably desire the hydrocarbon chains employed to be sufficiently long ($\geq C_{16}$) and to have them fairly densely packed in the surface, to make up a certain minimum thickness of the resulting surface-modifying layer. However, forcing them to adopt a crystalline state with tilted chains may cause complicating domain structures and grain boundaries to arise.

Concerning the *motional state*, one might preferably want the hydrocarbon chains to be in a "semi-frozen" solid (rather than close-packed crystalline) state, having a packing density between 0.22 and 0.25 nm^2 per single chain. For a less crowded amphiphile monolayer in a liquid-expanded state, one runs the risk of having the thermal motions make the water–hydrocarbon phase boundary fuzzy.

Clearly, when ionic head groups are attached to the bonding sites of opposite charge on a solid surface for which the thermal amplitudes are small, the chains may end up in some "frozen" gel state, hampering the equilibration of the adsorbate or at least making equilibration times exceedingly long. For mica and silica substrates there seems to be a large difference between C_{16} and C_{18} cationic surfactants in this regard, the latter approaching adsorption equilibrium at an extremely slow rate [23].

To summarize this discussion, we envisage an ideal, stable hydrophobic surface that exhibits an advancing contact angle against water of say 110° and which is in some solid state that maintains a certain degree of molecular mobility but is nevertheless smooth on the atomic scale. The fairly high mechanical tension acting in the water adjacent to the hydrocarbon film will further tend to dampen the thermal amplitudes in the interface. If properly balanced in terms of its mechanical properties so as to be able to withstand mechanical loads and being smooth enough from the outset, for such a surface contact angle hysteresis might be limited and the risk minimized of having bubbles attaching and bridging between approaching surfaces.

With the above background, let us now list some real surfaces which to a variable extent exhibit pronounced hydrophobic properties:

1. Cleaved mica with a deposited LB layer of a double-chain cationic surfactant, i.e. the hydrophobic surface employed by Claesson and Christenson [7, 8], and, more recently, by Meyer et al. [14] and Lin et al. [15]. It fulfils the quality criteria with respect to smoothness and motional state. It shows, however, lack of stability towards salt solutions and even towards passage through the water–air interface. One has to realize that the chemical potential of the attached amphiphile is raised by compressing the monolayer spread on the Langmuir trough to the desired packing density of about 0.25 nm^2 per single chain which, by the way, matches almost exactly the anionic site density of the mica surface (~0.50 nm^2). Consequently, upon screening (by adding salt) the attraction between unlike charges and also the electrostatic repulsion among the cationic head groups, there is a tendency for the surfactant to pass over to some less strained (bilayer) state. Further, the whole ion-exchange process is prone to be facilitated kinetically by the presence of small ions.

2. In essence, the same as above but instead relying on adsorption from a cyclohexane solution to attach the double-chain cationic surfactant to the mica. This scheme was successfully applied by Tsao and co-workers [12, 24], who obtained hydrophobized mica surfaces fairly similar to those made according

to point 1. In particular, by means of AFM they convincingly demonstrated that their surfaces were laterally homogeneous and free from defects over large distances, of the order of microns. Interestingly, a substantially smaller attraction force was recorded for C_{16} than for C_{18} chains and significant temperature effects were noted, the attraction becoming weaker and less long-range at higher temperature. However, stability towards salt solutions turned out to be less satisfactory than for the corresponding LB monolayers prepared according to point 1.

3. Self-assembling alkanethiols dissolved in e.g. alcohol are known to bind strongly to gold surfaces, a circumstance that was utilized by Ederth and co-workers [25, 26] to prepare hydrophobic surfaces on borosilicate glass substrates starting with forming (by electron beam evaporation) a 1-nm titanium layer, followed by a 10-nm gold layer. The gold surfaces made in this manner are slightly rough with peak-to-trough values of ~1.5 nm. The hydrocarbon chains become tilted and are tightly packed in a crystalline state. These surfaces exhibit a hydrophobic attraction of medium range but have been found to be plagued by sporadic bridging bubble steps in the surface force curves. Stability is, of course, not a problem.

4. Glass, silica or mica surfaces rendered hydrophobic by reaction with a silanation agent capable of reacting with surface OH groups such as (tridecafluoro-1,1,2,2-tetrahydrooctyl)dimethylchlorosilane used by Parker and Claesson [27]. For mica surfaces, a water plasma pretreatment is needed to introduce surface OH groups [28]. Stability towards salt solutions is as a rule obtained for this type of hydrophobic surface. However, to quote Christenson and Claesson [10]: “Experimental results with silylated surfaces have shown great variability depending on exact preparation conditions and further underscored the critical connection between details of surface chemistry, surface morphology and the measured forces”. Hence it seems that it is difficult to control fully a silanation reaction scheme so as to generate a sufficiently smooth, charge-free and laterally homogeneous hydrophobic surface.

5. Wood and Sharma [29] suggested that one might succeed in making a more stable and better characterized hydrophobic surface by first polymerizing octadecyltriethoxysilane at the air–water interface (pH 2) on a Langmuir trough, followed by LB deposition and chemical grafting to a mica surface. The surfaces obtained proved stable towards salt solutions but laterally inhomogeneous on the micron length scale. No long-range hydrophobic attraction was observed, presumably because of a lack of the molecular smoothness required.

6. Adsorption from water solution of a cationic surfactant such as CTAB (hexadecyltrimethylammonium bromide) or C_{18}TAC (octadecyltrimethylammonium chloride) on mica, glass or silica. Although easily prepared, surfaces of this nature have a major drawback: they are uncharged (or carry a minimal surface charge) only at one particular surfactant concentration, often denoted the charge neutralization concentration (c.n.c.). Furthermore, for the most desirable chain

length, C_{18}, adsorption equilibrium is approached only very slowly at room temperature. Hence, in spite of the circumstance that the phenomenon of long-range hydrophobic attraction is readily observed provided that the right surfactant concentration is chosen, fundamental studies are hampered by the variable state of the hydrophobic surfactant layer. The best thing to do is presumably to concentrate on those surface force curves which show the strongest hydrophobic attraction rather than to try to sort out exactly what electrostatic mechanisms operate at concentrations different from the c.n.c. The temperature dependence of the hydrophobic force, for instance, could perhaps be investigated in this manner by changing the temperature followed by adjustment of the surfactant concentration so as to recover a minimal surface charge.

The above listing, although certainly not complete, may serve as a ranking list of hydrophobic surfaces based on experiences in several laboratories during recent decades. Preparing hydrophobic surfaces of sufficient quality has turned out to be of crucial importance for drawing reliable conclusions about the strength, range and nature of the hydrophobic force. Although hydrophobic surfaces prepared by adsorption of surfactants from solution are at the bottom of the list, they are probably the most important in terms of applications, e.g. in flotation.

5.5 Hydrophobic Attraction Forces Under Ideal Conditions

For large separations between two interacting hydrophobic surfaces, an exponentially decaying negative tangential pressure component, p_T, gives rise to a likewise exponentially decaying surface force. However, for closer distances between the two hydrophobic surfaces, a more elaborate model is needed, in principle similar to that put forward by Cevc et al. [30], to account for repulsive hydration forces. In this type of model, changes of state arising in the residual thin water film itself are also taken into account.

A quasi-thermodynamic/structural model for hydrophobic attraction forces of the above kind was presented in 1989 by Eriksson et al. [31]. In contrast to most other tentative explanations of the hydrophobic force, this model has so far not been properly falsified in the Popper sense. It has rather been refuted by vaguely claiming that long-range structural effects are virtually excluded for (normal) liquids. As discussed above, however, water, with its ability to form a variety of hydrogen-bonding patterns, is still far from being a well-understood liquid at the molecular level. Hence it is worth bringing the ideas behind the water structure-based model into the open again, to scrutinize its basis and assess its predictive power by comparing it with the wider range of experimental results available today. It is not excluded, after all, that this model might serve as a valid point of departure for a deeper understanding of the hydrophobic force. While carrying through our discussion, we will, of course, also comment on alternative attempts to come to grips with the hydrophobic attraction. The premises are the following:

1. The hydrophobic solid surfaces which interact in a liquid water medium are *ideal*: they are geometrically *smooth* to within about 0.1 nm at 0 K, they exhibit small thermal amplitudes at room temperature and are laterally close to being homogeneous in every respect.

2. Relatively speaking, there is a free-energy *decrease* associated with the molecular reorganization of the first monolayer of water molecules next to the hydrophobic surfaces owing to minimization of the number of hydrogen bonds broken on contacting water with the hydrophobic surfaces.

3. The cooperatively enhanced tendency to avoid rupturing hydrogen bonds causes the surface-induced structure to be propagated (with a certain decay rate) towards the center of the thin film, resulting in a somewhat larger average number of hydrogen bonds per water molecule in the film than in the bulk. Hence, assuming short-range interaction forces only, there is a free-energy *increase* arising throughout the core of the thin film, owing to the imposed hydrogen-bond network formation.

The final (inhomogeneous) water state in the film reflects a balance between the *favorable* molecular reorganization occurring in the first (contact) layer of water molecules and the induced, *unfavorable* restructuring of the remainder of the film. In the following, these free energy changes are accounted for by making use of a dimensionless order parameter $s(x)$, which is a measure of the (average) *local* increase of the free energy density in the thin water film as compared with a corresponding slab of bulk water (x denotes the coordinate perpendicular to the thin film). Alternatively, we may assume that $s(x)$ mirrors the local increase of the average number of hydrogen bonds per water molecule in the film or the associated decrease in local density.

Let us now imagine that the final equilibrium state of the thin water film sandwiched between two hydrophobic surfaces is reached in a stepwise fashion. Starting from a thin film cut out of the bulk state, the first step involves establishing the proper molecular interactions at hydrocarbon–water interfaces while retaining the average spherical-symmetric orientation of all the water molecules in the film.

The second step implies a change in the packing and in the average orientation of the water molecules in each of the first molecular layers next to the hydrophobic surfaces, to yield a less dense and more ordered molecular state with an increased number of hydrogen bonds and a preference for tangential alignment of the HOH bisectors of the water molecules.

The film tension attained after these first two equilibration steps we denote γ_0^{f}. The third step comprises a reorganization of the hydrogen-bond network throughout the film, whereby the parameter s becomes a function of the x coordinate and the final equilibrium value of γ^{f} is reached.

In an approach similar to the so-called square gradient approximation, frequently used in the past to model liquid–vapor interfaces, we can express the model features invoked above in the following manner:

$$\gamma^{\mathrm{f}} = \gamma_0^{\mathrm{f}} - a s_0 + \int_{-h/2}^{h/2} \left[c_2 s^2 + \frac{1}{2} c_3 (\mathrm{d}s/\mathrm{d}x)^2 \right] \mathrm{d}x \tag{7}$$

The second term on the right-hand side of this expression (where a is a constant) accounts for an assumed linear free energy reduction due to changing the order parameter s_0 (and hence the packing density) for the water layers in direct contact with the hydrophobic surfaces, whereas the integral accounts for the free energy expense associated with structuring the core of the thin water film. Note in particular that including the squared gradient term is essential as it furnishes a mechanism of energetic coupling between successive layers of water molecules. Hence the constant c_3 reflects the (average) tendency for *cooperative* structure generation. Note also that no core term that is linear in s is included, the reason being that we are actually considering variations about a tensionless bulk water state.

Upon minimizing the film tension γ^{f} while taking into account the proper boundary conditions, we readily derive

$$s(x) = \left(\frac{a}{2 c_3 b} \right) \frac{\cosh(bx)}{\sinh(bh/2)} \tag{8}$$

and

$$\begin{aligned} \gamma^{\mathrm{f}}(h) &= \gamma_0^{\mathrm{f}} - (a^2/4 c_3 b) \coth(bh/2) = \gamma_0^{\mathrm{f}} - a s_0/2 \\ &= \gamma_0^{\mathrm{f}} - (B/2\pi) \coth(bh/2) \end{aligned} \tag{9}$$

The constant b stands for the quotient $\sqrt{2c_2/c_3}$, while the interaction constant B introduced in Eq. (9) can also be written in the form $B = \pi a^2/\sqrt{8 c_2 c_3}$.

For infinitely large film thicknesses, we obtain the film tension lowering caused by the imposed structuring due to the hydrophobic surfaces:

$$\gamma^{\mathrm{f}}(h = \infty) - \gamma_0^{\mathrm{f}} = -B/2\pi \tag{10}$$

implying that $B/4\pi$ is the corresponding reduction of the *interfacial tension* between water and a hydrophobic surface. It can be estimated as ~50 μN m^{-1}, that is, as about 0.1% of the total interfacial tension value of ~50 mN m^{-1}.

Now, by taking the difference between Eqs. (9) and (10) and making use of the Derjaguin approximation, one finds that the hydrophobic attraction force as measured by means of a surface force apparatus with cylindrically shaped hydrophobic surfaces having radii equal to R is given by the expression

$$F/R = 2\pi \Delta \gamma^{\mathrm{f}} = -B[\coth(bh/2) - 1] \tag{11}$$

For sufficiently large film thicknesses, the right-hand side of this equation becomes equal to $-2B\exp(-bh)$, i.e. in the (weak overlap) regime we predict $\ln(-F/R)$ to be a linear function of h. Moreover, in this range b^{-1} has the nature of a decay length.

We can put Eq. (11) in an approximate double exponential form:

$$F/R = -2B(e^{-bh} + e^{-2bh}) \tag{12}$$

where the second exponential accounts for the amplification of the surface force usually seen at medium separations. Unlike Eq. (11), however, although hardly astonishing, Eq. (12) does not extrapolate properly down to small h values. Hence it is understandable that it has become common practice to make use of a double exponential expression entailing four parameters instead of just two:

$$-F/R = C_1 e^{-h/\lambda_1} + C_2 e^{-h/\lambda_2} \tag{13}$$

which in many cases has been shown to represent experimental surface force data satisfactorily in a wide separation range. The first exponential yields the largest contribution for small separations, whereas the second dominates at large separations. Typical values derived are $C_1 = 0.2$ N m^{-1}, $\lambda_1 = 2$ nm, $C_2 = 1$ mN m^{-1} and $\lambda_2 = 20$ nm.

Yet another convenient, although somewhat less accurate, manner to represent experimental surface force data is to invoke an expression of the same mathematical form as for the van der Waals attraction:

$$\Delta\gamma^{\mathrm{f}} = -(K_{\mathrm{hph}}/12\pi)h^{-2} \tag{14}$$

resulting in a one-parameter surface force expression:

$$-F/R = (K_{\mathrm{hph}}/6)h^{-2} \tag{15}$$

Obviously, the strength of the hydrophobic attraction can easily be judged by comparing the value of the constant K_{hph} with the corresponding value of the Hamaker constant, which is usually on the order of 10^{-20} J. Typical K_{hph} values range between 10^{-18} and 10^{-19} J.

The disjoining pressure expression which can be derived from Eq. (9) by differentiation with respect to h is the following:

$$\pi_{\mathrm{D}} = -2c_2 s^2(x=0) = -(bB/2\pi)[\coth^2(bh/2) - 1] \tag{16}$$

Especially when displayed in this derivative mode, experimental hydrophobic interaction curves may appear to belong to two distinct regimes: below and above 10–20 nm, respectively (Fig. 5.4). The same holds true for the order parameter s in the middle of the thin film (where $x=0$)) as derived theoretically. Below about 15 nm, $s(x=0)$ rapidly diminishes with the film thickness h, whereas above about 15 nm, $s(x=0)$ decreases very slowly in an almost linear fashion with h (Fig. 5.5).

The hydrophobic surface forces arising within the large separation range are generally rather weak, at most about 1 mN m^{-1}, to be compared with

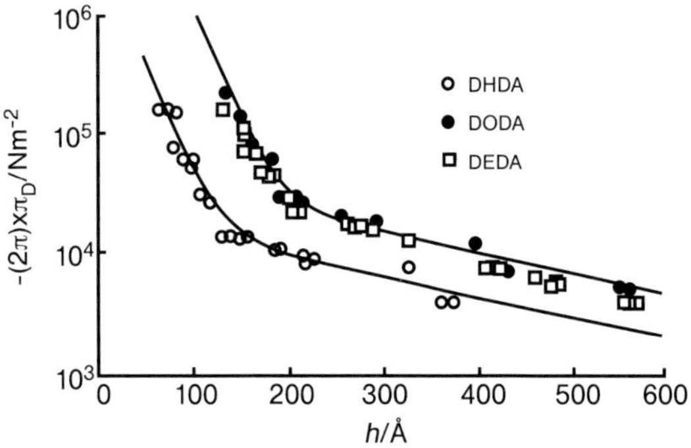

Fig. 5.4 Surface interaction pressures between mica surfaces modified with DHDA, DODA and DEDA monolayers in water at 25 °C as measured by Tsao et al. [24]. Note that the surface force curves are indistinguishable for DODA (C_{18} chains) and DEDA (C_{20} chains).

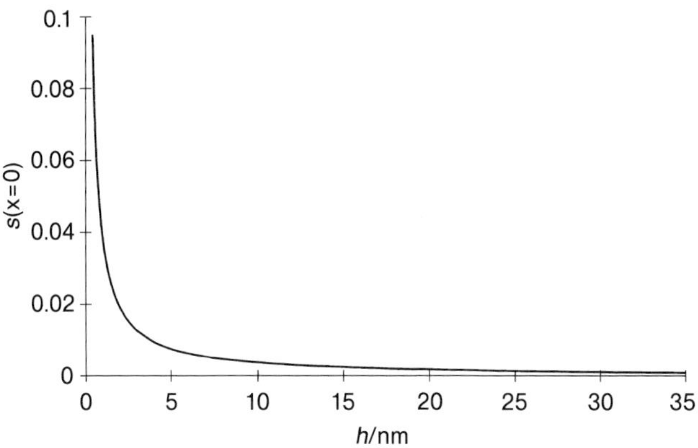

Fig. 5.5 Order parameter $s(x)$ in the midplane of a thin water film sandwiched between hydrophobic surfaces. According to Eq. (8) we have $s(x{=}0) = (a/2c_3b)\sinh^{-1}(bh/2)$ [31].

~500–600 mN m^{-1} for the maximum adhesion force at $h{=}0$ and can therefore easily be concealed, e.g. by a repulsive surface force of electrostatic origin.

Further, on the basis of Eq. (7), we can estimate the excess free energy per water molecule in the middle of the water film where the pressure tensor is likely to be approximately isotropic. Typically, in the medium separation range, $h \approx 10$ nm, we find this excess free energy to amount to about $4\times10^{-4}\ k_BT$ per molecule, to be compared with the energy of a hydrogen bond at room temperature of ~7 k_BT, showing that we are indeed studying very minute thermodynamic effects by means of a sensitive SFA or AFM set-up.

According to the theory summarized above, the strength of the hydrophobic attraction force is determined by the interaction constant $B = \pi a^2/\sqrt{8c_2c_3}$, which in turn is strongly dependent on the constant a related to the free energy change due to the reorganization of the contact layers of water molecules. Hence, B is expected to vary strongly with both the degree of hydrophobicity and the smoothness on the molecular scale of the solid (or liquid) surface.

The decay length in the large separation range, $b^{-1} = \sqrt{c_3/2c_2}$, on the other hand, should depend primarily just on the properties of the water in the thin film, becoming large if structuring occurs readily (as for bulk water at the freezing-point) or when there is a strong tendency to avoid rapid changes of the order parameter $s(x)$, that is, when cooperativity plays a significant role. If the hydrophobic surfaces are charged and (overlapping) electrostatic double layers are present, however, one has to expect a competition between the (tangential) polarization of the water molecules due to the hydrophobic surfaces and the (perpendicular) dipole alignment in the electrostatic (mean) field. Such a coupling might result in a larger constant c_2 and, hence, a shorter decay length.

Salt effects for uncharged or nearly uncharged surfaces are expected to be rather minor, provided, of course, that the hydrophobic surfaces themselves are stable in contact with salt solutions. By and large these features have been experimentally documented.

Furthermore, generalization of the water structure-based theory to the case of an unsymmetrical aqueous thin film between a strongly and a weakly hydrophobic surface has recently been accomplished [32]. The latter case has been studied experimentally by Claesson et al. [33] and by Yoon et al. [34].

To conclude this section, we have seen that the quasi-thermodynamic theory due to Eriksson et al. [31], which focuses on the rather minor free energy effects that are associated with restructuring of water in contact with hydrocarbon surfaces, is capable of systematizing several experimental findings concerning the attractive hydrophobic surface force. However, a fundamental problem related to this approach is that it does not really provide an understanding as to why the effect in some instances can be of such an extremely long range that it can be detected even for separations beyond 100 nm. One has to bear in mind, however, that the long-range hydrophobic surface forces are extremely weak and represent very minute thermodynamic effects and that hydrogen-bonded networks and chains of water molecules are known to be cooperatively stabilized, in other words, larger clusters are inherently more stable than small clusters. This may set the stage for extended water clusters of various shapes to arise, provided that the associated free energy expenses are small enough to be counterbalanced by corresponding mixing and size fluctuation entropies, a thermodynamic scenario that is familiar from the field of surfactant aggregation.

Nevertheless, there is definitely a need for more sophisticated models based on the concept of structure generation in water due to contact with a hydrophobic surface, and also of novel experimental methods by which one can investigate the detailed state of thin water films.

5.6 Bubble Attachment and Cavity Formation at Hydrophobic Surfaces

Upon observing more or less distinct steps in the surface force curves for hydrophobic surfaces (plasma-treated mica silylated with (tridecafluoro-1,1,2,2-tetraoctyl)dimethyldichlorosilane submerged in water at surface separations in the range of 100 nm, Parker et al. [35] suggested that these steps might actually demark the onset of the hydrophobic attraction, albeit an attraction totally different from the water structure-based one we have been discussing thus far. Hence they attributed the attraction seen to tiny adhering air bubbles which upon the approach of the hydrophobic surfaces suddenly form bridging (quasi-cylindrical) bubbles. Although the concept of bridging bubbles is in many ways an appealing one, there are a number of difficulties associated with claiming that a capillary mechanism of this kind is the chief reason for the long-range effects observed.

A primary difficulty is that very small air bubbles nucleated in a water phase are surprisingly short-lived. Because of the rather sizeable Laplace excess pressure, the air dissolves and the bubbles diminish in size at an accelerating rate. Straightforward calculations show that the lifetime of a small air bubble in water scales with the square of the bubble radius. It is about 10 ms for a radius of 1 μm and about 1 μs for a 10 nm bubble, whereas a bubble of mm size may subsist for days or even months [36, 37]. Although air bubbles as a rule are generated upon contacting two hydrophobic surfaces in an SFA device filled with water, followed by pulling them apart before starting the measurements proper, adhesion of a small air bubble is nonetheless likely to be a rare event already because of the limited life-span of these bubbles.

Moreover, unlike the situation for larger air bubbles which (because of the small excess pressure) can be treated as if they were thermodynamically *closed*, very small *open* air bubbles are not expected to adhere to an *ideal* hydrophobic surface in a stable manner [38–40]. Most *real* hydrophobic surfaces, however, contain defects which presumably play a crucial role in promoting bubble adhesion to occur to some minor extent.

What remains then, following the "bubble mechanism" line of reasoning, is the option that a limited number of small bubbles, which have survived (as they started out as relatively large bubbles), happen to adhere in an irreversible, defect-dependent manner to the hydrophobic surfaces and that a few of these adhering bubbles rather quickly form approximately cylindrical air bridges to the approaching hydrophobic surface. In this way the excess (Laplace) pressure will be efficiently cancelled and a bridging bubble can consequently subsist for a long time. To assume irreversibility for the wetting behavior is essential here. Otherwise, the formation of one large air-filled cavity, eventually giving rise to a huge surface force, would have to be inferred.

As is readily confirmed, the formation of a bridging bubble from an adhering bubble is likely to be advantageous free energy-wise. Nevertheless, it is significant, and also understandable in view of the short lifetime and submicroscopic size of such a bubble, that it has met with experimental difficulties to verify positively the presence of adhering air bubbles on hydrophobic surfaces [15, 41]. To judge from the evidence available today, the occurrence of adhering air bubbles depends strongly on the closer nature of the hydrophobic surface. This might be expected, of course, since non-ideal features in the hydrophobic surfaces themselves are likely to play a decisive role.

For a few years, it was widely held that the long-range part of the hydrophobic attraction is an artifact which can be attributed to a capillary bridging mechanism of the above character. More recently, however, several investigators have emphasized that the long-range hydrophobic attraction is actually present for a number of hydrophobic surfaces for which there are no signs of bubbles whatsoever. Hence the supposition has to be rejected that small adhering bubbles which collapse into bridging bubbles is the major cause of the "true" long-range hydrophobic attraction. Strong additional support for this conclusion has recently been obtained from careful degassing experiments [14, 23].

5.7
Electrostatic Correlation Forces

In the 1970s, it was realized that *correlation effects* may become significant especially for interacting double layers encompassing divalent counter-ions [42, 43]. An attraction of this origin is conceptually similar to the dispersion interaction as it relies on instantaneous, laterally inhomogeneous, counter-ion distributions outside the charged surfaces. On average, the fluctuations cause more weight to be given to attractive than repulsive configurations, resulting in an overall attraction. On this basis, one has been able to account for the large deviations from Poisson-Boltzmann (PB) theory seen for charged systems with divalent ions while at the same time verifying that the same PB theory as a rule is reasonably accurate for systems containing monovalent ions.

For the correlation interaction between surfaces carrying *mobile* charges, the Ninham-Parsegian expression holds for the disjoining pressure. At large separations it takes the following limiting form:

$$\pi_D = \frac{kT\kappa^3}{\kappa h} e^{-2\kappa h} \qquad h \gg \kappa^{-1} \tag{17}$$

The magnitude of the attractive interaction pressure predicted by means of this equation is, however, of the order of the van der Waals attraction. Hence it is evident that charge correlation attraction arising at the molecular (ion) length scale is not sufficiently strong to rationalize the experimentally measured hydrophobic attraction. This was realized by Forsman et al. [44], Miklavic et al. [45] and Kekicheff and Spalla [46], who argued that more sizeable contributions of this nature can be obtained for charged surfaces which are heterogeneous, containing (mobile) charged surfactant aggregates of some size, e.g. hemi-micelles.

In particular, Miklavic et al. [45] were able to demonstrate that, at relatively high salt concentration, the correlation attraction should decay with a decay length equal to $\kappa^{-1}/2$, i.e. half of the Debye length, upon increasing the surface separation. However, their calculations presuppose that the surface density of adsorbed micellar aggregates is practically independent of the concentration of added salt, a condition that for the most part is unlikely to be fulfilled.

Although these notions about ionic correlation attractions may be of some relevance for hydrophobic surfaces formed by adsorbing, e.g., a water-soluble cationic surfactant on mica, glass or silica, it seems out of the question to employ them to account for the archetypal long-range hydrophobic attraction documented for the LB monolayer-modified mica surfaces prepared by Claesson and Christenson [7, 8] or the corresponding mica surfaces prepared by adsorption from cyclohexane solution by Tsao et al. [24], at least insofar as no salt is added when making the surface force runs. The surface characterizations carried out and the preparation protocols employed seem to leave little room for speculation in this direction.

In a similar vein, it was suggested by Tsao et al. [12] and Pazhinur and Yoon [47] that correlations among colloidal-grained, ordered dipole domains across the thin film may give rise to sizeable attractions, far stronger than the van der Waals interaction. While these matters perhaps are not yet finally resolved, as already indicated above, it seems excluded that *all* observations regarding the hydrophobic attraction can be accounted for on the basis of correlation attractions alone.

Coming back to hemi-micelle formation, an aggregation of this kind is expected for hydrophobic surfaces prepared by surfactant adsorption from water solution, starting at the critical hemi-micelle concentration, where, however, the hydrophobic attraction already is at its peak. It was even demonstrated by Herder [48] several years ago that, on adsorbing a cationic surfactant on LB–DDOA-modified mica, the hydrophobic attraction rapidly vanishes and is replaced by electrostatic repulsion. Furthermore, a few years ago Craig et al. [13] thoroughly investigated the interaction between silica surfaces in dilute CTAB and CPC solutions in the presence of electrolytes and firmly concluded that a direct electrostatic mechanism for the hydrophobic attraction is hardly an option. Parker and Claesson [27] arrived at the same conclusion using the MASIF set-up and silanized glass spheres.

5.8 Surface Force Data Supporting the Water Structure Mechanism

After about two decades of research efforts world-wide, devoted to searching for the mechanism that gives rise to the hydrophobic attraction, it is probably time to turn back to where the surface force-based development phase started: the measurements by Israelachvili and Pashley [6]. These authors made use of mica surfaces submerged in dilute CTAB solution and interpreted the results rather straightforwardly in terms of the hydrophobic effect.

In the 1960s, the "mysteries" of water were still fashionable topics for study. A great deal of inspiration came, of course, from Pauling's investigations of the hydrogen bond [49] and from the fascinating field of clathrate physical chemistry [50]. Later, the tendency to restrict the discussion to structural aspects only, instead of employing the full statistical-mechanical machinery and the polywater affair, made this research direction fall into disrepute. More recently, however, it has become abundantly clear that the molecular understanding of liquid water is incomplete to an extent that renders it virtually impossible to make reliable theoretical predictions about (transient) H-bond-generated structures.

As we have already discussed earlier, in order to cope with the long-range hydrophobic attraction, a couple of approaches, totally different from the water structure approach, have been fairly thoroughly scrutinized. In particular, electrostatics beyond the standard PB mean-field description have been re-examined and, more trivial, although still fairly cumbersome, capillary effects have been investigated in detail. The latter phenomena include the formation of cavities and bridging bubbles between (hysteretic) hydrophobic surfaces submerged in water. In the end, however, both of these major alternative approaches have turned out to be untenable.

To elucidate further to what extent the water structure-based theory of the hydrophobic force compares with experimental data, let us go back to the perhaps most illuminating and even best documented experimental surface force study carried out so far that elucidated the hydrophobic attraction, namely that presented by Tsao et al. in 1991 [24]. These investigators made use of hydrophobized mica surfaces (mounted in an SFA) prepared by means of adsorption from cyclohexane solutions of cationic surfactants of different chain lengths: dihexadecyldimethylammonium (DHDA), dioctadecyldimethylammonium (DODA) and dieicosyldimethylammonium (DEDA), with bromide or acetate as the counter-ions. The resulting adsorption layers were characterized by AFM. Hence it was verified that the double-chain surfactant cations employed are electrostatically bonded, one to each anionic mica site, resulting in a packing density close to 0.50 nm^2 (0.25 nm^2 per single chain) and that they remain quantitatively bonded to their sites (although in a metastable state) even after contacting the surface with water and raising the temperature to 50 °C. Full stability towards salt solutions was, however, not achieved using this surface preparation method.

Moreover, the aforementioned authors demonstrated that while the DEDA monolayer preserves a frozen chain state at 50 °C, this is not the case for either the DHDA monolayer or the DODA monolayer. In fact, DHDA appears to be present in a melted chain state already at 25 °C whereas DODA melts in the range between 40 and 50 °C.

Correlating changes were observed in the surface force curves and it was concluded that a well-organized, smooth hydrocarbon monolayer of "frozen" hydrocarbon chains causes the strongest hydrophobic attraction. Later this finding was corroborated by Rabinovich et al. by means of FTIR measurements [51]. Concerning the chain length, it chiefly matters in the sense that it determines where on the temperature scale chain melting occurs. Moreover, for the same temperature, the decay length at large separations was found to be practically the same.

The observations just mentioned fit well into the theoretical model description presented earlier that presumes structural water effects. Accordingly, the strength of the hydrophobic force is determined primarily by the constant a of the (linear) response function $-as_0$, which participates in the expression for the constant B in Eq. (11):

$$B = \frac{\pi a^2}{\sqrt{8c_2c_3}} \tag{18}$$

In view of the cooperative nature of water structure generation, it seems very likely that a stronger response (larger a) in terms of free energy lowering should result for the contact monolayer insofar as it is fairly unperturbed by the thermal motions of the hydrocarbon surface. This dependence may suffice to account for the distinctly different surface force behaviors recorded by Tsao et al. for hydrocarbon surfaces in a frozen (larger B) or melted (smaller B) state.

In the same vein, raising the temperature should tend to make a smaller and, in addition, make the structuring of the film core more costly in terms of free energy, yielding a higher c_2, which should likewise tend to diminish B. Similarly, the decay length b^{-1}, which is given by $\sqrt{c_3/2c_2}$, depends inversely on c_2 and should therefore diminish with temperature, as is for the most part observed. Concerning c_3, just a modest dependence on temperature is expected.

On the other hand, the (unintentional) presence of a polarizing electrostatic field would tend to increase c_2 and hence shorten the decay length, whereas the presence of inert gas molecules in the form of clathrate guest molecules might conceivably make it less costly to restructure the water film core, implying a smaller c_2 and a longer decay length.

More importantly, however, we can carry through a virtually model-independent, surface-thermodynamic analysis of surface force data as complete as those of Tsao et al. [24]. Let us go back to the thermodynamic fundamental equation governing the present film case (cf. Appendix). For a pure water film in contact with water, we have

$$\mathrm{d}\gamma^{\mathrm{f}} = (V^{\mathrm{f,ex}}/A)\mathrm{d}p - (S^{\mathrm{f,ex}}/A)\mathrm{d}T - \pi_{\mathrm{D}}\mathrm{d}h \tag{19}$$

where $V^{\mathrm{f,\,ex}}$ is the volume in excess of the volume of a slab of bulk water that contains the same number of water molecules as the film of a certain thickness. The excess entropy $S^{\mathrm{f,\,ex}}$ is analogously defined. For an infinitely thick film, we accordingly obtain

$$\mathrm{d}\gamma^{\mathrm{f}}(\infty) = (V_{\infty}^{\mathrm{f,ex}}/A)\mathrm{d}p - (S_{\infty}^{\mathrm{f,ex}}/A)\mathrm{d}T \tag{20}$$

On subtracting Eq. (20) from Eq. (19) we next obtain

$$\mathrm{d}(\Delta\gamma^{\mathrm{f}}) = (\Delta V^{\mathrm{f,ex}}/A)\mathrm{d}p - (\Delta S^{\mathrm{f,ex}}/A)\mathrm{d}T - \pi_{\mathrm{D}}\mathrm{d}h \tag{21}$$

This equation is the appropriate thermodynamic relation for dealing with surface force data for thin films consisting of pure water. It obviously includes the partial derivative:

$$\left[\frac{\partial(\Delta\gamma^{\mathrm{f,ex}})}{\partial T}\right]_{\mathrm{p,h}} = -\frac{\Delta S^{\mathrm{f,ex}}}{A} \tag{22}$$

Therefore, if the surface force increases with temperature, which in the present case means that the attractive hydrophobic surface force becomes less negative when the temperature is raised, we must necessarily have that $\Delta S^{\mathrm{f,ex}}$ is a negative quantity. In turn, this evaluation result would imply that the thin water film is actually more ordered than an imaginary water film of the same thickness encompassing two non-interacting surface zones of water next to the hydrophobic surfaces. Note that provided the hydrophobic surfaces themselves do not suffer any changes in their intrinsic thermodynamic properties as the film thickness varies (which should not be the case, at least not if the surfaces are solid), there will be no contribution to $\Delta S^{\mathrm{f,ex}}$ other than that arising in the water film.

By integrating the disjoining pressure functions derived experimentally by Tsao et al. [24], one can obtain the corresponding relative film tensions (Fig. 5.6). It is

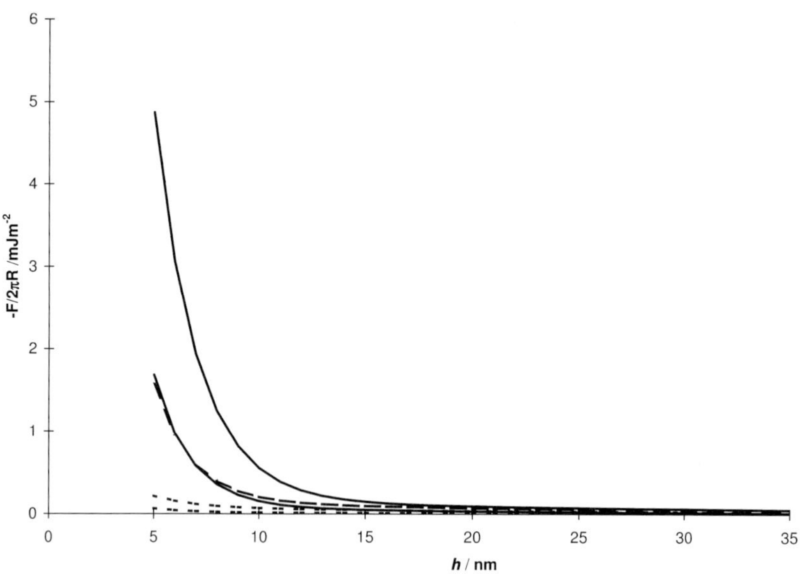

Fig. 5.6 Film tension changes for thin water films between hydrophobized mica surfaces at different temperatures derived from the disjoining pressure data presented by Tsao et al. [24]. Top curve: dieicosyldimethylammonium [DEDA(s)] and dioctadecyldimethylammonium [DODA(s)] at 25 °C. Middle curve: DEDA(s) at 50 °C (dashed) and DODA(s) at 40 °C (solid line). Bottom curves: dihexadecyldimethylammonium [DHDA(l)] at 25 °C (top) and 40 °C (bottom).

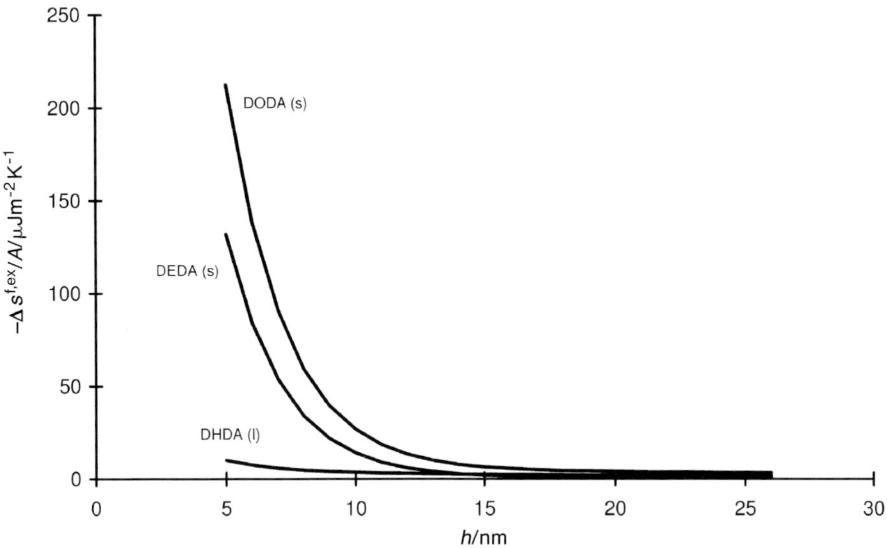

Fig. 5.7 Excess film entropies as functions of the separation between mica surfaces hydrophobized with DHDA, DODA and DEDA. The difference between the DODA and DEDA curves is presumably due to experimental uncertainty.

seen that these curves display the crucial feature discussed above: On diminishing the surface separation, the film tension reduction is larger at room temperature than above room temperature. Consequently, as soon as the surfaces interact, $\Delta S^{f,ex}$ is always a negative quantity. It is also worth noting that Eq. (11) with B and b values generated from the measurements of Claesson and Christenson [8] (who used similarly prepared hydrophobic surfaces), yields a reasonably good fit for the film tension curve obtained for DODA by Tsao et al. at room temperature.

The relative excess film entropies per unit area, $\Delta S^{f,ex}$, can be quantified using Eq. (22), resulting in the curves shown in Fig. 5.7. It is evident that the excess entropy becomes increasingly more negative when the water film becomes thinner, especially for the "frozen" hydrocarbon surfaces, DODA and DEDA.

Concerning the numerical values, it is illuminating to compare with the entropy reduction one would obtain for a corresponding slab of bulk water that undergoes freezing to ice. By invoking the entropy of fusion of ice (~22 J mol^{-1} K^{-1}), for a 5-nm water film we estimate $\Delta S^{f}_{freezing} = -6$ mJ m^{-2}K^{-1} which is about 30 times more than the entropy reduction quantified from the surface force measurements. In other words, the enhanced molecular order in the thin film is equivalent to introducing about 3% of ice ordering.

Furthermore, comparing with the relative film tensions shown in Fig. 5.6, we observe that for every thickness h it holds true that

$$|\Delta\gamma^f| \ll |T\Delta S^{f,ex}/A| \tag{23}$$

This inequality necessarily implies that the (negative) relative excess entropy of the film is approximately counterbalanced by a corresponding enthalpy term, that is

$$T\Delta S^{\mathrm{f,ex}} \approx \Delta H^{\mathrm{f,ex}} \tag{24}$$

Such an enthalpy–entropy compensation is perhaps the most distinguishing feature accompanying structural changes in water. For freezing of water at 0 °C we have, of course, that $\Delta G = 0$ which means that a relation similar to Eq. (24) is exactly fulfilled.

At this point, it is proper to emphasize that there is no need to invoke any mechanistic model whatsoever or to introduce any speculative assumptions about the nature of the thin film system, to make the above purely thermodynamic evaluation of the change in water film entropy. In addition, we note that it would be meaningless to propose an interaction mechanism incapable of giving rise to a significant reduction in the water film entropy. This circumstance apparently restricts the number of alternative mechanisms and, hence, lends support of the idea that we are really dealing with a water structuring effect proper caused by the contact with hydrophobic surfaces. As mentioned earlier, we envisage that the structure formation is most pronounced for smooth hydrophobic surfaces that are free from charges and made up of hydrocarbon chains in a frozen state.

The most important assumptions on which our conclusions above rest are the following:

1. The disjoining pressure data recorded by Tsao et al. [24] are representative of a water film state where the film itself is fully equilibrated whereas the (smooth) hydrophobic surfaces all the time remain in the same metastable frozen state.

2. No other conceivable mechanism except a water restructuring mechanism might reproduce the strong temperature dependence and the characteristic enthalpy–entropy compensation documented for a thin water film sandwiched between hydrophobic surfaces.

The reversibility demonstrated with respect to temperature changes and the close quantitative agreement with surface force data from other laboratories using similarly prepared hydrophobic surfaces lend support of point 1 above. Point 2 we shall return to below.

An additional advantage of the water restructuring model should be brought up explicitly: there is no need to assume that different physical mechanisms are operating at small and large separations. By making use of Eq. (11) and introducing just *two* parameters, the surface force strength parameter B and the decay length at large separations b^{-1}, the distinctly different behaviors seen in the two regimes can be accounted for within the same framework. Here we refer to Fig. 5.8, which includes a surface force function of the proposed form (Eq. 11) obtained by fitting to the data of Claesson and Christenson [8], viz.

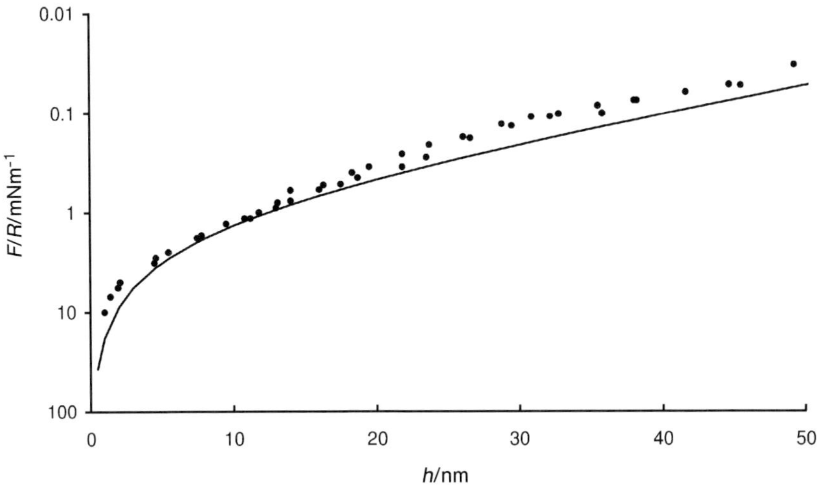

Fig. 5.8 Surface force data points recorded by Lin et al. [15] at 25 °C for DODA-modified mica surfaces in water, compared with the surface force function given by Eq. (11) upon inserting the parameter values $B = 0.600$ mJ m^{-2} and $b^{-1} = 15.8$ nm (yielding Eq. 25) as determined by fitting to the surface force data for the same experimental system due to Claesson and Christenson [8].

$$\frac{F}{R} = -0.600\left[\coth\left(\frac{h/2}{15.8}\right) - 1\right] \qquad (\text{mJ m}^{-2}) \tag{25}$$

(where h should be expressed in nm), together with the surface force data for the same experimental system recently published by Lin et al. [15]. It is evident that the above function produces a fairly good fit, especially for short separations (putting $B = 0.450$ yields an even better fit).

Correspondingly, the disjoining pressure function is given by

$$\pi_D = -6.04 \times 10^3\left[\coth^2\left(\frac{h/2}{15.8}\right) - 1\right] \qquad (\text{N m}^{-2}) \tag{26}$$

which, according to Fig. 5.9, represents fairly well the data recorded by Tsao et al. [24] for DODA and DEDA at 25 °C.

For small enough h values (less than about 2 nm), the [coth($bh/2$)–1] function can be approximated by $2B/bh$, that is the surface force is predicted to vary inversely with h. The good fit in the short separation range displayed in Fig. 5.8 is noteworthy. Still, it is hardly conceivable that the model concepts introduced to derive Eq. (11) would apply below ~2 nm; in other words, the agreement may appear reassuring but is presumably little else than fortuitous.

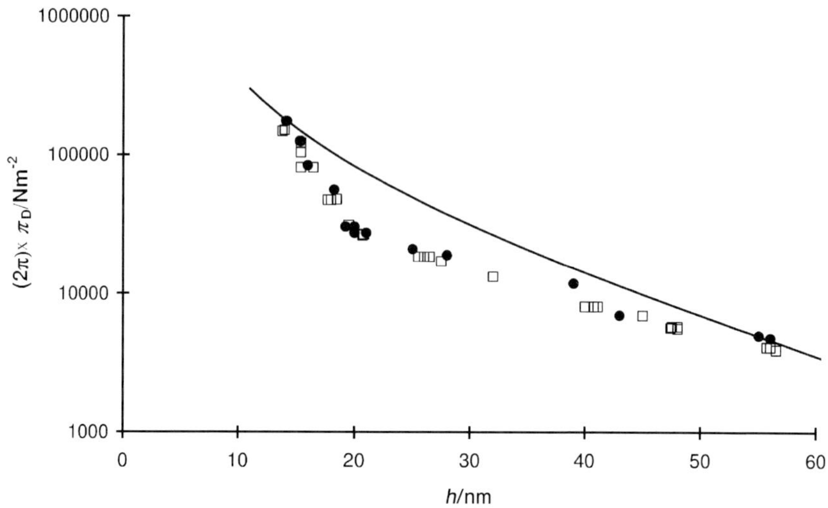

Fig. 5.9 Comparison between the disjoining pressure data reported by Tsao et al. [24] and the disjoining pressure function given by Eq. (26) obtained by inserting the parameter values $B = 0.600$ mJ m^{-2} and $b^{-1} = 15.8$ nm in Eq. (16).

5.9
The Effect of Solutes

Let us briefly make some comments on the effect of adding a solute to the water phase that is in equilibrium with the thin film. The main thermodynamic relation governing this case is the counterpart of the Gibbs surface tension equation, which at constant T and p reads

$$d\Delta\gamma^{f} = -\Delta\Gamma_{s}^{f,ex} d\mu_{s} - \pi_{D} dh \tag{27}$$

where μ_s, as before, denotes the chemical potential of the solute. Consequently, having access to equilibrium data on the surface force for a certain film thickness as a function of the solute concentration, we can obtain the change in solute excess, $\Delta\Gamma_{s}^{f,ex}$, for a film of that thickness as compared with an infinitely thick film.

Recently, interesting such SFA data were published by Meyer et al. [14] aiming to elucidate the effect of degassing (pure) water contacted with DODA-modified mica surfaces. Within most of the separation range below 50 nm, a reduction due to degassing of the magnitude of the hydrophobic force is clearly seen. Hence, referring rather to dissolution of the main air components N_2 and O_2, we recognize that the derivative

$$\left(\frac{\partial \Delta\gamma^{f}}{\partial \mu_{s}}\right)_{T,p,h} = -\Delta\Gamma_{s}^{f,ex} \tag{28}$$

has a negative sign. This obviously implies that $\Delta\Gamma_s^{f,ex}>0$ and that the air gases are enriched in a thin water film between hydrophobic surfaces in comparison with an infinitely thick film scaled down to the same thickness.

Taking 10–20-nm films as examples, by applying Eq. (28) we find that the film excesses of N_2 and O_2 are about the same as the amount of dissolved gases in corresponding (imaginary) thin films made up of bulk solution. This observation is in fair agreement with the idea of an enhanced water structure in thin water films delimited by hydrophobic surfaces. The solvent properties of the water change as additional water cage volume becomes available to the (clathrate-forming) gas molecules, in a similar manner as when reducing the temperature to below room temperature.

The higher content, relatively speaking, of the gas solutes for thin films will cause the film tension to become more negative in accordance with Eq. (28). Conversely, deaeration should tend to diminish the magnitude of the attractive hydrophobic force as observed by Meyer et al. [14]. Moreover, it seems most likely that the effects of deaeration documented by Pashley [52] on the stability of colloids and emulsions can also be understood, in a preliminary way at least, on this basis.

Concerning the addition of *alcohols* and *surfactants*, which readily adsorb on hydrophobic surfaces, in the dilute regime the situation is largely opposite: The surface force rises with μ_s, yielding a negative $\Delta\Gamma_s^{f,ex}$ that can be interpreted in terms of a smaller constant B and a diminishing decay length b^{-1}. Note that this holds for both alcohols and non-ionic surfactants. For a contact monolayer mixed with such surface-active species there is simply less free energy to be gained by restructuring (smaller a). Evidently, for ionic surfactants, in addition the electrostatic (double layer) repulsion has to be invoked.

Surface force curves for water and ethanol–water mixtures, recorded by Ederth [53] using C_{16}-alkanethiolated Au surfaces, are reproduced in Fig. 5.10. It

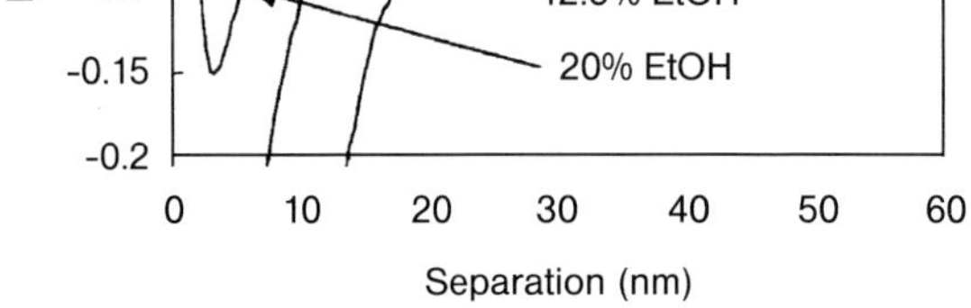

Fig. 5.10 Surface force isotherms presented by Ederth [53] for water and ethanol–water mixtures between hexadecanethiol surfaces, showing the pronounced effect on the hydrophobic attraction of adding alcohol. The advancing contact angles are 94° (12.5%) and 88° (20%), respectively.

appears that the surface force is substantially reduced in magnitude by adding alcohol. Using Eq. (28) we can estimate $\Delta\Gamma_s^{f,ex}$ as -4×10^{16} molecules m^{-2} for a 10-nm thick film in contact with ~16% alcohol solution to be compared with the alcohol content of a corresponding bulk solution film of $\sim 2\times10^{18}$ molecules m^{-2}. The reason why the surface force diminishes as alcohol is added is that the lowering of the film tension γ^f is larger for an infinitely thick film than for a thinner film. In turn, this is because adding alcohol to the thin film counteracts the favorable structure formation.

On the other hand, adding *salts* should not change the situation very much with respect to the hydrophobic force (provided that the metastable attachment of hydrocarbon chains remains unaffected), the chief reason being that small ions do not mix with the water in the contact monolayers adjacent to the hydrophobic surfaces. Hence the constant B should be left much the same and likewise b^{-1} insofar as the ions do not interfere significantly with the H-bond network formation. This agrees with a multitude of observations using hydrophobic surfaces which are sufficiently stable.

5.10 Conclusion

In his book *Intermolecular and Surface Forces* [54], Jacob Israelachvili writes: "It is the energy (or entropy) associated with the H-bonding network and proton hopping defects, which extends over much larger space than the molecular correlations, that is probably at the root of the long-range solvation interactions of water". After having scrutinized where the scientific issue stands today concerning the origin of the comparatively strong and long-range hydrophobic attraction, we do agree. No equally convincing approach to gaining a closer understanding of the hydrophobic surface force has so far been proposed. Various electrostatic interaction mechanisms have been subjected to critical tests and likewise the adhering-bridging-bubble track. Yet none of these approaches has been found to be applicable for the most ideal hydrophobic surfaces that one can prepare where the hydrophobic attraction is strongest. In the end, they have to be abandoned as possible starting points of a more elaborate theoretical modeling.

Two decades ago, the original proposition that the hydrophobic surface force is related to the structural response of water contacted with a solid hydrophobic surface appeared as a natural hypothesis. After some time, however, a paradigm shift occurred within the scientific community and eventually the notion of water restructuring was considered to be only a "bold conjecture", maybe even too bold to be taken seriously. Today, however, a vast amount of experimental results has accumulated, which, taken together, indicate that there is hardly any alternative route that might be fruitful when aiming at a molecular understanding of the hydrophobic attraction.

The versatility exhibited by water molecules to form clathrate cages for guest molecules of various size and shape indicates that reliable predictions of the structural state of water close to a real hydrophobic surface will be difficult to make, particularly, since the molecular description of liquid water is still much in dispute. For a long time, however, water has been known to be an associated liquid where three-dimensional networks play a significant role. Furthermore, the association process is of a cooperative nature, implying that the free energy per water molecule depends on the cluster size. Hence it ought to come as no great surprise that the introduction of an (infinite) hydrophobic *solid* surface can promote extensive structure generation in the adjacent water and that confinement of water in a thin film between two hydrophobic surfaces might strongly promote additional restructuring.

Given the background presented in this chapter, to attempt to resolve the long-standing issue about the origin of the hydrophobic force, it seems most urgent to

1. make several more studies of the temperature dependence of the hydrophobic force with the purpose of fully assessing the thermodynamic water film properties;
2. probe the state of the water next to carefully prepared and characterized hydrophobic surfaces, and also in thin water films sandwiched between hydrophobic surfaces, using e.g. the SFG or XAS technique;
3. make extensive experimental studies of the effect on the hydrophobic surface force of various clathrate-forming inert gases and other well-chosen solutes.

It may seem astonishing that we are still far from having reached a common understanding of the hydrophobic attraction force. In spite of all the labor spent, progress has been rather sluggish. For some years now the role of water merely as an ordinary solvent, and capillarity phenomena involving water and hydrophobic surfaces have been addressed in the first place. Hopefully to encourage research efforts along basically rather old-fashioned routes (which strange enough may now falsely seem almost novel), in this chapter we have focused on evidence in support of the original proposition that the hydrophobic force is linked with the special association properties of water, which express themselves e.g. in the formation of well-defined clathrates. However, the geometric and energetic constraints at a smooth, solid, planar (infinite) hydrophobic surface are evidently very different from those prevailing around a small guest molecule, apparently enabling extensive structures to form in a cooperative manner.

Appendix: Thermodynamics of a Planar Thin Aqueous Film Between Hydrophobic Solid Surfaces

To treat the thermodynamics of a (symmetric) thin aqueous solution film between two plane-parallel hydrophobic solid surfaces, let us start with writing down the thermodynamic fundamental equation of the film in the form

$$d\gamma^{f} = -(S^{f}/A)dT + hdp - \pi_{D}dh - \Gamma_{w}^{f}d\mu_{w} - \Gamma_{s}^{f}d\mu_{s} \tag{A.1}$$

Here we presuppose that the properties of the thin film are estimated by using the *equimolecular* dividing surfaces with respect to the solid component, that is by using the (solid)-dividing surfaces in the Gibbs notation. Thus, the film considered, which has thickness h (equal to the distance between the two dividing surfaces), only encompasses liquid matter in the form of water and a solute component denoted by subscript s. In the above expression, p stands for the ambient pressure, π_{D} for the disjoining (interaction) pressure and γ^{f} for the film tension. This equation is nothing but a straightforward extension of the Gibbs surface tension equation. As was considered by Gibbs [55], the latter equation holds only as long as the state of strain of the solid surfaces remains unchanged.

We emphasize that γ^{f} is *energetically* (rather than mechanically) defined, by means of the relation

$$\gamma^{f} = G^{f}/A - \Gamma_{w}^{f}\mu_{w} - \Gamma_{s}^{f}\mu_{s} \equiv G^{f,ex}/A = \Omega^{f}/A \tag{A.2}$$

Accordingly, γ^{f} is primarily the excess free energy per unit area compared to what we would have for a thin film of bulk-phase properties containing the same amounts of water and solute. To put it another way: it corresponds to the reversible cleavage work up to a certain surface separation h. In addition, it is formally the same as the Ω-potential per unit area of the thin film.

For the special case under discussion, however, of a thin film between hydrophobic surfaces, which is made out of a dilute water solution, γ^{f} has also a fairly clear mechanical significance. Since only minor contributions to the excess free energy and to the lateral tensions arise in the hydrocarbon part of the two interfaces, the integrated tangential pressure profile across the film thickness h to a good approximation corresponds to $-\gamma^{f}$.

Next, we have to bring Eq. (A.1) in harmony with the phase rule by invoking the Gibbs-Duhem relation for the bulk phase (denoted by superscript b):

$$S^{b}dT - V^{b}dp + n_{w}d\mu_{w} + n_{s}d\mu_{s} = 0 \tag{A.3}$$

On eliminating $d\mu_{w}$ from Eq. (A.1), we obtain

$$d\gamma^{f} = -(S^{f,ex}/A)dT + (V^{f,ex}/A)dp - \pi_{D}dh - \Gamma_{s}^{f,ex}\left(\frac{\partial \mu_{s}}{\partial \ln c_{s}}\right)_{T,p} d\ln c_{s} \tag{A.4}$$

where the coefficients of $\mathrm{d}T$, $\mathrm{d}p$ and $\mathrm{d}\mu_s$ are given by

$$S^{f,ex}/A = S^{f}/A - \Gamma^{f}_{w} s^{b}_{w} - \Gamma^{f}_{s} s^{b}_{s} \tag{A.5}$$

$$V^{f,ex}/A = h - \Gamma^{f}_{w} v^{b}_{w} - \Gamma^{f}_{s} v^{b}_{s} \tag{A.6}$$

and

$$\Gamma^{f,ex}_{s} = \Gamma^{f}_{s} - \Gamma^{f}_{w} c_s / c_w \tag{A.7}$$

In these expressions, s^b and v^b denote partial molar volumes in the bulk solution and c_w and c_s the bulk-phase concentrations of water and solute, respectively.

The (reversible) surface force measured in a surface force device is proportional to the film tension difference $\Delta\gamma^f = \gamma^f(h) - \gamma^f(\infty)$, for which we finally obtain, assuming an ideal solution of the solute and noting that $\pi_D(\infty) = 0$:

$$\mathrm{d}\Delta\gamma^{f} = -(\Delta S^{f,ex}/A)\mathrm{d}T + (\Delta V^{f,ex}/A)\mathrm{d}p - \pi_D \mathrm{d}h - \Delta\Gamma^{f,ex}_{s} RT \mathrm{d}\ln c_s \tag{A.8}$$

with obvious definitions of the coefficients of $\mathrm{d}T$, $\mathrm{d}p$ and $\mathrm{d}\mu_s$ as the differences for the properties in question between a thin and an infinitely thick film. Since practically all residual contributions from the solid surfaces to the thermodynamic film properties cancel upon forming these differences, the constraint of a constant surface strain can be disregarded and, furthermore, $\Delta\gamma^f$ corresponds almost exactly to the change in overall lateral mechanical tension of the film when comparing a thin and a very thick film; cf. the equivalent equation presented by Gibbs for the non-interacting case (Eq. 678 in Ref. [55]).

For the case of a 1:1 electrolyte being the solute, the last term of the above expression has, of course, to be multiplied by a factor of 2.

Hence, by plotting $\Delta\gamma^f$ as a function of the state variables T, p and c_s for fixed surface separations h, important information can be gathered as to the thermodynamic thin film (difference) properties $\Delta S^{f,ex}/A$, $\Delta V^{f,ex}/A$ and $\Delta\Gamma^{f,ex}_{s}$. By using Eq. (A.2) and the enthalpy definition, we furthermore have

$$\Delta G^{f,ex}/A = \Delta H^{f,ex}/A - T\Delta S^{f,ex}/A = \Delta\gamma^{f} \tag{A.9}$$

implying that the corresponding enthalpy difference is readily derived. Note, in particular, that gaining knowledge about $\Delta S^{f,ex}/A$ and $\Delta H^{f,ex}/A$ will be crucial for understanding the structural aspects of thin aqueous films, whereas the effect of solutes on the surface force is intimately related to $\Delta\Gamma^{f,ex}_{s}$. In all likelihood, producing the proper sets of data and analyzing them thermodynamically might contribute a great deal to resolving the current research issues about the nature and properties of thin water films between hydrophobic surfaces.

References

1 Blake, T.D., Kitchener, J.A., *J. Chem. Soc., Faraday Trans.*, **1979**, *68*, 1435–1442.

2 Derjaguin, B.V., Churaev, N.V., Muller, V.M., *Surface Forces*, Consultants Bureau, New York, **1987**, Chapter 7; Derjaguin, B.V., Churaev, N.V., *J. Colloid Interface Sci.*, **1974**, *49*, 249.

3 Israelachvili, J.N., Tabor, D., *Proc. Roy. Soc. London*, **1972**, *A331*, 19–38.

4 Parker, J.L., *Prog. Surf. Sci.*, **1994**, *379*, 205.

5 Yaminsky, V.V., Ninham, B.W., Stewart, A.M., *Langmuir*, **1996**, *12*, 3357.

6 Israelachvili, J.N., Pashley, R.M., *Nature*, **1982**, *300*, 341; *J. Colloid Interface Sci.*, **1984**, *98*, 500.

7 Christenson, H.K., Claesson, P.M., *Science*, **1988**, *239*, 390.

8 Claesson, P.M., Christenson, H.K., *J. Phys. Chem.*, **1988**, *92*, 1650.

9 Rabinovich, Ya.I., Derjaguin, B.V., *Colloids Surf.*, **1988**, *30*, 243.

10 Christenson, H.K., Claesson, P.M., *Adv. Colloid Interface Sci.*, **2001**, *91*, 391–436.

11 Spalla, O., *Curr. Opin. Colloid Interface Sci.*, **2000**, *5*, 5–12.

12 Tsao, Y., Evans, D.F., Wennerström, H., *Langmuir*, **1993**, *9*, 779–785.

13 Craig, V.S.J., Ninham, B.W., Pashley, R.M., *Langmuir*, **1998**, *14*, 3326–3332.

14 Meyer, E.E., Lin, Q., Israelachvili, J.N., *Langmuir*, **2005**, *21*, 256–259.

15 Lin, Q., Meyer, E.E., Tadmor, M., Israelachvili, J.N., Kuhl, T.L., *Langmuir*, **2005**, *21*, 251–255.

16 Miranda, P.B., Shen, Y.R., *J. Phys. Chem. B* **1999**, *103*, 3292–3307; Du, Q., Freysz, E., Shen, Y.R., *Science*, **1994**, *264*, 826–828.

17 Wernet, Ph., Nordlund, D., Bergmann, U., Cavalleri, M., Odelius, M., Ogasawara, H., Näslund, L.Å., Hirsch, T.K., Ojamäe, L., Glatzel, P., Pettersson, L.G.M., Nilsson, A., *Science*, **2004**, *304*, 995–999.

18 Cavalleri, M., Thesis, Stockholm University, 2004.

19 Ludwig, R., *Angew. Chem. Int. Ed.*, **2001**, *40*, 1808–1827.

20 Lee, C.Y., McCammon, J.A., Rossky, P.J., *J. Chem. Phys.*, **1984**, *80*, 4448.

21 Forsman, J., Jönsson, B., Woodward, C.E., *J. Phys. Chem.*, **1996**, *100*, 15005.

22 Derjaguin, B.V., *Kolloid-Z.*, **1934**, *69*, 155.

23 Zhang, J., Yoon, R.-H., Mao, M., Ducker, W.A., *Langmuir*, **2005**, *21*, 5831–5841.

24 Tsao, Y., Yang, S.X., Evans, D.F., Wennerström, H., *Langmuir*, **1991**, *7*, 3154–3159.

25 Ederth, T., Liedberg, B., *Langmuir*, **2000**, *16*, 2177.

26 Ederth, T., Claesson, P.M., Liedberg, B., *Langmuir*, **1998**, *14*, 4782–4789.

27 Parker, J.L., Claesson, P.M., *Langmuir*, **1994**, *10*, 635–639.

28 Parker, J.L., Cho, D.L., Claesson P.M., *J. Phys. Chem.*, **1989**, *93*, 6121.

29 Wood, J., Sharma, R., *Langmuir*, **1994**, *10*, 2307–2310.

30 Cevc, G., Podgornik, R., Zeks, B., *Chem. Phys. Lett.*, **1982**, *91*, 193.

31 Eriksson, J.C., Ljunggren, S., Claesson, P.M., *J. Chem. Soc., Faraday Trans. 2*, **1989**, *85*, 163–176.

32 Eriksson, J.C., Henriksson, U., Kumpulainen, A., *Colloids Surf. A: Physicochem. Eng. Aspects*, **2006**, *282/283*, 79–83.

33 Claesson, P.M., Herder, P.C., Blom, C.E., Ninham, B.W., *J. Colloid Interface Sci.*, **1987**, *118*, 68.

34 Yoon, R.-H., Flinn, D.H., Rabinovich, Ya.I., *J. Colloid Interface Sci.*, **1997**, *185*, 363–370.

35 Parker, J.L., Claesson, P.M., Attard, P., *J. Phys. Chem.*, **1994**, *98*, 8468.

36 Epstein, P.S., Plesset, M.S., *J. Chem. Phys.*, **1950**, *18*, 1505.

37 Ljunggren, S., Eriksson, J.C., *Colloids Surf. A: Physicochem. Eng. Aspects*, **1997**, *129/130*, 151–155.

38 Kralchevsky, P.A., *Langmuir*, **1996**, *12*, 5951.

39 Eriksson, J.C., Ljunggren, S., *Colloids Surf. A: Physicochem. Eng. Aspects*, **1999**, *159*, 159–163.

40 Ryan, W.L., Hemmingsen, E.A., *J. Colloid Interface Sci.*, **1993**, *157*, 312–317.

41 Mao, M., Zhang, J., Yoon, R.-H., Ducker, W. A., *Langmuir,* **2004**, *20,* 1843–1849.

42 Guldbrand, L., Jönsson, B., Wennerström, H., Linse, P., *J. Phys. Chem.*, **1984**, *80,* 2221.

43 Kjellander, R., Marcelja, S., *Chem. Phys. Lett.*, **1984**, *112,* 49.

44 Forsman, J., Jönsson, B., Åkesson, T., *J. Phys. Chem.*, **1998**, *102,* 5082–5087.

45 Miklavic, S. J., Chan, D. Y. C., White, L. R., Healy, T. W., *J. Phys. Chem.*, **1994**, *98,* 9022–9032.

46 Kekicheff, P., Spalla, O., *Phys. Rev. Lett.*, **1995**, *75,* 1851–1854.

47 Pazhinur, R., Yoon, R.-H., *Miner. Metall. Process.*, **2003**, *20,* 178–184.

48 Herder, P. C., *J. Colloid Interface Sci.*, **1990**, *134,* 336.

49 Pauling, L., *The Nature of the Chemical Bond,* Cornell University Press, Ithaca, NY, 1960.

50 Franks, F., ed., *Water, a Comprehensive Treatment,* Plenum Press, New York, 1973.

51 Rabinovich, Ya., Guzonas, D. A., Yoon, R.-H., *Langmuir,* **1993**, *9,* 1168–1170.

52 Pashley, R. M., *J. Phys. Chem. B,* **2003**, *107,* 1714–1720.

53 Ederth, T., Thesis (Appendix), Royal Institute of Technology, Stockholm, 1999.

54 Israelachvili, J. N., *Intermolecular and Surface Forces,* 2nd edn., Academic Press, New York, 1991, Chapter 15.

55 Gibbs, J. W., *The Scientific Papers of J. Willard Gibbs,* Vol. I, Dover, New York, 1961, pp. 328–329.

6
Long-range Surface Forces in Molecular Liquids: Trends in the Theory

Ludmila Boinovich and Alexandre Emelyanenko

Abstract

This chapter presents a review of theoretical approaches to the calculation of surface forces determining the stability of thin liquid interlayers separating colloid particles or macroscopic bodies. We analyze briefly the limitations in the bases of the classical DLVO theory and consider the trends of its further development. Special attention is given to approaches in which the peculiarities of the molecular structure and properties of liquids in interlayers are accounted for.

6.1
Introduction

The quantitative analysis of surface forces acting between two macroscopic bodies, separated by a liquid interlayer, is a key point in the solution of many problems in colloid and interfacial science. The consideration of the liquid interlayer is equivalent, in essence, to the withdrawal of the short-range concept and to the transition from 2D to 3D analysis of the thermodynamics of heterogeneous systems. The fruitfulness of such 3D approach was demonstrated in studies by Derjaguin and his school (see [1–3] and references cited therein). As early as the 1930s it was proven by Derjaguin and coworkers [4, 5] that the consideration of the interaction between the approaching surfaces is equivalent to the analysis of disjoining pressure as a function of interlayer thickness. The thickness dependence of disjoining pressure is the principal thermodynamic characteristic of the thin interlayer distinguishing the thin film and the bulk state of a liquid. Hence, if the interlayer thickness is such that its intermediate part possesses bulk properties and, therefore, there is no overlapping of subsurface zones adjoining the phase boundaries, then the disjoining pressure in such an interlayer vanishes. In contrast, the overlapping of subsurface zones leads to anisotropy of the pressure tensor inside the interlayer and to the appearance of a separation-dependent disjoining pressure:

Colloids and Interface Science Series, Vol. 1
Colloid Stability: The Role of Surface Forces, Part I. Edited by Tharwat F. Tadros

ISBN: 978-3-527-31462-1

$$\Pi(h) = P_N(h) - P_0 \tag{1}$$

where $P_N(h)$ is normal to the interface component of the pressure tensor in a thin interlayer with thickness h and P_0 is the pressure in the bulk isotropic liquid being in equilibrium with the interlayer under consideration. The type of dependence of disjoining pressure on interlayer thickness reflects the physical nature of forces determining the interaction between the surfaces confining the interlayer. In turn, the nature of the above forces itself is defined by properties of the confining phases, the peculiarities of interactions between the liquid and the confining phases and the properties of the liquid constituting the interlayer. It is worth noting that the latter are thickness dependent owing to the interaction of liquid with the confining phases.

In contrast to the atomic liquids, the picture of interactions between confining surfaces in molecular liquids is much more complex in view of the above-mentioned factors. In this chapter, proceeding from the simplest to more complex models of liquids, we shall consider different long-range surface forces determining the interactions in systems containing thin interlayers of molecular liquids.

6.2 Molecular Forces

The mechanism of van der Waals or dispersion interactions through thin liquid interlayers was one of the first to be developed theoretically. We will not delve here into historical details, for which we address the reader to the excellent monograph by Derjaguin, Churaev and Muller [1], but just note that since the publication of the general quantitative theory of molecular forces by London in 1930 [7], further improvements of the theory of interaction between two macroscopic bodies separated by a liquid interlayer were made on the basis of two main approaches – a microscopic and a macroscopic. In the former, developed by de Boer [8] and Hamaker [9], the molecular interaction between two similar macroscopic bodies was calculated as the resultant of London forces between all pairs of molecules constituting the bodies. Details of calculations can be found either in the original papers [8, 9] or in more recent monographs [1, 10, 11]. In the frame of this approach, the analytical expressions were derived for interaction forces between two spheres, between a sphere and a flat surface and between two flat surfaces through a vacuum gap, both in the limit of small separations (non-retarded limit) and that of large separations (with accounting for electromagnetic retardation). Application of numerical methods in the frame of the microscopic approach allows the interaction forces between arbitrary-shaped bodies to be computed. The summation rule, introduced by Hamaker [9], allows the application of the microscopic approach to the interaction of condensed media not only vacuum separated, but also through the condensed matter interlayer. However, it should be noted that the contemporary condensed matter physics considers the additivity of London forces as a very rough approximation for the calculation of interaction forces between dense media. Hence the simplicity of the microscopic method of calculations does not expiate the quantitative (and for absorbing interlayers even the qualitative) inexactness of computed force.

The exact macroscopic approach for calculations of dispersion interactions between bodies separated either by a vacuum [12] or by an absorbing condensed phase interlayer [13, 14] was developed in a series of papers by Lifshitz and coworkers (DLP theory). In this approach, the macroscopic bodies separated by a thin interlayer are considered as continuous media, interacting by means of electromagnetic field fluctuations. It is important to note that the thickness of the interlayer separating the bodies is supposed to be large compared with the molecular scale. The calculations in [12] were based on the fact that the interaction force between two bodies in vacuum equals the *zz*-component of the averaged Maxwell stress tensor. To compute the latter, Lifshitz introduced into the Maxwell equations the sources of fluctuation field and applied to these sources the fluctuation dissipation theorem for field averaging. Later, it was shown [15] that for both a vacuum and a medium-filled gap between two macroscopic bodies the problem of finding the stress tensor might be reduced to computation of appropriate Green functions. Since the theory is based on the exact Maxwell equations, the retardation effects are accounted for automatically. In general case the disjoining pressure $\Pi(h)$, equal to the specific force of interaction between two macroscopic bodies through a uniform plane-parallel liquid interlayer having thickness h, might be computed as follows [12, 13]:

$$\Pi(h) = -\frac{kT}{\pi c^3}\sum_{N=0}^{\infty}{}' \xi_N^3 \varepsilon_3^{3/2} \int_1^{\infty} p^2 \left\{ \left[\frac{(s_1+p)(s_2+p)}{(s_1-p)(s_2-p)} \exp\left(\frac{2p\xi_N h \varepsilon_3^{1/2}}{c}\right) - 1 \right]^{-1} + \left[\frac{(s_1+ap)(s_2+bp)}{(s_1-ap)(s_2-bp)} \exp\left(\frac{2p\xi_N h \varepsilon_3^{1/2}}{c}\right) - 1 \right]^{-1} \right\} dp \qquad (2)$$

where k is Boltzmann's constant, T is the absolute temperature, p is an integration variable, $\xi_N = 2\pi kTN/\hbar$, with natural number N, $\hbar$ is Planck's constant, $s_1 = (a-1+p^2)^{1/2}$, $s_2 = (b-1+p^2)^{1/2}$, $a = \varepsilon_1/\varepsilon_3$, $b = \varepsilon_2/\varepsilon_3$ and $\varepsilon_1(i\xi_N)$, $\varepsilon_2(i\xi_N)$ and $\varepsilon_3(i\xi_N)$ are the frequency dependences of the dielectric permittivities of phases 1, 2 and separating interlayer 3, respectively. The prime on the summation sign indicates that the first term (with $N=0$) has to be taken half-weighted. Note that straightforward substitution of $N=0$ into Eq. (2) leads to uncertainty since ξ_N turns out to be zero, while the integral over dp diverges. However, the appropriate replacement of variables [13] reveals the finiteness of the term under consideration.

As follows from Eq. (2), to compute the disjoining pressure at any separation the only information one needs to know is the frequency dependences of the dielectric permittivities $\varepsilon_1(i\xi_N)$ for all contacting media. The latter can be determined on the basis of the Kramers-Kronig relation [16] and data on the imaginary part of the dielectric permittivity, measured for a wide frequency range from hertz to exahertz. The analysis of Eq. (2) shows that the dispersion interaction between two similar bodies is always attractive. In the case of non-symmetric systems, for instance wetting films, the net interaction might be either attractive or repulsive, being dependent on the relation between $\varepsilon_1(i\xi_N)$, $\varepsilon_2(i\xi_N)$ and $\varepsilon_3(i\xi_N)$. Equation (2) can be simplified by formally proceeding from summation over N to integration over ξ. However, this procedure may cause an error in the calculated interaction force, connected with the ignored contribution of the first term (with $N=0$) of the sum during integration. This error might be essential for the interlayers of polar liquids, possessing considerable absorbance in the region of low frequencies of electromagnetic vibrations. In the limiting cases of small and large separations, compared with the wavelength characteristic of absorbance spectra of contacting media, Eq. (2) reduces to relations derived on the basis of the microscopic approach for non-retarded and retarded forces, respectively.

An alternative macroscopic approach for calculation of dispersion interactions was developed by van Kampen et al. [17]. The method is based on the calculation of the excess free energy of normal modes for surface oscillators (in harmonic approximation) depending on the thickness of the interlayer separating the interacting surfaces. This approach seems to be simpler and more demonstrative than that used in DLP theory, but is justly mentioned in the literature as "heuristic and intuitive" [18]. Later, the justification of van Kampen et al.'s method with simultaneous analysis of its applicability to arbitrary media and for solving new problems was performed by Barash and Ginzburg [19].

Experimental investigations of molecular forces were conducted for different model systems in vacuum, in ambient air and in liquids. The detailed analysis of the main results is given in [1]. Without entering into details, we just note here that generally the interactions between macroscopic bodies separated by vacuum or an air gap are well described by the macroscopic theory of van der Waals forces, within 15–30% error. The divergence between experimentally measured and theoretically predicted forces observed for small gap thicknesses is usually attributed to surface roughness or adsorption or formation of oxide layers on the surfaces of interacting solid bodies (see [1] and references cited therein). In contrast, the interactions between solids separated by liquid interlayers often demonstrate behavior essentially different from that predicted by existing theories of van der Waals forces (see, for example, Figs. 6.1 and 6.2 and [22–25]). The reasons of such failure of above theories are obviously related to the assumptions, explicitly or implicitly put on their basis for consideration of liquid interlayers.

The detailed analysis of the basic assumptions of the DLP theory of molecular forces between macroscopic bodies [13, 14] was performed by Barash and Ginzburg [19]. In [19], the Green functions for the Maxwell equations were applied to account for the contribution of equilibrium longwave electromagnetic fluctuations into the free energy of the system comprising two macroscopic bodies sep-

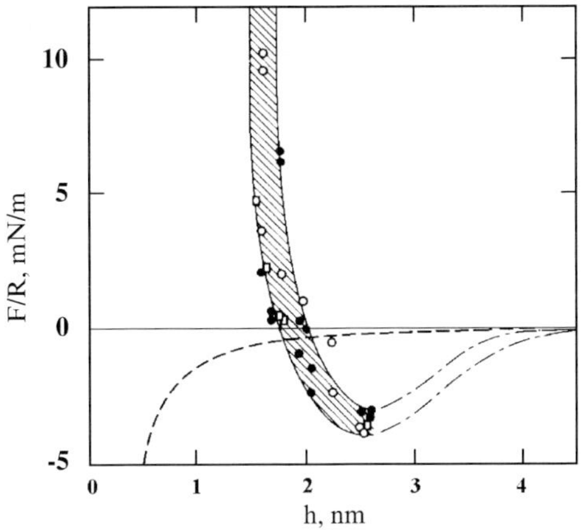

Fig. 6.1 Measured forces as a function of separation between mica sheets in white oil at 22 °C for different experimental conditions [20]. Closed symbols correspond to completely dry liquid in equilibrium with air dried over P_2O_5; open symbols represent measurements in dry nitrogen. Dashed line: van der Waals forces.

arated by a thin liquid interlayer. Note that a similar approach, based on Green functions, was also used in [14], but in [19] the authors employed the method of expanding the solutions of Maxwell equations over eigenfunctions of auxiliary systems. This procedure reduces the infinite series describing the system free energy to a fairly simple expression, suitable for further analysis. Such an analysis, thoroughly performed in [18], revealed the following limitations in the applicability of equations derived in [13] for forces of dispersion interaction between macroscopic bodies.

First, the results obtained in DLP theory are based on approximation of the negligible contribution of interaction between the longwave fluctuation field and the substance to the dielectric permittivity of the latter. That is, in essence, the DLP theory neglects the spatial dispersion for any arbitrary system. Second, the derivation of expressions for the system energy and the interaction forces, as given in DLP theory, implies that the densities and dielectric permittivities of bodies constituting the system do not change during their approach or separa-

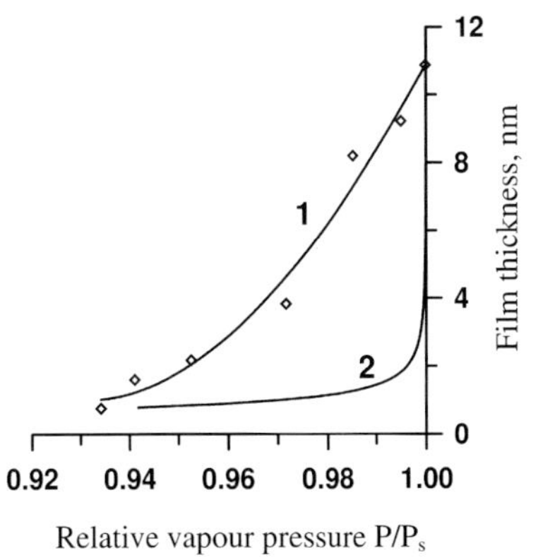

Fig. 6.2 Isotherm of adsorption of heptanol on optical glass K-8: (1) experimental data of Derjaguin and Zorin [21]; (2) theoretical curve for van der Waals interaction.

tion. This approximation corresponds to the usual formulation of the problem and is well satisfied for the interacting bodies themselves, whereas for thin interlayers not only it is not a strict one, but even ought to be inaccurate. Hence, just in the case when the dielectric permittivity of a liquid interlayer is dependent on its thickness (such a dependence should manifest, first of all, in the absorbance spectra of a liquid in the thin-film state) and/or in the case of considerable spatial dispersion in a liquid constituting the interlayer, the DLP approach in its classical form [13, 14] does not provide an adequate description of the real system stability. For complex molecular liquids, as a rule, both of above-mentioned factors apply, which leads to divergence between interaction forces measured experimentally and those predicted by the theory. Further, for interlayer thicknesses of several nanometers, the treatment of a molecular liquid as a continuous medium is not wholly correct, as the condition that the interlayer thickness is much larger than the molecular scale is no longer fulfilled. Finally, the fact is neglected that at the surfaces of interacting bodies there might exist subsurface layers with dielectric permittivities that are essentially different from the corresponding bulk properties used in calculations.

6.3 Ion–Electrostatic Interactions

The next mechanism we consider briefly is that determining the forces of electrostatic interactions in films of ionic liquids. These forces arise owing to overlapping of diffuse ionic atmospheres when two charged entities, separated by an interlayer of electrolyte solution, approach each other. The main difficulty in the calculation of those forces is the determination of the localization and concentration of ions in the gap between the approaching surfaces with account being taken of the deformation of ionic atmospheres when approaching each other. First rigorous calculations of the ion–electrostatic component of the disjoining pressure based on the solution of the Debye-Hückel equation for strong electrolytes were performed by Derjaguin [26]. In that work, the relations were derived for the forces of electrostatic interactions between two identical spheres or between two flat plates separated by a liquid interlayer. The results were obtained on the basis of the linearized Debye-Hückel equation and therefore were applicable for weakly charged sols only. The interaction of strongly charged particles in an electrolyte solution was considered by Derjaguin and Landau [27, 28]. They derived expressions for energies and forces of interactions for flat and convex surfaces under the condition of constant surface potential when they approached each other. On the basis of the solutions obtained, Derjaguin and Landau formulated the criterion of coagulation and validated the Hardy-Schultze rule, stating that the coagulating concentration of an electrolyte decays rapidly with increasing counter-ion valency. Almost at the same time, Verwey and Overbeek [29] independently considered the electrostatic interactions between similar flat surfaces for arbitrary potentials under the same condition of constant surface potential. Further, they had introduced the Stern correction, accounting for the existence of a dense part of the electric double layer (EDL). These studies [27–29] constituted the basis of the classical Derjaguin–Landau-Verwey-Overbeek (DLVO) theory of the stability of lyophobic colloids.

Since the mechanism of charge formation plays a key role in determining the boundary conditions for the electrostatic problem considered, further development of the theory of electrostatic forces between similar surfaces included both numerical and analytical solutions for different boundary conditions and for symmetric and non-symmetric electrolytes [1]. It is worth noting the useful feature of all solutions, which facilitates the evaluation of the magnitude of the ion–electrostatic component of the disjoining pressure for systems in which the mechanism of charge formation either is unknown or varies with the thickness of the interlayer. Namely, all the possible dependences of the disjoining pressure on the interlayer thickness for various boundary conditions fall within two limiting dependences, corresponding to the constant potential of boundaries of diffuse layers on the one hand and to the constant surface density of charge associated with the surface itself and the dense Stern monolayer on the other. The first condition provides a lower estimate of the attainable repulsive forces and the second a higher estimate.

Subsequently the calculations of the ion–electrostatic component of the disjoining pressure were generalized for the case of interactions between dissimilar particles. The role of such interactions is especially important for processes such as electrofiltration or electrophoretic precipitation in multicomponent suspensions, and also for interactions between an electrode and colloid particles. Derjaguin had proposed a graphical method of isodynamic curves [30], allowing one to assess the magnitude and the behavior of heterointeractions under conditions of constant potentials of diffuse ionic layers. The specific feature of the above interactions is that for two surfaces bearing charges of the same sign but of different surface density the ion–electrostatic forces are repulsive at large separations, but with the approach of two surfaces the repulsion goes through a maximum and then changes to attraction, which increases infinitely when the separation tends to zero. The analysis of interactions has shown that the maximal repulsion is determined by the lower of two surface potentials.

For the case of weakly charged surfaces, the analytical solution of the linearized Poisson–Boltzmann equation was obtained not only for the plane-parallel interlayers, but also for spherical particles with different radii. Thus, Hogg et al. [31] derived a simple relation for the energy of interaction between two spherical particles under the condition of constant surface potential, and Usui [32] solved the problem for the case of constant charge. The mixed condition of constant potential at one surface and constant charge at another for double-layer thicknesses that are small compared with the particle radii was analyzed by Kar et al. [33].

McCormack et al. [34] developed a general method for the numerical solution of the non-linear Poisson-Boltzmann equation, allowing one to perform computations of electrostatic interactions in plane-parallel interlayers of a symmetric electrolyte between dissimilar surfaces for any possible combination of constant charge/potential conditions at the approaching surfaces and without limitation on the magnitude of the surface charge density. It is worth emphasizing that the diversity of the force–distance dependences of the ion–electrostatic component of the disjoining pressure and the sensitivity of the above dependences to the boundary conditions are only manifested for those film thicknesses which are less than or comparable to the Debye screening length [1]. For larger interlayer thickness, the interaction decays exponentially with separation independent of the mechanism of charge formation, a fact which was repeatedly confirmed in the experiments (see, for instance, [35]).

The two above-considered mechanisms – of van der Waals forces and of electrostatic interactions – constituted the basis for constructing the classical DLVO theory of stability of lyophobic colloids. Namely, for lyophobic colloids, where the particles are presumed to interact weakly with the dispersing medium, the accounting for these two mechanisms is usually sufficient for the quantitative description of coagulation kinetics and the destabilizing influence of electrolyte additives. Seemingly the beginning of the 1960s, when the macroscopic DLP theory for van der Waals forces and the main approaches for calculations of electrostatic interactions of similar and dissimilar particles through uniform electrolyte interlayers were created, might be considered as the time of the final formulation of conventional DLVO theory. However, simultaneously with the development of the main statements of this theory, a number of experimental results were accumulated that did not quantitatively comply with, and sometimes even qualitatively contradicted, to the predictions of DLVO theory. As examples we can refer to the numerous studies of Derjaguin and his school [1], Exerowa and co-workers on studies of foam films [36–38], the results of Platikanov et al. on stability of wetting films [39, 40], unexpected signs in the interactions occurring between both dissimilar and identical interfaces obtained by Giesbers et al. [41], and many others.

6.4
Further Development of the Molecular Forces Theory

Nowadays it has become evident that main shortcomings of the DLVO theory follow directly from its advantages. In particular, just due to the continuum, without entering into details of molecular structure, consideration of liquids, the theory allows the calculation of surface forces on the basis of the macroscopic characteristics of the interacting media only. The attempts to improve the agreement between the theory and the experimental data led to the appearance of a variety of new mechanisms of long-range surface forces, which account for specific features of particular experimental systems and in the literature are frequently referred to as "non-DLVO forces".

Hence the non-uniformity of the liquid interlayer, arising due either to variation of density across the film or to monomolecular or diffuse adsorption of one of the solution components at the interfaces, might be accounted for in calculations of van der Waals interactions in the frame of multilayer models. Such an approach was adopted either with the approximation of constant dielectric properties within each layer [19, 42–45] or assuming different laws of their continuous variation across the layers [46, 47]. An alternative microscopic approach for the evaluation of the role of non-uniformity of the liquid interlayer was employed in [48–50] on the basis of molecular statistics. A method involving the thermodynamic consideration of diffuse adsorption layers and a macroscopic Lifshitz-type approach for describing the interaction between components of a solution and the confining phases was developed by Derjaguin and Churaev [51–53] and later generalized [54–58] for the case of solutions of polar solutes in non-polar solvents.

The spatial dispersion of dielectric permittivity arising in the interlayers of electrolyte solutions was considered in a series of papers by Smilga and Gorelkin [59–62] employing both the method of Bogolubov correlative functions and the method of Dzyaloshinsky et al. based on use of the Green temperature functions. An analytical solution was obtained for electrolytes modeled as a Debye plasma and as a collisionless plasma with mirror reflections at interfaces in the limiting case of uncharged interlayer surfaces and assuming a diagonal form of the dielectric permittivity tensor within the interlayer. The analysis of the solution obtained showed that the only significant deviation from Eq. (2) takes place with a zero frequency term. Further, the effect of spatial dispersion might be revealed in the appearance of the specific influence of boundaries on the dispersion law of interlayers. The latter circumstance is expected to be essential for films of specific solutions such as those of metals in ammonia for film thicknesses of a few nanometers [61, 62]. The zero frequency correction might noticeably change the molecular interaction forces for systems where the corresponding term is expected to be essential. Estimations [61, 62] show that the presence of dilute electrolyte solution leads to the screening of van der Waals interactions at distances of 10^{-7}–10^{-6} m. As a result, at such distances the molecular forces will decay exponentially, in contrast to conventional $1/h^4$ behavior as might be expected on the basis of Eq. (2).

It is interesting that the spatial dispersion in electrolyte solution emerges from the interactions of fluctuations of polarizations and ion–ion correlation-driven fluctuations of charges. The corresponding contribution of image charge forces to the interaction between two flat surfaces separated by an electrolyte interlayer was calculated [63–65] using anisotropic hypernetted chain closure for the Ornstein–Zernike equation.

6.5 Amendment of the Theory of Ion–Electrostatic Forces

The evolution of the conventional theory of electrostatic interactions between solid surfaces immersed in electrolyte solutions was directed to the relaxation of its main approximations such as continuity of charge distribution, condition of constant surface charge or potential at approaching surfaces and point-charge representation of ions in solution, to the consideration of other potentials (in addition to the Poisson potential), determining the ion distribution in the solution, and so on. In particular, Muller [66] has shown that the approaching of colloid particles is accompanied by the rearrangement of not only the diffuse charge, but also of the surface charge, with the exception of the chemisorbed part of the latter. He derived an exact parametric solution expressing the dependence of repulsive energy on the distance between two flat surfaces, using the Stern isotherm to describe the variation of ion adsorption with surface separation.

Buscall et al. [67] stressed that for small separations between charged surfaces, the condition of constant potential has to be replaced by a charge regulation model [68–71]. On the basis of the above model, Biesheuvel [72] calculated the correction to DLVO forces between hydrophilic surfaces in aqueous solution, assuming constant dielectric permittivity of the solvent and accounting for the existence of the Nernst layer of hydroxyl groups at the interface, which shifts the position of the plane of charge towards solution. This approach allowed the appearance of additional minimum between the DLVO-predicted primary and secondary minima on the energy–distance curve to be explained.

A general thermodynamic approach for considering the rearrangement of the charge distribution in a liquid interlayer with decreasing distance between two differently charged flat surfaces was presented by Lyklema and Duval [73]. In this approach, the charge regulation is assumed to occur at the Gouy-Stern layer. For different model systems, imitating the behavior of real complex systems, e.g. interactions of clay platelets and metal oxides, Lyklema and Duval obtained charge and potential distributions across plane-parallel interlayers and calculated the forces and energies of interaction between dissimilar surfaces. It was demonstrated that the approximation of a constant diffuse double-layer potential or charge works only at fairly large separation distances. At shorter distances, substantial deviations from both of the above limiting conditions were found. To describe the charge regulation processes taking place when the distance h separating the surfaces is reduced, Lyklema and Duval proposed to introduce the notion of regulation capacity, characterizing the ability of the interlayer to establish a new charge distribution without an essential change of surface potential. The higher are the regulation capacities, the better the interaction under the condition of constant potential is obeyed.

Another direction in the analysis of ion–electrostatic interactions includes accounting for different factors that determine the distribution of ions across the interlayer. The conventional theory of double-layer interaction, based on the classical form of the Poisson-Boltzmann equation, assumes that ions obey a Boltzmann distribution with a potential determined by the Poisson equation. This approach does not account for effects associated with the specific nature of ions in solution. In contrast, numerous experimental data on both the stability of disperse systems and measurements of surface forces through electrolyte interlayers indicate that electrostatic interactions depend not only on the ionic strength and ion valences, but also on the nature of cations and anions present in solution [74–76]. To explain the specific sequence of ions (Hofmeister series) according to their influence on the magnitude of double-layer interactions, it was proposed to include in the Poisson-Boltzmann equation additional potentials accounting for the nature of ion [77–83]. Thus, Ninham and Yaminsky [77] proposed to account for the potential of van der Waals forces between an ion and the rest of the system when analyzing the ion distribution across the system. Karraker and Radke [83] accounted for the ion-dispersion (van der Waals) potential and the potential of interaction between ions and their images in the vapor phase, inducing additional repulsion of ions from the vapor–liquid interface. Manciu and Ruckenstein [81, 82], in addition to the above-mentioned long-range ion-dispersion potential, considered the short-range ion-hydration potential, describing the difference in free energy of hydration of ions between the bulk solution and in the vicinity of the interface. The redistribution of ions in the interlayer under the influence of both the ion-dispersion and the ion-hydration potentials, being essentially dependent on the type of ions, leads to ion-specific changes of surface charges and potentials compared with those calculated on the basis of conventional theory and hence affects the double-layer interaction. It is interesting that despite the different choice of additional potentials, the authors of all the above-mentioned approaches were able to explain satisfactorily the experimental dependence of the surface tension of electrolyte solutions on their ionic strength [79, 81, 83].

Correlations between ions at opposite double layers and associated additional forces were considered in several studies [63, 84, 85]. The main contributions to excess energy, related to ion–ion correlations in the liquid interlayer, arise first of all due to correlations between fluctuations of charge density in double layers generated at the opposite surfaces of the interlayer. Second, accounting for the finite size of ions allows one to calculate the contribution to the energy evoked by additional ion core–core repulsion across the midplane. Third, image charge forces cause the deformation of diffuse ionic atmospheres with corresponding energetic effects. Finally, the deviation of ion concentration in the interlayer from that calculated on the basis of the conventional Poisson-Boltzmann approach, leads to a change in the Debye screening length. The analysis performed [63, 84, 85] indicates that the effect of ion–ion correlations usually manifests as an additional attraction between charged surfaces. This effect is most perceptible in solutions of di- and multivalent electrolytes, leading either to an essential decrease in double-layer repulsion or to a change in the sign of interaction. The size of the ion affects the interaction predominantly for small interlayer thicknesses, large ion radii and/or at high surface charges. The significance of ion–ion correlations for the stability of colloid solutions was proved experimentally by direct measurements of forces between mica surfaces in 0.15 M potassium chloride solution [86].

The renunciation of the model of uniformly distributed surface charge produced a series of models accounting for the discrete structure of the dense part of the double layer at the interface [87–92]. Thus, on the basis of a 2D modification of the Evald method, Zhiguleva and Smilga [90] analyzed the distribution of the electrostatic field strength and potential in EDL for different cases of regular charge distribution in a plain lattice. The results of their calculation of double-layer forces [90] have shown that for a wide class of problems the discreteness of charge distribution must be accounted for since the true interaction force might be as much as two orders of magnitude greater than the results obtained with the approximation of smeared charge. Derjaguin and Muller [93] have shown that in the case of non-localized adsorption of ions, the discreteness of charge distribution at the interface causes an electrostatically driven correlation of ion arrangement at the opposite film surfaces which should be taken into account. The calculations [93] indicate that such a correlation leads to the appearance of additional attractive forces, decaying as $1/h^4$. At small film thickness, these correlation forces might decrease the electrostatic repulsion and in some cases even lead to the transition from repulsion to attraction.

Basu and Sharma [94] have shown that additional repulsive forces, not accounted for in the DLVO theory, might originate from modified dielectric permittivity of the solvent near the charged interface. An attempt to account for the solvent structure in a thin interlayer through the analysis of the preferential orientation of dipole moments of solvent molecules (in continuum models such preferential orientation is interpreted as dielectric saturation) was made by Henderson and Lozada-Cassou [95]. They showed that such dielectric saturation is revealed in lowering of the static dielectric permittivity and leads to additional repulsion (above that predicted by DLVO theory) and hence it may be invoked to explain the hydration forces in electrolyte solutions. Paunov and coworkers [96–97] analyzed the role of the size of hydrated ions and demonstrated that the finiteness of ion size affects their distribution not only in the close vicinity of charged surface, but even beyond the EDL. Redistribution of ions across the film due to combined effects of the finiteness of ion size and the modified static dielectric permittivity of solvent requires the introduction of a correction in the calculations of the electrostatic component of the disjoining pressure. As was established in [96], with moderate overlapping of EDLs this correction provides additional repulsion at large separations and attraction for small separations.

The discreteness of the distribution of dipoles at interfaces has been considered by different workers [54–58, 87, 98–101]. Jonsson and Wennerstrom [98] analyzed the contribution of dipole image forces to the interactions between surfaces covered by surfactants with dipolar head groups in the framework of the continuum electrostatic model. They found that the above contribution caused the repulsive forces which could be responsible for the swelling of phospholipid bilayers in aqueous solutions. Kjellander and Marcelja [99], on the basis of the hypernetted chain approximation, showed that for dipoles oriented normally to the interface, correlation attraction between dipoles at opposite interlayer boundaries prevails over their repulsion due to image forces. Similar results were obtained by methods of perturbation theory by Attard and Mitchell [100] and Jonsson et al. [101] and on the basis of electrostatics methods by Martynov and Smilga [87]. Jonsson et al. [101] demonstrated that accounting for orientational rotation of dipoles in the frames of the longwave approximation of the perturbation theory allows one to obtain both the attraction and the repulsion between the dipole layers. They noted, however, that the magnitude of repulsion attainable is essentially lower than the van der Waals attraction in a symmetric system. The considerations in the above studies [87, 98–101] was restricted to the case of symmetric systems. Further, neither the structure of dipole molecules nor the dependence of adsorption on interlayer thickness was taken into account.

6.6 Electrostatic Interactions in Non-polar Media

A peculiarity of ion–electrostatic interactions in non-polar media is that usually bulk concentrations of ions are rather small and, consequently, the Debye screening length is much longer than the interlayer thickness. Therefore, the interaction between charged surfaces reduces to Coulomb forces. The detailed analysis of interactions between two charged colloid particles, between a particle and an electrode and between two polarizable particles in a non-uniform electric field, in the approximation of smeared surface charge, was considered in a recent review by Adamczyk [102] and therefore we shall not dwell on this question here. Instead, we shall discuss briefly the features of surface–surface interactions through a non-polar medium under the condition of a discrete dipole distribution. That case is of special importance when considering the stability of suspensions in solutions of polar substances in a non-polar solvent. The discrete character of the spatial distribution of dipole moments of polar molecules, both over the interface surfaces and across the interlayer, causes the emergence of different kinds of image forces and of dipole–dipole interactions inside the interlayer. The additional forces of interaction between confining phases, associated with the above-mentioned phenomena, were analyzed in a series of papers by Boinovich and Emelyanenko [54–58]. They considered both symmetric and non-symmetric systems, the latter being characterized by different amounts (and sometimes even types) of adsorption on the opposite boundaries of the interlayer. For evaluation of equilibrium adsorption, not only the specific potential of interaction with active centers was accounted for, but also the potentials of image forces and of the dispersion interaction between polar adsorbing molecules and confining phases. The last two potentials induce an essential dependence of the equilibrium amount of adsorbed molecules on the thickness of the liquid interlayer. It was found that even in the case of very dilute solutions the electrostatic interactions between the lattices of adsorbed dipoles and their images in confining phases can generate a significant contribution to the disjoining pressure. This contribution decays approximately as $1/h^4$ and might be either repulsive or attractive depending on the variation of adsorption versus interlayer thickness, the relation between dielectric permittivities of contacting media and the strength of the dispersion interaction between dipole molecules and confining media.

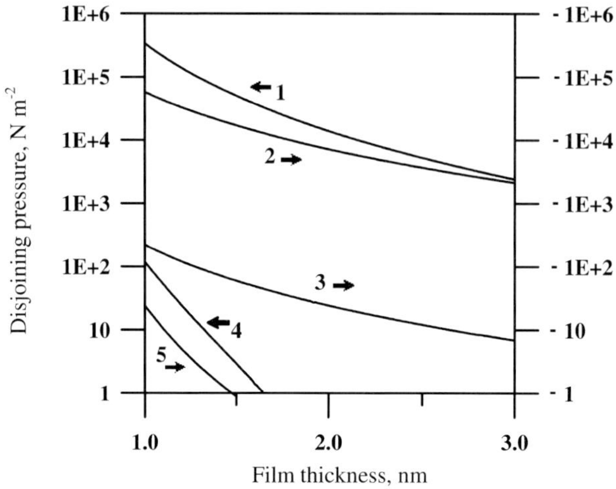

Fig. 6.3 Dependences of different components of disjoining pressure acting in wetting films of pentanol solution in pentane on a water surface versus the film thickness: (1) contribution due to interaction between the adsorbed monolayer of dipole molecules and its images; (2) molecular component; (3) adsorption component; (4) image forces contribution related to diffusely distributed pentanol molecules; (5) correlation attraction of adsorbed monolayers.

An analysis of the relations derived and the results of numerical calculation performed [54–58] for adsorption monolayers with different dipole moments and orientations of dipoles at the interface have shown that in asymmetric thin interlayers (for example, in wetting films) for both the planar and the normal to the interface orientation of dipole moments in the adsorbed monolayer the interaction between that dipole monolayer and its image can dominate over all the other types of interactions, including the interactions between diffuse adsorption layers, the forces of correlation attraction between the monolayers adsorbed on the opposite boundaries and the dispersion forces (Fig. 6.3). However, in view of the fact that the dispersion interaction decays with the film thickness approximately as $1/h^3$, it became to prevail over image forces at interlayers thicker than 2–3 nm.

6.7
Forces Due to Modified Structure of Liquid in the Interlayer

6.7.1
Treatment of Liquid Within a Continuum Approach

For the first time, the supposition that the structure of a liquid in a thin interlayer differs from that in the bulk phase and that such a difference might induce long-range surface forces, was put forward by Derjaguin and Kusakov in the 1940s [5]. Later, this supposition was repeatedly used with success in the analysis of the nature of different forces – hydration, solvation, structural, hydrophobic and so on. In such an analysis, different forces were associated with the deviation of the corresponding parameters of the structure in thin interlayers from their bulk values. One of the first theoretical attempts to consider processes occurring in a polar solvent under the influence of interlayer confinement was made by Marcelja and Radic [103]. They used Landau expansion of the density of free energy over the order parameter and showed that with moderate variations of the order parameter, accounting only for the quadratic with respect to the order parameter and its gradient terms of expansion, allows one to describe the exponentially decaying repulsion between the confining surfaces. In subsequent papers by many authors the order parameter was interpreted as a local water polarization [104–106].

The idea of surface-induced polarization of water molecules as a cause of hydration forces was further developed in [107–111]. Kornyshev and Leikin [109, 110] considered two mechanisms of polarization of water molecules near the interface. The first is associated with the contribution of interaction between oppositely charged groups of one surface of a lipid membrane (being electrically neutral as a whole) and their images in non-polar media, confining the water interlayer from the opposite side. The non-local electrostatics were used to take into account spatial correlations of polarization fluctuations in water under the influence of the charge distribution in neutral lipid membranes [109]. It was shown that the lateral ordering of polar groups along the interacting surfaces played a dominant role in the determination of interaction behavior. The chemical mechanism of water ordering considered by means of the free-field model with phenomenological order parameter allowed the difference in the water states near the polar surfaces and in the bulk to be analyzed [110]. Both the free-field and non-local electrostatic approaches predicted the exponentially decaying forces, determined by the lateral ordering of polar groups along the interacting surfaces of lipid membranes. Both the pre-exponential factors and the correlation lengths are dependent on the nature and the structural properties of surfaces confining the interlayer.

The combination of the Cahn approach [112], developed for the analysis of wetting transitions at an interface, with the ideas of Marcelja and Radic [103] was implemented in the framework of gradient theory by Mitlin and Sharma [113], who considered the interaction between two flat surfaces separated by a liquid interlayer. They employed a new analytical method allowing them to derive the asymptotically exact dependences of disjoining pressure, free energy and the profile of the order parameter as functions of interlayer thickness, for any arbitrary dependence of free energy on order parameter. In the general case, the disjoining pressure might be described by a set of exponential contributions with characteristic lengths proportional to the correlation length for the bulk liquid. The pre-exponential factors depend on the deviation of free energy density from the quadratic expansion and on the parameters of interaction between molecules of liquids and both of the confining surfaces. The net interaction forces might be either attractive or repulsive, with possible inversion of the interaction sign with change of the separation.

Another attempt to account for the interaction between the liquid and the confining surfaces in calculations of forces additional to DLVO was made by Kuklin [114]. In this approach, the free energy of the system was described by the harmonic approximation of the Landau–Ginzburg functional, dependent on the order parameter. In addition to those terms of expansion considered by Marcelja and Radic [103] and Kornyshev and Leikin [110], Kuklin had included functions describing the direct short-range interaction of the outermost molecular layers with confining solid surfaces. His analysis [114] indicated that in the general case of dissimilar confining surfaces their interaction might be either attractive or repulsive and is determined by four scaling parameters including two field parameters and two extrapolation lengths, characterizing the depth to which the electric field of each charged surface penetrates into the liquid interlayer. The dependence of interaction forces versus separation can be (for the specific choice of parameters and range of separations) close to the exponential one.

The approach developed by Eriksson et al. [115] conceptually is close to that of Marcelja and Radic [103]. However, in the former study, the role of the order parameter is a local relative increase in the number of hydrogen bonds per water molecule compared with that in a bulk solution. By expanding the free energy of the system (consisting of an aqueous interlayer confined between two hydrophobic surfaces) into a series over order parameter and performing minimization, Eriksson et al. concluded that an increase in the order parameter in the vicinity of the hydrophobic wall induced long-range attractive forces. However, the range of efficiency of the expected forces should be much shorter than that obtained experimentally for hydrophobic attraction.

Another possible mechanism of hydrophobic forces, with more long-range character and caused by the creation of a vapor- or air-filled cavity between the interacting lyophobic particles, was proposed by Yushchenko et al. [116]. The forces of hydrophobic attraction arising from such cavity formation are capillary. However, the above mechanism might be realized under the following conditions: the contact angle must be $>90^\circ$ and the cavity can be formed only after bringing the solid surfaces into contact. The evaluations made in [116] had shown that the cavity would be thermodynamically stable in the process of separation of contacted surfaces, whereas before first bringing them into contact the energetic barrier for cavity formation amounts to several tens of kT. Apparently just this circumstance explains the fact that in real experiments such a cavity was only observed after the preliminary contact of the surfaces in water [117]. It should be noted that hydrophobic attraction at large separation was also observed without detectable cavity formation [118, 119]. The mechanism invoked [118, 119] for the explanation of experimental data was based on the hypothesis that, prior to the formation of a vapor-filled cavity, the nucleus state might exist in the liquid interlayer. The different mechanisms leading to the appearance of long-range hydrophobic forces and associated with the formation of cavities or microbubbles between the interacting surfaces have also been discussed [120–123].

6.7.2
Accounting for a Discreteness of Liquid Structure

In the papers considered above, the solvent or the liquid constituting the liquid interlayer was considered in the continuum approximation. However, there is extensive literature [124–140] based on either the Lennard-Jones fluid model or the hard sphere model, where the main attention is given to the discreteness of the liquid in the interlayer. For these models, both the analytical procedures and the computer simulations reproduce well the oscillating regime of interaction forces, in agreement with experimental findings for liquid interlayers of different nature (see, for example, [11, 22]). The essential problem for the theoretical calculation of the interaction between two rigid walls immersed in a model liquid is the necessity to meet the condition of constant chemical potential of the liquid in the interlayer in the process of thickness variation. Whereas in numerical Monte Carlo simulations this condition is realized spontaneously in the frame of the grand canonic ensemble [124], it is difficult to be satisfied in calculations based on analytical theories. The rare exceptions are approximate theories, for instance, different versions of mean field theory [125, 126] or simple integral equation theories [127, 128] for non-uniform systems, in which the binary distribution functions of the bulk phase are used as a closure for the unary distribution function of the confined phase. However, the results of the latter are frequently incorrect because the isotropic bulk binary distribution function cannot account adequately for the anisotropic nature of the corresponding distribution in a thin interlayer. For that reason, several approaches providing a more reasonable description of anisotropy of the liquid in the interlayer were advanced. Kjellander and Sarman [131] proposed to apply anisotropic Percus-Yevick closure in the frame of integral equation theory based on the exact Ornstein-Zernike equation for an anisotropic binary correlation function. Attard and co-workers [132–134] have shown that different versions of the hypernetted chain (HNC) approximation applied to solve the Ornstein-Zernike equation allow adequate reproduction of oscillation forces in the frame of both the hard sphere model and the Lennard-Jones fluid model. It should be mentioned that some authors have express doubts about the correctness of the HNC approximation for dense fluids with short-range potentials [131].

Chandler and Andersen [135] developed the reference interacting site model (RISM), which can be considered as a molecular analog of the hard sphere Percus-Yevick approximation for simple fluids. In this model, designed for the investigation of molecular liquids with non-spherical molecules, it is assumed that equilibrium correlations between molecules in the dense phase are determined by the short-range repulsive part of the intermolecular potential. Grimson and Richmond [136] used the RISM model and the integral equation theory to compute the solvation forces for the interlayer of a diatomic liquid. Their results indicate an essential decrease in the amplitude of oscillations on transition from a hard sphere liquid to the RISM model, which explicitly accounts for the asymmetry of liquid molecules (Fig. 6.4).

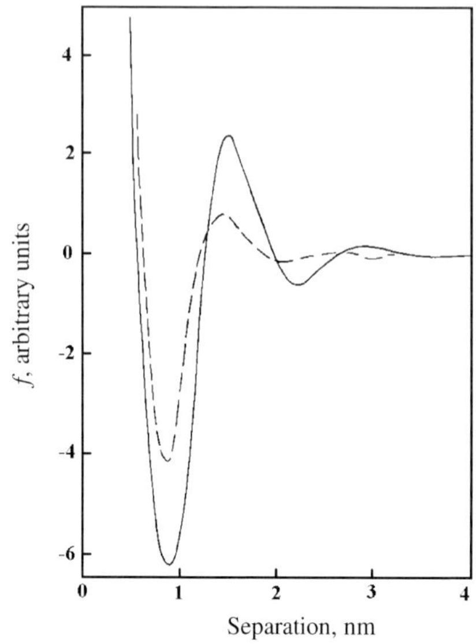

Fig. 6.4 Reduced force of interaction between two interfaces versus reduced separation for fluids with relative density 0.3584 [136]. Dashed line, for diatomic fluid with interatomic distance equal to 0.6σ (σ is the diameter of the atom); solid line, for hard sphere fluid with sphere diameter equal to 1.2146σ.

6.7.3
Phonon Mechanism of Long-range Forces

So far, the structuring of a liquid was interpreted as static deviations of parameters such as local polarization, intermolecular distances and orientations of molecules with respect to the interface from the corresponding bulk values. Additional surface forces, generated owing to that deviation, were attributed to the overlapping of boundary layers with the modified structure. Another approach to the problem of structuring of a liquid in a confined system was developed in a series of our studies [141–147]. This approach is based on the consideration of the influence exerted by confining phases on the dynamics of inter- and intramolecular motions of atoms and molecules in a thin interlayer.

The dynamic structure of a liquid is characterized by a spectrum of normal modes, associated with the acoustic waves in a thin interlayer. These modes depend essentially on both the interlayer thickness and the surface-induced alterations of intermolecular interactions. Each normal mode is determined by the parameters of the oscillator involved in the vibrational (rotational) motion, and also by its interaction with neighboring oscillators, whereas the probability of realization of a given mode, $\exp[-h/L(\omega)]$, depends on the ratio of the thickness, h, of the liquid interlayer to the free path length, L, of the corresponding phonon. The differences in dynamic structure in the bulk liquid and in the confined liquid are revealed explicitly, for example, in the absorbance spectra of the liquid in the corresponding state (Fig. 6.5). For thin liquid interlayers such differences lead to thickness-dependent contributions to the excess free energy of the liquid interlayer. The relation describing the corresponding contribution of

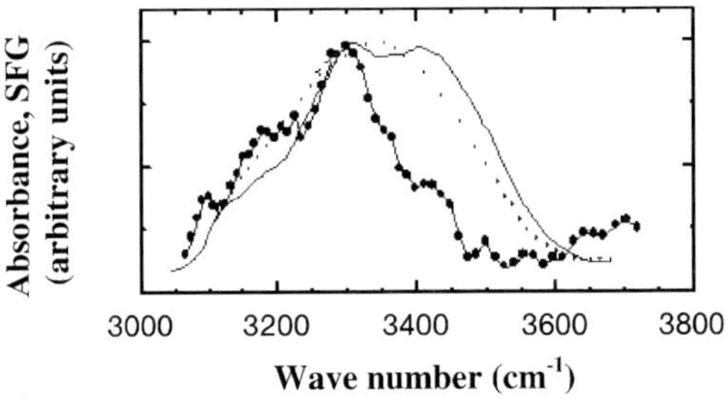

Fig. 6.5 Infrared spectra for ethanol in the OH stretching region. The dotted curve is the bulk liquid phase; the solid curve corresponds to a 25-nm film of liquid ethanol confined by fluorite plates [162]. The circles are sum frequenz generation (SFG) spectra for the ethanol–air interface [160].

collective vibrational excitations to the disjoining pressure of the liquid interlayer was derived [141] within the harmonic approximation, that is, neglecting the phonon–phonon interaction:

$$\Pi(h) = \sum_j \int_0^\infty \left[-\frac{dK_j}{dh} - \frac{K_j h}{L_j^2(\omega)} \frac{dL_j(\omega)}{dh} + \frac{K_j}{L_j(\omega)} \right] \exp\left[-h/L_j(\omega)\right] d\omega \tag{3}$$

where

$$K_j = Z_j(\omega) kT \ln \left[\frac{\mathrm{sh}(\hbar\omega/2kT)}{\mathrm{sh}(\hbar\omega_{j0}/2kT)} \right]$$

$Z_j(\omega)$ is the configuration average density of vibrational states for oscillators of the jth kind, obtained by the diagonalization of dynamic matrices for instantaneous states of liquid film with thickness h, ω_{j0} is eigenfrequency of corresponding vibration in bulk liquid, $L_j(\omega)$ is the free path of a j-type phonon with frequency ω and the summation is performed over all types of oscillators (j) associated with the branches of dispersion curve. The product $Z_j(\omega)\exp(-H/L_j)$ plays the role of the density of phonon states in the liquid film, taking into account the finite time of mode relaxation. It should be noted that the contribution of acoustic branches of collective vibrations, that is, waves of density, including the hydrodynamic waves, have been analyzed in several studies by different approaches [141, 148–150], starting from the classical work of Dzyaloshinsky et al. [14]. The numerical estimations made in these papers indicated negligibly small contributions of these types of phonons to both the excess free energy of the film and the interaction forces between the confining surfaces. In contrast, the contribution of optical phonons [141, 146–147] to the disjoining pressure for associated molecular liquids might dominate over contributions of other types of long-range surface forces.

The three terms in the first set of square brackets in Eq. (3) reflect three regimes on the disjoining pressure–distance curve. The first plays the prevailing role at small interlayer thickness, when the discreteness of liquid structure has to be accounted for. The discreteness causes oscillations of force constants of the dynamic matrix, which, in turn, generate oscillations of density of vibrational states and, as a result, oscillations of both the excess free energy and the disjoining pressure are observed. The invoking of the phonon mechanism for the analysis of oscillating (solvation) forces in thin films [145] allowed us to explain many experimental findings such as the non-sinusoidal character of oscillations, variation of the oscillation period with film thickness and close values of the oscillation period for different liquids. It is worth stressing that the phonon approach predicts the oscillations not around the zero force, but with respect to an exponentially decaying function, described by the last term in the brackets in Eq. (3).

In the intermediate thickness range when the contribution of oscillations of density of vibrational states vanishes, but it is not yet possible to neglect the variation of free path lengths of phonons with film thickness, the second term in the first set of brackets in Eq. (3) starts to dominate. Its sign and magnitude are determined by the derivative dL/dh, which depends mainly on the character of the interaction between the oscillator and the confining phases. The role of this factor was discussed in detail in [147]. In particular, it was shown that a change of surface wettability causes a redistribution of density of vibrational states over frequencies and may lead to the alteration of both the magnitude and the sign of the phonon component of the disjoining pressure.

The final term prevails for the films thick enough when values of L and K become almost independent of thickness. For that case, the phonon mechanism provides the exponential decay of disjoining pressure with film thickness. In the general case for a wide range of thicknesses, the phonon component of the disjoining pressure is expressed, as a rule, by the sum of several exponentially decaying contributions. This type of behavior correlates well with numerous experimental data (see, for example, [1] and references cited therein).

In the framework of the analysis of dynamic structure in films of molecular liquids [146], the various aspects of the temperature dependence of oscillating and exponentially decaying forces found a consistent explanation in accordance with numerous experimental and simulation data [151–157].

Hence the phonon mechanism of surface forces evidently follows from accounting for the peculiarities of the structure of the molecules, constituting the liquid interlayer. This mechanism relies on the consideration of variations in the frequency spectra of liquids under confinement. Such variations were repeatedly observed in both the real and numeric experiments (see, for instance, [158–162]). The approach described above provides a deeper understanding of the dynamic processes taking place in liquids due to confinement, allowing one to evaluate the role of the interaction between a liquid and the confining phases in the formation of dynamic structure. It is significant to note that in the frame of the phonon approach it is possible to describe successfully both the oscillating (solvation) forces and the different manifestations of exponentially decaying forces, such as the hydration repulsion between hydrophilic surfaces and the hydrophobic attraction.

6.8
Forces Due to High Molecular Weight Polymers and Chain Surfactants

One more mechanism of forces not accounted for in the DLVO theory but of great practical importance is associated with the forces arising from presence of high molecular weight polymers or chain surfactants in the thin liquid interlayer. The main factor determining the additional forces compared with the DLVO theory in that case is the character of the interaction between the polymer molecules and the surfaces of particles separated by a liquid interlayer. Three distinctive cases deserve consideration: polymers adsorb on the particles reversibly or irreversibly or do not adsorb at all. For the estimation of the sign, magnitude and range of action of the forces under consideration it is also important to take into account the features of the interaction between polymer and solvent, the surface coverage and the polymer concentration.

The models describing the interaction between particles covered by irreversibly adsorbed polymers were proposed by de Gennes [163] and by Fleer and co-workers [164–166]. The former approach is based on methods of scaling theory and is restricted to the case when polymer molecules occupy only part of the accessible adsorption sites on the particle surface (case of weak coupling), while the adsorption energy of each polymer chain is large. The latter approach employs self-consistent field theory and is not constrained by a low adsorption condition. Both approaches are valid in the case when the velocity of surface approach is high compared with the rate of establishing the adsorption equilibrium and at the same time is small compared with the rate of conformational equilibration within the adsorbed layer. Several mechanisms of forces are possible in that case. At low polymer concentration in a good solvent, the approach of the surfaces under the condition of constant surface coverage results in the appearance of repulsive forces decaying inversely as the cube of separation at distances comparable to double the thickness of the adsorption layer [167]. The increase in polymer concentration causes augmentation of the surface coverage and in the case of chemisorption leads to the formation of brushes with high grafting density and a predominantly normal to interface orientation of the polymer chains. The interaction of surfaces is determined here by two factors. First, the elastic energy of initially stretched chains of brushed structure decreases with decreasing separation, providing a negative (attractive) contribution to the disjoining pressure, which decays as $h^{3/4}$. Second, the osmotic contribution, associated with changes in mixing entropy and solvation energy of polymer segments, and also steric effects, related to the restriction in the number of possible polymer conformations in a narrow gap, induce a repulsion proportional to $h^{-9/4}$. The net force is determined by the competition between the above-mentioned factors and by the interlayer thickness, normalized to the thickness of the adsorption layer. The numerous experimental data indicate the dominant role of the steric effects (see, for example, [168, 169]). It is interesting that well before the formulation of theories of the steric component of the disjoining pressure, the idea on elastic repulsion of polymer- or surfactant-adsorbed layers was invoked by Rehbinder [170] for explanation of the effect of polymers and surfactants on colloid stability.

For the case of reversibly adsorbed polymers (as a rule, of homopolymers), both the de Gennes [167] and Fleer et al. [166] approaches predict the attractive forces due to bridging attraction, when under conditions of approaching equilibrium and the possibility of free desorption, the polymer chains from one particle may establish bridges with another one.

The effect of non-adsorbing polymers on the interaction between flat surfaces and between spherical particles has been considered in numerous studies (see, for example, [171–175]). The sign and strength of interaction depend on the separation, the presence of charge and the shape of polymer molecules. Thus, the approaches of Asakura and Oosawa [172–173], Feigin and Napper [174] and Walz and Sharma [175] have shown that a reduction in the distance between the macroscopic surfaces through the solution interlayer until separations shorter than the size of the free polymer molecules results in lowering of the polymer concentration in the interlayer compared with that in the bulk solution. The osmotic pressure arising gives rise to an attractive interaction referred to as depletion forces. The magnitude of these forces increases with increase in polymer concentration [166, 172, 174, 175], with transition from sphere-like to chain- or asymmetrically shaped polymer molecules [172, 173], with an increase in charge carried by the polymer molecules [173, 175] and when the polymer molecules with higher molecular weight than that of the free polymer are adsorbed at the particle–solution interface [174]. In contrast, at large distances between interacting particles, their approach causes an increase in polymer concentration in the interlayer due to squeezing of the solvent. This results in repulsive forces. It should be noted that repulsion at large separations is always weaker than attraction at small h. Nevertheless, the increase in long-range repulsion with increase in concentration of a charged polymer provides stability of suspensions at polymer concentrations >1 vol.% [175].

6.9 Conclusions

We have briefly reviewed the basic theoretical approaches for the calculation of long-range surface forces. On the one hand, following the established tradition, and on the other, for the sake of deeper insight into the physical nature of the analysed phenomena, different physical mechanisms determining these forces were considered separately. Evidently, such an approach for many systems could be oversimplified and rough [77, 80] and might lead to double accounting for the effect of the same characteristic feature on the stability of colloid system. At the same time, the different approaches described here frequently do not exclude but rather complement each other, as occurs in calculations of ion–electrostatic interactions. Therefore, in performing the additive summation of different components of the disjoining pressure one needs to have a clear comprehension of specific features of the system accounted for by each contribution. Thus, for example, the frequencies of inter- and intramolecular vibrations in a thin interlayer are directly related to parameters of static structure, e.g. mutual molecular orientations and intermolecular distances. Therefore, in the frame of the phonon mechanism of disjoining pressure the changes in static structure will be accounted for spontaneously.

Nevertheless, we believe that the main advantage of separate consideration is the possibility of efficient selection of a model system for testing the applicability of one or another theoretical approach for the explanation of the physical mechanism of forces acting in a colloid system. Finally, in the frame of their correctness, the different approaches for calculations of the same kind of forces should lead to equivalent results. This provides the researcher with a freedom of choice of an appropriate theoretical approach for calculations, depending on the available parameters of the system under investigation.

References

1 B.V. Derjaguin, N.V. Churaev, V.M. Muller, *Surface Forces*, Consultants Bureau, Plenum Press, New York, **1987**.
2 B.V. Derjaguin, N.V. Churaev, *Wetting Films*, Nauka, Moscow, **1984** (in Russian).
3 B.V. Derjaguin, *Theory of Stability of Colloids and Thin Films*, Plenum Press, New York, **1989**.
4 B.V. Derjaguin, E. Obukhov, *Colloid J.*, **1935**, *1*, 385–398
5 B.V. Derjaguin, M.M. Kusakov, *Izv. Acad. Nauk SSSR*, **1937**, *5*, 1119; B.V. Derjaguin, M.M. Kusakov, *Acta Physicochim. URSS*, **1939**, *10*, 25.
6 B.V. Derjaguin, M.M. Kusakov, *Acta Physicochim. URSS*, **1939**, *10*, 153.
7 F. London, *Z. Phys.*, **1930**, *63*, 245.
8 I.H. de Boer, *Trans. Faraday Soc.*, **1936**, *32*, 1638
9 H.C. Hamaker, *Physica*, **1937**, *4*, 1058.
10 J. Mahanty, B.W. Ninham, *Dispersion Forces*, Academic Press, London, **1976**.
11 J. Israelachvili, *Intermolecular and Surface Forces*, Academic Press, London, **1992**.
12 E.M. Lifshitz, *Zh. Eksp. Teor. Phys.*, **1955**, *29*, 94–110 (in Russian).
13 I.E. Dzyaloshinsky, E.M. Lifshitz, L.P. Pitaevsky, *Sov. Phys. Usp.*, **1961**, *73*, 381.
14 I.E. Dzyaloshinsky, E.M. Lifshitz, L.P. Pitaevsky, *Zh. Eksp. Teor. Phys.*, **1959**, *37*, 229.
15 M.L. Levin, S.M. Rytov, *Theory of Equilibrium Thermal Fluctuations in Electrodynamics*, Nauka, Moscow, **1967** (in Russian).
16 L.D. Landau, E.M. Lifshitz, *Electrodynamics of Continuous Media*, Gostekhtheorizdat, Moscow, **1957** (in Russian).
17 N.G. van Kampen, B.R.A. Nijboer, K. Schram, *Phys. Lett. A*, **1968**, *26*, 307.
18 B.W. Ninham, V.A. Parsegian, G.H. Weiss, *J. Stat. Phys.*, **1970**, *2*, 323.
19 Yu.S. Barash, V.L. Ginzburg, *Sov. Phys. Usp.*, **1975**, *116*, 5.
20 J. Israelachvili, S.J. Kott, M.L. Gee, T.A. Witten, *Macromolecules*, **1989**, *22*, 4247.
21 B.V. Derjaguin, Z.M. Zorin, *Zh. Fiz. Khim.*, **1955**, *29*, 1755.
22 H.K. Christenson, *J. Dispers. Sci. Technol.*, **1988**, *9*, 171.

23 T. Nakada, S. Miyashita, G. Sazaki, H. Komatsu, A.A. Chernov, *Jpn. J. Appl. Phys.*, **1996**, *35*, 52.
24 C.E. Herder, B.W. Ninham, H.K. Christenson, *J. Phys. Chem.*, **1989**, *90*, 5801.
25 R.G. Horn, J.N. Israelachvili, E. Perez, *J. Phys.*, **1981**, *42*, 39.
26 B.V. Derjaguin, *Bull. Acad. Sci. URSS*, **1937**, *5*, 1153.
27 B.V. Derjaguin, L.D. Landau, *Acta Physicochim. URSS*, **1941**, *14*, 633.
28 B.V. Derjaguin, L.D. Landau, *Zh. Eksp. Teor. Phys.*, **1945**, *15*, 663.
29 E.J.W. Verwey, J.Th.G. Overbeek, *Theory of the Stability of Lyophobic Colloids*, Elsevier, Amsterdam, **1948**.
30 B.V. Derjaguin, *Discuss. Faraday Soc.*, **1954**, *18*, 85.
31 R. Hogg, T.W. Healy, D.W. Fuerstenau, *Trans. Faraday Soc.*, **1966**, *62*, 1638.
32 S. Usui, *J. Colloid Interface Sci.*, **1973**, *44*, 107.
33 G. Kar, S. Chander, T.S. Mika, *J. Colloid Interface Sci.*, **1973**, *44*, 347.
34 D. McCormack, S.L. Carnie, D.Y.C. Chan, *J. Colloid Interface Sci.*, **1995**, *169*, 177.
35 S.G. Flicker, J.L. Tipa, S.G. Bike, *J. Colloid Interface Sci.*, **1993**, *158*, 317.
36 D. Exerowa, P.M. Kruglyakov, *Foam and Foam Films*, Elsevier, Amsterdam, **1998**.
37 R. Sedew, D. Exerowa, *Adv. Colloid Interface Sci.*, **1999**, *83*, 111.
38 R. Cohen, G. Ozdemir, D. Exerowa, *Colloids Surf. B*, **2003**, *29*, 197.
39 B. Diakova, C. Filiatre, D. Platikanov, A. Foissy, M. Kaisheva, *Adv. Colloid Interface Sci.*, **2002**, *96*, 193.
40 B. Diakova, M. Kaisheva, D. Platikanov, *Colloids Surf. A*, **2001**, *190*, 61.
41 M. Giesbers, J.M. Kleijn, M.A. Cohen Stuart, *J. Colloid Interface Sci.*, **2002**, *252*, 138.
42 B.W. Ninham, V.A. Parsegian, *J. Chem. Phys.*, **1970**, *53*, 3398.
43 N.V. Churaev, *Colloid Polym. Sci.*, **1975**, *253*, 120.
44 D. Langbain, *J. Phys. Chem. Solids*, **1971**, *32*, 133.
45 D. Langbain, *J. Adhesion*, **1974**, *6*, 1.
46 V.A. Parsegian, G.H. Weiss, *J. Colloid Interface Sci.*, **1972**, *40*, 35.
47 Yu.S. Barash, *Van der Waals Forces*, Nauka, Moscow, **1988**.
48 E.N. Trofimova, F.M. Kuni, A.I. Rusanov, *Colloid J.*, **1969**, *31*, 578.
49 F.M. Kuni, A.I. Rusanov, E.N. Brodskaya, *Colloid J.*, **1969**, *31*, 860.
50 D.Y.C. Chan, D.J. Mitchell, B.W. Ninham, B.A. Pailthorpe, *J. Colloid Interface Sci.*, **1979**, *68*, 462.
51 B.V. Derjaguin, N.V. Churaev, *Dokl. Akad. Nauk SSSR*, **1975**, *222*, 554.
52 B.V. Derjaguin, N.V. Churaev, *Colloid J.*, **1975**, *37*, 1075.
53 B.V. Derjaguin, N.V. Churaev, *J. Colloid Interface Sci.*, **1977**, *62*, 369.
54 L.B. Boinovich, A.M. Emelyanenko, *Adv. Colloid Interface Sci.*, **2003**, *104*, 93.
55 L.B. Boinovich, A.M. Emelyanenko, *Colloid J.*, **2003**, *65*, 672.
56 L.B. Boinovich, A.M. Emelyanenko, *Colloid J.*, **2003**, *65*, 678.
57 L.B. Boinovich, A.M. Emelyanenko, *Colloid J.*, **2004**, *66*, 11.
58 L.B. Boinovich, A.M. Emelyanenko, *Colloid J.*, **2004**, *66*, 18.
59 V.P. Smilga, V.N. Gorelkin, in *Research in Surface Forces* (Ed. B.V. Derjaguin), Nauka, Moscow, **1967**, p. 175 (in Russian).
60 V.N. Gorelkin, V.P. Smilga, *Colloid J.*, **1972**, *34*, 685.
61 V.N. Gorelkin, V.P. Smilga, *Dokl. Akad. Nauk SSSR*, **1973**, *208*, 635.
62 V.N. Gorelkin, V.P. Smilga, In *Surface Forces in Thin Films and Stability of Colloids* (Ed. B.V. Derjaguin), Nauka, Moscow, **1974**, p. 206 (in Russian).
63 P. Attard, D.J. Mitchell, B.W. Ninham, *J. Chem. Phys.*, **1988**, *89*, 4358.
64 P. Attard, R. Kjellander, D.J. Mitchell, *Chem. Phys. Lett.*, **1987**, *139*, 219.
65 R. Kjellander, S. Marcelja, *Chem. Phys. Lett.*, **1987**, *142*, 485.
66 V.M. Muller, in *Surface Forces in Thin Films and Stability of Colloids* (Ed. B.V. Derjaguin), Nauka, Moscow, **1974**, p. 245 (in Russian).
67 R. Buscall, R. Ettelaie, T.W. Healy, *J. Chem. Soc., Faraday Trans.*, **1997**, *93*, 4009.
68 B.W. Ninham, V.A. Parsegian, *J. Theor. Biol.*, **1971**, *31*, 405.

69 D. Y. C. Chan, J. W. Perram, L. R. White, T. W. Healy, *J. Chem. Soc., Faraday Trans.*, **1975**, *71*, 1046.

70 D. C. Prieve, E. Ruckenstein, *J. Theor. Biol.*, **1976**, *56*, 205.

71 T. W. Healy, L. R. White, *Adv. Colloid Interface Sci.*, **1978**, *9*, 303.

72 P. M. Biesheuvel, *Langmuir*, **2001**, *17*, 3553.

73 J. Lyklema, J. F. L. Duval, *Adv. Colloid Interface Sci.*, **2005**, *114–115*, 27.

74 M. G. Cacace, E. M. Landau, J. J. Ramsden, *Q. Rev. Biophys.*, **1997**, *30*, 241.

75 R. M. Pashley, *J. Colloid Interface Sci.*, **1981**, *83*, 531.

76 T. W. Healy, A. Homola, R. O. James, R. G. Hunter, *Faraday Discuss. Chem. Soc.* **1978**, *65*, 156.

77 B. W. Ninham, V. V. Yaminsky, *Langmuir*, **1997**, *13*, 2097.

78 B. W. Ninham, *Prog. Colloid Polym. Sci.*, **2002**, *12*, 1.

79 M. Bostrom, D. R. Williams, B. W. Ninham, *Langmuir*, **2002**, *18*, 6010.

80 W. Kunz, P. Lo Nostro, B. W. Ninham, *Curr. Opin. Colloid Interface Sci.*, **2004**, *9*, 1.

81 M. Manciu, E. Ruckenstein, *Adv. Colloid Interface Sci.*, **2003**, *105*, 63.

82 M. Manciu, E. Ruckenstein, *Adv. Colloid Interface Sci.*, **2003**, *105*, 177.

83 K. A. Karraker, C. J. Radke, *Adv. Colloid Interface Sci.*, **2002**, *96*, 231.

84 R. Kjellander, S. Marcelia, *J. Phys. (Paris)*, **1988**, *49*, 1009.

85 P. A. Kralchevsky, V. N. Panov, *Colloids Surf.*, **1992**, *64*, 265.

86 R. M. Pashley, R. Kjellander, S. Marcelia, J. P. Quirk, *J. Phys. Chem.*, **1988**, *92*, 6489.

87 G. A. Martynov, V. P. Smilga, *Colloid J.*, **1965**, *27*, 250.

88 J. E. Lennard-Jones, B. M. Dent, *Trans. Faraday Soc.*, **1928**, *24*, 92.

89 R. Guidelli, *J. Chem. Phys.*, **1990**, *92*, 6152.

90 I. S. Zhiguleva, V. P. Smilga, in *Surface Forces in Thin Films and Stability of Colloids* (Ed. B. V. Derjaguin), Nauka, Moscow, **1974**, p. 220 (in Russian).

91 N. V. Grigoriev, V. S. Krylov, *Electrochemistry*, **1968**, *4*, 763.

92 S. Levine, G. M. Bell, D. Calvert, *Can. J. Chem.*, **1962**, *40*, 518

93 B. V. Derjaguin, V. M. Muller, *Dokl. Acad. Nauk SSSR*, **1975**, *225*, 601.

94 S. Basu, M. M. Sharma, *J. Colloid Interface Sci.*, **1994**, *165*, 355.

95 D. Henderson, M. Lozada-Cassou, *J. Colloid Interface Sci.*, **1994**, *162*, 508.

96 V. N. Paunov, R. I. Dimova, P. A. Kralchevsky, G. Broze, A. Mehreteab, *J. Colloid Interface Sci.*, **1996**, *182*, 239.

97 V. N. Paunov, B. P. Binks, *Langmuir*, **1999**, *15*, 2015.

98 B. Jonsson, H. Wennerstrom, *J. Chem. Soc., Faraday Trans. 2*, **1983**, *79*, 19.

99 R. Kjellander, S. Marcelja, *Chem. Scr.*, **1985**, *25*, 112.

100 P. Attard, D. J. Mitchell, *J. Phys. Chem.*, **1988**, *88*, 4391.

101 B. Jonsson, P. Attard, D. J. Mitchell, *J. Phys. Chem.*, **1988**, *92*, 5001.

102 Z. Adamczyk, *Adv. Colloid Interface Sci.*, **2003**, *100–102*, 267.

103 S. Marcelja, N. Radic, *Chem. Phys. Lett.*, **1976**, *42*, 129.

104 D. W. R. Gruen, S. Marcelja, V. A. Parsegian, In *Cell Surface Dynamics* (Eds. A. S. Perelson, C. Delisi, F. W. Wiegel), Marcel Dekker, New York, **1984**, p. 59.

105 D. Schiby, E. Ruckenstein, *Chem. Phys. Lett.*, **1983**, *100*, 277.

106 B. W. Ninham, *J. Phys. Chem.*, **1980**, *84*, 1423.

107 G. Cevc, R. Podgornik, B. Zeks, *Chem. Phys. Lett.*, **1982**, *91*, 193.

108 N. Ostrovsky, D. Sornette, *Chem. Scr.*, **1985**, *25*, 108.

109 S. Leikin, A. A. Kornyshev, *J. Chem. Phys.*, **1990**, *92*, 6890.

110 A. A. Kornyshev, S. Leikin, *Phys. Rev. A*, **1989**, *40*, 6431.

111 D. W. R. Gruen, S. Marcelja, *J. Chem. Soc., Faraday Trans. 2*, **1983**, *79*, 225.

112 J. W. Cahn, *J. Chem. Phys.*, **1977**, *66*, 3667.

113 V. S. Mitlin, M. M. Sharma, *J. Colloid Interface Sci.*, **1993**, *157*, 447.

114 R. N. Kuklin, *Colloid J.*, **1997**, *59*, 330.

115 J. C. Eriksson, S. Ljunggren, P. M. Claesson, *J. Chem. Soc., Faraday Trans. 2*, **1989**, *85*, 163.

116 V. S. Yushchenko, V. V. Yaminsky, E. D. Shchukin, *J. Colloid Interface Sci.*, **1983**, *96*, 307.

117 H. K. Christenson, P. M. Claesson, *Science*, **1988**, *239*, 390.

118 H. K. Christenson, P. M. Claesson, R. M. Pashley, *Proc. Indian Acad. Sci. (Chem. Ser.)*, **1987**, *98*, 379.

119 P. M. Claesson, H. K. Christenson, *J. Phys. Chem.*, **1988**, *92*, 1650.

120 J. L. Parker, P. M. Claesson, P. Attard, *J. Phys. Chem.*, **1994**, *98*, 8468.

121 A. Carambassis, L. C. Jonker, P. Attard, M. W. Rutland, *Phys. Rev. Lett.*, **1998**, *80*, 5357.

122 V. V. Yaminsky, B. W. Ninham, *Langmuir*, **1993**, *9*, 3618

123 D. R. Berard, P. Attard, G. N. Patey, *J. Chem. Phys.*, **1993**, *98*, 7236.

124 M. Schoen, T. Gruhn, D. J. Diestler, *J. Chem. Phys.*, **1998**, *109*, 301.

125 B. C. Freasier, S. Nordholm, *J. Chem. Phys.*, **1983**, *79*, 4431.

126 P. Tarazona, *Phys. Rev. A*, **1985**, *31*, 2672.

127 J. W. Perram, L. R. White, *Discuss. Faraday Soc.*, **1975**, *59*, 29.

128 D. Henderson, F. F. Abraham, J. A. Barker, *Mol. Phys.*, **1976**, *31*, 1291.

129 D. E. Sullivan, D. Levesque, J. J. Weis, *J. Chem. Phys.*, **1980**, *72*, 1170.

130 M. Plischke, D. Henderson, *J. Chem. Phys.*, **1986**, *84*, 2846.

131 R. Kjellander, S. Sarman, *Chem. Phys. Lett.*, **1988**, *149*, 102.

132 P. Attard, G. N. Patey, *J. Chem. Phys.*, **1990**, *92*, 4970.

133 P. Attard, D. R. Berrard, C. P. Ursenbach, G. N. Patey, *Phys. Rev. A*, **1991**, *44*, 8224.

134 P. Attard, J. L. Parker, *J. Chem. Phys.*, **1992**, *96*, 5093.

135 D. Chandler, H. C. Andersen, *J. Chem. Phys.*, **1972**, *57*, **1930**.

136 M. J. Grimson, P. Richmond, *J. Chem. Soc., Faraday Trans. 2*, **1980**, *76*, 1478.

137 I. K. Snook, W. van Megen, *J. Chem. Phys.*, **1980**, *72*, 2907.

138 J. Gao, W. D. Luedtke, U. Landman, *J. Phys. Chem. B*, **1997**, *101*, 4013.

139 G. A. Martynov, *Colloid J.*, **2000**, *62*, 393.

140 E. N. Brodskaya, V. V. Zakharov, A. Laaksonen, *Colloid J.*, **2002**, *64*, 596.

141 L. B. Boinovich, A. M. Emelyanenko, *Z. Phys. Chem.*, **1992**, *178*, 229.

142 L. B. Boinovich, A. M. Emelyanenko, *Colloid J.*, **1992**, *54*, 5.

143 L. B. Boinovich, A. M. Emelyanenko, *Colloid J.*, **1993**, *55*, 27.

144 A. M. Emelyanenko, L. B. Boinovich, *Colloid J.*, **1994**, *56*, 362.

145 L. B. Boinovich, A. M. Emelyanenko, *Prog. Colloid Polym. Sci.*, **1999**, *112*, 64.

146 L. B. Boinovich, A. M. Emelyanenko, *Adv. Colloid Interface Sci.*, **2002**, *96*, 37.

147 L. B. Boinovich, *Prog. Colloid Polym. Sci.*, **2004**, *128*, 164.

148 R. B. Jones, *Physica A*, **1981**, *105*, 395.

149 J. Gunning, D. Y. C. Chan, *J. Colloid Interface Sci.*, **1994**, *163*, 100.

150 D. Y. C. Chan, L. R. White, *Physica A*, **1981**, *122*, 505.

151 L. J. Lis, M. McAlister, N. Fuller, R. P. Rand, V. A. Parsegian, *Biophys. J.*, **1982**, *37*, 657.

152 L. D. Gelb, R. M. Lynden-Bell, *Phys. Rev. B*, **1994**, *49*, 2058.

153 H. E. Christenson, J. N. Israelachvili, *J. Chem. Phys.*, **1984**, *80*, 4566.

154 T. Nakada, S. Miyashita, G. Sazaki, H. Komatsu, A. A. Chernov, *Jpn. J. Appl. Phys.*, **1996**, *35*, 52.

155 G. F. Ershova, Z. M. Zorin, N. V. Churaev, *Colloid J.*, **1975**, *37*, 208.

156 V. D. Perevertayev, M. S. Metzik, *Colloid J.*, **1966**, *28*, 254.

157 J. Marra, J. Israelachvili, *Biochemistry*, **1985**, *24*, 4608.

158 D. Quan, E. Freysz, Y. R. Shen, *Science*, **1994**, *264*, 826.

159 S. B. Zhu, G. W. Robinson, *J. Chem. Phys.*, **1991**, *94*, 1403.

160 P. B. Miranda, Y. R. Shen, *J. Phys. Chem.*, **1999**. *103*, 3292.

161 L. F. Scatena, M. G. Brown, G. L. Richmond, *Science*, **2001**, *292*, 908

162 L. B. Boinovich, I. A. Gagina, A. M. Emelyanenko, *Colloid J.*, **1995**, *57*, 897.

163 P. G. de Gennes, *Macromolecules*, **1982**, *15*, 492.

164 J. M. H. M. Scheutjens, G. J. Fleer, *J. Phys. Chem.*, **1979**, *83*, 1619.

165 O. A. Evers, J. M. H. M. Scheutjens, G. J. Fleer, *Macromolecules*, **1990**, *23*, 5221.

166 G. J. Fleer, M. A. Cohen Stuart, J. M. H. M. Scheutjens, T. Cosgrove, B. Vincent, *Polymers at interfaces*. Chapman & Hall, London, **1993**.

167 P. G. de Gennes, *Adv. Colloid Interface Sci.*, **1987**, *27*, 189.

168 P. M. Claesson, A. Dedinaite, O. J. Rojas, *Adv. Colloid Interface Sci.*, **2003**, *104*, 53.

169 R. Sedev, D. Exerowa, *Adv. Colloid Interface Sci.*, **1999**, *83*, 111.

170 P. A. Rehbinder, *Izv. Acad. Nauk SSSR, Ser. Khim.*, **1936**, *5*, 639.

171 B. Vincent, *Colloids Surf.*, **1990**, *50*, 241.

172 S. Asakura, F. Oosawa, *J. Chem. Phys.*, **1954**, *22*, 1255.

173 S. Asakura, F. Oosawa, *J. Polym. Sci.*, **1958**, *33*, 183.

174 R. I. Feigin, D. H. Napper, *J. Colloid Interface Sci.*, **1980**, *75*, 525.

175 J. Y. Walz, A. Sharma, *J. Colloid Interface Sci.*, **1994**, *168*, 485.

7
Hydrophobic Forces in Foam Films

Roe-Hoan Yoon and Liguang Wang

Abstract

Hydrophobic particles suspended in water are attracted to each other due to forces that are larger than the van der Waals force, a process generally referred to as hydrophobic interaction. An air bubble in water may be considered hydrophobic by virtue of its high interfacial tension, thus raising the possibility that the hydrophobic interaction also plays a role in bubble coalescence. This chapter presents evidences for the presence of attractive hydrophobic forces in the thin films of water confined between air bubbles. The hydrophobic forces are strong at low concentrations of surfactants and/or inorganic electrolytes, and decrease with increasing concentrations. Further, the rate of film thinning is accelerated in the presence of attractive hydrophobic forces. The magnitudes of the hydrophobic forces determined using different experimental techniques are compared with the stability of three-dimensional foams and the Gibbs elasticities of foam films. At low surfactant concentrations, the hydrophobic force plays a role in bubble coalescence and hence foam stability, while the film elasticity is important at higher concentrations. Possible origins of the long-range hydrophobic forces that may be present in foam films are discussed.

7.1
Introduction

A better understanding of the mechanisms involved in the stability of bubbles and foams is of crucial importance in many scientific and technological fields, such as surface and colloid chemistry, biology, biochemistry, tertiary oil recovery, detergency, foam fractionation and mineral flotation. In flotation, small air bubbles are introduced to an aqueous suspension (pulp) containing both hydrophilic and hydrophobic particles, so that the latter are selectively collected on the bubble surface. The bubbles laden with hydrophobic particles rise to the surface of the pulp, forming a three-phase foam (froth), which is subsequently removed

Colloids and Interface Science Series, Vol. 1
Colloid Stability: The Role of Surface Forces, Part I. Edited by Tharwat F. Tadros

ISBN: 978-3-527-31462-1

mechanically or by displacement. The rate of flotation increases sharply with decreasing bubble size. Therefore, the bubble size in the pulp phase is reduced by adding (non-ionic) surfactants and increasing the energy dissipation for bubble generation, while the bubble size in the froth phase is controlled by adjusting the froth height. The importance of controlling froth stability in flotation has become widely recognized in recent years (Ata et al., 2003; Neethling and Cilliers, 2003; Mathe et al., 1998).

When small air bubbles are dispersed in water, the stability of the bubble-in-water suspensions may be determined by the surface forces between air bubbles in the same manner as for colloidal suspensions. It has been shown that the stability of hydrophobic particle suspensions is controlled by the long-range hydrophobic force (Xu and Yoon, 1989, 1990), which was first measured by Israelachvili and Pashley (1982). Thermodynamically, air bubbles (or holes) in water can be considered hydrophobic in view of the high interfacial tension at the bubble/water interface ($\gamma_{lv}=72.6$ mJ m^{-2}). When air bubbles coalesce, the free energy gained ($-\Delta G$) is 145.2 mJ m^{-2}, which is larger than that from the coalescence of hydrocarbon liquids in water. In this regard, air bubbles may be considered more hydrophobic than oils. If air bubbles are indeed hydrophobic, it may be reasonable to expect that the hydrophobic force also plays a role in the coalescence of air bubbles. In this regard, the role of surfactant (or frothers) may be to decrease (or damp) the hydrophobic force and stabilize air bubbles.

Aqueous foam is another form of gas dispersion in water. When the gas fraction is increased, bubbles contact each other and deform, creating lamellae and plateau borders between the bubbles in contact. As a lamella becomes thinner by drainage, the surfaces of neighboring bubbles may be brought to a distance range of a net attractive force, which in turn can cause the thin film to rupture and the bubbles to coalesce. Hence foam stability is largely affected by the stability of the individual foam films (or lamellae). Many investigators have shown that the stability of foam films can be predicted using the DLVO theory (Derjaguin and Landau, 1941; Verwey and Overbeek, 1948), which assumes that the fate of a thin film is controlled by the sum of the attractive van der Waals force and the repulsive double-layer force. Angarska et al. (2004) showed, however, that the DLVO theory is applicable only at high surfactant concentrations where the hydrophobic force has disappeared completely. They showed that at low surfactant concentrations it is necessary to consider the contributions from the hydrophobic force. It has been shown that hydrophobic force also plays an important role in defoaming (Denkov, 2004), bubble coalescence (Craig et al., 1993) and breakup of thin evaporating water films (Sharma, 1998).

During the last decade, a considerable amount of work has been done to determine the magnitudes of the hydrophobic forces that may be present in foam films. A difficulty in exploring this possibility has been that the hydrophobic forces become stronger at lower surfactant concentrations but the films become unstable, which makes it difficult to carry out accurate measurements. The aim of this chapter is to review the work done recently, discuss the role of the hydrophobic force in film thinning and foam stability and discuss possible origin(s) of the hydrophobic forces in foam films.

7.2 Foam Films with Ionic Surfactants

7.2.1 Equilibrium Film Thickness

According to the DLVO theory, the disjoining pressure in a soap film can be expressed as the sum of the double-layer force (Π_{el}) and van der Waals force (Π_{vw}):

$$\Pi = \Pi_{el} + \Pi_{vw} \tag{1}$$

where the contribution from the double-layer force is usually given by

$$\Pi_{el} = 64 C_{el} RT \tanh^2\left(\frac{ze\psi_s}{4kT}\right) \exp(-\kappa H) \tag{2}$$

where C_{el} is electrolyte concentration, R the gas constant, T the absolute temperature, z ionic valency, e electronic charge, ψ_s double-layer potential, k Boltzmann's constant, κ the inverse Debye length and H the distance between two charged surfaces (film thickness), and the contribution from the van der Waals force is given by

$$\Pi_{vw} = -\frac{A_{232}}{6\pi H^3} \tag{3}$$

where A_{232} is the Hamaker constant for air (2) in water (3), which is usually of the order of 10^{-20} J.

As discussed in the forgoing section, hydrophobic force may play a role in foam films, in which case Eq. (1) may be extended as follows:

$$\Pi = \Pi_{el} + \Pi_{vw} + \Pi_{hb} \tag{4}$$

to include contributions from hydrophobic force (Π_{hb}), which may be expressed as a power law (Claesson et al., 1986; Yoon and Aksoy, 1999):

$$\Pi_{hb} = -\frac{K_{232}}{6\pi H^3} \tag{5}$$

where K_{232} is a hydrophobic force constant. An advantage of using Eq. (5) rather than an exponential form (Israelachvili and Pashley, 1982; Eriksson et al., 1989; Tsao et al., 1991) is that it is of the same form as the van der Waals force (Eq. 3). This will allow the magnitudes of the hydrophobic force to be compared directly with those of the van der Waals force by means of the two force constants, K_{232} and A_{232}.

When using the thin-film pressure balance (TFPB) technique to measure the thickness of a film (Scheludko and Exerowa, 1959; Scheludko, 1967; Exerowa and Kruglyakov, 1998), the disjoining pressure of a soap film should be equal to the capillary pressure (P_c) at equilibrium. Thus,

$$\Pi = P_c \tag{6}$$

where P_c is the capillary pressure at the meniscus of a horizontal film (Scheludko, 1967; Exerowa and Kruglyakov, 1998) is given by

$$P_c = \frac{2\gamma}{R_c} \tag{7}$$

where γ is surface tension and R_c is the inner radius (=2 mm) of a film holder. Substituting Eqs. (2), (3), (5) and (7) into Eq. (4), one obtains

$$64 C_{el} RT \tanh^2\left(\frac{ze\psi_s}{4kT}\right) \exp(-\kappa H_e) - \frac{A_{232}}{6\pi H_e^3} - \frac{K_{232}}{6\pi H_e^3} - \frac{2\gamma}{r_c} = 0 \tag{8}$$

in which H_e is the equilibrium thickness. Yoon and Aksoy (1999) used a cationic surfactant, dodecylammonium hydrochloride (DAH), to stabilize soap films, measured the equilibrium film thickness (H_e) using the microinterferometric technique (Scheludko, 1967) and calculated the double-layer potential (ψ_s) at the Stern plane. The double-layer potential was calculated by first calculating the adsorption density (Γ_+) of the ionic surfactant at the air/water interface, which in turn was used to calculate the surface potential (ψ_0) using the Gouy-Chapman theory. The surface potential was then corrected for the adsorp-

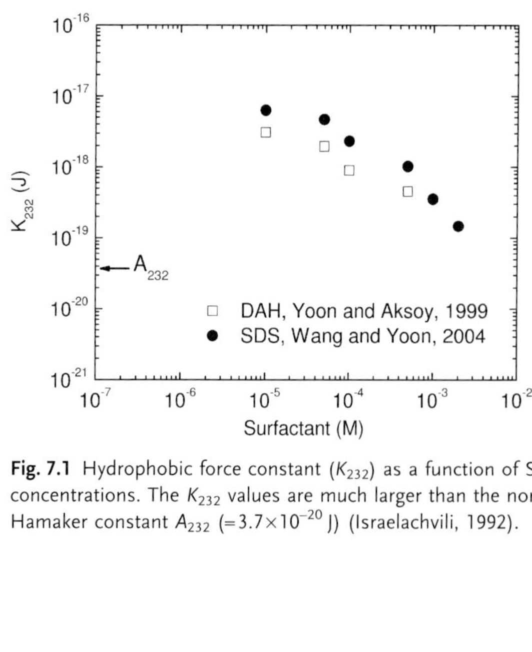

Fig. 7.1 Hydrophobic force constant (K_{232}) as a function of SDS and DAH concentrations. The K_{232} values are much larger than the non-retarded Hamaker constant A_{232} (=3.7×10^{-20} J) (Israelachvili, 1992).

tion of counter-ions at the air/water interface using the Stern model. Equation (8) can be solved for K_{232} using the values of H_e determined experimentally and the value of ψ_s obtained in the manner described above and the Hamaker constant (A_{232}) from literature. Israelachvili (1992), for example, gave the value of $A_{232}=3.7\times10^{-20}$ J. One should note that at low surfactant concentrations, the choice of A_{232} does not make a great difference in the calculation of K_{232} using Eq. (8) because the latter is much larger than the former.

Figure 7.1 shows the K_{232} values obtained using Eq. (8) from the equilibrium film thicknesses of the films stabilized with DAH (open squares) and sodium dodecyl sulfate (SDS) (closed circles). The K_{232} values for the DAH-stabilized films were obtained using the Stern model to correct the surface potentials for counter-ion adsorption, whereas those for the SDS-stabilized films were obtained using a more rigorous counter-ion binding model. The following is a counter-ion binding model derived by Wang and Yoon (2004):

$$\frac{\delta}{\varepsilon_0\varepsilon_s}\sqrt{8kT\varepsilon_0\varepsilon_r n_0}\,\sinh\left(\frac{e\psi_s}{2kT}\right)+\psi_s-\psi_0=0 \tag{9}$$

where δ is the distance between the polar head of a surfactant and the counter-ion in the Stern layer, ε_s the local dielectric constant in the Stern layer, ε_0 and ε_r are the permittivity of vacuum and the dielectric constant of water, respectively, and n_0 is the number density of the counter-ions in solution. This relation can be used to obtain the values of ψ_s from ψ_0. The *apparent* surface potential (ψ_0) can be calculated from the surface excess (Γ_s) of the surfactant ions (DS^-) using a similar method to that reported by Yoon and Aksoy (1999). The K_{232} values given in Fig. 7.1 for the SDS-stabilized films were obtained using Eq. (8) with the ψ_s values obtained using Eq. (9). The Stern potentials calculated using Eq. (9) are comparable to those obtained using other counter-ion binding models (Kalinin and Radke, 1996; Warszynski et al., 1998; Kralchevsky et al., 1999). It has been shown recently that the surface charge densities calculated using a counter-ion binding model are close to those measured using a film conductance technique (Yaros et al., 2003).

By extrapolating the K_{232} versus H plots shown in Fig. 7.1 to the lower concentration region, one can estimate the magnitude of K_{232} in the absence of surfactant and find that the hydrophobic force constant is larger than A_{232} by more than two orders of magnitude. This finding is consistent with thermodynamic considerations based on the changes in interfacial tensions at the air/water interface, which may be used as a measure of hydrophobicity. In the absence of surfactant, the interfacial tension is the largest (72.6 mN m^{-1}) and hence most hydrophobic. It decreases with increasing surfactant concentration, indicating a decrease in hydrophobicity. In this regard, an air bubble in water is most hydrophobic in pristine water and becomes less hydrophobic with increasing surfactant concentration. Hence the role of surfactant is to damp the hydrophobicity of air bubbles and to damp the hydrophobic force.

Note that the K_{232} values of the SDS-stabilized films are considerably larger than those of the DAH-stabilized films. This finding suggests that the head group of SDS may be less hydrophilic than that of the DAH group, which is likely in view of the fact that the $-SO_4^-$ group is considerably larger than the $-NH_3^+$ group. In general, the smaller the size of an ion is, the stronger the hydration energy becomes. Therefore, cations are more strongly hydrated than anions. On the other hand, the difference between the two sets of data shown in Fig. 7.1 may be due to the difference in calculating the Stern potentials.

7.2.2
Disjoining Pressure Isotherm

Although it is possible to determine K_{232} from equilibrium film thickness measurements, as shown in the foregoing paragraph, its accuracy depends critically on the validity of double-layer potential (ψ_s). Unfortunately, there are no reliable methods of directly measuring ψ_s at the air/water interface. Hence K_{232} by itself may not serve as direct evidence for the existence of the long-range hydrophobic force. As suggested by Tchaliovska et al. (1994), there is a need to measure surface forces (or disjoining pressures) as a function of film thickness (H), i.e. to obtain disjoining pressure isotherms. A challenge, however, is that it is difficult to measure disjoining pressures at low surfactant concentrations (owing to instability) where hydrophobic forces are discernible. This problem can be minimized by increasing the film stability by adding a small amount of an inorganic electrolyte. Further, the use of a bike-wheel film holder (Cascão Pereira et al., 2001) may stabilize the films during disjoining pressure measurement using the TFPB technique.

Figure 7.2 shows the disjoining pressure isotherms for the thin aqueous films stabilized in the presence of 10^{-4} M SDS and varying concentrations of NaCl (Wang and Yoon, 2004). A bike-wheel film holder, with an air/water entry pressure of $\sim$10 kPa, was used to measure the disjoining pressures above 150 Pa, whereas the lower pressures were measured using Scheludko cells with radii of

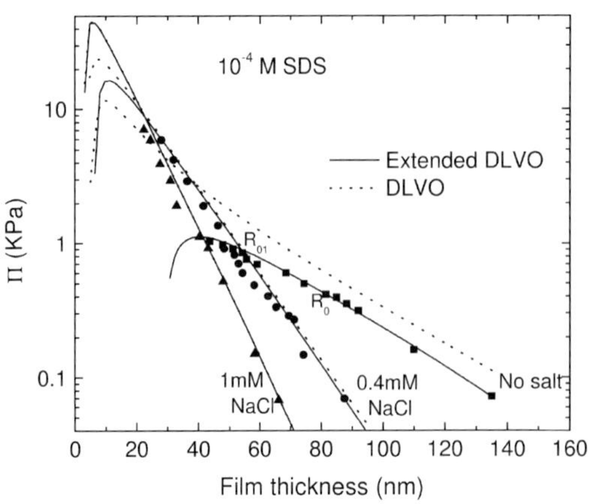

Fig. 7.2 Disjoining pressure isotherms obtained in 10^{-4} M SDS solutions at different NaCl additions (Wang and Yoon, 2004).

1.0 and 2.0 mm. The disjoining pressure isotherm obtained in the absence of NaCl (solid line) deviates significantly from the DLVO theory (Eq. 1) (dotted line) owing to the presence of the hydrophobic force. The experimental data can be fitted to the extended DLVO theory (Eq. 4), with $\psi_s = -114.3$ mV, $A_{232} = 3.7\times10^{-20}$ J, $K_{232} = 2.33\times10^{-18}$ J and κ^{-1} (Debye length) = 30.4 nm. The DLVO curve was obtained using the same set of parameters except that K_{232} was set to be zero. The value of ψ_s was obtained using the counter-ion binding model (Wang and Yoon, 2004) and the value of K_{232} that was used to fit the experimental data was obtained using the equilibrium film thickness method described in the foregoing section and presented in Fig. 7.1. That the K_{232} value obtained from one method can be used to fit the experimental data obtained using a completely different method is significant.

The isotherm obtained at 10^{-4} M SDS and in the absence of NaCl shows that the film ruptures at $H = 43$ nm and $\Pi = 1.1$ kPa. It should be noted here that this isotherm is the most typical of the many duplicate isotherms. In some cases, the rupture occurs at larger thicknesses and lower pressures. The lowest rupture pressure recorded is denoted R_0 on the isotherm. The measured rupture thicknesses are much larger than predicted by the DLVO theory as shown. Also, the measured disjoining pressures were considerably lower than predicted by the theory, indicating the presence of hydrophobic force. It may be of interest to note that these deviations from the DLVO theory are much larger than observed in the surface force measurements conducted with the mica surfaces coated with dodecylammonium hydrochloride (Yoon and Ravishankar, 1996). Furthermore, the experimental disjoining pressure isotherm begins to deviate from the DLVO theory at much larger separations (H) than observed in the force measurements conducted between macroscopic solid surfaces. These findings suggest that the hydrophobic forces operating in soap films are much longer ranged than observed with the amine-coated mica surfaces.

Note in Fig. 7.2 that the disjoining pressure isotherms deviate progressively less from the DLVO theory with increasing NaCl concentration. The isotherm obtained at 10^{-4} M SDS and 0.4 mM NaCl shows that the film ruptures at 6.0 kPa. In repeat measurements conducted at the same NaCl and SDS concentrations, the film ruptures at a pressure as low as 0.9 kPa, as indicated by R_{01}. At 1 mM NaCl, no rupture occurs at pressures as high as 8.5 kPa, which is the highest pressure attainable with the apparatus used for the measurements. With a better instrument, it may be possible to observe films rupturing at a pressure predicted by the DLVO theory. It seems that at 1 mM NaCl the classical DLVO theory can predict perfectly the disjoining pressure isotherm obtained, indicating that hydrophobic force disappears completely at such a high electrolyte concentration. This finding is consistent with the work of Craig et al. (1993), who showed that the coalescence of air bubbles in water begins to be mitigated at NaCl concentrations above ~1 mM. These authors noted that "as bubbles are highly hydrophobic ($\gamma_{23} = 72$ mJ m^{-2}), it is reasonable to assume that the hydrophobic force is present and acts to produce coalescence." Since bubble coalescence is controlled by long-range forces, the hydrophobic force in free films must also be a long-range force, as is clearly evidenced by the isotherms presented in Fig. 7.2.

The isotherms given in Fig. 7.2 show the classical behavior of colloidal systems; the films become thinner with increasing electrolyte concentration, which is due, of course, to double-layer compression. As the NaCl concentration is increased from 0 to 0.4 and 1 mM, the Debye length decreases from 30.4 to 13.6 and 9.2 nm, respectively.

7.2.3 Kinetics of Film Thinning

The lifetime of a foam film is controlled by drainage and rupture. For a horizontal foam film, the driving force for the drainage process is the capillary pressure, which is determined by the curvature at the plateau border and the surface tension. The drainage would stop when the capillary pressure becomes equal to the disjoining pressure of the film, which in turn gives rise to an equilibrium film thickness. Hence the initial process of film thinning is controlled by capillary pressure, while the subsequent process is controlled by surface forces. It would, therefore, be of interest to monitor the process of film thinning, which may shed light on the existence of hydrophobic force in thin aqueous films. Of particular interest would be to see if the later stages of a film thinning process are described by the classical (Eq. 1) or the extended DLVO theory (Eq. 4).

The rate of film thinning can be described by the Reynolds equation (Scheludko and Platikanov, 1961; Scheludko, 1967):

$$-\frac{dH}{dt} = \frac{2H^3\Delta P}{3\mu R_f^2} \tag{10}$$

where t is drainage time, μ dynamic viscosity, R_f film radius and ΔP the driving force for film thinning. In view of the discussions above, the driving force may be represented by the following relation:

$$\Delta P = P_c - \Pi \tag{11}$$

where P_c is the capillary pressure represented by Eq. (7) and Π is the disjoining pressure represented by Eq. (1) or (4).

The Reynolds equation was derived originally for film thinning that takes place under no-slip conditions. Therefore, it is applicable to foam films with tangentially immobile surfaces. It has been shown that this condition is met by using very small films stabilized by a surfactant (Exerowa and Kruglyakov, 1998; Langevin, 2000; Coons et al., 2005). It has also been shown that Eq. (10) can be used even at low surfactant concentrations (Angarska et al., 2004; Ivanov et al., 2005).

Figure 7.3 shows the results of the film thinning kinetic measurements conducted on two horizontal foam films stabilized at two different concentrations of SDS in the presence and absence of NaCl (Wang and Yoon, 2005). In both cases, the initial film thinning is fast and can be described by the Reynolds equation with

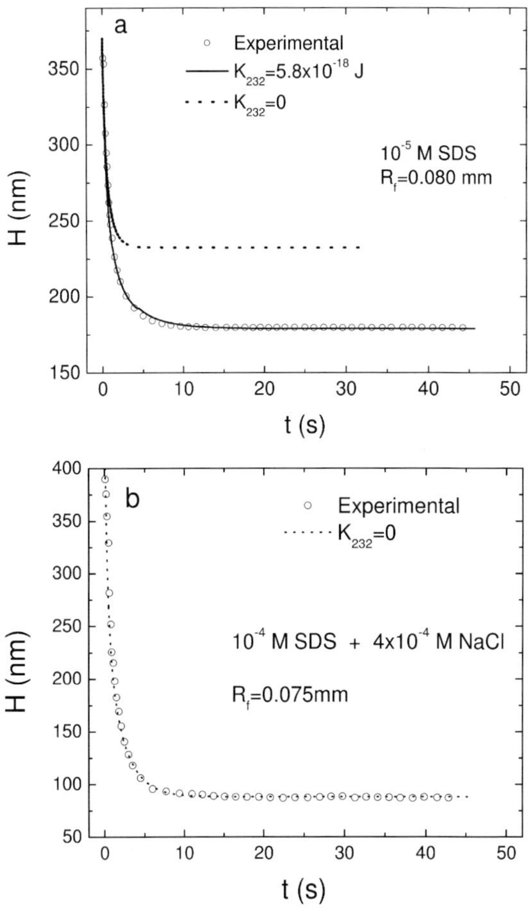

Fig. 7.3 Kinetics of film thinning at (a) 10^{-5} M SDS and (b) 10^{-4} M SDS and 4×10^{-4} M NaCl. The solid line represents the Reynolds equation (Eq. 10) with the extended DLVO theory and the dotted line represents the Reynolds equation with the DLVO theory (Wang and Yoon, 2005).

its driving force (ΔP) represented by Eq. (11) with $\Pi=0$, that is, the process is controlled solely by capillary force (Eq. 7). This is fully expected as the initial film thickness is too large for surface forces to affect the process. As the film thickness (H) is reduced to below $\sim$250 nm, however, surface forces begin to influence the film thinning process. The film thinning kinetic curve obtained at a relatively low SDS (10^{-5} M) concentration and in the absence of NaCl cannot be described by the Reynolds equation without considering the contribution from the hydrophobic force to ΔP, i.e. Eq. (11) in which Π is given by the DLVO theory (Eq. 1). The experimental data (open circles) can be fitted to the Reynolds equation when Π in Eq. (11) is represented by the extended DLVO theory (Eq. 4) with $K_{232}=5.8\times10^{-18}$ J, $\psi_s=-92$ mV and $A_{232}=3.7\times10^{-20}$ J. Not surprisingly, the K_{232} value is the same as obtained using the equilibrium film thickness technique and given in Fig. 7.1. (Note that equilibrium film thickness is reached during the later stage of the film thinning process.) On the other hand, the kinetic curve obtained at 10^{-4} M SDS in the presence of 4×10^{-4} M NaCl can be fitted to the Reynolds equation with the classical DLVO theory (Eq. 1) with $K_{232}=0$, $\psi_s=-110$ mV and $A_{232}=3.7\times10^{-20}$ J. The ψ_s value used for the fitting procedure has been obtained using the counter-ion binding model. Note that the hydrophobic force constant becomes zero owing to the presence of 0.4 mM NaCl. In the absence of NaCl, $K_{232}=5.8\times10^{-18}$ J at 10^{-4} M SDS (see Fig. 7.1). Hence hydrophobic force is damped by both surfactant and inorganic electrolyte. More importantly, the hydrophobic force constants determined using the static methods, i.e. the equilibrium film thickness and disjoining pressure isotherm methods, can be used to fit the dynamic film thinning data.

As has already been noted, the accuracy of determining the magnitudes of hydrophobic force depends critically on the double-layer potentials. Therefore, it would be of interest to study the kinetics of film thinning under conditions of $\Pi_{el}\approx0$, in which case the extended DLVO theory (Eq. 4) is reduced to

$$\Pi = \Pi_{vw} + \Pi_{hb} \tag{12}$$

By substituting this into Eq. (11) and then Eq. (10), one can predict the kinetics of film thinning using the Reynolds equation without the possible ambiguity associated with determining double-layer potentials.

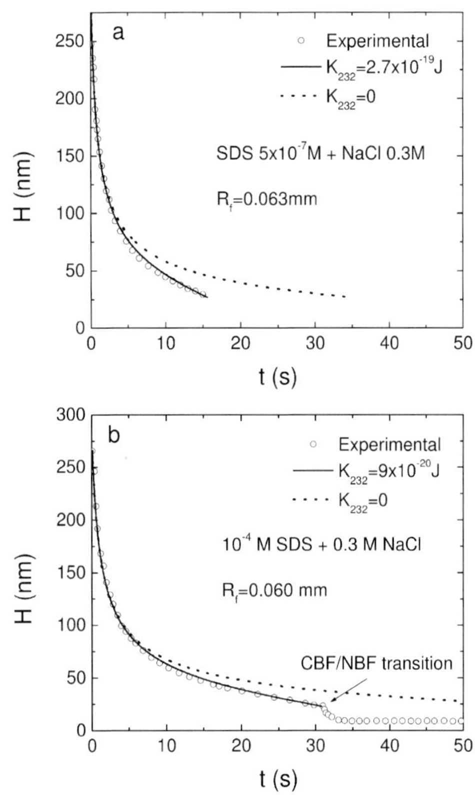

Fig. 7.4 Kinetics of film thinning at (a) 5×10^{-7} M SDS and 0.3 M NaCl and (b) 1×10^{-4} M SDS and 0.3 M NaCl. The solid line represents the Reynolds equation (Eq. 10) with the extended DLVO theory and the dotted line represents the Reynolds equation with the DLVO theory (Wang and Yoon, 2005).

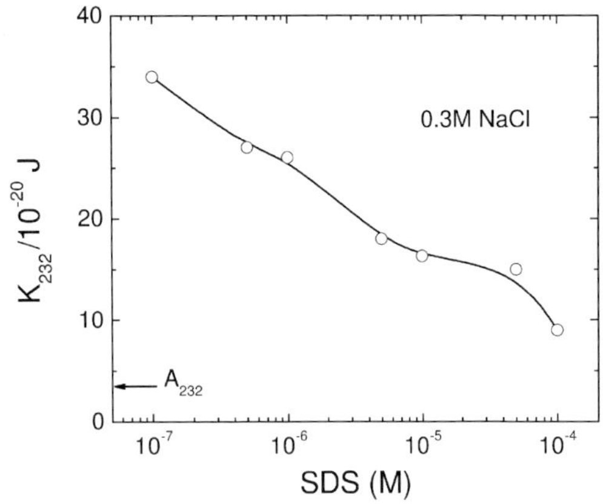

Fig. 7.5 Effect of SDS concentration on K_{232} at 0.3 M NaCl. The results were obtained by fitting the film thinning data to the Reynolds equation (Eq. 10) (Wang and Yoon, 2005). The arrow represents the non-retarded Hamaker constant ($=3.7\times10^{-20}$ J) (Israelachvili, 1992).

Figure 7.4 shows the results obtained in the kinetics studies conducted at 5×10^{-7} and 10^{-4} M SDS in the presence of 0.3 M NaCl. At the lower SDS concentration, the kinetics of film thinning are considerably faster than predicted (dotted line) without considering the hydrophobic force, i.e. $K_{232}=0$. The experimental data can be fitted to the Reynolds equation (solid line) with $K_{232}=2.7\times10^{-19}$ J and with the A_{232} values determined by considering the retardation effects (Russel et al., 1989). The film is shown to rupture at 28.2 nm in 15.7 s. At 10^{-4} M SDS and 0.3 M NaCl, the film becomes more stable owing to the higher surfactant concentration and ruptures at 23.3 nm in 30.9 s before the common black film (CBF) is transformed to a Newton black film (NBF). The kinetics curve can be fitted to the Reynolds equation with $K_{232}=9.0\times10^{-20}$ J before the CBF–NBF transition occurs.

The kinetics of film thinning was also studied at other SDS concentrations in the presence of 0.3 M NaCl and the K_{232} values were determined in the manner described in the foregoing paragraph are plotted as a function of SDS concentration in Fig. 7.5. At 10^{-7} M SDS, K_{232} is 3.4×10^{-19} J, which is 9.2 times larger than the non-retarded Hamaker constant (3.7×10^{-20} J). At 10^{-4} M SDS, K_{232} is 9×10^{-20}, which is 2.4 times larger than the Hamaker constant. Hence the hydrophobic forces are still observed under conditions of $\Pi_{el}\approx0$, demonstrating that hydrophobic forces in thin aqueous films are real. The differences between K_{232} and A_{232} are not as large as obtained in the absence of an inorganic electrolyte (see Fig. 7.1). The reason is that hydrophobic force is also damped by NaCl.

7.2.4 Critical Rupture Thickness

As the thickness of a foam film is reduced by drainage, the film ruptures catastrophically when the thickness reaches a critical thickness (H_{cr}). It is believed that a film surface is always in thermally or mechanically induced oscillation, causing the instantaneous distance between the two interfaces in a foam film to

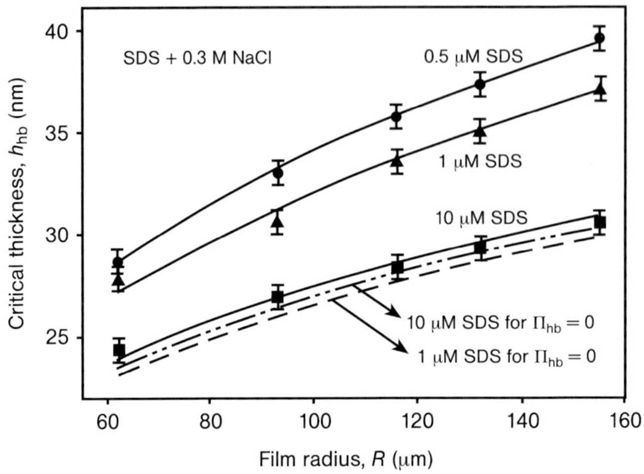

Fig. 7.6 Plots of H_{cr} versus R_f at 0.5, 1.0 and 10 μM SDS in the presence of 0.3 M NaCl. The solid lines represent model predictions by considering the presence of hydrophobic force in foam films ($\Pi_{hb} \neq 0$); $B=0.656$, 0.471 and 0.0334 μJ m^{-2} at 0.5, 1.0 and 10 μM SDS, respectively, and $\lambda=15.8$ nm. The dashed and dash-dotted lines represent the model predictions without considering the presence of hydrophobic force ($\Pi_{hb}=0$) (Angarska et al., 2004).

be smaller than the measured distance. The amplitude of the oscillation increases when the instantaneous distance between the two surfaces is within the range of an attractive force, i.e. van der Waals force. When the distance between the two surfaces reaches H_{cr}, the film ruptures spontaneously.

If a foam film ruptures at the H_{cr} predicted by the capillary wave model, there is no need to invoke the hydrophobic force. Indeed, many investigators have shown that, in the absence of repulsive forces, the model can predict the critical rupture thicknesses (Vrij, 1966; Vrij and Overbeek, 1968; Radoev et al., 1983; Valkovska et al., 2002). Recently, Angarska et al. (2004) showed that the predictions of the model developed by Valkovska et al. were in good agreement with the experimental data obtained by Manev et al. (1984). Note, however, that the rupture thicknesses were measured at high concentrations of SDS (4.3×10^{-4} M) and NaCl (0.25 M), under which conditions the air/water interface becomes hydrophilic, as shown in a previous section.

Angarska et al. (2004) showed, however, that the capillary wave model fails at low SDS concentrations. Figure 7.6 shows the H_{cr} values measured at 0.5×10^{-6}, 1.0×10^{-6} and 10×10^{-6} M SDS and 0.3 M NaCl concentrations at different film radii. At 1×10^{-6} M, the experimental values of H_{cr} were substantially larger than predicted by the capillary wave model derived by assuming that the disjoining pressure due to hydrophobic force (Π_{hb}) was zero. At 10^{-5} M, the discrepancy was much reduced; however, the experimental values were still considerably higher than predicted. Angarska et al. (2004) used the following form to represent the hydrophobic force (Eriksson et al., 1989):

$$\Pi_{\text{hb}} = -\frac{B}{2\pi\lambda}\left[\coth^2\left(\frac{H}{2\lambda}\right) - 1\right] \tag{13}$$

where the parameters B and λ represent the strength and decay length of the hydrophobic force, respectively. By assuming that $\lambda = 15.8$ nm at all SDS concentrations studied, Angarska et al. back-calculated the values of B to be 0.656, 0.471 and 0.0334 μJ m^{-2} at 0.5×10^{-6}, 1.0×10^{-6} and 10×10^{-6} M SDS, respectively and at 0.3 M NaCl. These results suggest that hydrophobic force decreases with increasing surfactant concentration, which is consistent with the results presented in the foregoing sections. It should be noted that the B values obtained by these investigators were small, possibly owing to the large decay length at all of the SDS concentrations investigated.

7.3
Foam Films with Non-ionic Surfactants

When using ionic surfactants, one can readily determine contributions from hydrophobic force (Π_{hb}) to the disjoining pressure (Π) in foam films. This can be done by determining the double-layer potentials (ψ_s) at the air/water interface by calculating the adsorption density of the charged species using the Gibbs adsorption isotherm, as described in the forgoing paragraph. When using non-ionic surfactants, however, the same technique cannot be used to determine the double-layer potentials. This problem can be overcome by conducting measurements at high electrolyte concentrations so that $\Pi_{\text{el}} \approx 0$. Under this condition, the disjoining pressure can be given as a sum of van der Waals force and hydrophobic force only, as shown in Eq. (12). By substituting this equation into Eq. (11), one obtains the driving force for film thinning (ΔP), which can be used in conjunction with the Reynolds equation (Eq. 10) to study the role of hydrophobic force in film thinning. One should note that the hydrophobic forces determined in this manner should be much smaller than those determined in the absence of electrolytes as the electrolyte by itself can damp the hydrophobicity and hydrophobic force.

7.3.1
Kinetics of Film Thinning

Figure 7.7 shows the film thinning kinetic curves obtained for the foam films stabilized by methylisobutyl carbinol (MIBC), a non-ionic surfactant used widely in the mining industry to produce bubbles and foams for flotation. The film thinning tests were conducted using the TFPB technique at 10^{-5}, 3×10^{-3} and 3×10^{-2} M MIBC in the presence of 0.1 M NaCl. At 10^{-5} M MIBC, the kinetics were considerably faster than predicted (dotted line) without considering the hydrophobic force, i.e. $K_{232} = 0$. The experimental data could be fitted to the Reynolds equation (solid line) with $K_{232} = 6\times10^{-19}$ J. In both cases, the A_{232} values were calculated by considering

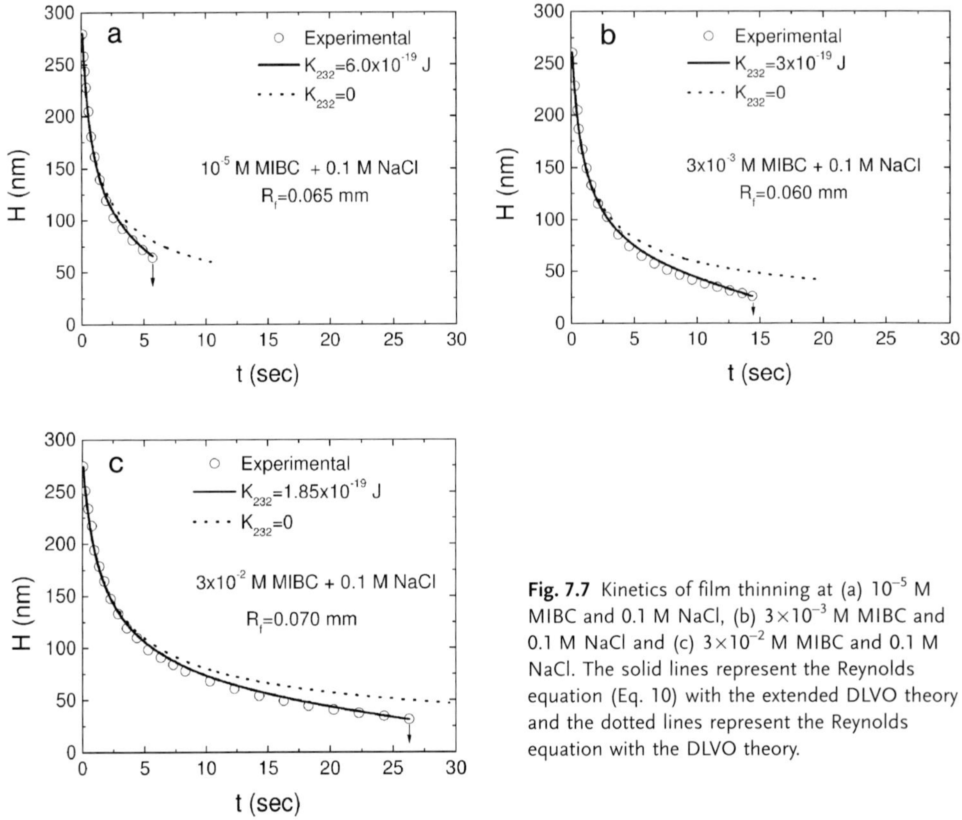

Fig. 7.7 Kinetics of film thinning at (a) 10^{-5} M MIBC and 0.1 M NaCl, (b) 3×10^{-3} M MIBC and 0.1 M NaCl and (c) 3×10^{-2} M MIBC and 0.1 M NaCl. The solid lines represent the Reynolds equation (Eq. 10) with the extended DLVO theory and the dotted lines represent the Reynolds equation with the DLVO theory.

the retardation effects as described by Russel et al. (1989). At 10^{-5} M MIBC, the film ruptured at $H=63.3$ nm in 5.7 s. At 3×10^{-3} M MIBC, the film ruptured at 26.1 nm in 14.4 s, indicating that the film had become more stable. At 3×10^{-2} M MIBC, the film lasted as long as 26.3 s, with the rupture occurring at 31.3 nm.

The kinetics of film thinning were also studied at other MIBC concentrations in the presence of 0.1 M NaCl and the results were used to determine the K_{232} values in the same manner as described in the foregoing paragraph. Figure 7.8 shows the K_{232} values determined as such as a function of MIBC concentration. The values of A_{232} calculated at H_{cr} are also plotted in Fig. 7.8. As shown, the K_{232} values determined from the kinetic data are $a\sim15$–90 times larger than the Hamaker constants. Since the hydrophobic force is represented in the same form as the van der Waals force, it is possible to compare the two by means of K_{232} and A_{232}. That the former is substantially larger than the latter even at a very high electrolyte concentration suggests that the hydrophobic force is real.

Note that the K_{232} values of the MIBC-stabilized films at 0.1 M NaCl are larger than those of the SDS-stabilized films at 0.3 M NaCl. This finding suggests that the hydroxyl group of MIBC may be less hydrophilic than that of the sulfate group.

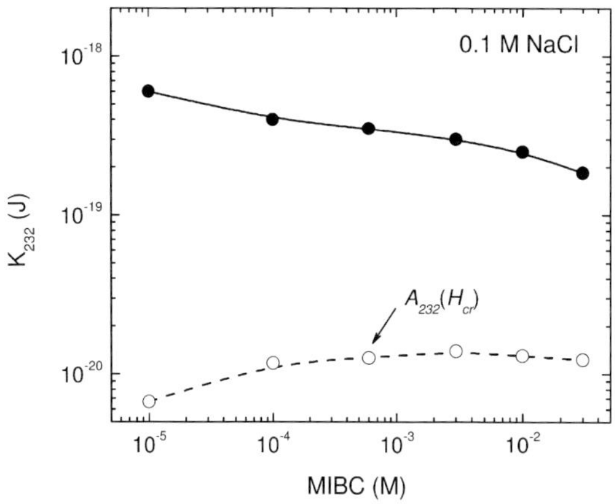

Fig. 7.8 Effect of MIBC concentration on K_{232} at 0.1 M NaCl. The results were obtained by fitting the film thinning data to the Reynolds equation (Eq. 10). The Hamaker constants (A_{232}) represent retarded Hamaker constants (Wang and Yoon, 2006).

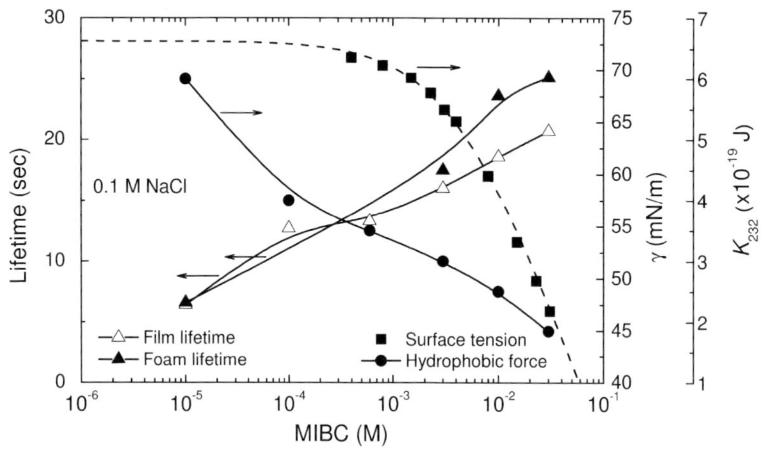

Fig. 7.9 Effect of MIBC concentration at 0.1 M NaCl on the lifetimes of foam (▲) and film (△). The surface tension data (■) are from Comley et al. (2002) and the dashed line represents the best fit of the surface tension data using the Langmuir-Szyszkowski equation (Eq. 14). Both the film and foam lifetimes increased with decreasing hydrophobic force constant K_{232} (●). The K_{232} values were calculated from the film thinning data using the Reynolds equation (Eq. 10) (Wang and Yoon, 2006).

On the other hand, the difference between the two sets of K_{232} values shown in Figs. 7.5 and 7.8 may be due to the difference in NaCl concentration. At 0.1 M NaCl, the hydrophobic force is damped to a less extent than at 0.3 M NaCl.

In Fig. 7.9, the K_{232} values given in Fig. 7.8 are re-plotted along with the lifetimes of the single foam films (film lifetime) and the lifetimes of the three-dimensional foams (foam lifetime) determined at different MIBC concentrations. The film lifetime represents the time elapsed prior to the rupture of a single foam film, whereas the foam lifetime represents the time it took for a three-dimensional foam to become single bubbles in the shake tests (Wang and Yoon, 2006). As shown, an increase in MIBC concentration caused both the film and the foam lifetimes to increase, which corroborates well with the changes in K_{232}.

According to Eqs. (7) and (10) to (12), the capillary pressure and, hence, the driving force (ΔP) for film thinning decrease with decreasing surface tension. Therefore, the decrease in surface tension should also be responsible for the increase in the film and foam lifetimes shown in Fig. 7.9. At low surfactant concentrations, however, the change in surface tension was, as depicted by the dashed line representing the Langmuir-Szyszkowski equation:

$$\gamma = \gamma_0 - RT\Gamma_{\mathrm{m}} \ln(1 + K_{\mathrm{L}} C_{\mathrm{s}}) \tag{14}$$

which was fitted to the surface tension data of Comley et al. (2002) shown in Fig. 7.9. In this equation, γ_0 is the surface tension of pure water, Γ_{m} ($=5\times10^{-6}$ mol m^{-2}) the monolayer coverage, K_{L} ($=230$ M^{-1}) the equilibrium constant and C_{s} the bulk MIBC concentration. Hence the data in Fig. 7.9 show that at low surfactant concentrations both the single foam film and the three-dimensional foam are destabilized by hydrophobic force, whereas at high surfactant concentrations they are stabilized by surface tension lowering.

It is well known that the stability of foams and foam films is determined not only by surface forces (or disjoining pressures) but also by film elasticities (Langevin, 2000; Fruhner et al., 2000; Stubenrauch and Miller, 2004). Therefore, the film

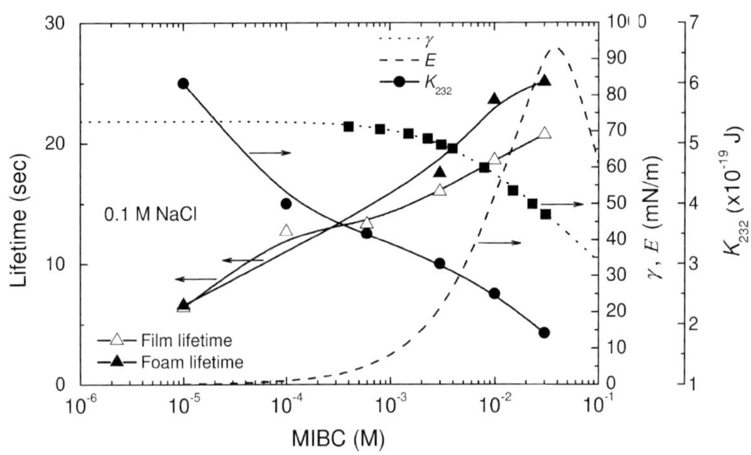

Fig. 7.10 Effects of MIBC concentration on hydrophobic force constant (K_{232}), surface tension (γ) film elasticity (E), film lifetime (△) and foam lifetime (▲) at 0.1 M NaCl. The surface tension data (■) are from Comley et al. (2002). The dotted line represents the best fit of the surface tension data using Eq. (14) and the dashed line represents E calculated using Eq. (17).

and foam lifetimes measured in the work of Wang and Yoon (2006) have been re-plotted in Fig. 7.10 as functions of MIBC concentration and are compared with film elasticities (E). The film elasticity is defined in the format of the Gibbs elasticity:

$$E = 2A\frac{d\gamma}{dA} = 2A\frac{d\gamma}{dc}\frac{dc}{dA} \tag{15}$$

where A is the film surface area and c is the bulk surfactant concentration. For a closed system, the volume of a foam film, $V = AH$, is constant, i.e. $dV = 0$, where H is film thickness. Also, the total number of the surfactant molecules is constant in a closed system. After several mathematical steps, including using the Gibbs adsorption isotherm, Wang and Yoon (2006) obtained an expression for E as follows:

$$E = \frac{4RT\Gamma^2}{c(H + 2d\Gamma/dc)} \tag{16}$$

where Γ is the surface excess of the surfactant. Equation (16) is similar to Christenson and Yaminsky's model (1995), except that the latter authors ignored the $d\Gamma/dc$ term, which is significantly large at low surfactant concentrations. Equation (16) can be transformed into a more useful form:

$$E = \frac{4cRT\Gamma_m^2 K_L^2}{H(1 + K_L c)^2 + 2\Gamma_m K_L} \tag{17}$$

by combining it with the Langmuir isotherm.

Figure 7.10 gives the Gibbs elasticities (E) calculated at H_{cr} using Eq. (17) (dashed line). In these calculations, the values of Γ_m and K_L were obtained by fitting the surface tension data to the Langmuir-Szyszkowski equation [Eq. (14)]. Also shown in Fig. 7.10 are (i) the film lifetimes and the foam lifetimes measured at different MIBC concentrations and (ii) the K_{232} values obtained from the film thinning kinetic data. The amounts of MIBC used in the flotation industry are usually in the range $(0.5\text{–}1.5)\times10^{-4}$ M. In this concentration range, the film elasticities (E) calculated using Eq. (17) are very low (<2 mN m^{-1}), as shown in Fig. 7.10. Therefore, it may be difficult to conclude that, at such low surfactant concentrations, bubbles and foams are stabilized by the increased elasticity. On the other hand, the hydrophobic force constant (K_{232}) decreases sharply with MIBC concentration. It may be stated, therefore, that the stability arises from the decrease in hydrophobic force caused by the small amount of the non-ionic surfactant added to the system. At higher surfactant concentrations, the bubbles and foams are stabilized by the increase in elasticity.

In the presence of 0.1 M NaCl, the values of K_{232} are in the range $(3.7\text{–}4.5)\times10^{-19}$ J, which are more than 25 times larger than A_{232}. In the absence of electrolyte, the bubbles should be more hydrophobic. That the bubbles remain hydrophobic at the frother concentrations where flotation is practiced should help increase the kinetics of bubble-particle interaction. At higher MIBC concentrations, the hydrophobic force decreases further, making the hydrophobic interaction less likely. It is well known that excessive frothers are detrimental to flotation (Klimpel and Hansen, 1987).

Although not shown in Fig. 7.10, the Gibbs elasticities calculated using Christenson and Yaminsky's model (1995), i.e. Eq. (16) without the $d\Gamma/dc$ term, are substantially larger than obtained using Eqs. (16) and (17). Because this term is in the denominator, their model over-predicts the elasticity at low surfactant concentrations. When the surfactant concentration is very high (i.e. above 0.5 M), however, the two models merge. Therefore, Christenson and Yaminsky's model is applicable only at very high concentrations.

7.3.2
Critical Rupture Thickness

Based on the capillary wave mechanism discussed earlier, Vrij (1966) and Vrij and Overbeek (1968) derived a model to predict the critical rupture thickness (H_{cr}) as follows:

$$H_{cr} \approx 0.207\left(\frac{A_{232}^2 R_f^2}{\gamma P_c}\right)^{1/7} \tag{18}$$

where R_f is the film radius in a TFPB. According to this model, H_{cr} should vary primarily with A_{232} and γ at a given R_f. Wang and Yoon (2006) measured the critical rupture thicknesses of foam films stabilized at different concentrations of MIBC and compared the results with the changes in A_{232} and γ with the surfactant concentration, as shown in Fig. 7.11. The measurements were conducted in 0.1 M NaCl solutions, with films whose radii (R_f) were kept relatively constant in the range 0.060–0.070 mm. As shown, H_{cr} decreases sharply at the

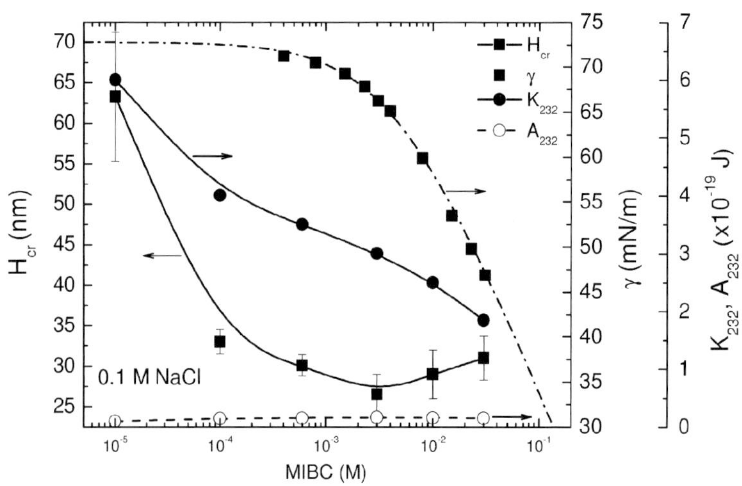

Fig. 7.11 Effects of MIBC concentration on H_{cr} (■), surface tension (γ), K_{232} (●) and A_{232} (○). H_{cr} varies closely with K_{232} (Wang and Yoon, 2006).

range 10^{-5}–8×10^{-4} M MIBC, whereas both A_{232} and γ vary little. On the other hand, K_{232} decreases sharply in the same concentration range. Therefore, the results in Fig. 7.11 suggest that the decrease in H_{cr} in the lower concentration range is largely due to the decrease in hydrophobic force caused by the increase in MIBC concentration. As the concentration continues to increase, the decrease in H_{cr} tapers off and reaches a minimum (26.5 nm) at 3×10^{-3} M MIBC. The increase in H_{cr} above 3×10^{-3} M MIBC can be attributed to the lowering of surface tension, which is more significant than the decrease in hydrophobic force.

7.4 Possible Origins of Hydrophobic Force

7.4.1 Adsorption

It has been shown in the foregoing sections that the hydrophobic force at the air/water interface is strongly a function of surfactant and NaCl concentrations. Angarska et al. (2004) suggested that these species adsorb at the air/water interface and hydrophilize the interface. They showed an excellent correlation between the hydrophobic force constant, i.e. the parameter B in Eq. (13), and the "inverse ionic adsorption" at the air/water interface. The latter quantity refers to $(\Gamma_0+\Gamma_s+\Gamma_c)^{-1}$, where Γ_s is the surface excess of surfactant (dodecyl sulfate or DS^-) ions and Γ_c and Γ_0 are the same for counter-ions (Na^+ ions) and hydroxyl ions, respectively. Γ_s can be obtained using the Gibbs adsorption equation and Γ_c from a counter-ion binding model.

Figure 7.12 shows a plot of the hydrophobic force constant, i.e. K_{232} in Eq. (5), versus the area, a, occupied by individual surfactant molecules adsorbed at the air/water interface. The parameter a is equivalent to Γ_s^{-1}, but it is probably easier to attach a physical meaning to the former. A high value of a means that a surfactant molecule has a large "parking area" and that a large part of the interface is unoccupied. In general, K_{232} is shown to increase with increasing a, indicating that hydrophobic force is strongest with the pristine air/water interface (or air bubbles in water) and becomes weaker with the adsorption of surfactant and/or electrolyte.

The K_{232} versus a curves in Fig. 7.12 can be subdivided into three different zones. In zone I, DS^- ions adsorb at the air/water interface, forming more or less a close-packed monolayer. In zone II, the surfactant adsorption is significant but the molecules are not well ordered. According to Eriksson and Ljunggren (1989), however, the surfactant may still adsorb as clusters as a means of minimizing free energy. In zone III, the surfactant adsorbs individually at the air/water interface, which is referred to as "the gaseous state" in the sense that the surfactant solution follows Henry's law.

The most striking feature of Fig. 7.12 is that NaCl has a profound impact on hydrophobic force. At a given a, the values of K_{232} obtained in the presence of

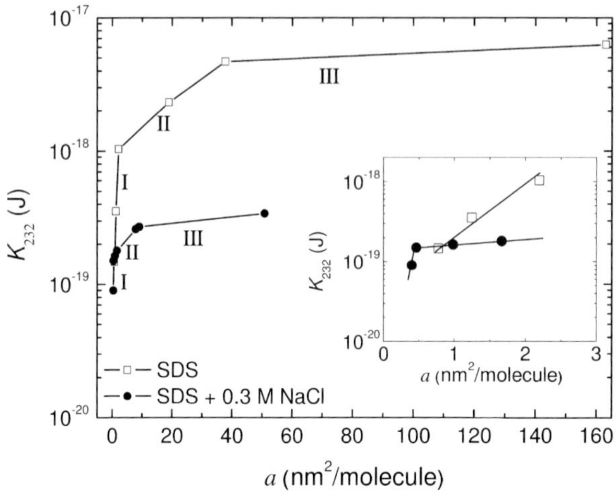

Fig. 7.12 Plots of the hydrophobic force constant (K_{232}) versus the area (a) occupied by individual DS^- ions at the air/water interface. The open squares represent the values of K_{232} for the SDS-stabilized foam films in the absence of NaCl and closed circles represent the same in the presence of 0.3 M NaCl (Wang and Yoon, 2005).

NaCl are substantially lower than those obtained in the absence of NaCl. The lowering of K_{232} in the presence of the electrolyte can be attributed to the presence of Na^+ ions at the interface as counter-ions. From Eq. (14), one can determine the monolayer coverage (Γ_m) to be 8.5×10^{-6} mol m^{-2} in the absence of NaCl, which in turn gives $a=0.20$ nm^2. In the presence of 0.3 M NaCl, Γ_m becomes much smaller (1×10^{-6} mol m^{-2}), with $a=1.67$ nm^2. That a is much larger in the presence of NaCl indicates that significant portion of the air/water interface is occupied by Na^+ ions, which renders the interface hydrophilic and, hence, reduces the hydrophobic force.

Table 7.1 shows the effect of NaCl on a and K_{232} at 10^{-4} M SDS. In the absence of NaCl, $a=19$ nm^2, which is reduced to 0.40 nm^2 in the presence of 0.3 M NaCl. Hence the presence of inorganic electrolyte allows the ionic surfactant moieties to come closer together owing to the screening effect, which gives rise to smaller a and, hence, lower K_{232} values. Therefore, NaCl has dual effects in damping the hydrophobic force: one is to decrease a and the other is to be

Table 7.1 Effect of NaCl on the area occupied by DS^- ions at the air/water interface at 10^{-4} M SDS.

NaCl concentration (M)	Area occupied (a) (nm^2 per molecule)	K_{232} (10^{-20} J)
0	19.0	233
0.3	0.40	9

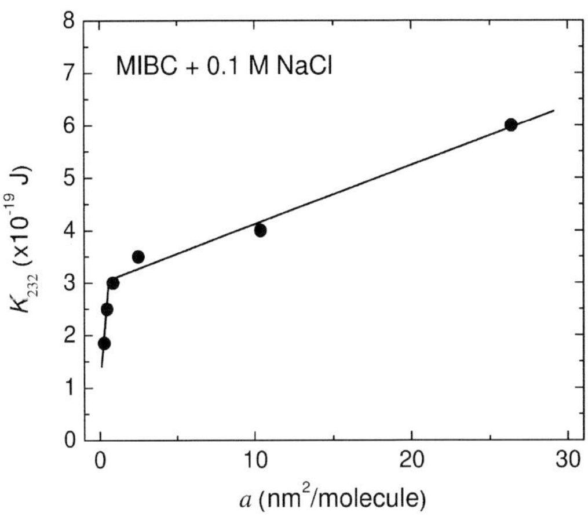

Fig. 7.13 Hydrophobic force constants (K_{232}) for the MIBC-stabilized foam films plotted versus the area (a) occupied by individual MIBC molecules at the air/water interface (Wang and Yoon, 2006).

co-present with the ionic surfactant at the air/water interface. The presence of either DS^- or Na^+ ions should cause a decrease in surface tension and, hence, a decrease in hydrophobic force.

Figure 7.13 shows a K_{232} versus a plot for the foam films stabilized by MIBC. The trend is the same as observed with the SDS-stabilized foam films. Hydrophobic force increases with increasing a, indicating that the pristine air/water interface is most hydrophobic. The plot has two distinct zones. In zone I, a close-packed monolayer of MIBC is formed at the interface, causing a precipitous decrease in K_{232} with decreasing a. In zone II, the adsorption of surfactant molecules is significant but they are not well ordered. That only two zones are observed with MIBC, unlike the three zones observed with the SDS-stabilized foam films, probably indicates that the hydrocarbon chains of this short-chain surfactant are less likely to form surface micelles than the C_{12}-chain surfactant.

7.4.2
Structure

It may be of interest to discuss here the work of Craig et al. (1993), who showed that coalescence of gas bubbles is critically affected by various electrolytes. This finding suggests that an electrolyte by itself (not as counter-ions of ionic surfactants) can damp the hydrophobic force. It is difficult, however, to explain this phenomenon with an adsorption mechanism. As is well known, an inorganic electrolyte undergoes negative adsorption due to the strong hydration energies associated with the ionic species. One possible explanation may be that electrolytes may cause a decrease in the cohesive energy of water and, hence, the hydrophobic force. As suggested by van Oss (1994), hydrophobic interaction arises from the strong cohesive energy of water (–102 mJ m^{-2}). Consider two nanosized gas bubbles suspended in water. Since water molecules cannot form strong hydrogen bonds with them, the bubbles will be pushed away from surrounding water molecules, which may be manifested as a hydrophobic interaction (or force) between the nanosized bubbles. This explanation is similar to the mechanism for the hydrophobic effect between hydrocarbon chains (Tanford, 1980; Lazaridis, 2001). In the presence of electrolytes, the cohesive energy of water may be reduced, resulting in a reduced hydrophobic force. However, the changes in cohesive energy and possibly in water structure may be discernible only at high electrolyte concentrations. Craig et al. (1993) observed the changes in bubble stability in the range 0.007–0.15 M NaCl. They suggested that the answer to their observations may reside in changes in "water structure" associated with solutes.

7.4.3
Long-range Force

The TFPB studies presented in the foregoing sections provided direct evidence that the stability of foam films is controlled by the long-range hydrophobic force. In earlier work, the magnitudes of the hydrophobic forces were estimated using ionic surfactants based on equilibrium film thickness measurements (Yoon and Aksoy, 1999; Wang and Yoon, 2004). The use of ionic surfactants made it possible to determine the double-layer potentials at the air/water interfaces, which in turn were used to determine the magnitudes of the hydrophobic forces using the extended DLVO theory (Eq. 4). However, there was a degree of uncertainty in determining the surface potentials despite the precautions taken in the experiments and appropriate corrections for changes in potential due to counter-ion binding.

More recently, the uncertainties in determining the double-layer potentials have been eliminated by monitoring the kinetics of film thinning at high electrolyte concentrations. The results of the kinetic studies still show the presence of the hydrophobic force, although its magnitude is much smaller than in the absence of electrolytes. That the hydrophobic force is much reduced in the presence of electrolytes does not mean that the long-range force is of electrostatic origin. It has been shown that the hydrophobic force is damped substantially by the presence of an electrolyte at the air/water interface. Similarly, Angarska et al. (2004) measured the critical rupture thicknesses in 0.3 M NaCl and detected the presence of hydrophobic force at low surfactant concentration. It appears, therefore, the hydrophobic force is real and that it is a long-range force. According to the film thinning kinetic studies conducted by Wang and Yoon (2005), the hydrophobic force begins to affect the process from ~250 nm at 10^{-5} M SDS and in the absence of electrolyte (see Fig. 7.3 a). This distance is much longer than that observed between hydrophobic surfaces of macroscopic solids. The strongest evidence that the hydrophobic forces present in foam films are long-range has been provided by the disjoining pressure isotherms. The isotherms obtained at 10^{-4} M SDS and in the absence of electrolyte shows long-range attraction at separation distances above 100 nm.

Meyer et al. (2005) measured long-range hydrophobic forces between mica surfaces coated with double-chain surfactants and suggested that the very long-range hydrophobic forces might be of electrostatic origin. It is difficult to trace its origin to electrostatic attraction when significant non-DLVO attraction is still observed at high electrolyte concentrations. An alternative would be to consider a molecular/structural origin. The hydrophobic force may be related to the strong cohesive energy of water, as has already been discussed. The free energy gained from the hydrogen bonding of water molecules is much larger than the energy of interaction between water and hydrophobic surfaces. This may cause the water film tension to diminish as the thickness shrinks, which is manifested as a long-range attractive force. Laskowski and Kitchener (1969) suggested that "the multi-molecular water layer on the surface of a hydrophobized silica is unstable, which is ascribed to a less favorable state of molecular association at a certain distance from the surface than in normal (bulk) water". These investigators were the first to recognize the existence of a long-range non-DLVO hydrophobic force and to suggest that the long-range character arises from the structural properties of water.

More recently, Eriksson and Yoon (2006) integrated the experimental disjoining pressure isotherms obtained at different temperatures by Tsao et al. (1991) and showed that the excess film entropies ($\Delta S^{f,ex}$) are negative and increase with decreasing distance separating hydrophobic surfaces. This finding suggests that water is more structured in the intervening layer, which may also be sup-

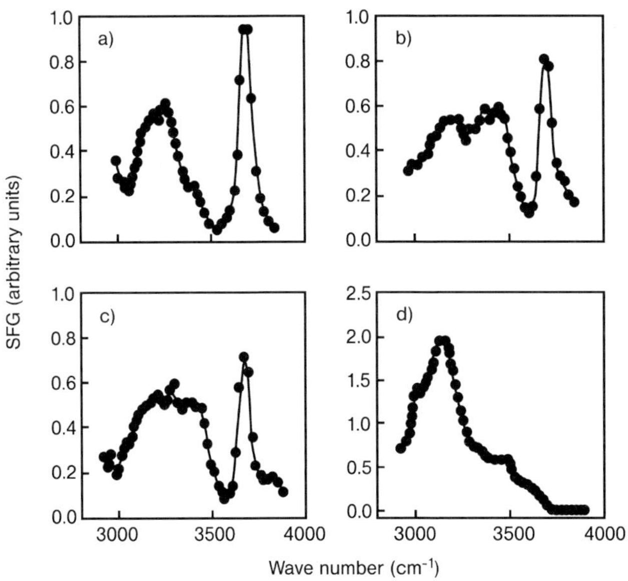

Fig. 7.14 Sum frequency generation (SFG) spectra of the quartz–OTS–water (A), air–water (B), hexane–water (C) and quartz–ice interface (D) (Du et al., 1994).

ported by the sum-frequency (SF) spectra of the water on hydrophobic surfaces. Figure 7.14 shows the SF spectra of four different interfaces (Du et al., 1994). It shows that the SF spectra of the interfacial water on the silica surfaces coated with octadecyltrichlorosilane (OTS) are similar to that of ice, whereas the water on the OTS-coated silica surface shows a peak at 3700 cm^{-1}, which is considered a signature of hydrophobic surfaces. The characteristic peak is due to the dangling (or free) OH groups oriented at the interface, which may indicate that the layer of water in the vicinity of a hydrophobic surface is well ordered compared with the bulk water. The characteristic peak is also observed at the air/water and air/hexane interfaces.

7.5 Implications for Flotation

Froth flotation is the most versatile and efficient separation technology used in the mining industry worldwide. It is used for separating coal from ash- and SO_2-forming minerals, copper sulfides from quartz, potash from halides and anatase from clay, to cite a few examples. Flotation is also used to remove oil from waste water. Sulman and Kirkpatrick-Picard (1905) were the first to obtain a US patent for using air bubbles to collect hydrophobic particles. Although the technology has evolved into many different forms over the years, the basic principle involved in the process remains the same, that is, attachment of hydrophobic particles to air bubbles. That only hydrophobic particles attach themselves to the surface of air bubbles is the basis for separating hydrophobic from hydrophilic particles. Flotation is a fast process despite the fact that both of the DLVO forces operating between bubbles and particles are often repulsive. One way of explaining the discrepancy between theory and practice would be to accept that hydrophobic force exists in the wetting films between bubbles and particles and that its role is to counterbalance the repulsive forces so that the energy barrier is reduced.

Although the hydrophobic force is essential for bubble–particle interactions, it destabilizes the bubbles and foams that are essential for flotation. In general, smaller air bubbles give higher flotation rates and, hence, higher recoveries. In the flotation industry, various non-ionic surfactants are used as frothers to produce small air bubbles. The amounts of the frothers used in the flotation industry are usually in the range 5–10 ppm. At such low concentrations, film elasticities (E) are low while hydrophobic force decreases sharply. It may be reasonable, therefore, to consider that bubbles are stabilized by the decrease in hydrophobic force rather than by the decrease in elasticity. At low concentrations of MIBC and 0.1 M NaCl, the values K_{232} are in the range $(3.7–4.5) \times 10^{-19}$ J, which are high enough to allow hydrophobic particles to be attracted to bubble surface via hydrophobic interaction. These values are more than 25 times larger than the Hamaker constant, A_{232}. The difference between K_{232} and A_{232} should be substantially larger in the absence of electrolyte. At higher MIBC concentrations, the hydrophobic force decreases further, making the hydrophobic interaction weaker. This may be the reason why the mining industry uses relatively weak frothers and almost never strong ionic surfactants such as SDS. It is well known that the excessive use of frothers is detrimental to flotation (Klimpel and Hansen, 1987).

7.6 Conclusion

The thin film pressure balance (TFPB) studies conducted at low surfactant concentrations show that hydrophobic force is present in foam films and that it is a long-range force. Since it is an attractive force, it serves as a major factor destabilizing the bubbles and foams produced at low surfactant concentrations or in the presence of weaker surfactants. It has been found that hydrophobic force decreases with increasing surfactant concentration, particularly in the presence of an inorganic electrolyte. At high concentrations, the stability is, therefore, controlled by the DLVO forces, i.e. double-layer and van der Waals forces.

The decrease in the hydrophobic force in surfactant solutions can be explained by the adsorption of surfactants and counter-ions at the air/water interface, which should in turn hydrophilize the interface. While the high interfacial tension at the air/water interface supports the idea that air bubbles are hydrophobic, there is no molecular and structural evidence that the hydrophobic force is a long-range force. Although the TFPB work discussed in this chapter provides evidences for the existence of the long-range hydrophobic force, further work is needed to elucidate its origin. A better understanding of its origin would be useful for developing more efficient surfactants for controlling the stability of bubbles and foams for different applications.

References

Angarska, J. K., Dimitrova, B. S., Danov, K. D., Kralchevsky, P. A., Ananthapadmanabhan, K. P., Lips, A., *Langmuir 20* (**2004**) 1799.

Ata, S., Ahmed, N., Jameson, G. J., *Int. J. Miner. Process. 72* (**2003**) 255.

Cascão Pereira, L. G., Johansson, C., Blanch, H. W., Radke, C. J., *Colloids Surf. A: Physicochem. Eng. Aspects 186* (**2001**) 103.

Christenson, H. K., Yaminsky, V. V., *J. Phys. Chem. 99* (**1995**) 10420.

Claesson, P. M., Blom, C. E., Herder, P. C., Ninham, B. W., *J. Colloid Interface Sci. 114* (**1986**) 234.

Comley, B. A., Harris, P. J., Bradshaw, D. J., Harris, M. C., *Int. J. Miner. Process. 64* (**2002**) 81.

Coons, J. E., Halley, P. J., McGlashan, S. A., Tran-Cong. T., *Colloids Surf. A: Physicochem. Eng. Aspects 263* (**2005**) 197.

Craig, V. S. J., Ninham, B. W., Pashley, R. M., *J. Phys. Chem. 97* (**1993**) 10192.

Denkov, N. D., *Langmuir 20* (**2004**) 9463.

Derjaguin, B. V., Landau, L., *Acta Physicochim. URSS 14* (**1941**) 633.

Du, Q., Freysz, E., Shen, Y. R., *Science 264* (**1994**) 826.

Eriksson, J. C., Ljunggren, S., *Colloids Surf. 38* (**1989**) 179.

Eriksson, J. C., Yoon, R.-H., in "A century of Flotation", Fuerstenau, M. C., Jameson, G. J. and Yoon, R.-H. (ed.), Society of Mining Engineers, Golden, CO, **2006**, in press.

Eriksson, J. C., Ljunggren, S., Claesson, P. M., *J. Chem. Soc., Faraday Trans. 2 85* (**1989**) 163.
Exerowa, D., Kruglyakov, P. M., *Foam and Foam Films*, Elsevier, Amsterdam, **1998**.
Fruhner, H., Wantke, K.-D., Lunkenheimer, K., *Colloids Surf. A: Physicochem. Eng. Aspects 162* (**2000**) 193.
Israelachvili, J. N., *Intermolecular and Surface Forces*, Academic Press, London, **1992**.
Israelachvili, J. N., Pashley, R. M., *Nature 300* (**1982**) 341.
Ivanov, I. B., Danov, K. D., Ananthapadmanabhan, K. P., Lips, A., *Adv. Colloid Interface Sci. 114/115* (**2005**) 61.
Kalinin, V. V., Radke, C. J., *Colloids Surf. A: Physicochem. Eng. Aspects 114* (**1996**) 337.
Klimpel, R. R., Hansen, R. D. in Somasundaran, P., Moudgil, B. M. (eds.), *Reagents in Mineral Technology, Surfactant Science Series*, Vol. 27, Marcel Dekker, New York, **1987**, p. 395.
Kralchevsky, P. A., Danov, K. D., Broze, G., Mehreteab, A., *Langmuir 15* (**1999**) 2351.
Langevin, D., *Adv. Colloid Interface Sci.* 88 (**2000**) 209.
Laskowski, J., Kitchener, J. A., *J. Colloid Interface Sci. 30* (**1969**) 391.
Lazaridis, T., *Acc. Chem. Res. 34* (**2001**) 931.
Manev, E. D., Sazdanova, S. V., Wasan, D. T., *J. Colloid Interface Sci. 97* (**1984**) 591.
Mathe, Z. T., Harris, M. C., O'Connor, C. T., Franzidis, J.-P., *Miner. Eng. 11* (**1998**) 397.
Meyer, E. E., Lin, Q., Hassenkam, T., Oroudjev, E., Israelachvili, J. N., *Proc. Natl. Acad. Sci. USA 102* (**2005**) 6839.
Neethling, S. J., Cilliers, J. J., *Int. J. Miner. Process. 72* (**2003**) 267.
Radoev, B. P., Scheludko, A. D., Manev, E. D., *J. Colloid Interface Sci. 95* (**1983**) 255.
Russel, W. B., Saville, D. A., Schowalter, W. R., *Colloidal Dispersions*, Cambridge University Press, Cambridge, **1989**.
Scheludko, A., *Adv. Colloid Interface Sci. 1* (**1967**) 391.
Scheludko, A., Exerowa, D., *Commun. Dept. Chem., Bulg. Acad. Sci. 7* (**1959**) 123.
Scheludko, A., Platikanov, D., *Kolloid Z. 175* (**1961**) 150.
Sharma, A., *J. Colloid Interface Sci. 199* (**1998**) 212.
Stubenrauch, C., Miller, R., *J. Phys. Chem. B 108* (**2004**) 6412.
Sulman, H. L., Kirkpatrick-Picard, H. F., *US Patent* 793808 (**1905**).
Tanford, C., *The Hydrophobic Effect: Formation of Micelles and Biological Membranes*, 2nd edn., Wiley-Interscience, New York, **1980**.
Tchaliovska, S., Manev, E., Radoev, B., Eriksson, J. C., Claesson, P. M., *J. Colloid Interface Sci. 168* (**1994**) 190.
Tsao, Y., Yang, S. X., Evans, D. F., Wennerström, H., *Langmuir 7* (**1991**) 3154.
Valkovska, D. S., Danov, K. D., Ivanov, I. B., *Adv. Colloid Interface. Sci. 96* (**2002**) 101.
van Oss, C. J., *Interfacial Forces in Aqueous Media*, Marcel Dekker, New York, **1994**.
Verwey, E. J. W., Overbeek, J. Th. G., *Theory of the Stability of Lyophobic Colloids*, Elsevier, Amsterdam, **1948**.
Vrij, A., *Discuss. Faraday Soc. 42* (**1966**) 23.
Vrij, A., Overbeek, J. Th. G., *J. Am. Chem. Soc. 90* (**1968**) 3074.
Wang, L., Yoon, R.-H., *Langmuir 20* (**2004**) 11457.
Wang, L., Yoon, R.-H., *Colloids Surf. A: Physicochem. Eng. Aspects 263* (**2005**) 267.
Wang, L., Yoon, R.-H., *Colloids Surf. A: Physicochem. Eng. Aspects 282–283* (**2006**) 84.
Warszynski, P., Barzyk, W., Lunkenheimer, K., Fruhner, H., *J. Phys. Chem. B 102* (**1998**) 10948.
Xu, Z., Yoon, R.-H., *J. Colloid Interface Sci. 132* (**1989**) 532.
Xu, Z., Yoon, R.-H., *J. Colloid Interface Sci. 134* (**1990**) 427.
Yaros, H. D., Newman, J., Radke, C. J., *J. Colloid Interface Sci. 262* (**2003**) 442.
Yoon, R.-H., Aksoy, B. S., *J. Colloid Interface Sci. 211* (**1999**) 1.
Yoon, R.-H., Ravishankar, S. A., *J. Colloid Interface Sci. 179* (**1996**) 403.

8
Surfactant Nanostructures in Foam Films

Elena Mileva and Plamen Tchoukov

Abstract

A survey of recent experimental and theoretical investigations of microscopic foam films containing self-assembled amphiphilic nanostructures is presented. The specific advantage of the microscopic film technique is that the fine control of system parameters allows the estimation of the respective changes in film properties for low surfactant contents and extremely small concentration changes. This gives a unique possibility of reaching amphiphile quantities when initial onset of self-assembly is to be observed. The film characteristics are investigated via a microinterferometric method, which operates with the measuring cell of Scheludko and Exerowa. The experimental setup is additionally improved by including video recording and consecutive image analysis. The results show the following: (1) unstable black patterns (dots and spots) are observed, which have very short lifetimes and the films that contain them rupture quickly; (2) several of the kinetic characteristics of the films display a sharp change within a narrow surfactant concentration range. The experiments are interpreted on the basis of the assumption that a series of self-assembled aggregates exist at the interfaces and inside the thin film. The proposed theoretical scheme suggests a mechanism connecting the formation of unstable black patterns (dots and spots) with the reorganization and destruction of the existing surfactant assemblies both in the bulk of the film and on its interfaces. The results suggest that the observed unstable black formations may serve as indicators for the presence of surfactant nanostructures in amphiphilic solutions and the microscopic foam film technique has major potential as prospective instrumentation in the study of amphiphilic self-assemblies.

Colloids and Interface Science Series, Vol. 1
Colloid Stability: The Role of Surface Forces, Part I. Edited by Tharwat F. Tadros

ISBN: 978-3-527-31462-1

8.1 Background

The investigation of the self-assembly of amphiphilic molecules into micellar aggregates is one of the oldest topics in colloid chemistry [1, 2]. Within the wide diversity of nanophenomena in fluid media, of particular importance is the onset of self-assembled nanostructures in aqueous solutions of surfactants [3–11].

For more than 40 years the microscopic foam film technique has been a major instrumentation mode in the study of surface forces and stability of colloid systems [12–17]. One of the major advantages of this instrumentation is the possibility of registering and following fine changes in these forces during the drainage of the liquid films. The films have radii in the range 10–500 μm and are formed in the Scheludko-Exerowa measuring cell of the thin liquid film microinterferometric technique. This instrumentation allows the direct measurement of most parameters that determine the properties and the behavior of foam films. The microscopic foam film was the first simple model of two interacting fluid interfaces related to the study of the classical theory of Derjaguin, Landau, Verwey and Overbeek (DLVO theory) [12–23] and also provided its first experimental verification [13–15, 23].

Recent investigations on foam films from aqueous amphiphilic solutions have proved that the film drainage kinetics contain important information about the presence and reorganization of self-assembled nanostructures [24–27]. These studies are related to some other surfactant solution properties and particularly to the data obtained in surface tension measurements [28, 29]. In earlier studies, Exerowa and Scheludko [30] and later Exerowa and coworkers [28, 29, 31] reported precise results concerning the surface tension isotherms for amphiphilic solutions. The basic outcome was that the surface tension curves of aqueous solutions of certain surfactants (alkyl sulfate homologues, etc.) in the presence of added electrolyte run in an unexpected manner.

In Fig. 8.1 are reproduced the data [28, 29] for the case of solutions of sodium dodecyl sulfate obtained by the spherotensiometric technique of Scheludko [32] (accuracy $\pm$0.005–0.01 mN m^{-1}). The curves contain distinct kinks and plateau sectors. These regions are situated within the range of amphiphilic quantities that are orders of magnitude lower than both the bulk critical micellar concentration (CMC) and the close packing values of the adsorption layers on the water/air interface. One plateau portion is distinctly observed when the quantity of the added electrolyte is high: $C_{\mathrm{el,cr}}=0.5$ M (curve 1). For a lower quantity of added electrolyte, $C_{\mathrm{el,cr}}=0.1$ M, there are two portions: a plateau at lower surfactant concentration and a kink at high surfactant concentration (curve 2). Exerowa and coworkers [28, 29] were the first to advance the idea that this odd behavior might be due to the presence of smaller self-assembled structures. They called them *premicelles* and introduced the concept of critical premicellar concentration (CPC). If premicellar entities could exist, their presence should inevitably show up in the results from other types of experiments also, as is the case with the usual micelles (at the CMC).

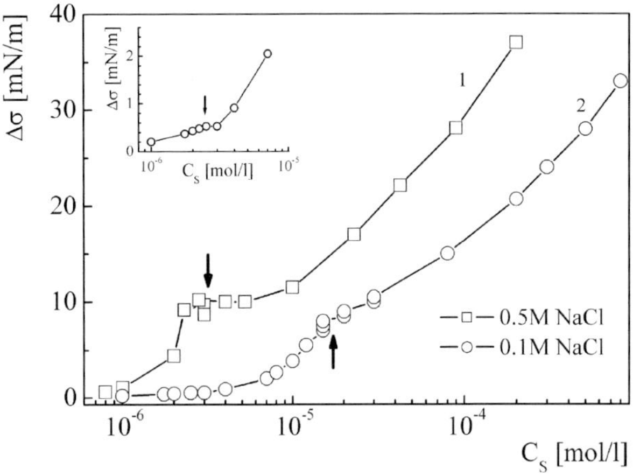

Fig. 8.1 Surface tension isotherms of sodium dodecyl sulfate solutions at different electrolyte (NaCl) concentrations and 22 °C.

One specific feature of the emerging surfactant self-assemblies in aqueous solutions is their labile character. This is allied to the dual structure of the amphiphiles and to the innate cause of the onset of these nanostructures, namely they appear as a result of weaker interactions summarized in the notion of "principle of opposing forces" [8, 10, 11]. Thus, although the amphiphilic aggregates are finite, they do not maintain permanent identities but are continuously exchanging fragments with each other [8, 33, 34]. For the same reason, these aggregates are usually very susceptible to every modification of the system's conditions: composition of the solution, temperature, proximity of interfaces, etc. Variation of any of the above affects the overall distribution of the amphiphilic structures. The labile character of the amphiphilic nanostructures, however, brings about serious difficulties in the identification of these species. There is a constant demand for more diverse approaches that could address such systems. The study of the reorganization and destruction of surfactant nanostructures using thin liquid film microinterferometric instrumentation has promoted a new option of this kind [24–27].

Generally, the presence of a phase boundary influences the specific conditions for the formation of self-assembled structures and induces changes in their size distribution compared with the bulk solution. Although the overall self-assembling principles are fully operative, the proximity of an interface induces specific demands for the form of the aggregates and the size distribution of the emerging surfactant aggregates in its proximity [35]. These reorganization possibilities potentially contain valuable structural information about the mechanism of surfactant self-assembly. An important issue is the accurate differentiation between the specific adsorption effects and the self-assembling phenomena. This distinction demands "milder" experimental conditions that are provided by fluid interfaces. The water/air interface in particular ensures an alternative "hydrophobic option" that might interfere in the self-assembling process near this interface.

The specific kinetic and thermodynamic properties of thin liquid films and especially the disjoining pressure create more additional options for the impact on existing micellar entities. The role of the disjoining pressure might be regarded as an influence of an "outer" background field. Thus, upon drainage of foam films, which originate from solutions containing self-assembled nanostructures, the initial bulk and interface size distributions are altered in comparison with the respective values in the primary solution. The existing aggregates are reorganized and destroyed. The effect of these events is evidenced through the specific run of common parameters that characterize the thinning of foam films [24–27, 36, 37]. Hence foam films can serve as very suitable instrumentation for model studies of micellar solutions. The results obtained via this "dynamic" tool, coupled with static and dynamic investigations of adsorption behavior, supply more detailed information about the existence and structure of surfactant assemblies in amphiphilic solutions.

The aim of this chapter is to give a survey of recent experimental data and model considerations that illustrate the impact of amphiphilic nanostructures on the drainage of microscopic foam films from aqueous surfactant solutions. The results obtained support the notion that the microinterferometric setup equipped with a Scheludko-Exerowa cell can serve as a valuable tool for the registration and investigation of self-assembled structures in surfactant solutions.

8.2
Drainage of Microscopic Foam Films

The experiments are performed with sodium dodecyl sulfate (SDS), specially synthesized for us by Henkel KGaA, Germany, that does not display a minimum in the surface tension isotherm. The experimental conditions are exactly the same as in the case of surface tension measurements [28, 29], namely, aqueous solutions of concentration range within $C_S \approx 10^{-6}$–10^{-4} M SDS~ ($C_S < C_{CMC}$). The added electrolyte is sodium chloride (NaCl, Merck), heated at 600 °C, and two concentrations are investigated: $C_{el} = 0.5$ and 0.1 M. The temperature is maintained strictly at 22 ± 0.1 °C. Triply-distilled water is used with electrical conductivity $k = 1.0$–$1.1 \times 10^{-6}\ \Omega^{-1}\ \mathrm{cm}^{-1}$.

The foam films are studied by the microinterferometric method, which operates with the measuring cell of Scheludko and Exerowa [13–15]. It is shown schematically in Fig. 8.2a. Microscopic films with radii of about 100 μm are formed in the middle of a biconcave drop, situated in a glass tube of diameter 0.4 cm, by withdrawing the liquid from it. In the case of the lowest concentrations and the blank probe a variant with a supplementary reservoir next to the meniscus is also used [14, 15]. The classical experimental scheme is additionally modified with video registration via a CCD camera (Sony, DXC-107P) (Fig. 8.2 b). The digitized image is processed with a powerful PC using a capture video card.

A particular advantage of the film techniques is that the microscopic dimensions of the thin liquid layer allow the possibility of dealing with very low sur-

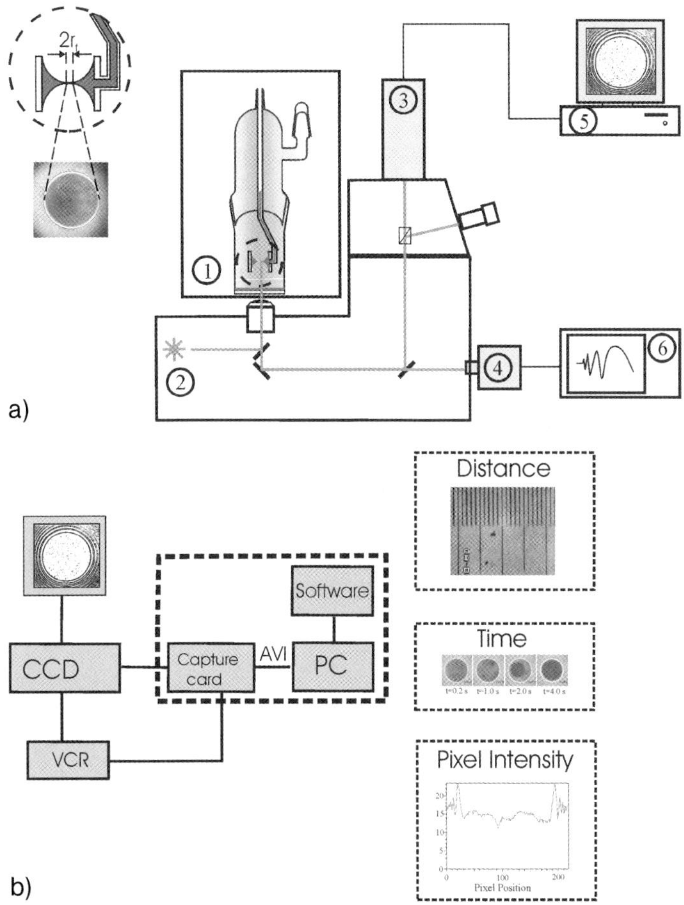

Fig. 8.2 (a) Microinterferometric experimental setup with the measuring cell of Scheludko and Exerowa. (1) Measuring cell in thermostating device; (2) microscope; (3) CCD video camera; (4) photomultiplier; (5) PC with capture video card; (6) $Y(t)$ recorder. (b) Schematic diagram of the stages in processing of the video recording.

factant concentrations of the primary amphiphilic solutions. This gives the possibility of tracing back to the conditions where the initial onset of (pre)micellar structures might be registered. Therefore, the general aim to outline the impact of surfactant assemblies on the drainage of microscopic foam films was reduced to the following specific tasks: (1) to juxtapose the foam film drainage properties on the previously obtained surface tension results and (2) to find the film characteristics that are sensitive to the presence of amphiphilic nanostructures.

The investigation of the foam film drainage with respect to surfactant concentration provides two types of results: (1) qualitative observations (unstable black patterns – dots and spots – are observed in the background thinning film);

(2) specific drainage characteristics (evolution of film thickness values, film drainage times).

8.2.1 Black Patterns

The observed black patterns may formally be classified as of two types. The first type (spots, black films) is characteristic of higher concentration ranges of the amphiphile used and are precursors of the classical black films [common black films (CBFs) and Newton black films (NBFs)] [13–15, 38, 39]. These patterns generally evolve into films that survive within longer time intervals (minutes, hours). They are very well described and have been related to the lifetimes and the stability of liquid films [13–15]. The common feature of the surfactant systems where these precursors of black films are observed is that the adsorption surfactant coverage on the liquid/air interface of the initial solutions is almost close packed. From this point of view, these black formations mark a specific stage of the evolution of black (plane-parallel) films, namely the transition to a new thickness of relatively stable configuration: CBF or NBF. This thickness transition has been denoted the critical thickness h_{cr} and the surfactant concentration at which the onset of this type of black patterns is observed is C_{bl} [14, 15, 39].

With certain amphiphilic solutions, however, black patterns of a second type, dots and unstable black spots, are observed [24–29]. The unstable black formations appear when there is some deficiency of the surfactant in the adsorption coverage on the interfaces for miscellaneous reasons. These formations have lifetimes of no more than several seconds. The respective foam films drain quickly and survive for at most a minute or two. Their onset is also closely related to the specific structure of the surfactant. There are no systematic data in this sense, but it seems that, as a rule, the stabilizing amphiphile should have a long hydrophobic tail as opposed to a comparatively small hydrophilic head [40, 41]. This surfactant structure ensures the formation of foam films at lower surfactant concentrations, which, although of shorter lifetime, still live long enough for it to be possible to investigate them.

The black dots (Fig. 8.3) appear at the lowest surfactant concentrations [at about $(2–3)\times10^{-6}$ M SDS]. They live for 3–10 s and do not grow in size. It should be noted that the minimum concentration for the first onset of these black patterns does not depend on the electrolyte concentration. Within the range of electrolyte concentrations added here, the "critical" value is at about $(2.0–2.5)\times10^{-6}$ M SDS [27]. The spots (Fig. 8.4) are characteristic of higher amphiphile concentrations. They live for less than 1 s but quickly grow in size. Upon increase of the amphiphile concentration both their number and the probability of their onset rise (Figs. 8.5 and 8.6).

In Figs. 8.3 and 8.4 are presented typical pictures of the foam films and results from the respective image analysis. The morphological difference between the two black formations is well distinguished. The "black dots" (Fig. 8.3 b) are shallow local thinnings with irregular thickness. The "spots" are thinner black

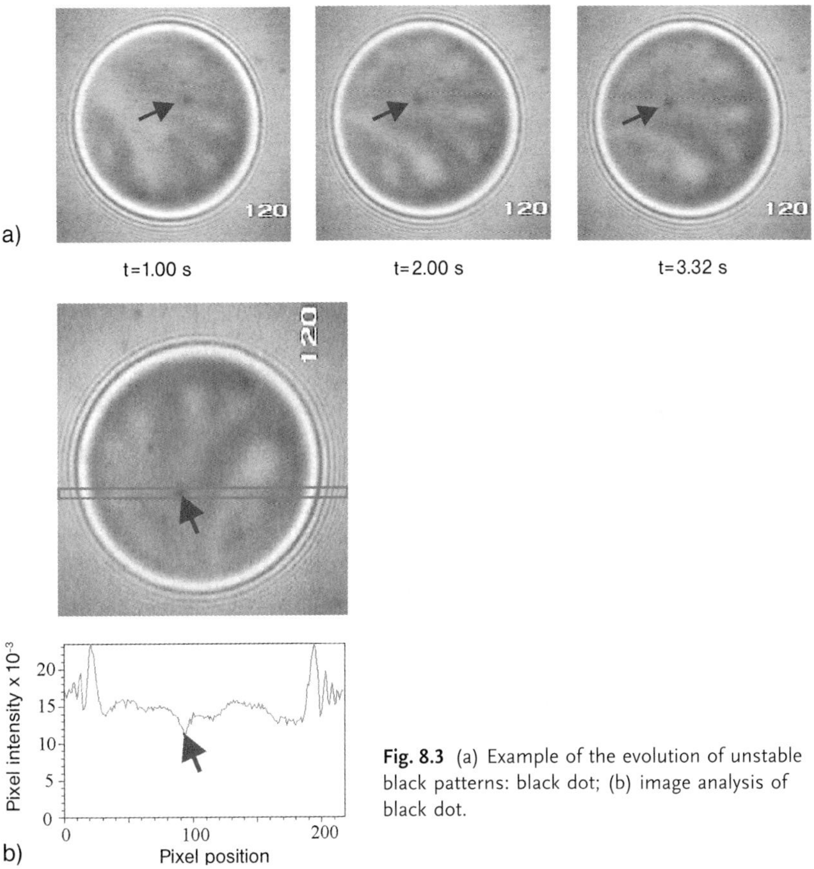

Fig. 8.3 (a) Example of the evolution of unstable black patterns: black dot; (b) image analysis of black dot.

patterns and may be viewed as plane-parallel portions (microfilms) within the background foam film (Fig. 8.4 b).

Although the surfactant concentration range is the same, there are some nuances in their lifetimes with respect to the added salt quantity. Namely, at higher electrolyte concentration (0.5 M NaCl), the dots live for a longer time and are well observed. At lower electrolyte concentration (0.1 M NaCl), there is a more subtle difference between the dots and the spots. The onset of a dot is first observed when the surfactant concentrations are from the range of the plateau portions of the surface tension isotherm (see the inset in Fig. 8.1). With a rise in the surfactant concentration, the increase in spot number is sometimes accompanied by the onset of new dots, which live for several seconds without change in size, but then grow and form larger, unstable spots.

The basic result of the qualitative observations is the concentration synchrony of the onset and properties of the unstable black patterns (black dots and spots) in the draining foam films and the kink and plateau regions in the surface tension isotherms. The appearance of black dots might be considered as an indica-

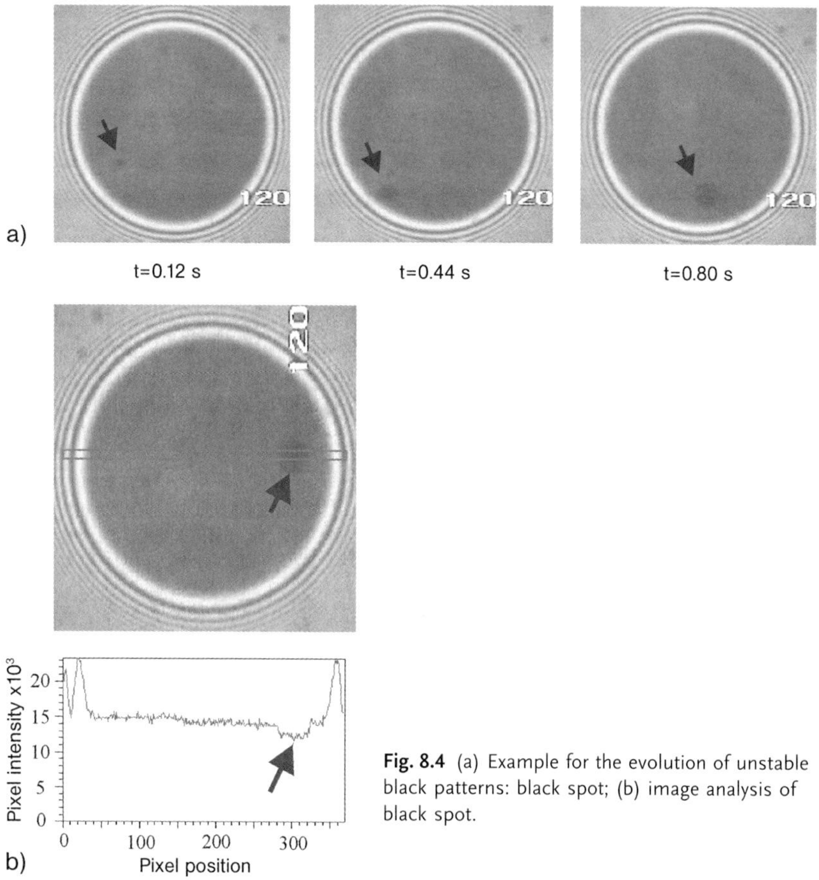

Fig. 8.4 (a) Example for the evolution of unstable black patterns: black spot; (b) image analysis of black spot.

tor for the presence of amphiphilic structures in the initial surfactant solutions from which the respective foam films are formed [37], while the onset and evolution of the black spots mark the crucial change in the interrelation between the bulk-film drainage properties and the change in the adsorption layers on the interfaces [25, 27]. The persistent emergence of, although unstable, clearly distinguished black patterns (dots and spots) influences the drainage behavior of the foam films.

8.2.2
Drainage Characteristics

The video recording and the image analysis permit closer examination of the drainage evolution of the foam films within the entire concentration interval that is investigated. The foam films studied do not reach equilibrium thickness [25, 26]. Provided that the investigated foam films originate from surfactant so-

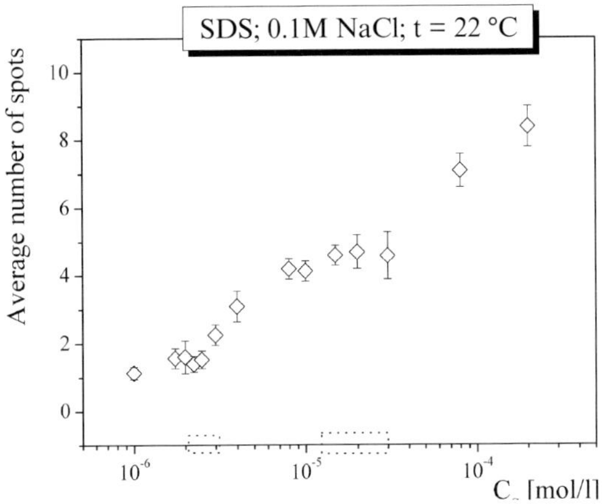

Fig. 8.5 Average number of spots versus surfactant concentration. The dashed rectangles denote the concentration intervals where the kinks in the surface tension curve are observed.

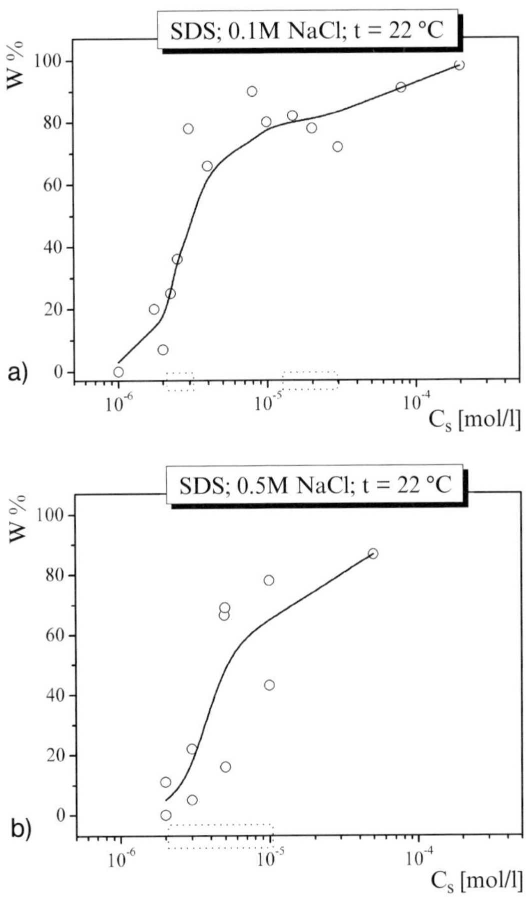

Fig. 8.6 Probability of observing unstable black patterns versus surfactant concentration. The dashed rectangle denotes the concentration interval where the plateau portion of the surface tension curve is observed. C_{el} = (a) 0.1, (b) 0.5 M NaCl.

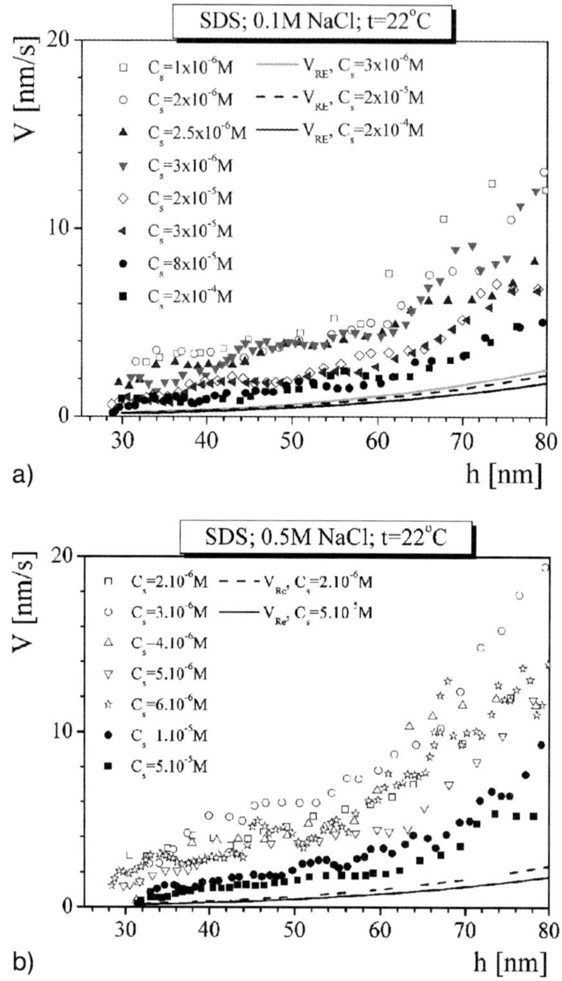

Fig. 8.7 Experimental thinning velocity results versus actual film thickness for certain surfactant concentrations. C_{el} = (a) 0.1 and (b) 0.5 M NaCl. The lines denote the calculated values for the Reynolds drainage in the respective cases.

lutions of concentrations that are lower than both the close packing value and the CMC values, they drain quickly and rupture within 1 min. The film drainage itself is performed in a regime of increased tangential mobility of the film interfaces, as can be seen from the representative results for the drainage velocity against the film thickness shown in Fig. 8.7.

Particularly informative are the following drainage characteristics of the studied films: (1) the evolution of film thickness with time and (2) film drainage times against the surfactant concentration of the initial solutions.

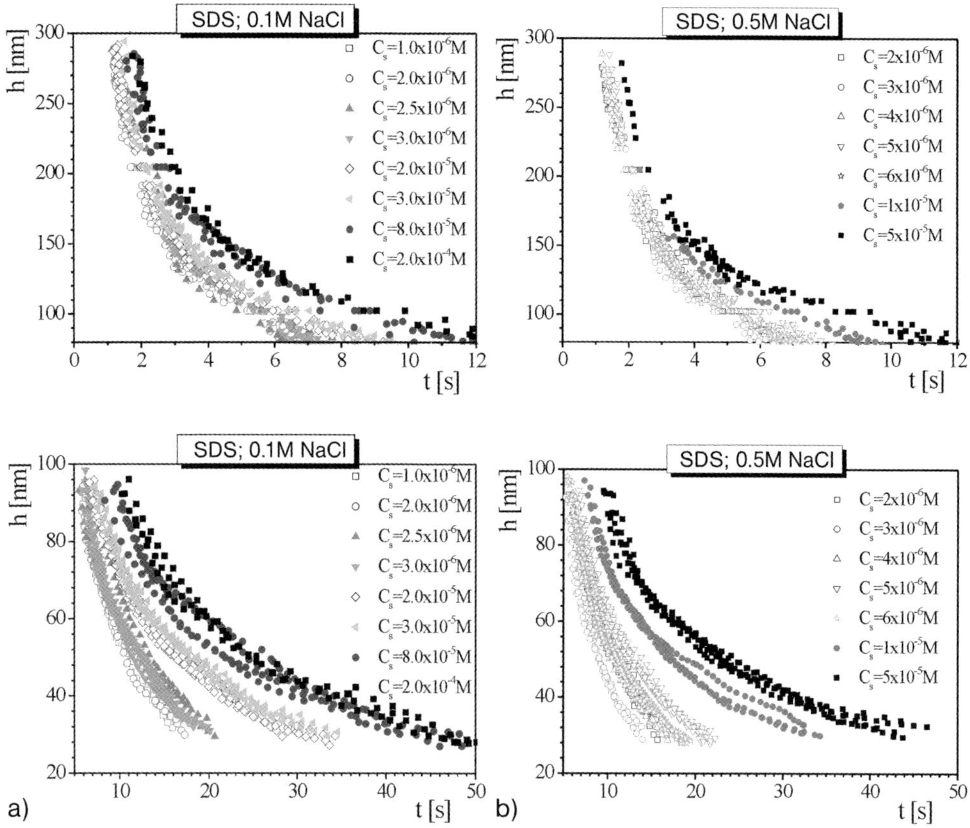

Fig. 8.8 Mean thickness evolution with time for C_{el} = (a) 0.1 and (b) 0.5 M NaCl.

The evolution of the film thickness with time is presented in Fig. 8.8. Depending on the surfactant concentration, the curves run in such a way that a characteristic "broom" of curves is formed. At high thickness values when the disjoining pressure is not still operative, the curves run in a bunch together. At lower thickness values, when a notable impact of the disjoining pressure is to be expected, the "broom" separates into several distinct bunches, that stretch out wide. These bunches are grouped in such a way that indicates the launch of the concentrations where the peculiarities in the surface tension curves are observed. Thus, in the case of the lower electrolyte concentration, the "broom" consists of three distinct bunches, grouped so as to indicate the onsets of the plateau and the kink regions (Fig. 8.8 a). The same is observed for higher electrolyte concentrations. Insofar as there is only one distinct plateau region in the surface tension isotherms, the respective bunches are just two (Fig. 8.8 b).

The film drainage times are specified from the moment of formation of a thin film until the onset of black spot or rupture. The cumulative results for the dependence of the mean drainage times (τ_m) for film radii of 100 μm are shown

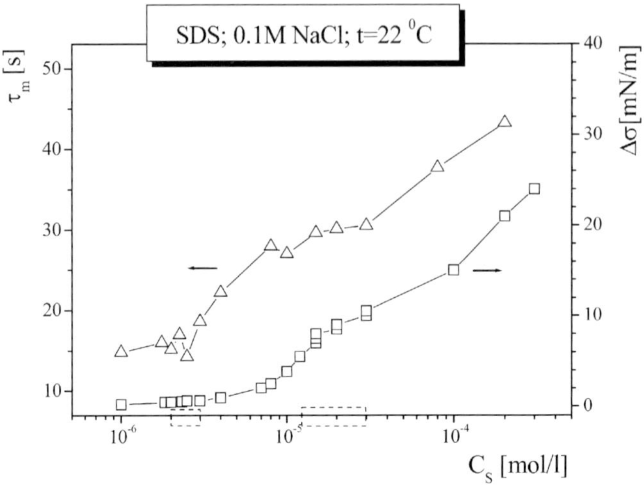

Fig. 8.9 Mean drainage times of foam films against the surfactant concentration of initial sodium dodecyl sulfate solutions at C_{el}=0.1 M NaCl.

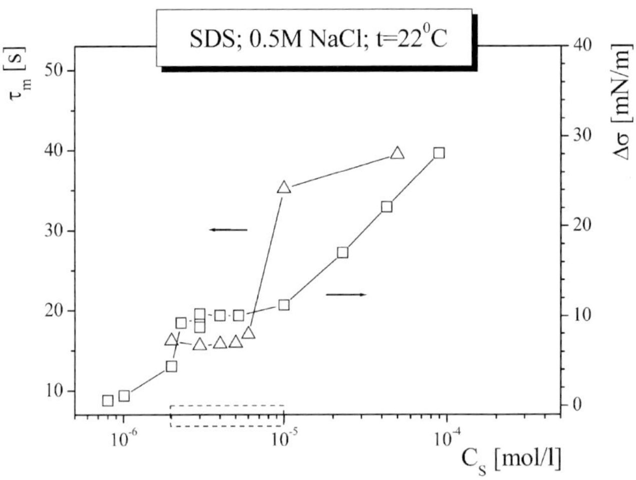

Fig. 8.10 Mean drainage times of foam films against the surfactant concentration of initial sodium dodecyl sulfate solutions at C_{el}=0.5 M NaCl.

in Figs. 8.9 and 8.10. The experimental results for the kinetics of foam film drainage are juxtaposed on the concentration peculiarities of the surface tension isotherms for the same surfactant systems. As one can see, there is a marked concentration coincidence of the run of foam film drainage times and the kink and plateaus in the respective surface tension isotherms.

8.3 Understanding the Experimental Results

The major results from the systematic foam film experiments are the following. First, all the film drainage peculiarities (black patterns, $h(t)$ curves, lifetimes) show a remarkable concentration coincidence with the odd properties of the surface tension curves. Second, within the film experiments, the number of the black patterns and the probability of their onset on the one hand and the lifetime of the films on the other rise in a synchronized manner with respect to the concentration. To summarize, the "irregular" run of the surface tension curves is a clear sign of a change in the state of the adsorption layer on the air/solution interface upon increase of the surfactant concentration. Further, in the film experiments, stepwise changes in the drainage characteristics appear persistently within the same concentration intervals, where the "irregularities" of the surface tension isotherms are observed.

In an attempt to seek an explanation of this "abnormal" behavior, the concept of the possible existence of surfactant nanostructures in the initial solutions was first abandoned. Therefore, systematic measurements of the critical thickness of the films were made. The results for both electrolyte concentrations are shown in Fig. 8.11. The critical thickness shows no distinct peculiarity and remains virtually constant within the investigated surfactant concentration interval. The run of the curves does not exhibit any trait that might be related to the respective surface tension and the foam film drainage results.

There is always a close relationship between the black pattern formation and the state of the adsorption layer on the film interfaces. In our particular case, the quantity of the stabilizing surfactant is insufficient to ensure the formation of equilibrium films, the foam films are unstable and they drain continuously and quickly within 1 min or so. Therefore, the changes in the surfactant coverage due the Marangoni effect must have an important effect on the drainage [42, 43] and one should account for the coupling of film hydrodynamics and the mass transfer of the stabilizing surfactant [44–46]. This fact alone, however,

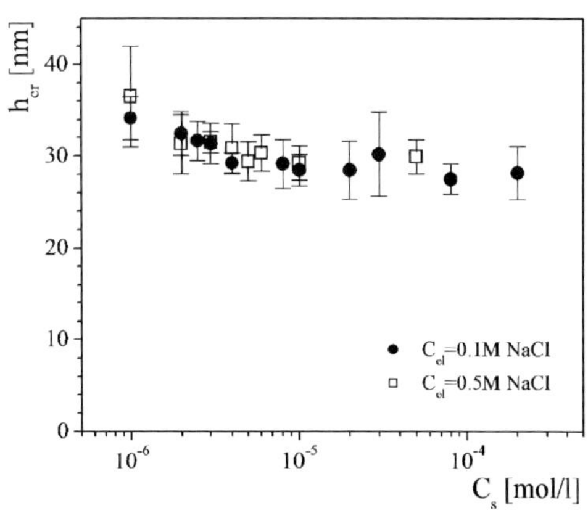

Fig. 8.11 Critical thickness of foam films as a function of the surfactant concentration.

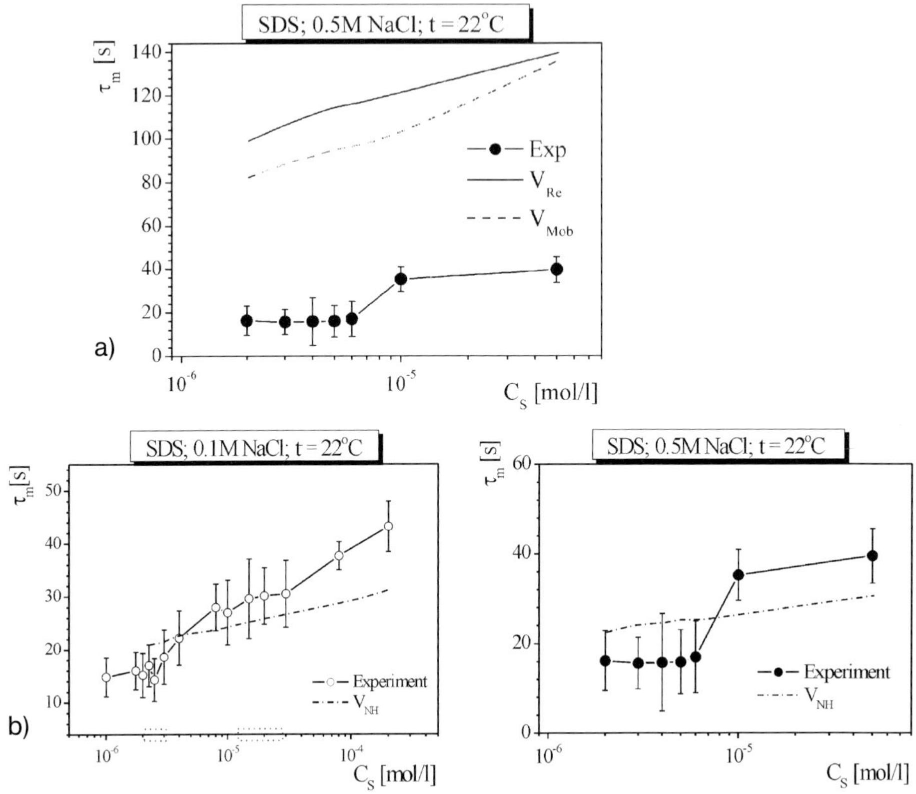

Fig. 8.12 Experimental data and drainage time according to conventional film hydrodynamics. (a) Influence of Marangoni effect (notations according to [43b]): $V_{mob}/V_{Re}=1+b+h_s/h$; $b=-(3\eta D)/\Gamma_0(\partial\sigma_0/\partial c)$; $h_S=[6\eta(\partial\Gamma_0/\partial c)/\Gamma_0(\partial\sigma_0/\partial c)]$; $V_{Re}=(2h^3\Delta P/3\eta R^2)$. (b) Influence of non-homogeneities in film thickness; V_{NH} is the velocity according to the law in [47].

would just back up the general tendency of the overall rise of the drainage time against the surfactant concentration. It cannot explain the stepwise changes in the foam film drainage experiments and their concentration coincidence with the peculiarities of the surface tension results with increase in the overall amphiphile quantity. The inadequacy of the common film hydrodynamic approach is illustrated by an attempt to relate the drainage times to the known hydrodynamic factors that are fully operative in our case also, namely the influence of the Marangoni effect [43] and the impact of thickness non-homogeneities on the film hydrodynamics (Fig. 8.12) [47].

Hence the coupled peculiarities of the "static" surface-tension measurements and of the "dynamic" foam film drainage characteristics cannot be understood within the premise of the monomolecular state of the amphiphile inside the film bulk. Therefore, a different notion is needed to correlate these two types of experiments.

To start, three key moments are mostly essential for the interpretation and analysis of the foam film drainage results:

1. amphiphilic nanostructures can exist in the initial surfactant solutions;
2. surface forces are dominated by the van der Waals disjoining pressure;
3. the films are draining in a regime of high interfacial mobility of the film interfaces and thickness non-homogeneities.

8.3.1
Premicellar Concept

The starting point of the interpretation of the experimental observations is the understanding that under the particular experimental conditions in the initial surfactant solutions smaller self-assembled structures (premicelles) may appear at concentrations lower than the CMC value. These nanostructures are somewhat different from the true micellar entities in related cases of stratified films originating from surfactant solutions at concentrations above the CMC [48]. Generally, the ordinary theory of micelle formation does not in itself pose any *a priori* restrictions on the minimum size of the aggregates [9]. It just states that the onset of self-assembled species in amphiphilic solutions is a result of the action of weaker "opposing forces". Therefore, if premicellar aggregates could exist, the general self-assembling principles should not be altered. Hence one possibility for modeling the geometry of the premicellar aggregate is to present them as symmetrical crumbly aggregates like the so-called plateau bodies [41]. An attempt to construct the size distribution of amphiphilic nanostructures based on this idea shows that such a notion is energetically plausible [41].

8.3.2
Surface Forces in the Films and Surfactant Self-assemblies

Applied to the foam film, the theory of amphiphilic self-assembly accounts for its most important characteristic, namely the disjoining pressure [36, 37, 40]. Foam films are regarded as obtained from a surfactant solution that initially contains self-assembled entities characterized by definite size distributions. The film itself is modeled as a bulk phase and two 2D surface phases (air/liquid interfaces) [23, 49]. It is presumed that the amphiphile self-assembly is possible in all the phases [36, 37, 40]. The latter are in thermal, mechanical and chemical equilibrium.

The specific bulk quantities are denoted with a superscript b. The size distribution curve of the micellar species for the bulk of the film may be presented as

$$X_n^b = X_n^{0,b} X_n^{int,b} X_n^{ad,b} X_n^{f,b} \tag{1}$$

where

$$X_n^b = \frac{N_n^b}{N_w^b + \sum_i iN_i^b}, \quad X_w^b = \frac{N_w^b}{N_w^b + \sum_i iN_i^b} \tag{2}$$

and

$$X_n^{0,b} = (X_1^b)^n \exp\left(\frac{n\mu_1^{0,b} - \mu_n^{0,b}}{kT}\right) \tag{3 a}$$

$$X_n^{int,b} = \exp\left[\frac{\sum_j \left(nu_{1j}^b - u_{nj}^n\right) N_j^b}{kTV_f}\right] \tag{3 b}$$

$$X_n^{ad,b} = \exp\left(\frac{-2\sigma_b}{kT}\right) \tag{3 c}$$

$$X_n^{f,b} = \exp\left\{\frac{2}{kT}[2\sigma_f(h) + \Pi(h)h]\left(n\frac{\partial A_f}{\partial N_1^b} - \frac{\partial A_f}{\partial N_n^b}\right)\right\}. \tag{3 d}$$

In the above, X_n^b and X_w^b are the mole fractions of the bulk solution aggregates and the solvent, respectively. The details of the theoretical scheme and the first two multipliers in Eq. (1) are reported elsewhere [33, 34, 36, 50, 51]. The most important advantage of Eqs. (1) to (3) is that the bulk micelle formation and interaction influence are effectively decoupled from the interface adsorption and specific film effects. In the above expressions, $\mu_1^{0,b}$ and $\mu_n^{0,b}$ are the standard chemical potentials of a monomer and an *n*-mer, respectively.

The size distribution curve of the 2D amphiphilic structures is obtained in complete analogy with the bulk micellization [35, 36, 52, 53]. All quantities characterizing the interfaces are denoted with a superscript s. The result is

$$X_n^s = X_n^{0,s} X_n^{int,s} X_n^{f,s} \tag{4}$$

where

$$X_n^s = \frac{N_n^s}{N_w^s + \sum_j iN_j^s} \tag{5}$$

$$X_n^{0,s} = (X_1^s)^n \exp\left(-\frac{\mu_n^{0,s} - n\mu_1^{0,s}}{kT}\right) \tag{6 a}$$

$$X_n^{int,s} = \exp\left[\frac{\sum_j (nu_{1j}^s - u_{nj}^s) N_j^s}{kTA_f} - \frac{\Delta a_n^s}{2kTA_f^2}\sum_i N_i^s \sum_j N_j^s u_{ij}^s\right], \quad \Delta a_n^s = na_1^s - a_n^s \tag{6 b}$$

$$X_{\rm n}^{\rm f,s} = \exp\left[-\frac{\Delta a_{\rm n}^{\rm s}}{kT}\sigma_{\rm f}(h)\right]\exp\left[\frac{v_{\rm a}\Pi(h)}{kT}\left(n\frac{\partial N_1^{\rm b}}{\partial N_1^{\rm s}} - \frac{\partial N_{\rm n}^{\rm b}}{\partial N_{\rm n}^{\rm s}}\right)\right]$$
$$\times \exp\left[-\frac{\Delta a_{\rm n}^{\rm s}}{kT}\int_{\infty}^{h}\left(\frac{\partial \Pi}{\partial h'}\right)h'{\rm d}h'\right] \qquad (6c)$$

Again, the interspecies interactions and the specific influence of the film bulk are effectively decoupled. In the above equations $X_{\rm n}^{\rm s}$ is the surface mole fraction of the interface n-mers and $\mu_1^{0,\rm s}$ and $\mu_{\rm n}^{0,\rm s}$ are the standard chemical potentials of a monomer and an n-aggregate in the 2D-phases, respectively. They carry the information about the intrinsic surface self-assembling properties of the amphiphile on the interface of a bulk solution before a thin film is formed from it.

For thinning foam films [$\Pi(h) \neq 0$], the effect of the disjoining pressure is very important (see Eqs. 3d and 6c). In all the cases studied the electrolyte concentration is high, the electrostatic component of the disjoining pressure is effectively depressed and the leading term of the surface forces is the van der Waals constituent $\Pi_{\rm vw}$. As shown in previous papers [36, 37], the van der Waals disjoining pressure acts in the direction of enhanced destruction of the existing self-assembled structures both in the film bulk and at the film interfaces. Hence in thinner films the fraction of the larger surfactant entities is decreased in comparison with thicker films and the number of the free monomers increases as the film drainage proceeds.

8.3.3 Foam Film Hydrodynamics

The films drain quickly in a regime of substantial interfacial mobility. The cause of this is that the overall amphiphile quantity is insufficient for the onset of equilibrium film thickness. From the extensive film drainage investigations, it is well known that this fact results in a specific coupling of film hydrodynamics and the mass transfer of the stabilizing surfactant [42–46]. Unlike the case of the monomolecular state of the surfactant, however, here there is an additional source of amphiphilic molecules: the *premicelles*.

Further, the unstable black patterns are a clear sign of thickness non-homogeneities in the course of the foam film drainage. This creates an extra option for the onset of local differences in the general coupling of film hydrodynamics and the amphiphile mass transfer. The onset of these non-homogeneities (black patterns) is accompanied by a slowing down of the overall film thinning velocity and by an increase in film lifetimes (Figs. 8.9 and 8.10). The observed unstable black formations visualize the places of smaller gap widths (microfilms) within the background thinning plane-parallel foam film. In these microfilms, more opportunities are created for the launch of local differences in the coupling of the film hydrodynamics and the surfactant mass transfer.

Hence in thinner regions, the local flow is retarded. The mechanism of this retardation is the same as in large plane-parallel draining film (Fig. 8.13) [42,

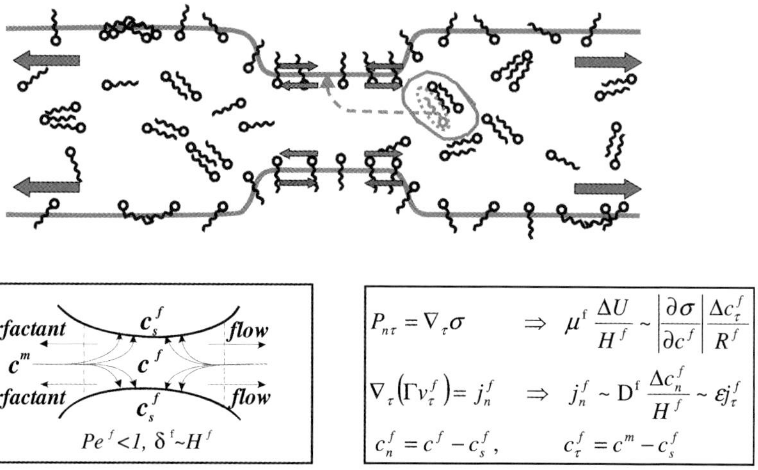

Fig. 8.13 Film hydrodynamics is coupled with the mass transfer of the stabilizing surfactant. Schematic diagram of the impact of surfactant nanostructures on the drainage of foam film with unstable black pattern.

43]: The flow sweeps the surfactant molecules outside the black pattern, creating locally a surface tension gradient that causes the onset of a tangential force that acts in a direction opposite to the fluid outflow, resulting in effective immobilization of the black pattern interfaces. The existing nanostructures in the microfilm bulk are already destroyed and it is depleted of surfactant molecules. In the neighboring regions, however, there are still self-assemblies. Upon thinning they are further on destroyed, providing an additional amount of monomers. The latter could participate in the feed-up of the black pattern interfaces (Fig. 8.13). This extra surfactant flow gives a certain additional time to the emerged local surface tension gradient to be maintained for a while and the interface flow to be retarded. This results in retardation of the further thinning within the black pattern region. The increase in the overall concentration of the initial solution leads to the onset of more of these black patterns (Fig. 8.5). The larger their number, the sharper is the rise of the overall drainage time of the film.

The proposed slow-down mechanism couples the presence of amphiphilic nanostructures with the specific film hydrodynamics and the mass transfer of the surfactant molecules in thinning foam films. It allows a unified explanation of the film drainage behavior and the surface tension measurements (Fig. 8.14). The juxtaposition of the presented results, related to two types of experiments (static and dynamic), shows a synchronized onset of specific peculiarities in the run of the surface tension isotherms and the drainage characteristics of foam films against the surfactant concentration. Therefore, it may be regarded as experimental evidence of amphiphilic self-assembly at surfactant concentrations lower than the CMC values.

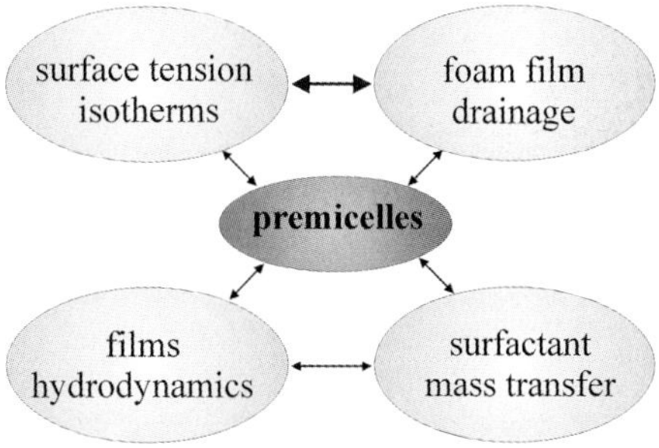

Fig. 8.14 Schematic diagram of the interpretation of the experimental results.

8.4 Conclusion

This chapter has attempted to draw attention towards the unexploited potential of the microinterferometric foam film technique equipped with a Scheludko-Exerowa measuring cell for the study of the surfactant self-assembly. It is shown that specific film parameters can be extracted from the systematic foam film experiments with initial solutions of surfactant concentration lower than the usual CMC. These parameters reflect the coupling of film dynamics and the reorganization of existing surfactant assemblies both in the film and in the adsorption layers on its interfaces. The run of the kinetic foam film characteristics against the amphiphile concentration may be considered as experimental evidence for the presence of amphiphilic structures in the primary solutions.

The reported results suggest that it seems encouraging that microscopic thin liquid film instrumentation might serve as a very appropriate and promising tool in the study of surfactant nanostructures in fluid media.

References

1 J. McBain, *Kolloid Z.*, *12* (**1913**) 256.
2 J. McBain, W. Dye, S. Jonson, *J. Am. Chem. Soc.*, *61* (**1939**) 321.
3 *Curr. Opin. Colloid Interface Sci.*, *1* (3) (**1996**) whole issue.
4 *Curr. Opin. Colloid Interface Sci.*, *2* (4) (**1997**) whole issue.
5 R. Rajagopalan, *Curr. Opin. Colloid Interface Sci.*, 6 (**2001**) 357.
6 W. Gelbart, A. Ben-Shaul, *J. Phys. Chem.*, *100* (**1996**) 13169.
7 P. Mukerjee, K. Mysels, *Critical Micelle Concentrations of Aqueous Surfactant Systems*, NSRDS, Washington, **1971**.
8 W. Gelbart, A. Ben-Shaul, D. Roux (Eds.), *Micelles, Membranes, Microemulsions and Monolayers*, Springer-Verlag, New York, **1994**.
9 A. Rusanov, *Micellization in Surfactant Solutions*, Khimiya, St. Petersburg, **1992**.
10 C. Tanford, *The Hydrophobic Effect: Formation of Micelles and Biological Membranes*, Wiley, New York, **1980**.
11 S. Chen, J. Huang, P. Tartaglia (Eds.), *Structure and Dynamics of Strongly Interacting Colloids and Supramolecular Aggregates in Solution*, NATO ASI Series, Series C: Mathematical and Physical Sciences, Vol. 369, Kluwer, Dordrecht, **1992**.
12 K. Mysels, K. Shinoda, S. Frankel, *Soap Films*, Pergamon Press, New York, **1959**.

13 A. Scheludko, *Adv. Colloid Interface Sci.*, *1* (**1967**) 391.

14 D. Exerowa, P. Kruglyakov, *Foam and Foam Films*, Elsevier, Amsterdam, **1998**.

15 D. Platikanov, D. Exerowa, in *Soft Colloids, Fundamentals of Interface and Colloid Science*, Vol. V (Ed. J. Lyklema), Elsevier, The Netherlands, **2005**, Chapter 6.

16 B. Derjaguin, N. Churaev, V. Muller, *Surface Forces*, Consultants Bureau, New York, **1987**.

17 B. Derjaguin, *Theory of Stability of Colloids and Thin Films*, Consultants Bureau, New York, **1989**.

18 D. Exerowa, T. Kolarov, Khr. Khristov, *Colloids Surf.*, *22* (**1987**) 171.

19 T. Kolarov, R. Cohen, D. Exerowa, *Colloids Surf.*, *42* (**1989**) 49.

20 D. Exerowa, D. Kashchiev, D. Platikanov, B. Toshev, *Adv. Colloid Interface Sci.*, *49* (**1994**) 303.

21 D. Exerowa, M. Zacharieva, R. Cohen, D. Platikanov, *Colloid Polym. Sci.*, *257* (**1979**) 1089.

22 D. Exerowa, D. Kashchiev, D. Platikanov, *Adv. Colloid Interface Sci.*, *40* (**1992**) 201.

23 J. de Feijter, in *Thin Liquid Films* (Ed. I.B. Ivanov), Marcel Dekker, New York, (**1988**) 1–47.

24 E. Mileva, D. Exerowa, *Adv. Colloid Interface Sci.*, *100–102* (**2003**) 547.

25 P. Tchoukov, E. Mileva, D. Exerowa, *Langmuir*, *19* (**2003**) 1215.

26 P. Tchoukov, E. Mileva, D. Exerowa, *Colloids Surf. A*, *238* (**2004**) 19.

27 E. Mileva, P. Tchoukov, D. Exerowa, *Adv. Colloid Interface Sci.*, *114–115* (**2005**) 47.

28 A. Nikolov, G. Martynov, D. Exerowa, *J. Colloid Interface Sci.*, *81* (**1981**) 116.

29 A. Nikolov, G. Martynov, D. Exerowa, W. Kaishev, *Kolloid Zh.*, *62* (**1980**) 672.

30 D. Exerowa, A. Scheludko, *Commun. Inst. Phys. Chem. Bulg. Acad. Sci.*, *3* (**1963**) 79.

31 D. Exerowa, A. Nikolov, in *Surfactants in Solution*, Vol. 4 (Ed. K.L. Mittal), Plenum Press, New York, **1984**, p. 1313.

32 A. Scheludko, A. Nikolov, *Colloid Polym. Sci.*, *253* (**1975**) 404.

33 D. Blankschtein, G. Benedek, G. Thurston, *J. Chem. Phys.*, *85* (**1986**) 7268.

34 P. Missel, N. Mazer, G. Benedek, C. Young, M. Carey, *J. Phys. Chem.*, *84* (**1980**) 1044.

35 J. Israelachvili, *Langmuir*, *10* (**1994**) 3774.

36 E. Mileva, D. Exerowa, *Colloids Surf. A*, *149* (**1999**) 207.

37 E. Mileva, D. Exerowa, P. Tchoukov, *Colloids Surf. A*, *186* (**2001**) 83.

38 A. Scheludko, D. Exerowa, *Commun. Inst. Phys. Chem. Bulg. Acad. Sci.*, *1* (**1960**) 203.

39 D. Exerowa, A. Nikolov, M. Zacharieva, *J. Colloid Interface Sci.*, *81* (**1981**) 419.

40 E. Mileva, D. Exerowa, In *Emulsions, Foams and Thin Films* (Eds. K.L. Mittal, P. Kumar), Marcel Dekker, New York, **1999**, Chapter 15.

41 E. Mileva, unpublished results.

42 J. Lee, T. Hodgson, *Chem. Eng. Sci.*, *23* (**1968**) 1375.

43 (a) I. Ivanov, D. Dimitrov, B. Radoev, *Colloid J.*, *41* (**1979**) 36; (b) I. Ivanov, *Pure Appl. Chem.*, *52* (**1980**) 1241.

44 E. Mileva, B. Radoev, *Colloids Surf. A*, *74* (**1993**) 259.

45 E. Mileva, L. Nikolov, *Colloids Surf. A*, *74* (**1993**) 267.

46 E. Mileva, B. Radoev, in *Emulsions: Structure, Stability and Interactions* (Vol. Ed. D.N. Petsev), Vol. 4 of *Interfacial Science and Technology* (Series Ed. A. Hubbard), Elsevier, London, **2004**, Chapter 6, pp. 215–256.

47 E. Manev, R. Tsekov, B. Radoev, *Dispers. Sci. Technol.*, *18* (**1997**) 769.

48 D. Langevin, A. Sonin, *Adv. Colloid Interface Sci.*, *51* (**1994**) 1.

49 A. Rusanov A, *Phasengleichgewichte und Grenzflächenerscheinungen*, Akademie Verlag, Berlin, **1978**.

50 E. Mileva, *J. Colloid Interface Sci.*, *178* (**1996**) 10.

51 E. Mileva, *J. Colloid Interface Sci.*, *232* (**2000**) 2110.

52 V. Fainerman, R. Miller, *Langmuir*, *12* (**1996**) 6011.

53 V. Fainerman, E. Aksenenko, R. Miller, *J. Phys. Chem.*, *104* (**2000**) 5744.

9
Nanoparticles in Confined Structures: Formation and Application

Alexander Kamyshny and Shlomo Magdassi

Abstract

Current approaches to the synthesis of nanoparticles in confined structures are described. The syntheses of inorganic and organic nanoparticles in nanometric confined structures, such as reverse micelles, water-in-oil and oil-in-water microemulsions, water-in-supercritical liquid microemulsions, micelles of amphiphilic block copolymers, miniemulsions, dendrimers, polymeric capsules, pore channels of mesoporous solids and nanoporous membranes and liquid crystals are discussed. Examples of practical applications of the nanoparticles obtained are presented.

9.1
Introduction

Nanoparticles are defined as particles having at least one dimension in the 1–100-nm range [1]. In recent years, nanoparticles have attracted considerable scientific and practical interest owing to their unique properties (mechanical, optical, electrical, thermal, magnetic, catalytic), which differ from those of atomic and bulk species [1–15]. These differences originate from the extremely large ratio of surface area to volume and in very small size ranges (1–10 nm) also from quantum confinement effects. Virtually nanosized materials are well known products with novel properties and promising applications. As a result, nanoscience and nanotechnology are now developing as new boundary fields of knowledge, which use the approaches and tools of physics, chemistry, biology and materials science. Functional nanosized materials of various composition and physicochemical properties (inorganic salts and oxides, metals, semiconductors, organic compounds, polymers) and morphologies (spheres, rods, wires, tubes, cubes, hexagons, triangles, etc.) and their assemblies pave the way to new applications and possibilities of making products such as catalysts, targeted drug delivery and implants, biosensors, bioseparation, magnetic resonance im-

Colloids and Interface Science Series, Vol. 1
Colloid Stability: The Role of Surface Forces, Part I. Edited by Tharwat F. Tadros

ISBN: 978-3-527-31462-1

aging, new electronic, optoelectronic and magnetic devices, small and very effective energy devices, nanoelectromechanical systems, ceramics and coatings and pigments for ink-jet inks [4–8, 16–36].

Since the electrical, optical, magnetic and catalytic properties of nanoparticles depend strongly on their dimensions and morphology, the development of methodologies for obtaining nanoparticles of controlled size, shape and polydispersity is of key importance for many applications and is currently a very active and challenging topic of research [14, 37–51].

Most of the reported methods for nanoparticle preparation were developed for inorganic materials, such as metals, semiconductors, magnetic materials and dielectrics, and may be roughly divided into two groups: physical and chemical methods. *Physical methods* are usually considered to be those in which bulk material is dispersed in a suitable medium with formation of nanosized particles and include mechanical grinding, mechanical alloying, spray coating, laser ablation, rapid quenching of supersaturated metal vapor by cold inert gas and solidifying droplets of molten metal by rapid cooling [9, 30, 45, 48]. *Chemical methods* include precipitation and redox reactions, thermal decomposition of organometallic compounds and electrochemical and sonochemical syntheses [5, 7, 14, 30, 45, 52–54]. Among the most widely used and most convenient are *wet chemical processes*, which are based on the approaches and tools of colloid chemistry (sol–gel processes, redox and precipitation reactions in solution) [5, 9, 35, 36, 38–43, 45, 48, 49, 51, 52, 55]. Essentially, solutions of various precursors are mixed in well-defined quantities and under controlled conditions such as reagent and additive concentrations, solvent polarity and viscosity, temperature and pH to promote the formation of colloidal dispersions or insoluble compounds, which precipitate out of solution. The advantage of wet chemical processes is that a large variety of compounds can be fabricated on essentially cheap equipment and in significant quantities.

To be of practical importance, dispersions of nanoparticles should be stable. To prevent agglomeration and precipitation and to obtain stable dispersions, the agglomeration process must be arrested in the early stages of particle formation. This objective is achieved by the addition of stabilizers to the reaction mixture. A number of substances, such as ionic and non-ionic surfactants, ionic and non-ionic polymers are used as nanoparticle stabilizers [5, 7, 9, 14, 35, 36, 41]. In general, the stabilizing effect is achieved by electrostatic and/or steric mechanisms. Electrostatic stabilization is based on Coulombic repulsion between particles caused by the electrical double layer formed by adsorbed ions [7]. Steric stabilization prevents agglomeration due to the formation of a protective shield between nanoparticles and is especially useful in the case of concentrated dispersions of nanoparticles [5, 7]. Therefore, charged polymers are the most effective stabilizers, since they combine both electrostatic and steric effects, that is, involving an electrosteric mechanism.

Particle size in wet chemical processes is usually controlled by adjustment of the kinetics of nucleation and growth [7, 14, 38, 40–42, 52, 56]. In homogeneous solutions, the nucleation process is affected by various parameters, such as the nature and concentration of reagents, temperature, the type and concentration of protecting agents and solvent viscosity. High supersaturation of reagents and a fast nucleation rate tend to produce small and numerous particles. To produce larger particles, a smaller number of nuclei should be generated and their growth should be the dominant process. This condition can be realized by slow reactions, the use of added seeds or slow addition of nanoparticle precursor to the system. Another possibility for particle growth is via aggregation of already formed particles. The growth mechanism favors the development of crystals, whereas the aggregation mechanism produces mostly spherical and polycrystalline particles. In reality, both mechanisms can be involved in nanoparticle formation.

The formation of nanoparticles having a desired morphology is difficult to achieve in homogeneous solutions because it requires controlling the growth rate of the specific crystalline plane of colloid particles. Nevertheless, the shape of nanoparticles (spheres, cubes, rods, wires, disks, prisms, hexagons, etc.) can also be controlled by the composition of the reaction mixture, relative rates of nucleation and growth and by the presence of preformed seeds, surfactants, polymers and other additives [14, 39, 43, 50, 52, 57–64].

Another approach for obtaining monodisperse nanoparticles of definite size and shape is to perform the formation of particles within confined nanometric structures [39, 45, 52, 57, 65–67]. These confined structures may be considered as nanoreactors, while the reaction volume is restricted to a cavity of "nanometric reaction vessel", which is a zone of reaction involving the interacting reagents. In addition to obtaining uniform nanoparticles of desired shape, this approach can be promising for producing assemblies of nanoparticles with a new collective physical behavior [68–72].

There are several types of confined systems that are used for nanoparticle formation: reverse micelles and water-in-oil (W/O) microemulsions [56, 69, 73–84], oil-in-water (O/W) microemulsions [36], micelles of amphiphilic block copolymers [65, 85–90], water-in-supercritical fluid (W/SCF) microemulsions [57, 91–94], interior cavities of dendritic molecules [50, 95–98], polymeric capsules [67, 99, 100], pore channels of mesoporous solids [45, 101–103], nanoporous membranes [104–106] and liquid crystals [44, 107, 108].

In this chapter, we survey the current approaches to the synthesis of nanoparticles in confined nanometric structures, and also practical applications of the nanoparticles obtained in various fields of science and technology. One-dimensional (1D) nanostructured materials, such as nanotubes and nanowires, and two-dimensional (2D) assemblies, which are of special interest in nanoscience and nanotechnology, will be mentioned here only if required for comprehensive discussion.

9.2 Synthesis of Nanoparticles in Nanoreactors

9.2.1 Micelles and Emulsions

Micelles and microemulsions are thermodynamically stable isotropic systems consisting of two immiscible liquids and an amphiphilic surfactant. At the macroscopic level, micelles and microemulsions appear to be homogeneous, transparent solutions, but at the microscopic level they are heterogeneous.

9.2.1.1 Reverse Micelles and W/O Microemulsions

Inorganic Nanoparticles

The most widely used templates for nanoparticles formation are reverse micelles and W/O microemulsions. Reverse micelles and W/O microemulsion droplets consist of an inner cavity containing water molecules surrounded by polar head groups of amphiphilic surfactant molecules, while hydrophobic chains are directed outwards, to the external oil phase. These inner cavities serve as nanoreactors for the production of nanoparticles, mostly spherical.

The difference between reverse micelles and microemulsions is not well defined and the most useful definition is related to the size of their inner cavity, which is determined by the molar ratio between water and surfactant, W_0. For example, the diameter of AOT micelles [Aerosol OT, sodium 1,4-bis(ethylhexyl)-sulfosuccinate], can by described by D (nm) $= 0.3\,W_0$ ($W_0 = [H_2O]/[AOT]$) [69, 76]. It is generally accepted that aggregates with radii of several nanometers ($W_0 < 10$–15) are reverse micelles (their diameter is determined mainly by the length of a surfactant molecule), whereas aggregates with radii of tens of nanometers and a surfactant-to-water volume ratio of <4 are microemulsions [76, 109–112] (Fig. 9.1 A, B). Water in the inner cavity of small micelles is considered to be structured owing to interaction with the surfactant polar head groups and counterions, whereas in microemulsion droplets it can be considered as a pseudo-phase (bulk water) [69, 109, 110]. In surfactant systems containing large amounts of both oil and water, the shape and dimensions of the surfactant aggregates may change with formation of interconnected water channels [51, 56, 69, 74, 76, 113]. Such structures are known as bicontinuous (Fig. 9.1 C).

A wide range of surfactants are used for the formation of reverse micelles and W/O microemulsions, which serve as nanoreactors for nanoparticle synthesis: anionic surfactants such as AOT, potassium oleate, sodium dodecyl sulfate (SDS), perfluoropolyethercarboxylic acid (PEPE) and dioleylphosphoric acid (DOLPA); cationic surfactants such as cetyltrimethylammonium chloride (CTAC) or bromide (CTAB), cetyldimethylbenzylammonium bromide (CDBA) and didodecyldimethylammonium bromide (DDAB); non-ionic surfactants such as polyoxyethylene derivatives (Triton N-42, N-57, N-60, N-101, X-15, X-35, X-45

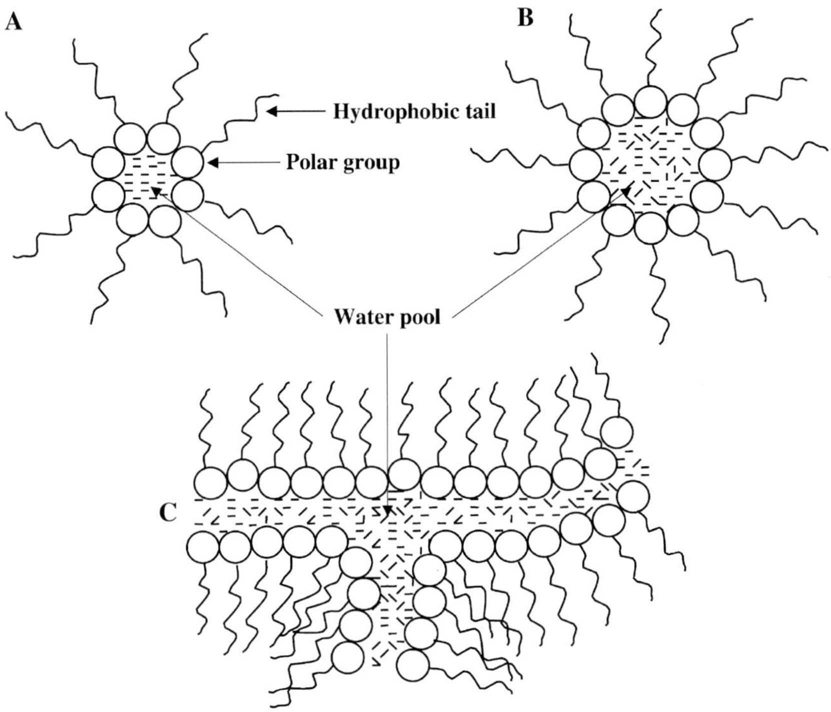

Fig. 9.1 Schematic presentation of reverse micelles (A), microemulsion droplets (B) and bicontinuous structures (C).

and X-100, Brij 30, Brij 97), sorbitan monooleate (Span 80), polyethylene glycol (PEG); and zwitterionic lecithins [78, 82]. The double-chained anionic AOT is the most widely used surfactant, since it forms relatively monodisperse and spherical reverse micelles in a wide range of concentrations, which are capable of solubilizing a large amount of water (up to W_0 as large as 40–60) [60, 76, 78, 79, 82, 110, 111] and can produce nanoparticles in the size range 5–50 nm [84] with polydispersity $<10\%$ [69]. In many cases, microemulsion compositions contain, in addition to surfactants, cosurfactants, such as medium-chain alcohols, acids or amines, which affect the fluidity of the interfacial films and the mean micellar size and facilitate transport of various entities through the interface [82, 114].

There are two basic approaches to nanoparticle synthesis in micelles and microemulsion droplets. One of them is based on their dynamic nature. Since the aqueous droplets of micelles and microemulsions continuously collide, their coalescence results in a very fast exchange of their water pools [82, 113], which makes possible chemical reactions between reactants solubilized in two different micelles [60, 74, 78, 79, 83, 115, 116] as presented in Fig. 9.2A. In the second approach (Fig. 9.2B), one of the reactants (e.g. metal salt) is solubilized in the inner (e.g. aqueous) cavity of a micelle or a microemulsion droplet while an-

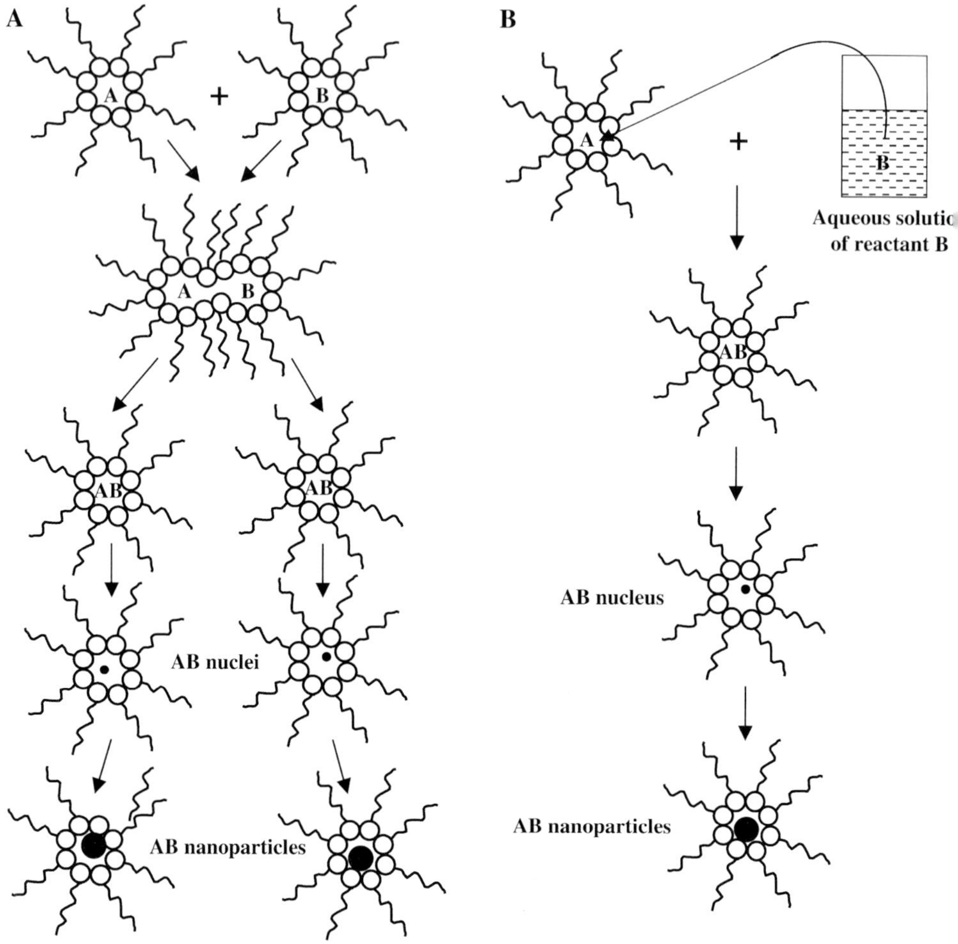

Fig. 9.2 Schematic presentation of two approaches to nanoparticle synthesis in micelles and microemulsion droplets. (A) Exchange of water pools followed by reaction between reagents and formation of nuclei and nanoparticles. (B) Interaction between reagent in water cavity of micelle and external reagent followed by formation of nuclei and nanoparticles.

other reactant (e.g. reducing agent) is added to the system and diffuses into this cavity, where the reaction occurs [78, 83, 115]. In addition, nanoparticle formation in micelle can be affected externally, for example by initiation of monomer polymerization by UV irradiation [73]. The use of functionalized surfactants, such as various salts of AOT, as one of the reagents improves significantly the synthesis and "quality" (e.g. crystallinity, stability) of nanoparticulate products (such salts can be obtained from Na-AOT by ion exchange [117]). For example, the synthesis of nanoparticles of a number of metals, metal salts, oxides and semiconductors has been performed with the use of $Cu(AOT)_2$ [73, 74, 76],

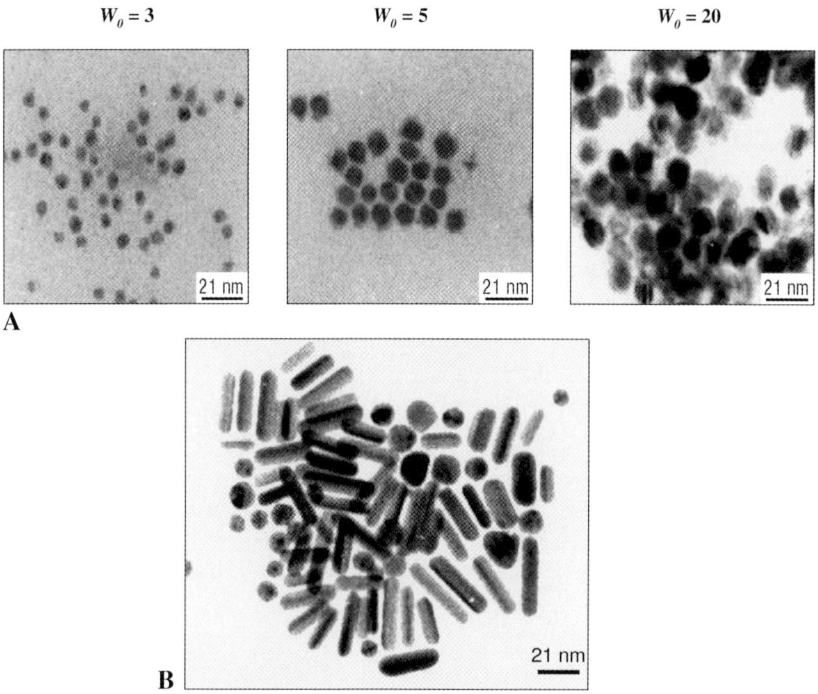

Fig. 9.3 TEM images of copper nanoparticles produced in reverse micelles of Na(AOT) and $Cu(AOT)_2$ (molar ratio 1:10) in isooctane at different W_0 values (A) and in bicontinuous structures formed in a $Cu(AOT)_2$–isooctane–H_2O system. (Reproduced with permission from [76], © 2003 Nature Publishing Group).

$Cd(AOT)_2$ [74, 117–119], $Zn(AOT)_2$ [119], $Mn(AOT)_2$ [74, 119], $Co(AOT)_2$ [73, 75, 120], $Ba(AOT)_2$ [121] and Ag(AOT) [70, 73, 122].

The possibility of controlling the size and shape of the inner cavity (nanoreactor) by changing the water content and the type of surfactant allows the nanoparticle shape and size to be controlled [69, 76]. As an example, Fig. 9.3 presents TEM images of spherical copper nanocrystals formed in reverse micelles at various W_0 values [76]. It is obvious that control of the inner cavity size by changing the W_0 value permits the control of the spherical nanoparticle size (Fig. 9.3 A). Synthesis in bicontinuous structures results in the formation of cylindrical, rod-shaped copper nanoparticles (Fig. 9.3 B). The other approach to the synthesis of rod-shaped metallic nanoparticles is the use of preformed seeds. These seeds are then added to a solution containing more metal salt, a reducing agent and rod-like micelles formed by a cationic surfactant, CTAB [37, 72].

In addition, many other factors, such as droplet concentrations, reactant concentrations, type of bulk solvent, rate of exchange of water pools between droplets, adsorption of micelle-forming surfactant molecules on nanoparticles (stabilizing effect) and ionic strength of the water pool also affect the size and shape of nanoparticles synthesized in reverse micelles and in W/O microemulsions [47, 69, 76, 78, 81–84, 115, 116, 123, 124]. These parameters are often interdependent and their precise control is still a complicated problem. Therefore, the synthesis of nanoparticles with a desired size and shape in micellar media is, to a considerable extent, a matter of art.

Since the first application of micellar aggregates for the preparation of metallic nanoparticles in 1982 [125], a large number of inorganic nanoparticles have been synthesized in reverse micelles and in W/O microemulsions: metals such as Ag [70, 84, 122, 126, 127], Au [47, 84, 115, 126], Cu [56, 69, 73, 74, 76, 84, 115, 128], Ni [84, 116, 124], Co [56, 75, 120], Pt [84, 115, 129], Pd [84, 115, 130, 131], Rh [115], Ir [115], bimetallic Au/Pd, Au/Ag, Pd/Pt [82], Pt/Co [132] and Pt/Ru [133, 134] and trimetallic Pt/Ru/Mo compounds [135]; salts such as $CaCO_3$ [123, 136], $CaSO_4$ [137], $BaSO_4$ [138–140], $BaCrO_4$ [121, 141, 142], $BaMoO_4$ [143], AgCl [79, 144], AgBr [79], AgI [69], Ag_2S [69], ZnS [145], MoS_x [146], Prussian Blue [147]; metal oxides and ceramics [78, 148–153]; metal borides [79, 115]; various semiconductors [74, 117–119, 154–160]; magnetic materials [73, 78, 116, 135, 149, 152, 161–163]; and high-temperature superconductors [116].

The use of W/O microemulsions for the preparation of inorganic nanoparticles requires, in many cases, their recovery from reverse micelles and removal of the organic solvent from the products. To remove the solvent and to obtain a powder of nanoparticles, the microemulsion can be simply dried in a vacuum oven [140]. Another approach for the recovery of inorganic nanoparticles is the addition of alcohols, such as ethanol or methanol [70, 73, 75, 119, 124, 126, 144, 154–156, 159, 163], or acetone [134, 149, 152], which induce phase separation. Sometimes capping agents, such as thiols (dodecanethiol, thiophenol) [70, 73, 119, 126, 144, 154–156] or fatty acids (e.g. lauric acid [75]) are added before phase separation. In general, the precipitates obtained can be easily redispersed in organic solvents, such as toluene or alkanes [70, 73, 119, 126, 149, 155, 156, 163]. In addition, inorganic nanoparticles synthesized in micelles can be separated from solution by immobilization on solid supports. For example, CdS nanoparticles formed in reverse micelles were incorporated into thiol-modified mesoporous silica [164] or bound to thiol-modified polystyrene particles [165] by simple addition of these supports to micellar solution.

Organic Nanoparticles

There are only few reports on the formation of organic nanoparticles in micelles and in microemulsions (excluding polymerization). Several types of organic nanoparticles were also produced in W/O microemulsions by direct precipitation in aqueous cores of droplets [79]. The method consists in addition of water-insoluble organic compounds dissolved in a suitable organic solvent to the microemulsion. This compound diffuses into the aqueous cavity of the reverse micelle by crossing the interfacial layer and precipitates there with formation of nanoparticles. The production of cholesterol, Rhodiarome and Rhovanil nanoparticles with average diameter 5–7 nm was successfully performed in AOT–

heptane–water, Triton–decanol–water and CTAB–hexanol–water microemulsions [79, 80]. Nanoparticles of Nimesulide (a non-steroidal anti-inflammatory drug) were obtained by the same method with the use of microemulsions composed of lecithin–isopropyl myristate–water–*n*-butanol (or 2-propanol) [166].

Polymeric Nanoparticles

The preparation of dispersions of polymeric particles by emulsion polymerization has been well known for many years and extensively applied in industry [167]. The concept of polymerization in microemulsions arose in the 1980s [168] and is now also a widely used method for the preparation of stable polymeric nanoparticles in the size range 5–50 nm [168, 169]. In the case of W/O microemulsions, a water-soluble monomer (e.g. acrylamide, acrylic acid), which may also act as a cosurfactant, is dissolved or dispersed in the water phase and polymerization is initiated thermally, photochemically or under γ-irradiation [168, 170]. An additional approach to the preparation of nanosized latexes is the use of a polymerizable surfactant, such as didodecyldimethylammonium methacrylate, which forms reverse micelles [73].

9.2.1.2 W/SCF Microemulsions

In recent years, starting from the first report on the synthesis of silver nanoparticles [171], W/SCF microemulsions are often used for various nanoparticle preparations [57, 92–94, 172].

A supercritical fluid (SCF) is a substance present at a pressure and temperature that are above their critical values. Under these conditions, the SCF exists as a single phase with a liquid-like density, but with mass transfer properties similar to those of gases [172, 173]. The most extensively used "oil" solvent in W/O microemulsions is supercritical CO_2, which offers several advantages over conventional solvents, such as fast reaction speed, rapid separation and easy removal of the solvent from nanoparticles by pressure reduction [82, 84]. In addition, CO_2 is relatively inert and environmentally friendly in comparison with organic solvents and has a low critical point at 73.8 bar and 31.1 °C, permitting processing under ambient conditions [172].

Most ionic surfactants with long hydrocarbon chains, such as the widely used AOT, are insoluble in CO_2 and cannot form microemulsions. Non-ionic surfactants have a greater affinity for CO_2 but, in general, they display little ability to self-aggregate and dissolve water owing to the weak interaction between their head groups. Therefore, surfactants with fluorinated tails or mixtures of surfactants (e.g. perfluoropolyethers [91, 93, 94], fluorinated AOT [174], mixture of AOT with fluorinated cosurfactant [93, 171, 175–178]) are usually used for the formation of W/CO_2 microemulsions [91, 94].

There are two basic methods of separation and collection of nanoparticles synthesized in W/CO_2 microemulsions: the *reducing pressure method*, in which the nanoparticles can be recovered by a slow pressure decrease in the reactor, allowing the surfactant to collapse on the bottom of the reaction vessel, or the *rapid expansion method*, in which the microemulsion is rapidly expanded via a capillary nozzle of small diameter into a collection solution [49, 94].

W/CO_2 microemulsions have been extensively used in recent years for the synthesis of metal and semiconductor nanoparticles. The following metal nanoparticles were obtained: Ag [57, 82, 171, 177, 178], Cu [57, 82, 91, 176], Pt and Pd [82, 94, 179]. Among semiconductors obtained by this method are CdS [57, 82, 179, 180], ZnS [57, 82], Ag_2S [175] and silver halides [176, 178].

9.2.1.3 **Micelles of Amphiphilic Block Copolymers**

Amphiphilic block copolymers represent a class of functional polymers with molecules consisting of at least two parts: hydrophilic and hydrophobic blocks. Such a composition leads to self-aggregation with the formation of micelles in dilute solutions in solvents that are selective to one of the blocks. Micelles comprised of a dense core formed by the insoluble block and a shell or corona formed by the soluble block take a spherical shape in most cases, but cylindrical, plate-like and other morphologies are also formed while varying the copolymer structure, its molecular weight and composition and the type of solvent [87, 88, 90, 181, 182]. Therefore, block copolymer micelles may also serve as nanometric "reaction vessels" (10–100 nm) or "nanocontainers" for the synthesis of nanoparticles [9, 86–90, 181, 182].

Micelles of amphiphilic block copolymers are widely used for the preparation of inorganic, especially metallic, nanoparticles [9, 86–88]. The principal approach to the preparation of inorganic nanoparticles is based on metal ion chelating or binding of positively charged fragments of a labile organometallic compound by the functional groups of a core-forming hydrophilic block (e.g. carboxylic), followed by reaction with a suitable reagent. Depending on the type of this reagent, nanoparticles of metal oxide, zerovalent metal, chalcogenide and other compounds can be formed within the micelle core. In the case of metal nanoparticles, two major types of morphology are possible: "cherry morphology" (one nanoparticle formed) and "raspberry morphology" (several smaller nanoparticles formed within the micelle core) [9, 65, 86, 88], as presented in Fig. 9.4. The type of morphology is largely determined by the reduction method: slow reduction favors cherry morphology, whereas fast reduction favors raspberry morphology (Fig. 9.4 A and B show TEM micrographs of gold nanoparticles obtained by slow reduction of $HAuCl_4$ with triethylsilane and by fast reduction with $LiAlH_4$, respectively, in micelles of polystyrene-*block*-poly-4-vinylpyridine [87]). In certain cases it is possible to bind metal compounds to functional monomers or to use organometallic monomers, which after polymerization form a micelle core filled with metal precursor [87, 183]. Numerous examples of amphiphilic block copolymers and nanoparticles of various inorganic materials, such as metals (Au, Pt, Pd, Co) and semiconductors (ZnF_2, ZnO, CdS, PbS), formed within the cores of polymer micelles have been reviewed [9, 86–88, 181, 183].

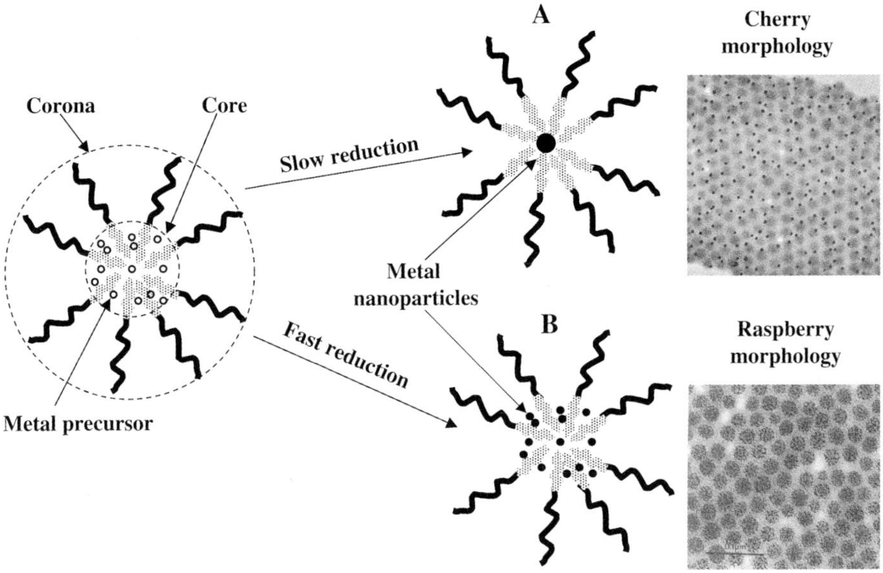

Fig. 9.4 Schematic presentation of two types the metal nanoparticles, which can be formed in micelles of block copolymer: with "cherry morphology" (one nanoparticle formed) and with "raspberry morphology" (several smaller nanoparticles are formed within the micelle core). TEM micrographs present gold nanoparticles obtained by slow reduction of $HAuCl_4$ with triethylsilane (A) and by fast reduction with $LiAlH_4$ (B), respectively, in micelles of polystyrene-*block*-poly-4-vinylpyridine. TEM micrographs are reproduced with permission from [87].

9.2.1.4 O/W Emulsions and Microemulsions

Organic Nanoparticles

A widely used method for the preparation of organic micro- and nanoparticles is the *solvent evaporation method* [49, 170, 184–186]. In this method, the preparation of nanoparticles is carried out by dissolving the organic compound and a protective polymer in a volatile water-immiscible organic solvent (e.g. methylene chloride, chloroform, ethyl acetate) followed by emulsifying the mixture with an aqueous solution in the presence of a suitable emulsifying agent. As a result, the organic compound is "locked" inside the oil droplets of the O/W emulsion. Subsequent removal of the solvent by evaporation results in formation of polymer-stabilized organic particles [186]. The size of these particles is determined mainly by the size of the emulsion droplets. Therefore, the use of emulsions with nanosized droplets results in the formation of organic particles in the nanometric size range (down to 100 nm) [186].

Conversion of oil droplets into organic nanoparticles has recently been reported while evaporation is performed by deposition of oil droplets on solid substrate by ink-jet printing. In this case, the use of O/W microemulsions composed of a suitable surfactant, water-insoluble solvent with dissolved functional organic material (e.g. hydrophobic dye) and cosolvent allows solid organic nanoparticles with an average size as small as 40–60 nm to be obtained [36, 187, 188].

The solvent evaporation method is extensively used for the preparation of biodegradable polymeric nanoparticles (nanospheres and nanocapsules) containing entrapped water-insoluble drugs [185, 186, 189, 190]. Poly(lactic acid) (PLA), poly(glycolic acid) (PGA) and poly(lactic/glycolic acid) (PLGA) are the most widely used biodegradable polymers, since they provide non-toxic, long-acting, injectable delivery systems with high drug loading and lack of significant side-effects [189–199].

Polymeric Nanoparticles

Most studies on obtaining polymeric nanoparticles in O/W microemulsions have dealt with the polymerization of hydrophobic monomers (e.g. styrene, methyl methacrylate) present within the oil droplets stabilized by cationic and non-ionic surfactants [168]. Polymeric nanoparticles can be functionalized by incorporation into a microemulsion of polymerizable cosurfactants with desired functionality, such as hydroxyalkyl methacrylates [200, 201], methyl acrylate [201], vinylbenzyl chloride, *N*-acryloyloxysuccinimide, vinylpyridine and methacrylic acid [202]. Such nanoparticles can be further modified by surface reactions with various chemical agents [202]. The use of regular CTAB micelles in water for the synthesis of fluorescent polymer/mesoporous silica nanocomposite has recently been reported (2-naphthol was enzymatically polymerized inside such micelles and then tetramethyl orthosilicate was condensed around polymer-filled micelles) [203]. The fabrication of photochromic nanomaterial composed of a conductive polymer and organic nanoparticles was performed by polymerization of a mixture of pyrrole monomer with pyrene in the hydrophobic interior of DeTAB micelles in water [204].

9.2.1.5 **Miniemulsions**

Emulsions with droplet sizes between those of conventional emulsions and microemulsions (i.e. 50–500 nm) are named miniemulsions, although other terms (nanoemulsions, ultrafine emulsions, submicron emulsions and translucent emulsions) are also used [205–208]. In contrast to microemulsions, miniemulsions are metastable (kinetically stable), but owing to their small droplet size may appear transparent. Miniemulsions may be prepared mechanically by high-shear stirring, high-pressure homogenization and ultrasound emulsification [205, 206, 208] or by low-energy methods, such as phase inversion temperature (PIT) [207–209] or emulsion inversion point (EIP) at constant temperature [207] methods. The concept of the PIT method is based on the fact that O/W emulsions stabilized by non-ionic emulsifiers containing ethoxylated groups can show a temperature-induced phase inversion to W/O emulsion, since these surfactants undergo a dehydration process during heating and become more hydrophobic. The EIP method is based on dropwise addition of water to a mixture of oil and surfactant, that results in changing the spontaneous curvature of the surfactant from initially stabilizing a W/O emulsion to an O/W emulsion [207].

Miniemulsion droplets have been successfully used as nanoreactors for polymerization reactions and the formation of nanosized latexes (numerous examples can be found in [206]). Miniemulsions can also be used for the preparation of functionalized polymeric nanoparticles. As reported recently, the polymerization of oil droplets in an O/W miniemulsion (ethoxylated non-ionic surfactants, Brij 96V and Brij 92V, monomeric lauryl acrylate with dissolved fluorescence probe, pyrene, as an "oil" phase) at temperatures lower than the PIT results in the formation of polymeric nanoparticles with an average size of 46 nm, which contain an embedded fluorescence probe. This probe changes the emission color from blue to violet after polymerization owing to the absence of excimers, which is an indication of a drastic viscosity change [209].

9.2.2 Dendrimers

Dendrimers represent a class of macromolecules with a highly branched regular structure, which is characterized by a central core, internal repeating units and a large number of terminal functional groups. Dendrimer molecules are capable of taking up guest molecules between the branches and therefore can serve as templates for nanoparticle synthesis. The synthesis of dendrimers generally involves grafting of repetitive branch units on the dendritic core and, according to the number of reaction cycles, dendrimers are subdivided into generations. During the past two decades, a wide variety of dendrimers with various cores, branches and terminal groups have been synthesized and applied in different areas including drug delivery, catalysis, harvesting and sensors [50, 97, 98, 210–212]. At present, poly(amidoamine) dendrimer (PAMAM, Fig. 9.5 A) and its derivatives with various lengths of branches are available commercially and are extensively used for nanoparticles synthesis [50, 98]. Most publications deal with the use of dendrimers as templates for the synthesis of metallic nanoparticles, such as Au [50, 52, 95–98, 213–216], Ag [50, 52, 95–98, 215, 217], Cu [50, 96–98], Pt [50, 96–98], Pd [50, 96–98, 218, 219], Ru [96] and Ni [96]. Mixing of metal ions with dendrimers in aqueous solution results in complex formation between ions and the dendrimer's functional groups (e.g. amino or carboxyl) that might occur at the surface or/and in the central core. As a result, after reduction of metal ions, the nanoparticles formed with a size of a few nanometers may be located on the surface or/and in the core of a dendrimer molecule, as shown in Fig. 9.5 (for comparison, the average diameter of the generation-4 PAMAM with a molecular weight of ~20 000 is about 4.5 nm [97, 220]). In the case of a dendrimer molecule with a diameter smaller than that of the nanoparticle, it acts simply as a stabilizing polymer [50, 95–98, 213]. The formation of metallic nanoparticles can be performed not only in aqueous but also in non-aqueous solutions (toluene, chloroform, ethyl acetate, formamide) and the use of hydrophobically modified PAMAM dendrimers (e.g. surface methyl ester, stearyl groups) [88, 98].

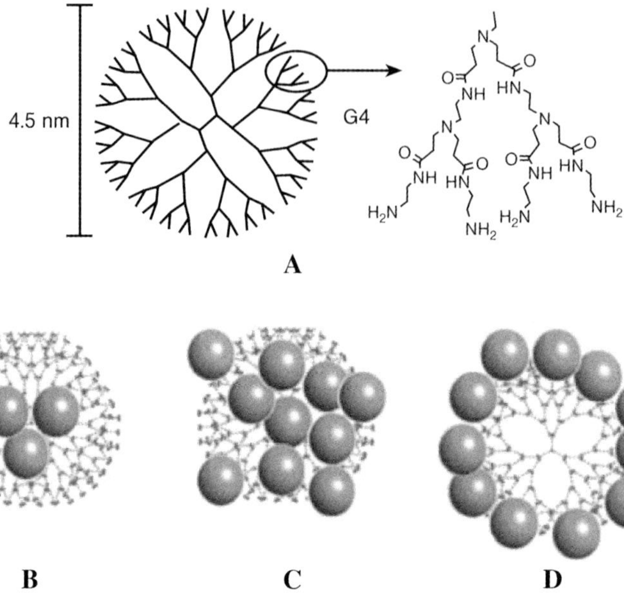

Fig. 9.5 Structure of G4 PAMAM dendrimer (A) (Reproduced with permission from [97], © 2001 Springer) and possible topologies of dendrimer/metal nanocomposites: internal (B), mixed (C) and external (D) location of nanoparticles. (Reproduced with permission from [95], © 1999 Kluwer Academic Publishers).

Dendrimers are also used as nanoreactors for the synthesis of semiconductor nanoclusters (quantum dots). For example, CdS nanoclusters with a diameter of <3 nm were prepared by reaction of Cd^{2+} with S^{2-} in aqueous or methanolic solutions of PAMAM [97, 98, 220].

9.2.3 Porous Matrices

Solid porous inorganic materials, such as zeolites, silica, alumina, aluminosilicates and glasses, can be used for nanoparticle synthesis within their pore channel [45]. According to the IUPAC classification [102, 221], porous solids are divided into three main categories according to their pore size: micro- (<2 nm), meso- (2–50 nm) and macroporous (>50 nm).

Among porous materials, mesoporous silicas are widely applied in nanoparticle synthesis. They are characterized by large specific surface areas and ordered systems of pores with fairly uniform size distribution. Nanostructures synthesized inside pores may be obtained in the form of particles, wires and tubes [45].

There are two basic approaches to nanoparticle synthesis in pores of silica: impregnation of the sorbent with metal salt solution followed by chemical reaction and modification of silica with a nanoparticles precursor [e.g. complex formation between metal ion and surface atoms (silicon, oxygen)] also followed by chemical transformation [45]. With the use of these approaches, a number of metallic and bimetallic nanoparticles (Ag [222–224], Cu [222], Au [225], Pd [226], Pt [226], Rh [226], Fe [45], Au/Ag [103] and Pd/Au [227]) and nanoparticles of semiconductors (ZnO [228], TiO_2 [229], CdSe [230]) have been prepared in mesoporous silica.

The main method of nanoparticle synthesis in zeolites is based on exchange of counterions (H^+, Na^+, NH_4^+) for the cation of the desired metal followed by its reduction with a suitable reducing agent [45]. Microporous zero-dimensional zeolites, characterized by open porosity with an ordered arrangement of spherical pores, were used for the preparation of metallic (Cu, Au, Pd, Fe) [101, 231–234] and bimetallic nanoparticles and nanoclusters (Pt/Pd, Au/Pd) [101, 235, 236] and also nanoparticles of semiconducting materials (CdS, NiS) [237, 238].

Nanoporous polymeric matrices in the form of beads, membranes, films, etc. are also used for the synthesis of nanosized materials. For example, nanoparticles of metals (Au, Ag, Pt, Pd, Rh, Cu, Co, Ni) were synthesized in pores of poly(styrene sulfonate) and poly(*N*-isopropyl acrylamide/acrylic acid/2-hydroxyethyl acrylate) microgels [88, 239, 240], hyper-cross-linked polystyrene [88, 241], cellulose fibers [242], perfluorinated ionomer membranes [243], poly(acrylic acid)/poly(allylamine hydrochloride) multilayer films [244, 245] and films of various block copolymers [246]. Nanoporous polymer materials were also used for the formation of semiconducting (CdS, Ag_2S, PbS) [240, 243, 245] and superparamagnetic (Fe_3O_4) [240] nanoparticles.

An elegant method for the preparation of porous, high-surface-area nanotubes has been reported recently. This method is based on the assembly of metal (Au, Ag) nanoparticles on the pore walls of nanoporous aminosilane-modified alumina membrane into a tubular structure spanning the entire pore length. Dissolution of the alumina template results in the release of solid, self-sustained metallic nanotubes [106, 247].

9.2.4
Polyelectrolyte Micro- and Nanocapsules

Polyelectrolyte microcapsules and nanocapsules can be fabricated by sequential layer-by-layer adsorption of oppositely charged polyelectrolytes from aqueous solution on the surface of colloidal template core particles of 0.1–10 μm diameter (Fig. 9.6 A). The polyelectrolyte shell can be assembled by employing a great variety of components: synthetic and natural polyelectrolytes, proteins, nucleic acids, inorganic nanoparticles and lipids [67, 99, 100, 248]. Hollow polyelectrolyte microcapsules (HPMs) can be prepared by removal of the core material after the coating layer has been formed. The removal of the core material (polystyrene, calcium carbonate) can be achieved by dissolution or chemical decomposition and the HPM can be refilled again with a desired core material [186]. Such a refilling is possible, since HPMs are permeable towards inorganic and organic molecules [67, 186]. The structure of HPMs allows the synthesis of na-

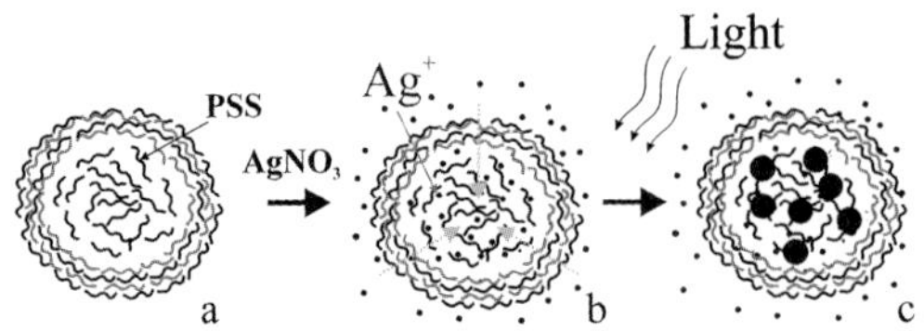

Fig. 9.6 Schematic illustration of photoinduced formation of silver nanoparticles inside HPM: (a) HPM loaded with electron-donating poly(styrene sulfonate) (PSS) molecules; (b) incubation with $AgNO_3$ solution; (c) silver nanoparticles formed inside HPM core after polychromatic irradiation. (Reproduced from [99]).

noparticles in the hollow cavity and also in the polyelectrolyte multilayer, which acts as protective medium for nanoparticles.

Diffusion of $NaWO_4$ solution into hollow capsules built of poly(styrene sulfonate) (PSS) and poly(allylamine hydrochloride) (PAH) and loaded with PSS results in polycondensation of tungstate ions in the capsule interior with low pH (a pH gradient across the capsule shell is induced by negatively charged PSS molecules) followed by formation of WO_3 [67, 249]. Nanosized Fe_2O_3 and Fe_3O_4 were formed in the same capsules, but loaded with PAH [67, 250, 251]. Entrapped PSS or polyaniline molecules can act as electron donors for the reduction of photoinduced silver ions in both the capsule core and the shell with the formation of nanosized particles, as shown in Fig. 9.6 [99]. Metal ions such as Pd^{2+} and Cu^{2+} can also be reduced inside PAH/PSS capsules by a photocatalytic process mediated by TiO_2 incorporated into the capsule shell [67]. Microcapsules made by layer-by-layer deposition of poly(acrylic acid) and polyacrylamide on amine-functionalized polystyrene as a core followed by cross-linking of the deposited polymeric layers were shown to be suitable templates for reduction of metal (Ag, Pd) ions in the capsule's shell. Subsequent dissolution of the core in a suitable solvent resulted in the formation of hollow microcapsules reinforced with metallic nanoparticles [252].

9.2.5
Liquid Crystals

Liquid crystals (LCs) are thermodynamically stable phases with an order intermediate between that of an anisotropic solid crystal and that of a liquid. The substances that form LCs are typically of organic origin and are characterized by rod-shaped molecules (e.g. surfactants, amphiphilic molecules consisting of hydrophobic and hydrophilic blocks). When such molecules are mixed with water, they are self-organized into periodic lattices consisting of nanometric domains (Fig. 9.7). The water cavities (hydrophilic domains) of the LC lattice might be considered as nanoreactors, which confine the reaction medium for the formation of various particles as presented in Fig. 9.7 D for lamellar LC systems. Thus, by controlling the geometry of the liquid crystalline phase, it is possible to control the size and morphology of the synthesized nanoparticles.

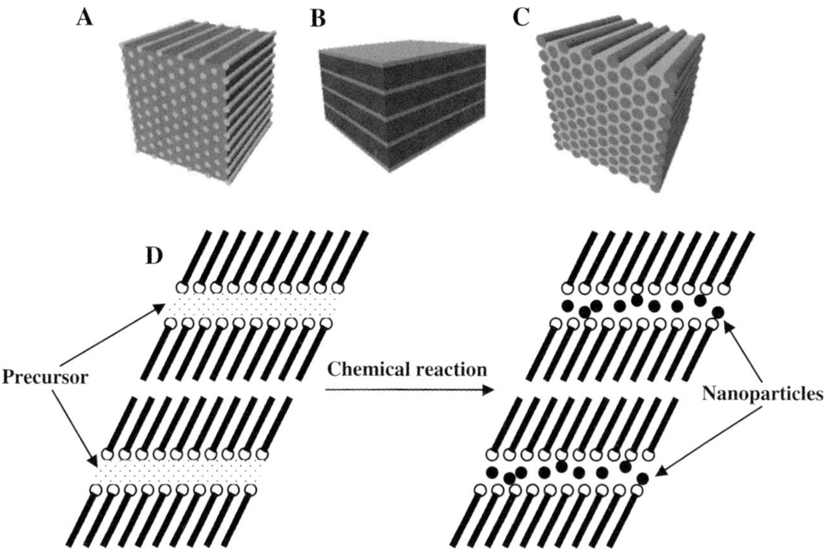

Fig. 9.7 Schematic presentation of lyotropic LCs with hexagonal (A), lamellar (B) and inverse hexagonal (C) liquid crystalline phases (hydrophilic and hydrophobic domains are dark and light colored, respectively) and scheme of preparation of metallic nanoparticles in water cavities of lamellar LCs. (Reproduced with permission from [44], © 2004 American Chemical Society.

Until now, LCs have been used mainly for the preparation of metallic and semiconducting nanoparticles. The synthesis can be performed by using a precursor aqueous solution (e.g. metal salts) as an aqueous phase to which a suitable reagent is then added. This approach was used for obtaining Ag nanoparticles in lamellar LCs by reduction of $AgNO_3$ with $NaBH_4$ [107], catalytic Pd nanoparticles in lyotropic LCs by reduction of Pd(II) salt with H_2 gas [253] and BiOCl nanoparticles in lyotropic LCs by reaction of $BiCl_3$ with NH_4OH [254].

Another approach, which is based on mixing of two preformed liquid crystalline systems containing interacting reagents dissolved in the aqueous phase, was successfully used for the preparation of metallic Bi nanoparticles by reduction of $BiCl_3$ with $CrCl_2$ in lyotropic LCs [44], in order to obtain semiconducting PbS nanoparticles by reaction of water-soluble lead salts with Na_2S in lyotropic and lamellar LCs [44, 255] and for the synthesis of Ag_2S from $AgNO_3$ and Na_2S [108] and ZnS from zinc acetate and thioacetamide in lamellar LCs [256].

In the case of metallic nanoparticles preparation by chemical reduction of corresponding ions, the LC-forming surfactants or polymers containing oxidizable functional groups facing the water domain are often used as *in situ* reductants. For example, silver and gold nanoparticles were prepared by this method in lamellar LCs formed by Triton X-100–$C_{10}H_{21}OH$–H_2O [257] and tetraethylene glycol monododecyl ether [258] and in hexagonal and reverse hexagonal LCs formed by poly(ethylene oxide)-*block*-poly(propylene oxide)-*block*-poly(ethylene oxide) (PEO–PPO–PEO) block copolymers (Pluronics) [259–262].

9.3 Applications

At present, the practical use of nanomaterials is at early stages and has not yet found extensive industrial applications, although the great potential of nanomaterials in various fields of science, technology and medicine is obvious. The use of nanoparticles is expected not only to provide materials with enhanced and even new suitabilities by re-engineering the existing materials to the nanoscale, but also to create more efficient approaches to manufacturing low-cost multifunctional materials for optics, electronics, optoelectronics, information and communication technologies [6, 8, 19, 24, 30, 31, 49, 50, 88], catalysis [5, 7, 11, 17, 20, 27, 30, 52, 88, 96], biotechnology and medicine (targeted drug delivery and implants, biosensors, bioseparation, magnetic resonance imaging) [21, 25, 30, 31, 33, 34, 49, 52, 161, 172, 190, 263–265], semiconducting, magnetic and superconductive devices [30, 31, 33, 34, 52, 88, 161], oxide-based thin-film power devices [17], ceramics and coatings [30, 78, 148–153] and pigments for ink-jet inks [35, 36, 187, 188]. We present here a number of examples of current and potential applications of nanoparticles formed in confined structures with special focus on catalysis, drug delivery and ink-jet printing.

9.3.1 Catalysis

The effectiveness of catalytic and photocatalytic reactions may be, in general, significantly improved and reaction rates may be increased by a factor of 100 by using nanoparticulate catalysts, which are characterized by high specific surface areas and quantum confinement effects [17]. Such catalysts can be successfully employed in many industrial catalytic processes.

For example, nanoparticles of metals, such as Pt, Pd, Rh, and their alloys formed in reverse micelles and W/O and W/SCF microemulsions were shown to be effective catalysts in reactions of isomerization, hydrogenation, selective ring opening, reduction and oxidation and hydrogen–deuterium exchange [83, 129–131, 133, 179]. Metal oxide nanoparticles prepared by the microemulsion technique were used for sulfur removal from gasoline [83, 266] and for pollution abatement in automobiles [267, 268]. Pt nanoparticles stabilized in micelles of polystyrene-*block*-poly-4-vinylpyridine and poly(ethylene oxide)-*block*-2-vinylpiridine were used as catalysts in cross-coupling reactions between aryl halides and alkenes [269] and in selective hydrogenation reactions of 2-butane-1,4-diol [270].

Dendrimer-encapsulated Pd nanoparticles were found to be effective catalysts in reactions of alkene hydrogenation in aqueous media and in organic solvents [96, 219] and in the cross-coupling reaction between phenylboronic (or 2-thienylboronic) acid and iodobenzene in aqueous medium [218].

Pt nanoparticles embedded in pores of hyper-cross-linked polystyrene catalyze the direct oxidation of L-sorbose to 2-keto-L-gulonic acid [241] and Pd nanoparticles incorporated into lyotropic LCs catalyze the hydrogenation of benzaldehyde and the Heck coupling reaction (olefination of aryl halides) [253].

9.3.2
Nanoparticles in Drug Delivery

The small size of nanoparticles allows them to penetrate and pass through biological barriers and to dissolve faster than micron-size particles. In addition, high loadings of various therapeutic and protective agents can be achieved inside non-toxic nanoparticles with various suitabilities and release characteristics. Therefore, nanoparticles may serve as effective carriers of drugs and physiologically active compounds. Solid nanoparticles made of biodegradable polymers by the solvent evaporation technique, such as the hydrophobic PLAs with degradation rates of several months, the more hydrophilic PGAs with degradation rates of several days and their copolymers PLGA are extensively used in various clinical applications, especially for drug delivery [89, 185, 189, 190, 263, 264]. For example, the antimicrobial agent ciprofloxacin [191], AG-1295 (a tyrphostin compound with proven antirestenotic activity) [193, 195], the anti-inflammatory drug nimesulide [197] and the anti-ischemic drug N^6-cyclopentyladenosine [199] were embedded into PLA nanospheres and pilocarpine hydrochloride was included in PLGA nanospheres [196]. Another type of biodegradable nanoparticles, which are able to entrap the model drug, Cytarabine, was prepared by UV irradiation of W/O microemulsions obtained by mixing an aqueous solution of an acryloylated polymer PHG with propylene carbonate–ethyl acetate [271].

Micelles of self-assembled amphiphilic block copolymers stabilized by intravesicular cross-linking were shown to possess great potential for the delivery of drugs embedded in their hydrophobic cores. They can be loaded with various physiologically active compounds and their permeability can be controlled either by attachment of functional groups or by reconstitution of membrane proteins within the micellar shell [89, 263, 272]. For example, drug delivery systems based on micelles of amphiphilic block copolymers loaded with antitumor drugs, cisplatin [273] and taxol [274], a cytotoxic agent, doxorubicin [275], and an antifungal agent, amphotericin B [276], have been reported.

9.3.3
Patterning of Organic Nanoparticles by Ink-jet Printing

Recently, O/W microemulsions containing hydrophobic colorants dissolved in volatile solvent droplets were evaluated as water-based ink-jet inks, which may provide a new route for low-cost fabrication of various products that requires fine patterning of functional molecules and nanocomposites [36, 187, 188]. The reported microemulsions were composed of a volatile solvent (e.g. toluene), a gemini-type surfactant (e.g. didodecyldiphenyl ether disulfonate) and a hydrophobic dye (Sudan IV, Blue GL or Nile Red) [36, 277]. In fact, organic nanoparticles are formed by evaporation of the microemulsion droplets in the size range

3–20 nm after their deposition on the solid substrate by ink-jet printing. The average size of the nanoparticles formed was found to be 40–60 nm [188, 277]. The larger size of the nanoparticles after printing and evaporation compared with that of the initial droplets might be a result of their spreading on the substrate and/or coalescence of the primary droplets during evaporation.

It is expected that microemulsions containing functional organic molecules will find many applications as ink-jet inks for organic nanoparticle patterning.

References

1 J.H. Fendler, *Chem. Mater.* **1996**, *8*, 1616–1624.
2 A. Henglein, *J. Phys. Chem.* **1993**, *97*, 5457–5471.
3 A.P. Alivisatos, *Science* **1996**, *271*, 933–937.
4 I. Bica, *Mater. Sci. Eng. B* **1999**, *68*, 5–9.
5 J.D. Aiken III, R.G. Finke, *J. Mol. Catal. A* **1999**, *145*, 1–44.
6 A.N. Shipway, E. Katz, I. Willner, *ChemPhysChem* **2000**, *1*, 18–52.
7 H. Bönnemann, R.M. Richards, *Eur. J. Inorg. Chem.* **2001**, 2455–2480.
8 G. Schmid, *Adv. Eng. Mater.* **2001**, *3*, 737–743.
9 A.B.R. Mayer, *Polym. Adv. Technol.* **2001**, *12*, 96–106.
10 D.L. Feldheim, C.A. Foss Jr., in *Metal Nanoparticles: Synthesis, Characterization and Applications* (D.L. Feldheim, C.A. Foss Jr., Eds.), Marcel Dekker, New York, **2002**, 1–15.
11 R. Kho, C.L. Torres-Martínez, in *Encyclopedia of Surface and Colloid Science* (P. Somasundaram, Ed.), Marcel Dekker, New York, **2002**, 3629–3642.
12 Q. Jiang, S. Zhang, M. Zhao, *Mater. Chem. Phys.* **2003**, *82*, 225–227.
13 C.N.R. Rao, A. Müller, A.K. Cheetham, in *The Chemistry of Nanomaterials: Synthesis, Properties and Applications* (C.N.R. Rao, A. Müller, Eds.), Wiley-VCH, Weinheim, **2004**, 1–11.
14 L.M. Liz-Marzán, *Mater. Today* **2004**, *7*, 26–31.
15 M.A. El-Sayed, *Acc. Chem. Res.* **2004**, *37*, 326–333.
16 S.V. Sonti, A. Bose, *J. Colloid Interface Sci.* **1995**, *170*, 575–585.
17 W.A. Zeltner, M.A. Anderson, in *Fine Particles Science and Technology: from Micro to Nanoparticles* (E. Pelizzetti, Ed.), Kluwer, Dordrecht, **1996**, 643–655.
18 D. Bockelmann, M. Lindner, D. Bahnemann, in *Fine Particles Science and Technology: from Micro to Nanoparticles* (E. Pelizzetti, Ed.), Kluwer, Dordrecht, **1996**, 675–689.
19 R.F. Service, *Science* **2000**, *287*, 1902–1903.
20 N.R. Jana, Z.L. Wang, T. Pal, *Langmuir* **2000**, *16*, 2457–2463.
21 A.P. Alivisatos, *Sci. Am.* **2001**, *285*, 67–73.
22 G.M. Whitesides, J.C. Love, *Sci. Am.* **2001**, *285*, 39–47.
23 C.M. Lieber, *Sci. Am.* **2001**, *285*, 58–64.
24 S.A. Maier, M.L. Brongersma, P.G. Kik, S. Meltzer, A.A.G. Requicha, H.A. Atwater, *Adv. Mater.* **2001**, *13*, 1501–1505.
25 C.M. Niemeyer, *Angew. Chem. Int. Ed.* **2001**, *40*, 4128–4158.
26 Y. Li, E. Boone, M.A. El-Sayed, *Langmuir* **2002**, *18*, 4921–4925.
27 N. Toshima, Y. Shiraishi, in *Encyclopedia of Surface and Colloid Science* (P. Somasundaram, Ed.), Marcel Dekker, New York, **2002**, 879–886.
28 P.V. Kamat, *J. Phys. Chem. B* **2002**, *106*, 7729–7744.
29 E. Katz, I. Willner, *Angew. Chem. Int. Ed.* **2004**, *43*, 6042–6108.
30 M.J. Pitkethly, *Nano Today* **2004** (December), 20–29.
31 A.P. Dowling, *Nano Today* **2004** (December), 30–35.
32 M. Mandal, S. Kundu, S.K. Ghosh, S. Panigrahi, T.K. Sau, S.M. Yusuf, T. Pal, *J. Colloid Interface Sci.* **2005**, *286*, 187–194.

33 Y. Wang, Z. Tang, N.A. Kotov, *Nano Today* **2005** (May), 20–31.

34 L. LaConte, N. Nitin, G. Bao, *Nano Today* **2005** (May), 32–38.

35 S. Magdassi, A. Bassa, Y. Vinetsky, A. Kamyshny, *Chem. Mater.* **2003**, *15*, 2208–2217.

36 A. Kamyshny, M. Ben-Moshe, S. Aviezer, S. Magdassi, *Micromol. Rapid Commun.* **2005**, *26*, 281–288.

37 G.-J. Lee, S.-I. Shin, Y.-C. Kim, S.-G. Oh, *Mater. Chem. Phys.* **2004**, *84*, 197–204.

38 D.V. Goia, E. Matijević, *New J. Chem.* **1998**, *22*, 1203–1215.

39 J.H. Adair, E. Suvaci, *Curr. Opin. Colloid Interface Sci.* **2000**, *5*, 160–167.

40 J. Park, E.V. Privman, E. Matijević, *J. Phys. Chem. B* **2001**, *105*, 11630–11635.

41 I. Sondi, D.V. Goia, E. Matijević, *J. Colloid Interface Sci.* **2003**, *260*, 75–81.

42 D.V. Goia, *J. Mater. Chem.* **2004**, *14*, 451–458.

43 L. Suber, I. Sondi, E. Matijević, *J. Colloid Interface Sci.* **2005**, *288*, 489–495.

44 T.M. Dellinger, P.V. Braun, *Chem. Mater.* **2004**, *16*, 2201–2207.

45 Yu.D. Tretyakov, A.V. Lukashin, *Russ. Chem. Rev.* **2004**, *73*, 899–921.

46 M.P. Pileni, *J. Phys. Chem.* **2001**, *105*, 3358–3371.

47 C.-L. Chiang, M.-B. Hsu, L.-B. Lai, *J. Solid State Chem.* **2004**, *177*, 3891–3895.

48 N. Toshima, T. Yonezawa, *New J. Chem.* **1998**, *22*, 1179–1201.

49 D. Horn, J. Rieger, *Angew. Chem. Int. Ed.* **2001**, *40*, 4330–4361.

50 T. Goodson III, O. Varnavski, Y. Wang, *Int. Rev. Phys. Chem.* **2004**, *23*, 109–150.

51 M.-P. Pileni, *C.R. Chimie* **2003**, *6*, 965–978.

52 G.B. Sergeev, *Russ. Chem. Rev.* **2001**, *70*, 809–825.

53 J. Asenjo, R. Amigo, E. Krotenko, F. Torres, J. Tejada, E. Brillas, G. Sardin, *J. Nanosci. Nanotechnol.* **2001**, *1*, 441–449.

54 A. Gedanken, *Ultrasonics Sonochem.* **2004**, *11*, 47–55.

55 G. De, A. Licciulli, C. Massaro, L. Tapfer, M. Catalano, G. Battaglin, C. Meneghini, P. Mazzoldi, *J. Non-Cryst. Solids* **1996**, *194*, 225–234.

56 I. Lisiecki, *Colloids Surf. A* **2004**, *250*, 499–507.

57 P.S. Shah, T. Hanrath, K.P. Johnston, B.A. Korgel, *J. Phys. Chem. B* **2004**, *108*, 9574–9587.

58 X. Peng, L. Manna, W. Yang, J. Wickham, E. Scher, A. Kadavanich, A.P. Alivisatos, *Nature* **2000**, *404*, 59–61.

59 G. Nizri, S. Magdassi, J. Schmidt, Y. Cohen, Y. Talmon, *Langmuir* **2004**, *20*, 4380–4385.

60 T. Teranishi, in *Encyclopedia of Surface and Colloid Science* (P. Somasundaram, Ed.), Marcel Dekker, New York, **2002**, 3314–3327.

61 A. Huczko, *Appl. Phys. A* **2000**, *70*, 365–376.

62 S. Chen, D.L. Carroll, *Nano Lett.* **2002**, *2*, 1003–1007.

63 C. Salzemann, I. Lisiecki, J. Urban, M.-P. Pileni, *Langmuir* **2004**, *20*, 11772–11777.

64 Y. Sun, Y. Xia, *Analyst* **2003**, *128*, 686–691.

65 A. Mayer, M. Antonietti, *Colloid Polym. Sci.* **1998**, *276*, 769–779.

66 M. Mandal, S.K. Ghosh, S. Kundu, K. Esumi, T. Pal, *Langmuir* **2002**, *18*, 7792–7797.

67 D.G. Shchukin, G.B. Sukhorukov, *Adv. Mater.* **2004**, *16*, 671–682.

68 G.B. Khomutov, V.V. Kislov, M.N. Antipina, R.V. Gainutdinov, S.P. Gubin, A.Yu. Obyydenov, S.A. Pavlov, A.A. Rakhnyanskaya, A.N. Sergeev-Cherenkov, E.A. Soldatov, D.B. Suyatin, A.L. Tolstikhina, A.S. Trifonov, T.V. Yurova, *Microelectron. Eng.* **2003**, *69*, 373–383.

69 M.P. Pileni, *Langmuir* **1997**, *13*, 3266–3267.

70 A. Taleb, C. Petit, M.P. Pileni, *J. Phys. Chem. B* **1998**, *102*, 2214–2220.

71 M.-P. Pileni, in *Metal Nanoparticles: Synthesis, Characterization and Applications* (D.L. Feldheim, C.A. Foss Jr., Eds.), Marcel Dekker, New York, **2002**, 207–236.

72 C.J. Murphy, N.R. Jana, *Adv. Mater.* **2002**, *14*, 80–82.

73 M.P. Pileni, A. Hammouda, I. Lisiecki, L. Motte, N. Moumen, J. Tanori, in *Fine Particle Science and Technology* (E. Pelizzetti, Ed.), Kluwer, Dordrecht, **1996**, 413–429.

74 M.P. Pileni, in *Nanoparticles and Nanostructured Films: Preparation, Characterization and Applications* (J.H. Fendler, Ed.), Wiley-VCH, Weinheim, **1998**, 71–100.

75 I. Lisiecki, M.P. Pileni, *Langmuir* **2003**, *19*, 9486–9489.

76 M.-P. Pileni, *Nat. Mater.* **2003**, *2*, 145–150.

77 G. Kästle, H.-G. Boyen, F. Weigl, G. Lengl, T. Herzog, P. Ziemann, S. Riethmüller, O. Mayer, C. Hartmann, J.P. Spatz, M. Möller, M. Ozawa, F. Banhart, M.G. Garnier, P. Oelhafen, *Adv. Funct. Mater.* **2003**, *13*, 853–861.

78 K. Osseo-Asare, in *Handbook of Microemulsion Science and Technology* (P. Kumar, K.L. Mittal, Eds.), Marcel Dekker, New York, **1999**, 549–603.

79 L. Jeunieau, F. Debuigne, J.B. Nagy, in *Reactions and Synthesis in Surfactant Systems* (J. Texter, Ed.), Marcel Dekker, New York, **2001**, 609–631.

80 F. Debuigne, L. Jeunieau, M. Wiame, J.B. Nagy, *Langmuir* **2000**, *16*, 7605–7611.

81 M.A. López-Quintela, *Curr. Opin. Colloid Interface Sci.* **2004**, *8*, 137–144.

82 M.A. López-Quintela, C. Tojo, M.C. Blanco, L. García Rio, J.R. Leis, *Curr. Opin. Colloid Interface Sci.* **2004**, *9*, 264–278.

83 S. Eriksson, U. Nylén, S. Rojas, M. Boutonnet, *Appl. Catal. A* **2004**, *265*, 207–219.

84 I. Capek, *Adv. Colloid Interface Sci.* **2004**, 49–74.

85 G. Carrot, J.C. Valmalette, C.J.G. Plummer, S.M. Scholz, J. Dutta, H. Hofmann, J.G. Hilborn, *Colloid Polym. Sci.* **1998**, *276*, 853–859.

86 L. Bronstein, M. Antonietti, P. Valetsky, in *Nanoparticles and Nanostructured Films: Preparation, Characterization and Applications* (J.H. Fendler, Ed.), Wiley-VCH, Weinheim, **1998**, 145–171.

87 S. Förster, M. Antonietti, *Adv. Mater.* **1998**, *10*, 195–217.

88 L.M. Bronstein, S.N. Sodorov, P. Valetsky, *Russ. Chem. Rev.* **2004**, *73*, 501–515.

89 M. Sauer, W. Meier, *ACS Symp. Ser.*, **2004**, *879*, 224–237.

90 M. Vamvakaki, L. Papoutsakis, V. Katsamanis, T. Afchoudia, P.G. Fragouli, H. Iatrou, N. Hadjichristidis, S.P. Armes, S. Sidorov, D. Zhirov, V. Zhirov, M. Kostylev, L.M. Bronstein, S.H. Anastasiadis, *Faraday Discuss.* **2005**, *128*, 129–147.

91 K.P. Johnston, C.T. Lee, G. Li, P. Psathas, J.D. Holmes, G.B. Jacobson, M.Z. Yates, in *Reactions and Synthesis in Surfactant Systems* (J. Texter, Ed.), Marcel Dekker, New York, **2001**, 349–358.

92 G.D. Holmes, D.M. Lyons, K.J. Ziegler, *Chem. Eur. J.* **2003**, *9*, 2144–2150.

93 X. Ye, C.M. Wai, *J. Chem. Educ.* **2003**, *80*, 198–204.

94 J. Liu, Y. Ikushima, Z. Shervani, *Curr. Opin. Colloid Interface Sci.* **2003**, *7*, 255–261.

95 L. Balogh, R. Valluzzi, K.S. Laverdure, S.P. Gido, G.L. Hagnauer, D.A. Tomalia, *J. Nanopart. Res.* **1999**, *1*, 353–368.

96 R.M. Crooks, M. Zhao, L. Sun, V. Chechik, L.K. Yeung, *Acc. Chem. Res.* **2001**, *34*, 181–190.

97 R.M. Crooks, B.L. Lemon III, L. Sun, L.K. Yeung, M. Zhao, *Top. Curr. Chem.* **2001**, *212*, 81–135.

98 K. Esumi, *Top. Curr. Chem.* **2003**, *227*, 31–52.

99 D.G. Shchukin, I.L. Radtchenko, G.B. Sukhorukov, *ChemPhysChem.* **2003**, *4*, 1101–1103.

100 X. Shi, M. Shen, H. Möhwald, *Prog. Polym. Sci.* **2004**, *29*, 987–1019.

101 J.B. Nagy, I. Hannus, I. Kiricsi, in *Nanoparticles and Nanostructured Films: Preparation, Characterization and Applications* (J.H. Fendler, Ed.), Wiley-VCH, Weinheim, **1998**, 389–427.

102 G.J. de A.A. Soler-Illia, C. Sanchez, B. Lebeau, J. Patarin, *Chem. Rev.* **2002**, *102*, 4093–4138.

103 J.F. Hund, M.F. Bertino, G. Zhang, C. Sotiriou-Leventis, N. Leventis, *J. Non-Cryst. Solids* **2004**, *350*, 9–13.

104 J.C. Hulteen, C.R. Martin, in *Nanoparticles and Nanostructured Films: Preparation, Characterization and Applications* (J.H. Fendler, Ed.), Wiley-VCH, Weinheim, **1998**, 235–262.

105 M. Wirtz, M. Parker, Y. Kobayashi, C. R. Martin, *Chem. Eur. J.* **2002**, *8*, 3573–3578.

106 M. Lahav, T. Sehayek, A. Vaskevich, I. Rubinstein, *Angew. Chem. Int. Ed.* **2003**, *42*, 5576–5579.

107 R. Patakfalvi, I. Dekany, *Colloid Polym. Sci.* **2002**, *280*, 461–470.

108 Y. Ding, B. Xu, R. Guo, M. Shen, *Mater. Res. Bull.* **2005**, *40*, 575–582.

109 M. Zulauf, H.-F. Eicke, *J. Phys. Chem.* **1979**, *83*, 480–486.

110 N. M. Correa, M. A. Biasutti, J. J. Silber, *J. Colloid Interface Sci.* **1995**, *172*, 71–76.

111 J. Faeder, B. M. Ladanyi, *J. Phys. Chem. B* **2000**, *104*, 1033–1046.

112 D. M. Togashi, S. M. B. Costa, *Phys ChemChemPhys* **2000**, *2*, 5437–5444.

113 M. J. Rosen, *Surfactants and Interfacial Phenomena*, Wiley, New York, **2004**, 105–177.

114 B. Lindman, N. Kamenka, B. Brun, P. G. Nilsson, in *Microemulsions* (I.D. Robb, Ed.), Plenum Press, New York, **1982**, 115–129.

115 J. B. Nagy, In *Handbook of Microemulsion Science and Technology* (P. Kumar, K. L. Mittal, Eds.), Marcel Dekker, New York, **1999**, 499–547.

116 M. A. López-Quintela, J. Rivas, *J. Colloid Interface Sci.* **1993**, *158*, 446–451.

117 C. Petit, P. Lixon, M. P. Pileni, *J. Phys. Chem.* **1990**, *94*, 1598–1606.

118 M. P. Pileni, L. Motte, C. Petit, *Chem. Mater.* **1992**, *4*, 338–345.

119 M. P. Pileni, *Catal. Today* **2000**, *58*, 151–166.

120 J. Eastoe, S. Stebbing, J. Dalton, R. K. Heenan, *Colloids Surf. A* **1996**, *119*, 123–131.

121 M. Li, H. Schnablegger, S. Mann, *Nature* **1999**, *402*, 393–395.

122 C. Petit, P. Lixon, M.-P. Pileni, *J. Phys. Chem.* **1993**, *97*, 12974–12983.

123 M. Li, S. Mann, *Adv. Funct. Mater.* **2002**, *12*, 773–779.

124 D. E. Zhang, X. M. Ni, H. G. Zheng, Y. Li, X. J. Zhang, Z. P. Yang, *Mater. Lett.* **2005**, *59*, 2011–2014.

125 M. Boutonnet, J. Kizling, P. Stenius, *Colloids Surf.* **1982**, *5*, 209–225.

126 P. Barnickel, A. Wokaun, *Mol. Phys.* **1990**, *69*, 1–9.

127 J. Zhang, B. Han, M. Liu, D. Liu, Z. Dong, J. Liu, D. Li, J. Wang, B. Gomg, H. Zhao, L. Rong, *J. Phys. Chem.* **2003**, *107*, 3679–3683.

128 F. Chen, G.-Q. Xu, T. S. A. Hor, *Mater. Lett.* **2003**, *57*, 3282–3286.

129 O. P. Yadav, A. Palmqvist, N. Cruise, K. Holmberg, *Colloids Surf. A* **2003**, *221*, 131–134.

130 M. Spiro, D. M. de Jesus, *Langmuir* **2000**, *16*, 2464–2468.

131 D. M. de Jesus, M. Spiro, *Langmuir* **2000**, *16*, 4896–4900.

132 X. Zhang, K.-Y. Chan, *J. Mater. Chem.* **2002**, *12*, 1203–1206.

133 X. Zhang, K.-Y. Chan, *Chem. Mater.* **2003**, *15*, 451–459.

134 J. Solla-Gullón, F. J. Vidal-Iglesias, V. Montiel, A. Aldaz, *Electrochim. Acta* **2004**, *49*, 5079–5088.

135 X. Zhang, F. Zhang, K.-Y. Chan, *J. Mater. Sci.* **2002**, *39*, 5845–5848.

136 F. Rauscher, P. Veit, K. Sundmacher, *Colloids Surf. A* **2005**, *254*, 183–191.

137 G. D. Rees, R. Evans-Gowing, S. J. Hammond, B. H. Robinson, *Langmuir* **2000**, *16*, 4896–4900.

138 L. Qi, J. Ma, H. Cheng, Z. Zhao, *Colloids Surf. A* **1996**, *108*, 117–126.

139 J. D. Hopwood, S. Mann, *Chem. Mater.* **1997**, *9*, 1819–1828.

140 J. Koetz, J. Bahnemann, G. Lucas, B. Tiersch, S. Kosmella, *Colloids Surf. A* **2004**, *250*, 423–430.

141 C. J. Johnson, M. Li, S. Mann, *Adv. Funct. Mater.* **2004**, *14*, 1233–1239.

142 Z. Li, J. Zhang, J. Du, B. Han, T. Mu, Y. Gao, Z. Liu, *Mater. Chem. Phys.* **2005**, *91*, 40–43.

143 Z. Li, J. Du, J. Zhang, T. Mu, Y. Gao, B. Han, J. Chen, J. Chen, *Mater. Lett.* **2005**, *59*, 64–68.

144 M. Husein, E. Rodil, J. Vera, *Langmuir* **2003**, *19*, 8467–8474.

145 V. T. Liveri, M. Rossi, G. D'Arrigo, D. Manno, G. Micocci, *Appl. Phys.* **1999**, *69*, 369–373.

146 K. E. Marchand, M. Tarret, J. P. Lechaire, L. Normand, S. Kasztelan, T. Cseri, *Colloids Surf. A* **2003**, *214*, 239–248.

147 S. Vaucher, M. Li, S. Mann, *Angew. Chem. Int. Ed.* **2000**, *39*, 1793–1796.

148 F. J. Arriagada, K. Osseo-Asare, *J. Colloid Interface Sci.* **1995**, *170*, 8–17.

149 D.-W. Kim, S.-G. Oh, J.-D. Lee, *Langmuir* **1999**, *15*, 1599–1603.

150 Y. Huang, T. Ma, J.-L. Yang, L.-M. Zhang, J.-T. He, H.-F. Li, *Ceram. Int.* **2004**, *30*, 675–681.

151 X. Sui, Y. Chu, S. Xing, M. Yu, C. Liu, *Colloids Surf. A* **2004**, *251*, 103–107.

152 K. D. Kim, S. H. Kim, H. T. Kim, *Colloids Surf. A* **2005**, *254*, 99–105.

153 X. Fang, C. Yang, *J. Colloid Interface Sci.* **1999**, *212*, 242–251.

154 A. R. Kortan, R. Hull, R. L. Opila, M. G. Bawendi, M. L. Steigerwald, P. J. Carrol, L. E. Brus, *J. Am. Chem. Soc.* **1990**, *112*, 1327–1332.

155 N. Pinna, K. Weiss, H. Sack-Kongehl, W. Vogel, J. Urban, M. P. Pileni, *Langmuir* **2001**, *17*, 7982–7987.

156 N. Pinna, K. Weiss, J. Urban, M.-P. Pileni, *Adv. Mater.* **2001**, *13*, 261–264.

157 D. Ingert, M.-P. Pileni, *Adv. Funct. Mater.* **2001**, *11*, 136–139.

158 B. A. Simmons, S. Li, V. T. John, G. L. McPherson, A. Bose, W. Zhou, J. He, *Nano Lett.* **2002**, *2*, 263–268.

159 B. Zhang, G. Li, J. Zhang, Y. Zhang, L. Zhang, *Nanotechnology* **2003**, *14*, 443–446.

160 T. Hirai, Y. Bando, *J. Colloid Interface Sci.* **2005**, *288*, 513–516.

161 P. Tartaj, M. del Puerto Morales, S. Veintemillas-Verdaguer, T. González-Carreño, C. J. Serna, *J. Phys. D: Appl. Phys.* **2003**, *36*, R182–R197.

162 V. Uskoković, M. Drofenik, I. Ban, *J. Magn. Magn. Mater.* **2004**, *284* 294–302.

163 Y. Lee, J. Lee, C. J. Bae, J.-G. Park, H.-J. Noh, J.-H. Park, *Adv. Funct. Mater.* **2005**, *15*, 503–509.

164 T. Hirai, T. Sato, I. Komesawa, *J. Colloid Interface Sci.* **2001**, *235*, 358–364.

165 T. Hirai, H. Okubo, I. Komesawa, *J. Phys. Chem.* **2001**, *105*, 9711–9714.

166 J. F. Debuigne, J. Cuisenaire, L. Jeunieau, B. Masereel, J. B. Nagy, *J. Colloid Interface Sci.* **2001**, *243*, 90–101.

167 K. Tauer, in *Reactions and Synthesis in Surfactant Systems* (J. Texter, Ed.), Marcel Dekker, New York, **2001**, 429–453.

168 F. M. Pavel, *J. Dispers. Sci. Technol.* **2004**, *25*, 1–16.

169 M. Antonietti, R. Basten, S. Lohmann, *Macromol. Chem. Phys.* **1995**, *196*, 441–466.

170 J. Texter, in *Reactions and Synthesis in Surfactant Systems* (J. Texter, Ed.), Marcel Dekker, New York, **2001**, 577–607.

171 M. Ji, X. Chen, C. M. Wai, J. L. Fulton, *J. Am. Chem. Soc.* **1999**, *121*, 2631–2632.

172 P. J. Ginty, M. J. Whitaker, K. M. Shakesheff, S. M. Howdle, *Nano Today* **2005** (August), 42–48.

173 J. L. Fulton, in *Handbook of Microemulsion Science and Technology* (P. Kumar, K. L. Mittal, Eds.), Marcel Dekker, New York, **1999**, 629–650.

174 H. Ohde, M. Ohde, F. Bailey, H. Kim, C. M. Wai, *Nano Lett.* **2002**, *2*, 721–724.

175 J. Liu, P. Raveendran, Z. Shervani, Y. Ikushima, *Chem. Commun.* **2004**, 2582–2583.

176 H. Ohde, J. M. Rodriguez, X.-R. Ye, C. M. Wai, *Chem. Commun.* **2000**, 2353–2354.

177 H. Ohde, F. Hunt, C. M. Wai, *Chem. Mater.* **2001**, *13*, 4130–4135.

178 J. Liu, P. Raveendran, Z. Shervani, Y. Ikushima, Y. Hakuta, *Chem. Eur. J.* **2005**, *11*, 1854–1860.

179 M. Ohde, H. Ohde, C. M. Wai, *Langmuir* **2005**, *21*, 1738–1744.

180 J. D. Holmes, P. A. Bhargava, B. A. Korgel, K. P. Johnston, *Langmuir* **1999**, *15*, 6613–6615.

181 W. Hamley, *Nanotechnology* **2003**, *14*, R39–R54.

182 L. Zhang, K. Yu, A. Eisenberg, *Science* **1996**, *272*, 1777–1779.

183 R. E. Cohen, *Curr. Opin. Solid State Mater. Sci.* **1999**, *4*, 587–590.

184 R. Arshady, in *Microspheres, Microcapsules and Lipisomes*, Vol. 1 (R. Arshady, Ed.), Citus Books, London, **1999**, 279–326.

185 S. Desgouilles, C. Vauthier, D. Bazile, J. Vacus, J.-L. Grossiord, M. Veillard, P. Couvreur, *Langmuir* **2003**, *19*, 9504–9510.

186 A. Kamyshny, S. Magdassi, in *Encyclopedia of Surface and Colloid Science* (P. Somasundaram, Ed.), Marcel Dekker, New York, **2004**, 1–15.

187 S. Magdassi, M. Ben-Moshe, L. Berenstein, A. Zaban, *J. Imaging Sci. Technol.* **2003**, *47*, 357–360.
188 S. Magdassi, M. Ben-Moshe, *Langmuir* **2003**, *19*, 939–942.
189 H. Okada, H. Toguchi, *Crit. Rev. Ther. Drug Carr.* **1995**, *12*, 1–99.
190 D. Quintanar-Guerrero, E. Allemann, H. Fessi, E. Doelker, *Drug Dev. Ind. Pharm.* **1998**, *24*, 1113–1128.
191 W. P. Yu, J. P. Wong, T. M. S. Chang, *Artif. Cells Blood Subst. Immob. Technol.* **1999**, *27*, 263–278.
192 T. Chandy, G. S. Das, G. H. R. Rao, *J. Microencapsulation* **2000**, *17*, 625–638.
193 I. Fishbein, M. Chorny, S. Banai, A. Levitzki, H. D. Danenberg, J. Gao, X. Chen, E. Moerman, I. Gati, V. Goldwasser, G. Golomb, *Arterioscler. Thromb. Vasc. Biol.* **2001**, *21*, 1434–1439.
194 Y. Liu, X. Deng, *J. Control. Release* **2002**, *83*, 147–155.
195 M. Chorhy, I. Fishbein, H. D. Danenberg, G. Golomb, *J. Control. Release* **2002**, *83*, 389–400.
196 J. Vandervoort, K. Yoncheva, A. Ludwig, *Chem. Pharm. Bull.* **2004**, *52*, 1273–1279.
197 M. N. Freitas, J. M. Marchetti, *Int. J. Pharm.* **2005**, *295*, 201–211.
198 A.-M. Layre, R. Gref, J. Richard, D. Requier, H. Chacun, M. Appel, A. J. Domb, P. Couvreur, *Int. J. Pharm.* **2005**, *298*, 323–327.
199 A. Dalpiaz, E. Leo, F. Vitali, B. Pavan, A. Scatturin, F. Bortolotti, S. Manfredini, E. Durini, F. Forni, B. Brina, M. A. Vandelli, *Biomaterials* **2005**, *26*, 1299–1306.
200 C. Larpent, E. Bernard, J. Richard, S. Vaslin, *Macromolecules* **1997**, *30*, 354–362.
201 Y. Zhang, T. Guo, G. Hao, M. Song, B. Zhang, *J. Appl. Polym. Sci.* **2003**, *90*, 3625–3630.
202 C. Larpent, E. Bernard, J. Richard, S. Vaslin, *React. Funct. Polym.* **1997**, *33*, 49–59.
203 C. Ford, M. Singh, L. Lawson, J. He, V. John, Y. Lu, K. Papadopoulos, G. McPherson, A. Bose, *Colloids Surf. B* **2004**, *39*, 143–150.
204 J. Jang, J. H. Oh, *Adv. Mater.* **2003**, *15*, 977–980.
205 A. Forgiarini, J. Esquena, C. González, C. Solans, *Prog. Colloid Polym. Sci.* **2000**, *115*, 36–39.
206 M. Antonietti, K. Landfester, *Prog. Polym. Sci.* **2002**, *27*, 689–757.
207 P. Fernandez, V. André, J. Rieger, A. Kühnle, *Colloids Surf. A* **2004**, *251*, 53–58.
208 T. Tadros, P. Izquierdo, J. Esquena, C. Solans, *Adv. Colloid Interface Sci.* **2004**, *108/109*, 303–318.
209 L. Spernath, S. Magdassi, Patent Application PCT/IL2005/000416, **2005**.
210 F. Vögtle, S. Gestermann, R. Hesse, H. Schwierz, B. Windisch, *Prog. Polym. Sci.* **2000**, *25*, 987–1041.
211 P. E. Froehling, *Dyes Pigm.* **2001**, *48*, 187–195.
212 J. W. Lee, K. Kim, *Top. Curr. Chem.* **2003**, *228*, 111–140.
213 R. West, Y. Wang, T. Goodson III, *J. Phys. Chem. B* **2003**, *107*, 3419–3426.
214 F. Grohn, B. J. Bauer, Y. A. Akpalu, C. L. Jackson, E. J. Amis, *Macromolecules* **2000**, *33*, 6042–6050.
215 A. Manna, T. Imae, K. Aoi, M. Okada, T. Yogo, *Chem. Mater.* **2001**, *13*, 1674–1681.
216 K. Satoh, T. Yoshimura, K. Esumi, *J. Colloid Interface Sci.* **2002**, *255*, 312–322.
217 H. W. Lu, S. H. Liu, X. L. Wang, X. F. Qian, J. Yin, Z. K. Zhu, *Mater. Chem. Phys.* **2003**, *81*, 104–107.
218 Y. Li, M. A. El-Sayed, *J. Phys. Chem. B* **2001**, *105*, 8938–9843.
219 M. Ooe, M. Murata, T. Mizugaki, K. Ebitani, K. Kaneda, *Nano Lett.* **2002**, *2*, 999–1002.
220 L. H. Hanus, K. Sookal, C. J. Murphy, H. J. Ploehn, *Langmuir* **2000**, *16*, 2621–2626.
221 K. S. W. Sing, D. H. Everett, H. Raw, L. Moscou, R. A. Pierotti, J. Rouquerol, T. Siemieniewska, *Pure Appl. Chem.* **1985**, *57*, 603–619.
222 G. De, L. Tapfer, M. Catalano, G. Battaglin, F. Caccavale, F. Gonella, P. Mazzoldi, R. F. Haglund Jr., *Appl. Phys. Lett.* **1996**, *68*, 3820–3822.

223 W. Chen, J. Zhang, *Scr. Mater.* **2003**, *49*, 321–325.
224 X.-G. Zhao, J.-L. Shu, B. Hu, L.-X. Zhang, Z.-L. Hua, *Mater. Lett.* **2004**, *58*, 2151–2156.
225 G. Fu, W. Cai, C. Kan, C. Li, Q. Fang, *J. Phys. D: Appl. Phys.* **2003**, *36*, 1382–1387.
226 B.-H. Han, S. Polarz, M. Antonietti, *Chem. Mater.* **2001**, *13*, 3915–3919.
227 C. Li, W. Cai, C. Kan, G. Fu, *Scr. Mater.* **2004**, *50*, 1481–1486.
228 W.-H. Zhang, J.-L. Shi, L.-Z. Wang, D.-S. Yan, *Chem. Mater.* **2000**, *12*, 1408–1413.
229 S.I. Seok, J.H. Kim, *Mater. Chem. Phys.* **2004**, *86*, 176–179.
230 H. Parala, H. Winkler, M. Kolbe, A. Wohlfart, R.A. Fischer, R. Schmechel, H. von Seggern, *Adv. Mater.* **2000**, *12*, 1050–1055.
231 D. Guillemot, M. Polisset-Thfoin, J. Frassard, *Catal. Lett.* **1996**, *41*, 143–148.
232 Y.-M. Kang, B.-Z. Wan, *Catal. Today* **1997**, *35*, 379–392.
233 A. Seidel, J. Loos, B. Boddenberg, *J. Mater. Sci.* **1999**, *9*, 2495–2498.
234 N. Nishimiya, T. Kishi, T. Mizushima, A. Matsumoto, K. Tsutsumi, *J. Alloys Compd.* **2001**, *319*, 312–321.
235 T. Rades, M. Polisset-Thfoin, J. Fraissard, *Top. Catal.* **2000**, *11/12*, 283–287.
236 A.-M. Diamy, Z. Randriamanantenasoa, J.-C. Legrand, M. Polisset-Thfoin, J. Fraissard, *Chem. Phys. Lett.* **1997**, *269*, 327–332.
237 P.W. de Bont, M.J. Vissenberg, E. Boellaard, V.H.J. de Beer, J.A.R. van Veen, R.A. van Santen, A.M. van der Kraan, *Hyperfine Interact.* **1998**, *111*, 39–44.
238 H.V. García, M.H. Vélez, O.S. Garrido, J.M.M. Duart, J. Jiménez, *Solid-State Electron.* **1999**, *43*, 1171–1175.
239 N.T. Whilton, B. Berton, L. Bronstein, H.-P. Hentze, M. Antonietti, *Adv. Mater.* **1999**, *11*, 1014–1018.
240 J. Zhang, S. Xu, E. Kumacheva, *J. Am. Chem. Soc.* **2004**, *126*, 7908–7914.
241 L.M. Bronstein, G. Goerigk, M. Kostylev, M. Pink, I.A. Khotina, P.M. Valetsky, V.G. Matveeva, E.M. Sulman, M.G. Sulman, A.V. Bykov, N.V. Lakina, R.G. Spontak, *J. Phys. Chem. B* **2004**, 18234–18242.
242 J. He, T. Kunitake, A. Nakao, *Chem. Mater.* **2003**, *15*, 4401–4406.
243 H.W. Rollins, F. Lin, J.U. Johnson, J.-J. Ma, J.-T. Liu, M.H. Tu, D.D. DesMarteau, Y.-P. Sun, *Langmuir* **2000**, *16*, 8031–8036.
244 T.C. Wang, M.F. Rubner, R.E. Cohen, *Langmuir* **2002**, *18*, 3370–3375.
245 S. Joly, R. Kane, L. Radzilowski, T. Wang, A. Wu, R.E. Cohen, E.L. Thomas, M.F. Rubner, *Langmuir* **2000**, *16*, 8031–8036.
246 S. Horiuchi, T. Fujita, *Langmuir* **2003**, *19*, 2963–2973.
247 T. Sehayek, M. Lahav, R. Popovitz-Biro, A. Vaskevich, I. Rubinshtein, *Chem. Mater.* **2005**, *17*, 3743–3748.
248 F. Caruso, *Adv. Mater.* **2001**, *13*, 11–22.
249 D.G. Shchukin, W. Dong, G.B. Sukhorukov, *Micromol. Rapid Commun.* **2003**, *24*, 462–466.
250 I.L. Radtchenko, M. Giersig, G.B. Sukhorukov, *Langmuir* **2002**, *18*, 8204–8208.
251 D.G. Shchukin, I.L. Radtchenko, G.B. Sukhorukov, *Mater. Lett.* **2003**, *57*, 1743–1747.
252 D. Lee, M.F. Rubner, R.E. Cohen, *Chem. Mater.* **2005**, *17*, 1099–1105.
253 J.H. Ding, D.L. Gin, *Chem. Mater.* **2000**, *12*, 22–24.
254 T.M. Dellinger, P.V. Braun, *Scr. Mater.* **2001**, *44*, 1893–1897.
255 R. Guo, T.Q. Liu, *Colloids Surf. A* **1997**, *123/124*, 587–591.
256 D. Zhang, L. Qi, H. Cheng, J. Ma, *J. Colloid Interface Sci.* **2002**, *246*, 413–416.
257 M.-H. Lee, S.-G. Oh, K.-D. Suh, D.-G. Kim, D. Sohn, *Colloids Surf. A* **2002**, *210*, 49–60.
258 L. Qi, Y. Gao, J. Ma, *Colloids Surf. A* **1999**, *157*, 285–294.
259 L. Wang, X. Chen, J. Zhao, Z. Sui, W. Zhuang, L. Xu, C. Yang, *Colloids Surf. A* **2005**, *257/258*, 231–235.
260 M. Andersson, V. Alfredsson, P. Kjellin, A.E.C. Palmqvist, *Nano Lett.* **2002**, *2*, 1403–1407.

261 L. Wang, X. Chen, J. Zhan, Z. Sui, J. Zhao, Z. Sun, *Chem. Lett.* **2004**, *33*, 720–721.

262 L. Wang, X. Chen, J. Zhan, Y. Chai, C. Yang, L. Xu, W. Zhuang, B. Jing, *J. Phys. Chem. B* **2005**, *109*, 3189–3194.

263 S. K. Sahoo, V. Labhasetwar, *Drug Discov. Today* **2003**, *8*, 1112–1120.

264 T. M. Fahmy, P. M. Fong, A. Goyal, W. M. Saltzman, *Nano Today* **2005** (August), 18–26.

265 G. J. Kim, S. Nie, *Nano Today* **2005** (August), 28–33.

266 P.-O. F. Andersson, M. Pirjamali, S. G. Järås, M. Boutonnet-Kizling, *Catal. Today* **1999**, *53*, 565–573.

267 T. Masui, K. Fujiwara, K. Machida, G. Adachi, T. Sakata, H. Mori, *Chem. Mater.* **1997**, *9*, 2197–2204.

268 A. Trovarelli, M. Boaro, E. Rocchini, C. de Leitenburg, G. Dolcetti, *J. Alloy Compd.* **2001**, *323/324*, 584–591.

269 S. Klingelhöfer, W. Heitz, A. Greiner, S. Oestreich, S. Förster, M. Antonietti, *J. Am. Chem. Soc.* **1997**, *119*, 10116–10120.

270 N. Semagina, E. Joannet, S. Parra, E. Sulman, A. Renken, L. Kiwi-Minsker, *Appl. Catal. A* **2005**, *280*, 141–147.

271 E. F. Craparo, G. Cavallaro, M. L. Bondi, G. Giammona, *Macromol. Chem. Phys.* **2004**, *205*, 1955–1964.

272 K. Kataoka, A. Harada, Y. Nagasaki, *Adv. Drug Deliv. Rev.* **2001**, *47*, 113–131.

273 N. Nishiyama, Y. Kato, Y. Sugiyama, K. Kataoka, *Pharm Res.* **2001**, *18*, 1035–1041.

274 V. P. Torchilin, A. N. Lukyanov, Z. Gao, B. Papahadjopoulos-Sternberg, *Proc. Natl. Acad. Sci. USA* **2003**, *100*, 6039–6044.

275 K. Kataoka, T. Matsumoto, M. Yokoyama, T. Okano, Y. Sakurai, S. Fukushima, K. Okamoto, G.S. Kwon, *J. Control. Release* **2000**, *64*, 143–153.

276 M. L. Adams, D. R. Andes, G. S. Kwon, *Biomacromolecules* **2003**, *4*, 750–757.

277 S. Magdassi, M. Ben-Moshe, Y. Talmon, D. Danino, *Collids Surf. A* **2003**, *212*, 1–7.

10
Colloid Stability Using Polymeric Surfactants

Tharwat Tadros

Abstract

This overview starts with a general classification of polymeric surfactants with particular reference to A–B, A–B–A block and BA_n or AB_n graft types (where B is the anchor chain and A is the stabilizing chain). The solution properties of polymeric surfactants was described with particular reference to the A chain–solvent interaction as determined by the Flory-Huggins interaction parameter χ. For A chains to be in good solvent conditions (an important criterion for steric stabilization), χ should remain <0.5 under all conditions (such as temperature changes and/or in the presence of electrolytes). The adsorption and conformation of polymeric surfactants were briefly described at the solid–liquid interface. For a full description of polymer adsorption one needs to obtain three main parameters: the amount of polymer adsorbed Γ as a function of equilibrium concentration in the bulk solution, the fraction of segments p in direct contact with the surface (which determines the strength of adsorption) and the segment density distribution $\rho(z)$ or hydrodynamic thickness δ_h. The theories of polymer adsorption are briefly mentioned. The use of polymeric surfactants of the A–B–A, BA_n and AB_n types for stabilization of various disperse systems was described. A graft copolymer of poly(methyl methacrylate)/(polymethacrylic acid) backbone with poly(-ethylene oxide) (PEO) side-chains, Atlox 4913 and Hypermer CG-6, was applied for stabilization of preformed polystyrene latex dispersions (that were prepared using surfactant-free emulsion polymerization). The stability of the system was investigated using rheological measurements. Force–distance curves for two mica cylinders containing an adsorbed graft copolymer of Atlox 4913 were obtained and this confirmed the strong steric repulsion between the adsorbed polymer layers. A hydrophobically modified graft copolymer with an inulin (polyfructose) backbone with several alkyl side-chains (INUTEC SP1) was applied to the preparation of polystyrene and poly(methyl methacrylate) latexes. Stable systems could be produced using low concentrations of INUTEC SP1. The stability of the resulting latexes was assessed by measuring the critical coagulation concentration (CCC) using $CaCl_2$. The CCC increased rapidly on post-addition of INUTEC

Colloids and Interface Science Series, Vol. 1
Colloid Stability: The Role of Surface Forces, Part I. Edited by Tharwat F. Tadros

ISBN: 978-3-527-31462-1

SP1 to the latex dispersions. The high stability of the latexes was accounted for by multi-point attachment of the graft copolymer (by several alkyl chains) and the strong hydration of the polyfructose loops and tails. Evidence for this strong repulsion was obtained using atomic force microscopy (AFM) measurements using a hydrophobized glass sphere and plate both containing an adsorbed layer of INUTEC SP1. The latter was also used for the stabilization of oil-in-water (O/W) emulsions and the results showed very high stability both at high temperature (50 °C) and in the presence of high electrolyte concentrations. This high stability was due to the multi-point attachment of the graft copolymer to the oil droplets and the strong hydration of the polyfructose loops and tails both at high temperature and in the presence of high electrolyte concentrations. Evidence for this strong hydration was obtained from cloud point measurements of inulin solutions. This strong steric repulsion was also confirmed by measuring the disjoining pressure as a function of film thickness between two oil drops. Stable water-in-oil (W/O) emulsions could also be prepared using an A–B–A block copolymer of polyhydroxystearic acid (PHS) (A) and PEO (B) (Arlacel P135). Stable nano-emulsions could be produced using INUTEC SP1. The latter could reduce Ostwald ripening when compared with conventional surfactants. Finally, polymeric surfactants of Aralacel P135 and INUTEC SP1 could be applied for the preparation of very stable multiple emulsions. Both W/O/W and O/W/O multiple emulsions could be prepared and their stability was investigated by taking optical micrographs at several periods for systems stored at 50 °C.

10.1 Introduction

Polymeric surfactants are ideal systems for the colloid stability of many dispersions, e.g. solid–liquid (suspensions), liquid–liquid (emulsions), nano-emulsions and multiple emulsions. The most commonly used polymeric surfactants are those of the A–B, A–B–A block and BA_n or AB_n graft copolymer type. Here B represents the "anchor" chain that is insoluble in the medium and has a strong affinity to the surface. This ensures strong adsorption and lack of desorption on approach of the particles. The A chains, referred to as the stabilizing chains, are chosen to be highly soluble in the medium and strongly solvated by its molecules. This ensures effective steric stabilization [1] as discussed in Chapter 1 of this volume.

For a proper choice of the polymeric surfactant, one has to consider the adsorption and conformation of the polymeric surfactant at the interface [2]. It is also essential to consider the effect of the medium on this adsorption process and the systems variables such as temperature changes and addition of electrolytes and non-electrolytes. These variables can affect the overall steric repulsion and under some conditions this repulsive interaction may change to an attractive interaction, resulting in flocculation of the dispersion.

This chapter starts with a section on the general classification of polymeric surfactants, giving examples of systems that can be used to stabilize various disperse systems. This is followed by a section on the solution properties of polymeric surfactants and the effect of some variables, e.g. temperature. The adsorption and conformation of polymeric surfactants at the solid–liquid interface are briefly described. The final part summarizes some of the model systems that have been stabilized with polymeric surfactants and the techniques that may be used to study this stability.

10.2 General Classification of Polymeric Surfactants

The simplest type of polymeric surfactant is a homopolymer, which is formed from the same repeating units: for example, poly(ethylene oxide) (PEO) or polyvinylpyrrolidone (PVP). Homopolymers have little surface activity at the oil/water (O/W) interface and hence they are seldom used for stabilization of emulsions. However, homopolymers may adsorb significantly at the solid/liquid (S/L) interface either by specific interaction, e.g. hydrogen bonding (PVP on silica), or simple adsorption of few polymer segments on the surface without any specific interaction. Even if the adsorption energy per monomer segment χ^s is small (a fraction of kT, where k is Boltzmann's constant and T is the absolute temperature), the total adsorption energy per molecule may be sufficient (several monomer units are adsorbed at the surface) to overcome the unfavorable entropy loss when a polymer chain approaches the surface. However, this type of non-specific adsorption does not confer sufficient stability on the dispersion owing to the possible desorption on approach of a particle.

Homopolymers are modified by introducing short blocks of segments which show strong adsorption on the surface. A good example is partially hydrolyzed poly(vinyl alcohol) (PVA) which contains short blocks of poly(vinyl acetate) (PVAc). These short blocks give the molecule its amphipathic character and hence the molecule becomes surface active at the O/W interface. In addition, on a hydrophobic surface such as polystyrene latex, strong adsorption occurs by hydrophobic bonding between the PVAc blocks and the surface. Adsorption in this case becomes irreversible and a high-affinity isotherm is produced [3]. The PVA loops and tails dangling in solution provide effective steric stabilization.

The most effective polymeric surfactants are those of the A–B, A–B–A block, BA_n or AB_n types. As mentioned in Chapter 1 of this volume, these satisfy the criteria of effective steric stabilization. The B chain adsorbs strongly on the surface, forming small loops with multi-point attachment, whereas the highly soluble and strongly solvated A chains provide steric stability. One of the earliest A–B–A block copolymers were those based on PEO (the A chains) and poly(propylene oxide) (PPO) (the B chain). These are commercially available under the trade name Pluronics (BASF), Synperonic PE and Poloxamers (UNIQEMA, ICI). Molecules with various PEO and PPO compositions are available and the commercial name is followed by the letter L (liquid), P (paste) and F (flake). This is followed by two numbers that represent the composition – the first digit represents the PPO molecular mass and the second digit represents the percentage of PEO, Pluronic F68 (PPO Mol Wt 1501–1800 + 140 mol EO) and Pluronic L62 (PPO Mol Wt 1501–1800+15 mol EO). These block copolymers were used for the stabilization of emulsions and suspensions. However, they are not the most effective stabilizers since the PPO chain is not sufficiently hydrophobic to provide a strong "anchor" to the droplet or particle surface. They probably adsorb by a process referred to as "rejection anchoring"; the PPO chain which is not soluble in water or oil adsorbs by rejection from the oil and the water. For suspensions, the PPO chain is also not strongly adsorbed on a hydrophobic surface. For that reason, several other di- and tri-blocks were synthesized with the B chain being polystyrene (PS) or poly(methyl methacrylate) (PMMA) and the A chains being PEO or PVA. The PS or PMMA B chain is strongly adsorbed on a hydrophobic surface. However, these blocks of PEO–PS–PEO and PEO–PMMA–PEO are not commercially available.

A commercially available graft copolymer of PMMA (with some polymethacrylic acid, PMMAc) and several PEO chains grafted on the backbone is available (Hypermer CG-6 or Atlox 4913, UNIQEMA) have been successfully applied for stabilization of suspensions (see below). This graft copolymer is sometimes referred to as a "comb" polymeric surfactant and on adsorption it produces a "brush" that is very effective for steric stabilization.

Recently, an AB_n has been synthesized [4, 5] consisting of an inulin (linear polyfructose) backbone on which several alkyl chains have been grafted. This graft copolymer strongly adsorbs on hydrophobic surfaces with multi-point attachment points (by several alkyl chains) leaving loops and tails of polyfructose dangling in solution, thus providing effective steric stabilization. This graft copolymer has been applied to the stabilization of emulsions, nano-emulsions, multiple emulsions, suspensions and latex dispersions, as will be discussed below.

10.3 Solution Properties of Polymeric Surfactants

To understand the solution properties of block and graft copolymers, one should first consider the case of homopolymers, which was discussed in detail using the Flory-Huggins theory [6]. The latter considers the free energy of mixing of pure polymer with pure solvent, ΔG_{mix}, in terms of two contributions: an enthalpy of mixing term, ΔH_{mix}, and an entropy of mixing term, ΔS_{mix}:

$$\Delta G_{mix} = \Delta H_{mix} - T\Delta S_{mix} \tag{1}$$

The entropy of mixing is given by

$$\Delta S_{mix} = -k(n_1 \ln \phi_1 + n_2 \ln \phi_2) \tag{2}$$

where k is Boltzmann's constant, T is the absolute temperature, n_1 is the number of moles of solvent with volume fraction ϕ_1 and n_2 is the number of moles of polymer molecules with volume fraction ϕ_2.

The enthalpy of mixing is given by

$$\Delta H_{\text{mix}} = n_1 \phi_2 \chi kT \tag{3}$$

where χ is a dimensionless interaction parameter and χkT expresses the difference in energy of a solvent molecule in pure solvent compared with its immersion in pure polymer (or equivalently a measure of transfer of a segment from pure polymer to pure solvent); χ is usually referred to as the Flory–Huggins interaction parameter, which is a measure of excess affinity of segments to each other over that of the solvent. If the polymer segment and solvent molecule have the same polarity and polarizability, $\chi=0$, and this is referred to as athermal solvent. However, χ usually has a small value (<0.5) and this is referred to as a good solvent. Under this condition, the volume exclusion implies that the walk through space has to avoid itself and this leads to strong coil expansion (swelling). Polymer segments in a good solvent avoid each other. Similarly, polymer coils avoid each other and they are reluctant to interpenetrate. This is the origin of stabilization by the A chains of an A–B, A–B–A block and BA_n or AB_n graft copolymers.

The condition $\chi=0.5$ refers to ideal mixing of the polymer with the solvent. In this case, the net excluded volume and the second virial coefficient (see below) are zero. This condition is referred to as the θ-point. Under these conditions, the effective segment–segment repulsion vanishes and this denotes the point of incipient phase separation. Clearly, when $\chi>0.5$, phase separation occurs and this leads to incipient flocculation (see Chapter 1 in this volume).

A useful analysis for describing the non-ideality of mixing of a polymer with a solvent is to consider the osmotic pressure of the solution π in terms of its concentration c_2 and its volume fraction:

$$\frac{\pi}{c_2} = RT\left[\frac{1}{M_2} + \left(\frac{v_2^2}{V_1}\right)\left(\frac{1}{2} - \chi\right)c_2 + \ldots\right] \tag{4}$$

where v_2 is the partial specific volume of the polymer with molecular weight M_2 ($v_2 = V_2/M_2$) and V_1 is the molar volume of the solvent.

The second term on the right-hand side of Eq. (4) is the second virial coefficient B_2, which is a measure of non-ideality of mixing of the polymer with the solvent:

$$B_2 = \left(\frac{v_2^2}{V_1}\right)\left(\frac{1}{2} - \chi\right) \tag{5}$$

From Eq. (5), $B_2=0$ when $\chi=½$ and this represents the θ-condition described above. When $\chi<½$, B_2 is positive and this represents good solvent conditions whereby the polymer chains are strongly solvated and they resist interpenetration. In contrast, when $\chi>½$, i.e. the solvent is worse than a θ-solvent, B_2 is negative and the polymer chains are attracted to each other and precipitation of the polymer may take place.

Since the solvency of the medium for the polymer chain depends on temperature, one can also define a θ-temperature at which $\chi = ½$. The θ-temperature can be easily defined if one considers the enthalpy (κ_1) and entropy (ψ_1) with respect to [½–χ]:

$$\left(\frac{1}{2} - \chi\right) = \kappa_1 - \psi_1 \tag{6}$$

θ is related to κ_1 and ψ_1 by the expression

$$\theta = \left(\frac{\kappa_1}{\psi_1}\right) T \tag{7}$$

Combining Eqs. (6) and (7):

$$\left(\frac{1}{2} - \chi\right) = \psi_1 \left[\left(\frac{\theta}{T}\right) - 1\right] \tag{8}$$

From Eq. (8), it is clear that if $\theta = T$, $\chi = ½$ and this defines the θ-temperature. The latter is the temperature at which the polymer chains in solution have no repulsion or attraction.

The Flory-Huggins theory fails to explain a number of experimental results, e.g. the dependence of χ on polymer concentration and the phase separation of many polymer solutions upon heating, e.g. aqueous solutions of poly(ethylene oxide) (PEO). The Flory-Huggins theory predicts phase separation only on cooling [according to Eq. (8) only if $T < \theta$, $\chi > ½$]. Phase separation on heating could be accounted for by introducing the concept of free volume. Near the critical point, where phase separation occurs, there are no bonds between the molecules constraining the separation of solvent molecules. Hence upon heating a polymer solution the increase in free volume of the solvent is large and much larger than that of the polymer. This free volume difference creates a large difference in thermal expansion between the polymer and solvent and leads to phase separation on heating. Hence many polymer–solvent mixtures show a lower critical solution temperature (LCST) below which a clear solution is obtained and an upper critical solution temperature (UCST) above which a clear solution is also obtained. Above the LCST, phase separation occurs on heating (clouding of polymer solution), whereas below the UCST, phase separation occurs on cooling.

The solution behavior of block and graft copolymers is much more complicated than that of homopolymers. In dilute solutions in solvents that are good solvents for both components A and B, the polymer exhibits similar behavior characteristics to photopolymer chains resulting from interactions with solvent molecules and each other. Two possible models have been suggested: the first assumes that there are only a few heterocontacts (the segregated model) and the two components A and B behave like homopolymer chains. The second model assumes a random structure that takes into account some overlap between different blocks creating heterocontacts between unlike segments. Results using PS–*block*–PMMA copolymers in toluene [7] showed a tightly coiled conformation surrounded by a slightly expanded PS shell.

In selective solvents, whereby one of the components (B) is in poor solvent conditions (i.e. insoluble in that solvent) whereas the second component (A) is in a good solvent (i.e. soluble in that solvent), separation of the amphipathic part (B) of the block or graft copolymer into a distinct phase will occur, leaving the soluble (A) part in solution. The insoluble part of the amphipathic copolymers will aggregate, forming "micelle-like" structures. In very dilute solution, the block or graft copolymer is monomolecular with the insoluble part of the chain forming the core and the A chains forming a solvated shell [8]. This is represented schematically in Fig. 10.1 a. At a critical concentration, sometimes referred to as the critical aggregation concentration (CAC), aggregation of the molecules takes place as represented in Fig. 10.1 b. The CAC of most block and graft copolymers is very low in comparison with the critical micelle concentration (CMC) for surfactants.

Several experimental procedures can be applied to obtain the size and shape of the aggregate units in block and graft copolymers. These include X-ray scattering (SAXS), neutron scattering, dynamic light scattering, gel permeation chromatography, osmotic pressure and viscosity. The results showed that for A–B–A-type triblocks is a solvent that preferentially dissolves the A blocks, uniform spherical micelles are obtained which are in equilibrium with molecularly dissolved molecules in solution. These micelles show a core–shell structure, where the core is made up of the insoluble inner block (B) surrounded by solvated outer blocks (A).

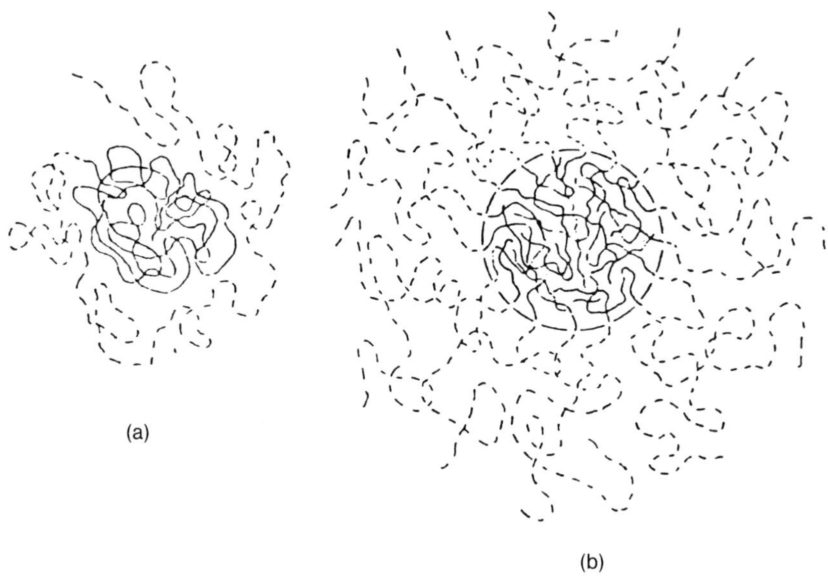

Fig. 10.1 Block copolymers in selective solvents: (a) monomolecular, (b) multimolecular.

10.4
Adsorption and Conformation of Polymeric Surfactants at Interfaces

Understanding the adsorption and conformation of polymeric surfactants at interfaces is key to knowing how these molecules act as stabilizers. Most basic ideas on the adsorption and conformation of polymers have been developed for the solid/liquid interface [9, 10]. The same concepts can be applied to the liquid/liquid interface with some modifications where some parts of the molecule may reside within the oil phase (for O/W emulsions) or the water phase (for W/O emulsions). Such modification does not alter the basic concepts, particularly when dealing with stabilization by these molecules.

The process of polymeric surfactant adsorption involves a number of interactions that must be considered separately. Three main interactions must be taken into account, namely the interaction of the solvent molecules with the surface (or oil in case of O/W emulsions) that need to be displaced for the polymer segments to adsorb, the interaction between the stabilizing chain A and the solvent (that can be described by the Flory–Huggins interaction parameter χ) and the interaction between the segments of the B chain with the surface (that can be described by an adsorption energy per segment χ^s). The balance of these interactions determines the adsorption and conformation of the polymeric surfactant molecule at the interface.

Figure 10.2 shows a schematic representation of the conformation of different polymers on a solid surface. For a hompolymer to adsorb, the adsorption energy per segment χ^s must exceed a critical value so that the total energy of adsorption per molecule is sufficient to compensate for the entropy loss when a polymer chain approaches the surface. This explains why homopolymers with no specific adsorption on the surface are not efficient for stabilization of dispersions. For enhanced adsorption, one can introduce a short block in the molecule which shows specific adsorption on the surface, as represented in Fig. 10.2b. However, the most effective stabilizers for dispersions are those of the A–B, A–B–A block and BA_n grafts, as represented in Fig. 10.2d–f.

For full characterization of polymer adsorption and conformation, one needs to obtain three parameters: the amount of adsorption Γ (mg m^{-2} or mol m^{-2}) as a function of equilibrium concentration C_2 (i.e. the adsorption isotherm), the fraction of segments in direct contact with the surface p and the extension of the layer from the surface $\rho(z)$ or the adsorbed layer thickness δ. These parameters need to be investigated as a function of the system variables such as temperature, adsorption energy per segment χ^s and solvency of the medium for the A chains that is determined by the Flory-Huggins interaction parameter χ.

Measurement of the adsorption isotherm is, in general, straightforward. A given amount of particles with mass m of surface area A is equilibrated with polymer solutions with various initial concentration C_1 and after equilibrium has been reached (which may take several hours depending on the molecular weight of the polymer), the particles are separated and the equilibrium concentration C_2 is determined analytically:

$$\Gamma = \frac{C_1 - C_2}{mA} \tag{9}$$

The adsorption isotherm is obtained by plotting Γ versus C_2. For sufficiently high molecular weight of the polymer, a high-affinity isotherm is obtained. The first polymer molecules added are virtually completely adsorbed, giving a high Γ at an equilibrium concentration approaching zero, and this is followed by a nearly horizontal part, the pseudo-plateau. In this case no desorption can be detected and adsorption is referred to as being irreversible. The amount adsorbed depends on the difference in interaction between a polymer segment and a solvent molecule with the surface. Hence the net adsorption energy depends on the nature of the solvent and the surface.

For A–B, A–B–A block and BA_n graft copolymers, large adsorbed amounts are obtained compared with homopolymers. In this case, only one type of block (B) is at the surface (the "anchor") and the other block (A) extends away from the surface (the "buoy"). Obviously, the anchor block should consist of a minimum number of adsorbing monomers to hold the molecules tethered to the surface. It has been found that Γ is highest when the length of the adsorbing block is in the range 10–20% of the total chain length.

The bound fraction p or train density gives an indication of how well anchored the chain is at the interface. The strength of anchoring is important as

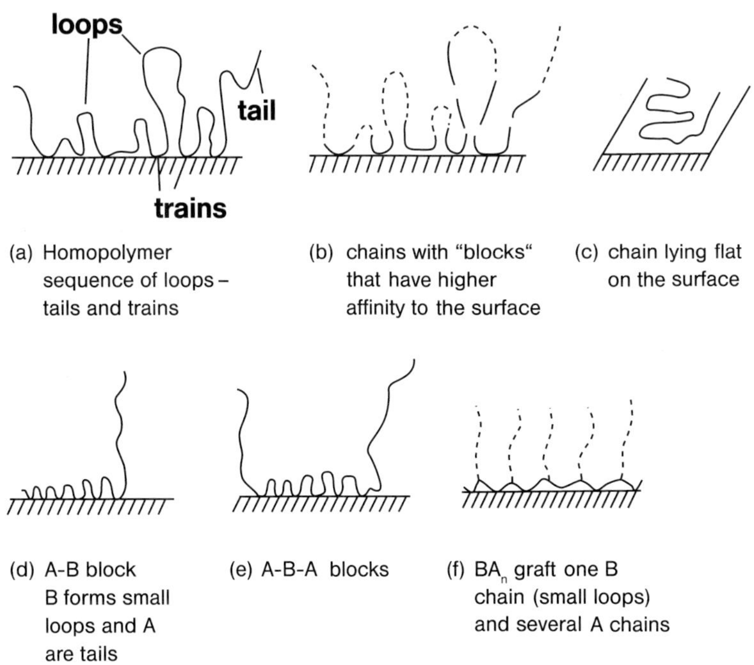

Fig. 10.2 Schematic representation of the conformation of various polymeric surfactants on a solid surface.

it determines the difficulty in removing the polymer from the surface and hence its effectiveness as a stabilizer in colloidal systems (see Chapter 1 in this volume). Also, it gives some idea of the conformation of a homopolymer at the interface. A high p corresponds to a flat conformation with small loops, whereas a low p corresponds to an adsorbed layer with large loops and tails. For block and graft copolymers, a high p value for the anchor chain B ensures strong adsorption leaving the A chains extending away from the surface and this ensures effective steric stabilization.

In order to measure p, some characteristic change in the system must occur to differentiate between segments that are not in contact with the surface and those in the interfacial region. For copolymer systems, the bound fraction can be decomposed into contributions from two chemically different segments (of components A and B) and techniques that can resolve chemical structure are required. Several techniques have been applied for measuring p and these have been summarized by Fleer et al. [9]. These include infrared (IR), nuclear magnetic resonance (NMR) and electron spin resonance (ESR) spectroscopy and microcalorimetry.

The adsorbed layer thickness δ is one of the most important factors in determining the effectiveness of the adsorbed polymer layer in stabilizing colloidal dispersions (see Chapter 1 in this volume). There are several techniques for measuring δ, which for convenience may be classified into three main categories [9]: static, hydrodynamic and disjoining pressure (or direct measurement of surface forces as a function of separation distance between surfaces or particles). The static technique is based on the measurement of the volume fraction profile or segment density distribution $\delta(z)$ that can be obtained using, for example, small-angle neutron scattering [11]. The hydrodynamic methods (which give a hydrodynamic thickness δ_h) are based on the fact that the presence of the adsorbed layer impedes the flow of solvent across the surface or it limits the diffusion of coated particles. The most convenient method to determine δ_h is the quasi-elastic light scattering technique (which is referred to as photon correlation spectroscopy, PCS) [12]. It is based on measurement of the intensity fluctuation of scattered light by the particles as they undergo Brownian diffusion. This allows one to obtain the diffusion coefficient of the particles from which the hydrodynamic radius R_h can be obtained using the Stokes-Einstein equation ($D = kT/6\pi\eta R_h$, where k is Boltzmann's constant, T is the absolute temperature and η is the viscosity of the medium). By measuring R_h in the presence and absence of an adsorbed layer, one can obtain δ_h by difference.

Several theories are available for the description of polymer adsorption and conformation and these has been described in detail in various texts [9, 10]. The polymer conformation can be described as walks in continuous space or on a lattice. This can be obtained using either exact enumeration or Monte Carlo methods. The lattice model can be applied using a mean field approximation. A quasi-lattice model was developed by Scheutjens and Fleer [12], who described all chain conformations as a step-weighted random walk on a quasi-crystalline lattice which extends in parallel layers away from the surface. Each step in the random walk was assigned a weighting factor p_i that was considered to consist of three contributions, namely the adsorption energy χ^s (only in the first layer near the surface), the configurational entropy of mixing and the segment–solvent interaction parameter χ (the Flory-Huggins interaction parameter). A full description of the theory is beyond the scope of this chapter.

As an alternative to the mean-field approach, it is possible to take into account the excluded volume correlations occurring in polymers in a good solvent. In this picture, the polymer chain is considered to be made of "blobs" that can interact as hard spheres. It is relevant to the case of graft copolymers that are considered to produce a polymer "brush". This approach was considered by de Gennes [13] and is referred to as the scaling theory. Details of this theory are beyond the scope of this chapter.

10.5 Stabilization of Solid–Liquid Dispersions Using Graft Copolymers

As a model for solid–liquid dispersions (suspensions), preformed polystyrene (PS) latex was used [14]. The latex dispersions were prepared using surfactant-free emulsion polymerization (with a z-average diameter of 427 and 867 as measured by PCS) and they were stabilized using graft copolymers of a poly(methyl methacrylate)/poly(methacrylic acid) (PMMA/PMMAc) backbone (B) on which several PEO chains ($M = 750$) were grafted. Two graft copolymers were used, namely Atlox 4913 and Hypermer CG-6 (similar to Atlox 4913 but containing a higher proportion of PMMAc). The adsorption isotherms of the two graft copolymers on the two lattices were determined at 20 °C and are shown in Figs. 10.3 and 10.4. The amount adsorbed (expressed in mg m^{-2}) was independent on the particle size. The extent of adsorption was higher for Atlox 4913 than Hypermer CG-6. This probably reflects the nature of the backbone of the graft copolymer. Hypermer CG-6 contains a higher proportion of PMAc, which probably makes it more polar, thus reducing its adsorption compared with Atlox 4913 with a

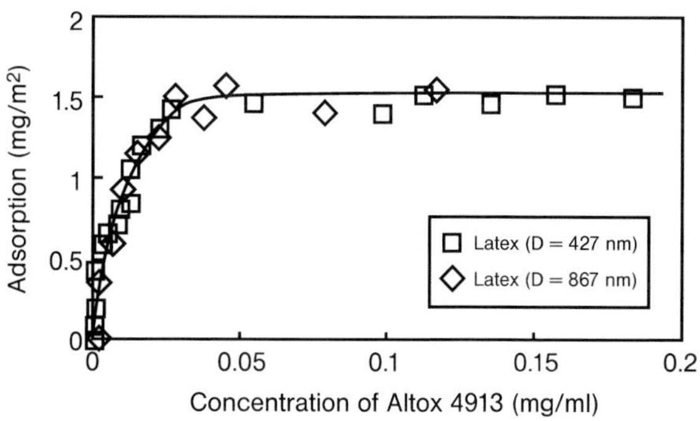

Fig. 10.3 Adsorption isotherm of Atlox 4913 on PS latex.

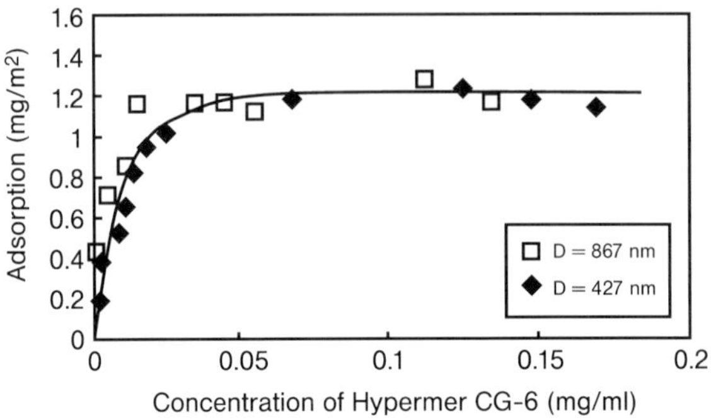

Fig. 10.4 Adsorption isotherm of Hypermer CG-6.

less polar backbone. An increase in temperature showed an increase in the amount of adsorption and this is related to the lower solubility of the polymer at higher temperatures.

The stability of the latex dispersions was investigated using rheological measurements, which could also be used to obtain information on the adsorbed layer thickness as a function of volume fraction of the dispersion [14]. Two types of measurements were carried out, namely steady-state shear stress (σ)–shear rate (γ) and dynamic or oscillatory techniques. In the first case, σ is plotted as a function of γ and this allows one to obtain the yield stress σ_β and the viscosity η using the Bingham equation:

$$\sigma = \sigma_\beta + \eta\gamma \,. \tag{10}$$

As an illustration, Fig. 10.5 shows the variation of σ_β with the core volume fraction ϕ of the latexes for Atlox 4913. Similar plots were obtained for Hypermer CG-6. It can be seen that σ_β increases above a critical ϕ and this denotes the onset of the start of interaction between the PEO layers of the graft copolymer.

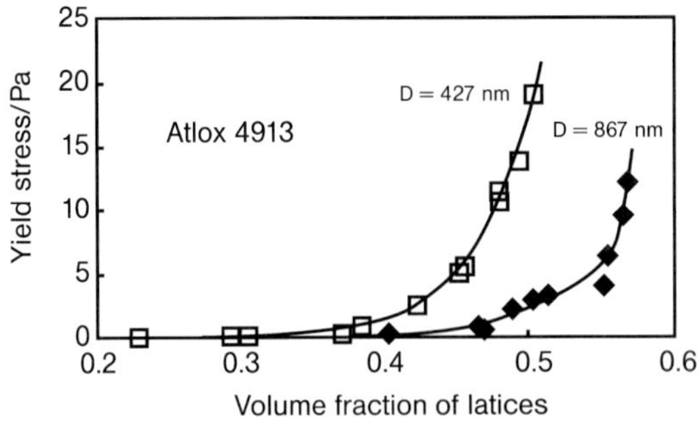

Fig. 10.5 Variation of yield stress with core volume fraction for two latexes stabilized with Atlox 4913.

The value of ϕ_{cr} above which σ_β increases is lower for the smaller particle size latex. This can be understood since the effective volume fraction ϕ_{eff} is determined by the ratio of the adsorbed layer thickness to the particle radius R:

$$\phi_{eff} = \phi\left(1 + \frac{\delta}{R}\right)^3 \tag{11}$$

It is clear from Eq. (11) that, at any given core volume fraction ϕ, the effective volume fraction ϕ_{eff} is higher for the smaller size latex. From η, one can obtain ϕ_{eff} using the Dougherty-Krieger equation [15]:

$$\eta = \left[1 - \left(\frac{\phi_{eff}}{\phi_p}\right)\right]^{-[\eta]\phi_p} \tag{12}$$

where $[\eta]$ is the intrinsic viscosity, which is equal to 2.5 for hard spheres, and ϕ_p is the maximum packing fraction, which could be obtained from the following expression:

$$\frac{\eta_r^{1/2} - 1}{\phi} = \frac{1}{\phi_p}(\eta_r^{1/2} - 1) + 1.25 \tag{13}$$

ϕ_p was found to be 0.6 for the latex with $D = 427$ nm and 0.64 for the latex with $D = 867$ nm. This is consistent with random packing.

Using Eqs. (11) to (13), one can obtain δ as a function of ϕ. The results showed that, for Atlox 4913, δ decreases from 17.5 to 6.5 nm as ϕ increases from 0.36 to 0.57. For Hypermer CG-6, δ decreases from 11.8 to 6.0 nm as ϕ increases from 0.49 to 0.58. This reduction in δ with increase in ϕ may be due to the compression of the PEO layers on close approach of the particles.

In dynamic or oscillatory measurements, a sinusoidal strain with amplitude γ_0 is applied on the system and the resulting stress oscillation with amplitude σ_0 is simultaneously measured. The strain and stress sine waves oscillate with the same frequency ω (rad s^{-1}) but for a viscoelastic system such as the latex dispersions described above the sine waves are shifted by a time Δt. The phase angle shift δ is simply $\Delta t\omega$. From σ_0, γ_0 and δ one can obtain the complex modulus $|G^*| = \sigma_0/\gamma_0$, the elastic modulus $G' = |G^*|\cos\delta$ and the viscous modulus $G'' = |G^*|\sin\delta$. Initially the strain amplitude is gradually increased at constant frequency ω (say 6 rad s^{-1}) to obtain the linear viscoelastic region at which $|G^*|$, G' and G'' are independent of γ_0. The frequency is then changed at constant amplitude (in the linear viscoelastic region) to obtain the dependence of $|G^*|$, G' and G'' on frequency. Details of the analysis have been given elsewhere [16]. As an illustration, Fig. 10.6 shows the variation of $|G^*|$, G' and G'' (at low amplitude and frequency of 6.2 rad s^{-1}) with the latex ($D = 427$ nm) core volume fraction. At ϕ below 0.3, $G'' > G'$, which indicates weak interaction between the PEO layers. However, at $\phi > 0.3$, $G' > G''$ and G' increases rapidly with increase in ϕ, whereas G'' remains low. At sufficiently high ϕ, G' approaches $|G^*|$ very

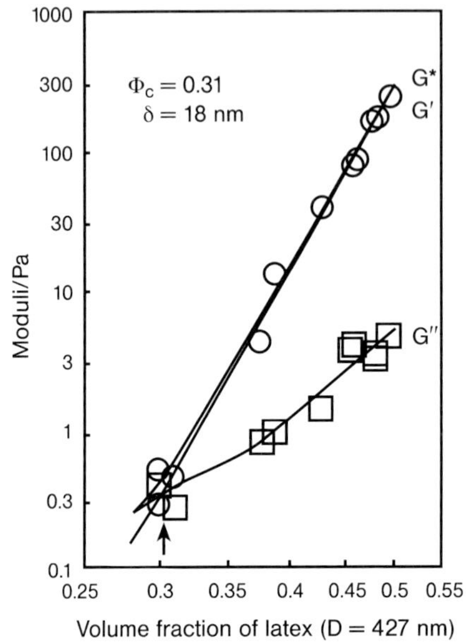

Fig. 10.6 Variation of moduli with core volume fraction of latex stabilized with Atlox 4913.

closely. This behavior reflects the steric interaction between the polymer layers. When $\phi<0.3$, the surface-to-surface distance h is large compared with twice the adsorbed layer thickness, 2δ. Under these conditions, the interaction between the PEO chains of the graft copolymer is weak and the system shows more viscous than elastic response. At $\phi>0.3$, the PEO chains approach very closely and steric repulsion sets in since h becomes comparable to 2δ. At sufficiently high ϕ values, such steric interaction becomes very strong and the PEO layers may undergo some compression and G' approaches $|G^*|$ very closely (i.e. the system behaves as a near elastic "gel"). The critical volume fraction at which $G'=G''$ (i.e. the cross-over point) can be used to obtain the adsorbed layer thickness. From the viscosity results, the maximum packing fraction was found to be 0.6, which would be the effective volume fraction at the cross-over point. Using Eq. (11), $\delta=18$ nm, which is comparable to the value obtained from viscosity measurements.

The above steric interaction can be studied directly using surface force measurements. Basically, one measures the energy $E(D)$–distance D curves for two smooth mica cylinders each containing an adsorbed layer of the graft copolymer. Figure 10.7 shows the energy–distance curve for the mica surfaces with the adsorbed graft copolymer. This curve shows a monotonic and approximately exponential decrease in $E(D)$ with increasing surface separation. The exponential nature of the decay makes it difficult to assess precisely the point at which interaction begins, although it falls below the detection limit of the instrument at about 25 nm, indicating a layer thickness in the region of 12.5 nm.

Using the de Gennes scaling theory [13], the energy of interaction between polymer layers can be calculated:

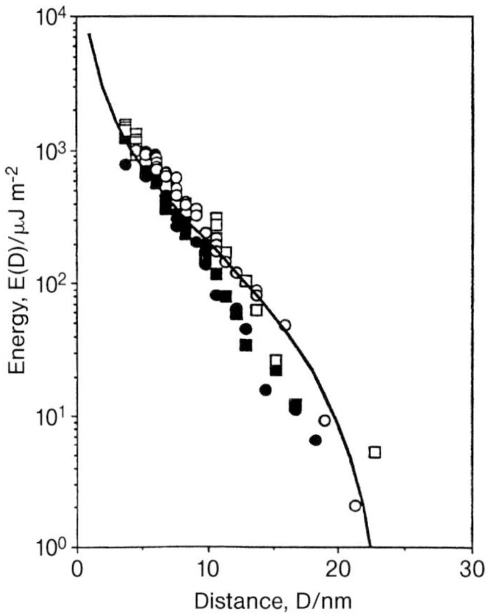

Fig. 10.7 Interaction energy $E(D)$ versus surface separation D for mica surfaces containing adsorbed layers of Atlox 4913: ■, □ decompression; ●, ○ second decompression; (—) calculated from Eq. (14).

$$E(D) = \frac{\beta kT}{s^3}\left[\frac{(2L)^{2.25}}{1.25(D)^{1.25}} + \frac{D^{1.75}}{1.75(2L)^{0.75}}\right] - \left(\frac{2L}{1.25} + \frac{2L}{1.75}\right) \tag{14}$$

where L is the stabilizer thickness on each surface (taken to be equal to 12.5 nm), s is the distance between side attachment points and β is a numerical factor.

The stability of the dispersions can be assessed by measuring the critical flocculation temperature in the presence and absence of electrolytes. This can be assessed by following the storage modulus G′ as a function of temperature. As an illustration, Fig. 10.8 shows the results for latex dispersions stabilized with Atlox 4913 in aqueous solution and in the presence of 0.1, 0.2 and 0.3 mol dm^{-3} Na_2SO_4.

In the absence of electrolyte, G′ remains constant up to 65 °C (the highest temperature studied), indicating stability of the dispersion. This is not surprising since the PEO chains of Atlox 4913 remain hydrated up to this temperature and even higher. Cloud point measurements (see below) showed that PEO solutions remain clear up to 100 °C. In the presence of 0.1 mol dm^{-3} Na_2SO_4, G′ started to increase above 40 °C, which is the critical flocculation temperature (CFT) at this electrolyte concentration. Under this condition, the PEO chains become dehydrated and at and above 40 °C PEO–PEO contact is possible, resulting in flocculation. With further increase in Na_2SO_4 concentration, the CFT decreases, as is clearly shown in Fig. 10.8.

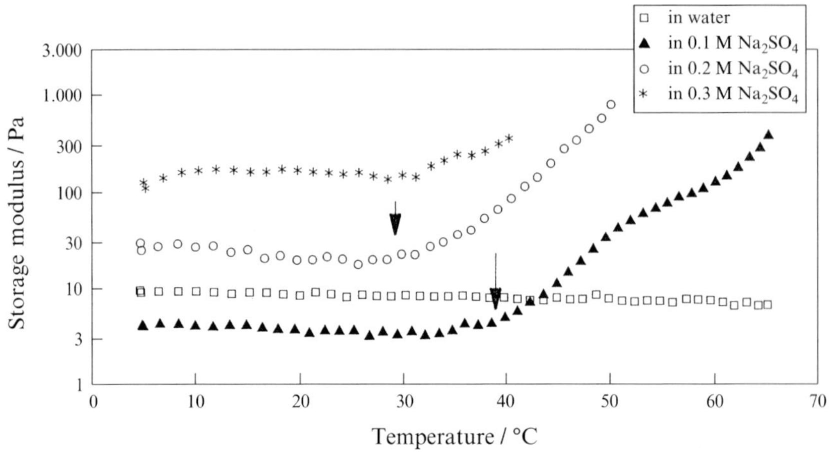

Fig. 10.8 Variation of storage modulus with temperature for a latex dispersion ($\phi = 0.4$) stabilized with Atlox 4913: □ in water; ▲ in 0.1 mol dm^{-3} Na_2SO_4; ○ in 0.2 mol dm^{-3} Na_2SO_4; ☆ in 0.3 mol dm^{-3} Na_2SO_4.

10.6 Emulsion Polymerization Using Graft Copolymer (INUTEC SP1) and Stability of the Resulting Latex

The graft copolymer of hydrophobically modified inulin (INUTEC SP1, ORAFTI, Belgium) was used in emulsion polymerization of styrene and methyl methacrylate. The latexes were prepared in a semi-spherical batch reactor and the dispersions were stirred continuously during preparation. The reactions were carried out for 24 h for polystyrene (PS) and for 6 h for poly(methyl methacrylate) (PMMA), at a constant temperature of 80 °C under a nitrogen atmosphere [16]. The *z*-average particle size of the latex was determined using PCS. The stability of the latex was determined by measuring the turbidity as a function of time. The initial slope gives a measure of the rate of coagulation k and when the critical coagulation concentration (CCC) is reached this slope becomes constant and this gives the rate constant of fast flocculation k_0. The ratio k_0/k gives the stability ratio W. The latter decreases linearly with increase in electrolyte concentration and at and above the CCC, when $W = 1$ (fast flocculation), the slope becomes constant.

Emulsion polymerization of styrene and methyl methacrylate using INUTEC SP1 and potassium persulfate as initiator showed an optimum weight ratio of [INUTEC]/[monomer] of 0.003 for PS and 0.001 for PMMA. Such a low polymeric surfactant concentration for producing latex up to 40 wt% latex is remarkable and it shows the effectiveness of INUTEC SP1 in stabilizing the latex. This, as discussed above, is due to the multi-anchor attachment of the polymeric surfactant and the strong hydration of linear polyfructose loops and tails which

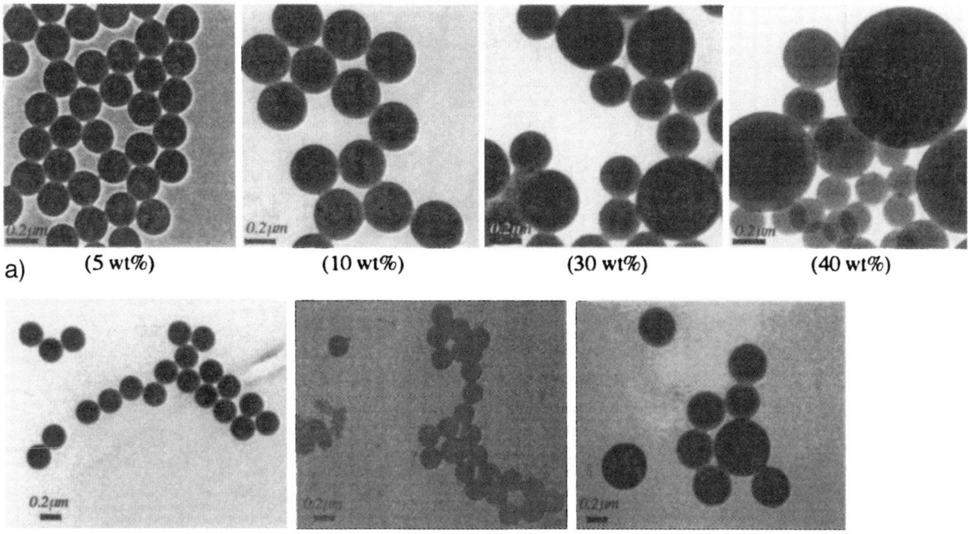

a)

b)

Fig. 10.9 TEM pictures of PS (a) and PMMA (b) latex at various weight fractions of latex: (a) from left to right, 5, 10, 30 and 40 wt%; (b) from left to right, 10, 20 and 30 wt%.

provide effective steric stabilization. As an illustration, Fig. 10.9 shows transmission electron micrographs of PS and PMMA latex at various weight fractions of the latex. This clearly shows that the latex size and its polydispersity increase with increase in the weight fraction of the latex. However, both latexes were very stable, showing no flocculation over very long periods (several months), and this confirms the effective steric stabilization produced using INUTEC SP1 (enhanced steric stabilization).

The stability of the latex was further investigated by measuring the CCC using $CaCl_2$. As an illustration, Fig. 10.10 shows the $\log W$ versus $\log C$ plots for PS latex (5 wt%) and also with post-addition of INUTEC SP1. Without any extra addition of INUTEC SP1, the CCC is 0.028 mol dm^{-3}. However, on post-addition of INUTEC SP1, the CCC increases, becoming >0.4 mol dm^{-3} on addition of 0.05% INUTEC SP1 ([INUTEC]/[PS] ratio=0.01). This enhanced adsorption is due to the increased adsorption of INUTEC SP1. The latex prepared at an [INUTEC]/[monomer] ratio of 0.003 is not fully covered by the graft copolymer, hence there are several uncovered patches of the surface that contain sulfate groups from the initiator, and this explains the low CCC which is normally obtained with electrostatically stabilized dispersions. However, the CCC obtained in the presence of INUTEC SP1 in the emulsion polymerization is still higher than that obtained in its absence. The partial coverage by the graft copolymer is sufficient for stabilization in the absence of added electrolyte. Indeed, comparison of the results with those obtained using other conventional surfactants at

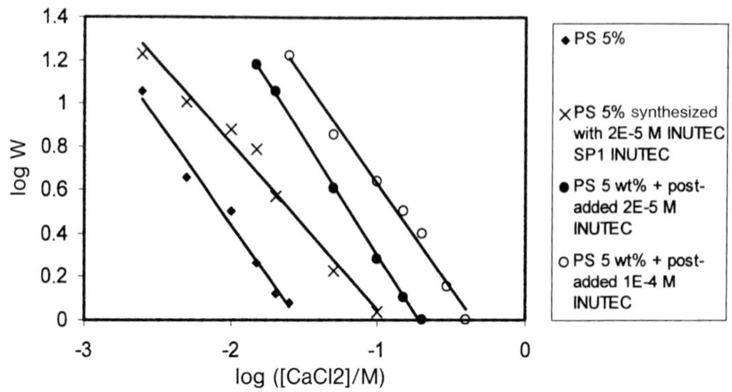

Fig. 10.10 Log *W* versus log *C* curves for PS latex at various INUTEC SP1 concentrations.

comparable concentrations to INUTEC SP1 showed a lack of stability using these systems [15, 16]. Only when using INUTEC SP1 could a stable latex be produced at high monomer concentration (>20 wt%).

The steric stability using INUTEC SP1 has recently been investigated [17] using force–distance measurements for two surfaces containing adsorbed layers of the graft copolymer. For that purpose, the atomic force microscopy (AFM) technique was used. In this method, a hydrophobically modified (using dichlorodimethylsilane) glass sphere (~30 μm in diameter) that was attached to the cantilever of the AFM was used. The plate was also made of glass that was hydrophobically modified using dichlorodimethylsilane. Both the sphere and the plate were coated with an adsorbed layer of INUTEC SP1. As an illustration, Fig. 10.11 shows the force–distance curves at two INUTEC SP1 concentrations in solution, namely 0.08 and 0.1%, which correspond to full coverage of the sur-

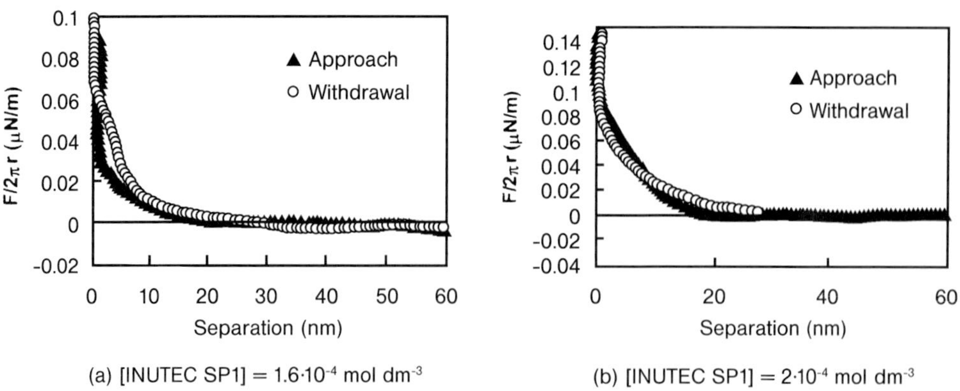

Fig. 10.11 Force–distance curves between hydrophobized glass surfaces containing adsorbed layers of INUTEC SP1: (a) 0.08% (1.6×10^{-4} mol dm^{-3}) in solution; (b) 0.1% (2×10^{-4} mol dm^{-3}) in solution.

faces by the graft copolymer. Upon retraction, some small degree of hysteresis in the interaction is observed, but essentially the decompression curve follows the compression curve back to equilibrium.

It is reasonable to assume that the layer thickness is half the separation where the interaction between the two surfaces occurs. This gives a layer thickness in the region of 9 nm when the polymer concentration is 1.6×10^{-4} mol dm^{-3} and it does not increase with further increase in polymer concentration. This is consistent with a single layer of molecules attached to the surface (i.e. there is no multilayer adsorption).

The force–distance curves were obtained at various Na_2SO_4 concentrations (0.3–1.5 mol dm^{-3}) and the results showed a gradual decrease in adsorbed layer thickness from 9 nm in water to 3 nm in 1.5 mol dm^{-3} Na_2SO_4. However, the interaction was always repulsive even at such high electrolyte concentration. This clearly shows that INUTEC SP1 would be an effective stabilizer over a large range of electrolyte concentrations.

10.7
Emulsions Stabilized Using Polymeric Surfactants

10.7.1
Oil-in-Water Emulsions Stabilized Using INUTEC SP1

Several oils, e.g. Isopar M and silicone oil, have been used for the preparation of O/W emulsions and the polymeric surfactant concentration was varied between 0.5 and 2% based on the oil phase [18, 19]. The latter was varied between 10 and 50% (v/v). It was found that the stability of the emulsion against coalescence was maintained at INUTEC SP1 concentrations ≥1.0% based on the oil phase. For example, with 50:50 (v/v) Isopar M, the total polymeric surfactant necessary for stabilization of the emulsion is only 0.5% (based on the total composition). This low emulsifier concentration required for stabilization of the emulsion is due to its strong adsorption at the O/W interface. The graft copolymer adsorbs with multi-point attachment (several alkyl groups are anchored leaving linear polyfructose loops and tails dangling in solution and providing an effective steric barrier). These emulsions are stable both in aqueous solution and in the presence of high electrolyte concentrations (up to 4 mol dm^{-3} NaCl and 1.5 mol dm^{-3} $MgSO_4$). As an illustration, Fig. 10.12 shows optical micrographs of diluted Isopar O/W emulsions that were stored for 1.5 and 15 weeks at 50 °C. Similar results were obtained in the presence of high electrolyte concentrations. This high stability at high temperatures and high electrolyte concentrations is due to the strong hydration of the linear polyfructose loops and tails. This could be demonstrated by measuring the cloud point of the polymeric surfactant backbone, namely inulin (INUTEC N25). The results showed no cloudiness of the polymer solution up to 100 °C both in water and in the presence of NaCl solutions as high as 4 mol dm^{-3} and $MgSO_4$ solutions as high as

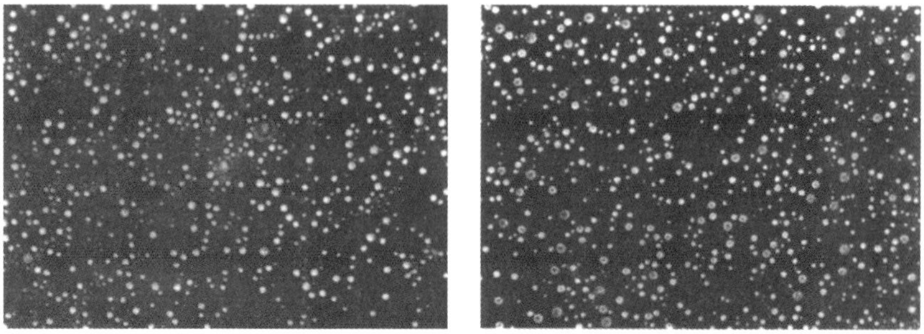

(a) (b)

Fig. 10.12 Optical micrographs of dilute 50:50 (v/v) IsoparM-in-water emulsions containing 2% Inutec SP1 (based on the oil phase) that were stored at 50 °C for (a) 1.5 and (b) 14 weeks.

1 mol dm^{-3}. Such high cloud points were not obtained with poly(ethylene oxide) (PEO), indicating the dehydration of these polymers under such conditions of high temperature and high electrolyte concentration. This shows the superior stabilizing effect of INUTEC SP1 compared with polymers based on PEO such as the Pluronics described above.

The high stability of the emulsions against coalescence is due to the steric interaction between the polyfructose layers which produce high elasticity at the interface. This prevents any thinning and disruption of the liquid film between the droplets. Evidence for this strong elastic interaction has been obtained recently using disjoining pressure $\Pi(h)$ versus h (where h is the film thickness) between the oil measurements between oil droplets [19, 20]. As an illustration, Fig. 10.13 shows the $\Pi(h)$–h isotherms at an INUTEC SP1 concentration of 2×10^{-5} mol dm^{-3} at various NaCl solutions (0.05–2 mol dm^{-3}).

It can be seen from Fig. 10.13 that the initial thicknesses are within the range 9–11 nm, after which starts the transition zone, corresponding to capillary pressure of 0.6–1 kPa. In this zone, all films transform to a Newton black film (NBF) with a jump and the pressure increases rapidly while the film thickness remains constant at ~7 nm. The jump indicates a low barrier at such separation distances, indicating absence of an electrostatic component in the disjoining pressure. This is due to the non-ionic nature of INUTEC SP1. The NBF does not rupture up to the highest pressure of 45 kPa. This clearly indicates the very high stability of the liquid film. These results are direct proof of the expected emulsion stability shown above. With droplets of 10 μm, the capillary pressure is of the order of 3 kPa, whereas for 1-μm droplets the capillary pressure is 30 kPa, both of which are lower than the highest pressure of 45 kPa that is obtained with emulsion films and showing no rupture. This strong steric repulsion and the hydration of the polyfructose loops and tails at high electrolyte concentrations demonstrates the unique performance of INUTEC SP1 as a stabilizer for O/W emulsions.

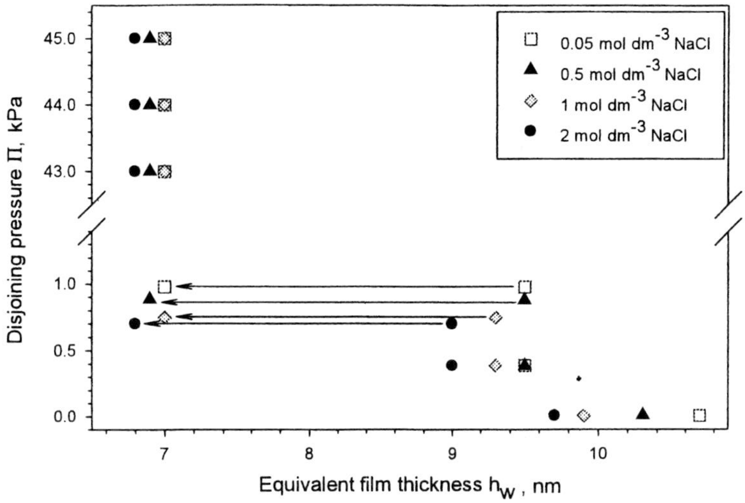

Fig. 10.13 $\Pi(h)$ versus h isotherms for emulsion films at 2×10^{-5} mol dm^{-3} INUTEC SP1 and various NaCl concentrations.

10.7.2
Water-in-Oil (W/O) Emulsions Stabilized with Arlacel P135

W/O emulsions were prepared using an A–B–A block copolymer of polyhydroxystearic acid (PHS, A chains) and PEO (B chain). This PHS–PEO–PHS block copolymer is commercially available (UNIQEMA, ICI) and has been applied for stabilization of W/O emulsions in many personal care formulations. It has a weight-average molecular weight of 6809 and a number-average molecular weight of 3500. The polymeric surfactant adsorbs with the PEO chain residing in the water droplets, leaving the PHS chains extended in the oil phase. A systematic investigation of the adsorption of the polymer at the O/W interface was carried out using a specially designed Langmuir trough. The polymeric surfactant film was spread at the O/W interface and the area of the interface was gradually decreased using moving barriers. The surface pressure Π was simultaneously measured using a Wilhelmy plate [21]. As an illustration, Fig. 10.14 shows the variation of surface pressure with area per molecule A of the block copolymer at the water/air (W/A) and water/oil (W/O) interfaces. As A decreases, Π increases and, below a critical value A_c, it increases very sharply with further decrease in A. At the W/O interface, Π reached ~50 mN m^{-1}, indicating an interfacial tension γ close to zero (note that γ_0 for a clean interface is ~50 mN m^{-1} and $\Pi=\gamma_0-\gamma$). This low interfacial tension explains the ease of emulsification of water in an oil solution of Arlacel P135.

The oil film thickness between two water droplets was measured using an interferometric technique and the results for Isopar M (the oil used for preparation of the W/O emulsion) showed a film thickness of 15.8 ± 0.8 nm. This gives

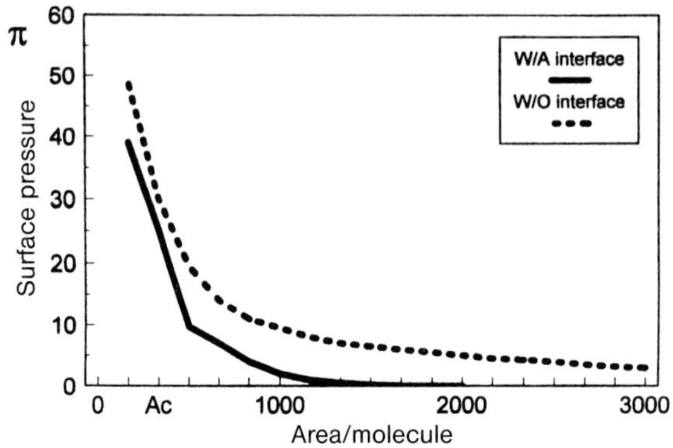

Fig. 10.14 Variation of surface pressure with area per molecule of the block copolymer at the water/air (W/A) and water/oil (W/O) interfaces.

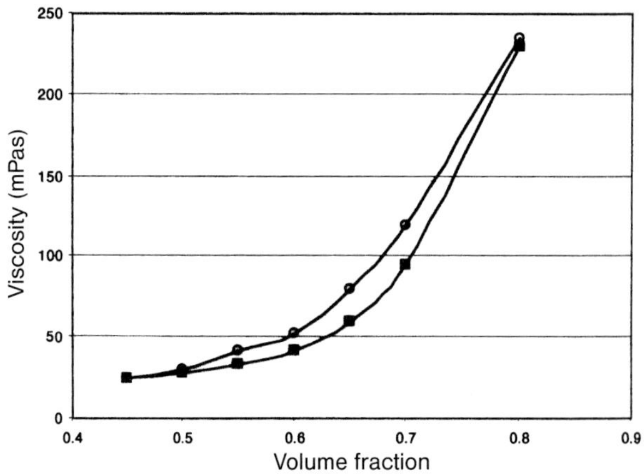

Fig. 10.15 Viscosity–volume fraction curves for W/O emulsions stabilized with Arlacel P135.

an adsorbed layer thickness in the region of 8 nm, which is consistent with other types of measurements, e.g. by the rheological technique (see below).

W/O emulsions were prepared using Isopar M as the oil and Arlacel P135. The water volume fraction was gradually increased and it was possible to prepare fluid emulsions with a water volume fraction ϕ exceeding 0.75. They have a z-average radius (as measured by PCS) of 183 nm. These emulsions were very stable, showing no coalescence for several months both at room temperature and 50 °C. As an illustration, Fig. 10.15 shows the variation of viscosity with water volume fraction ϕ [22]. It can be seen that the viscosity of the emulsion remains low (< 100 mPa s) up to $\phi = 0.65$, above which there is a slow increase in viscosity. Even at $\phi = 0.75$ the viscosity is still low (< 300 mPa s).

The viscosity–volume fraction curves can be used to obtain the adsorbed layer thickness δ. The data can be fitted to the Dougherty-Krieger equation (15) to obtain the effective volume fraction ϕ_{eff}:

$$\eta_r = \left[1 - \left(\frac{\phi_{\text{eff}}}{\phi_p}\right)\right]^{-[\eta]\phi_p} \tag{15}$$

where ϕ_p is the maximum packing fraction that could be obtained from a plot of $1/\eta^{\frac{1}{2}}$ versus ϕ and this was found to be 0.84 (which is higher than the theoretical maximum packing of 0.74), and this high value is due to the polydispersity of the emulsion; $[\eta]$ is the intrinsic viscosity that is equal to 2.5 for hard spheres. From ϕ_{eff} and ϕ one can obtain the adsorbed layer thickness δ:

$$\phi_{\text{eff}} = \phi\left[1 + \left(\frac{\delta}{R}\right)\right]^3 \tag{16}$$

The results showed that at $\phi = 0.4$, $\delta = 10$ nm and this is the fully extended PHS chain length. It is comparable to the value of 8 nm obtained using liquid film measurements described above. The thickness δ showed a linear decrease with increase in ϕ and this could be attributed to interpenetration and/or compression of the PHS chains on close approach of the water droplets. Evidence of this interaction was also obtained using viscoelastic measurements, which showed a rapid increase in the elastic modulus G' when the volume fraction of the emulsion exceeded 0.67. This reflects the elastic interaction between the PHS chains, thus producing strong steric repulsion and hence high stability of the W/O emulsion.

10.8 Stabilization of Nano-emulsions Using INUTEC SP1

Nano-emulsions are systems that cover the size range 50–200 nm that have long–term kinetic stability and sometimes are referred to as "approaching thermodynamic stability" [23, 24]. With a sufficiently small droplet size and small refractive index difference between the droplets and the medium, they can appear transparent or translucent. The inherently high colloid stability of nano-emulsions can be understood from consideration of their steric stabilization, e.g. when using non-ionic surfactants and/or polymers. This can be easily illustrated if one considers the energy–distance curve for a sterically stabilized system with increasing the ratio of adsorbed layer thickness to particle radius (δ/R), as represented schematically in Fig. 10.16. It can be seen than that the depth of the minimum, G_{min}, decreases as δ/R increases. This is the basis of the high kinetic stability of nano-emulsions. For example, for a system with a radius of 50 nm and an adsorbed layer thickness of 10 nm (which is common with many polymeric surfactants), δ/R is large (0.2) and G_{min} becomes very shallow (which could be less than kT). Under these conditions, the Brownian

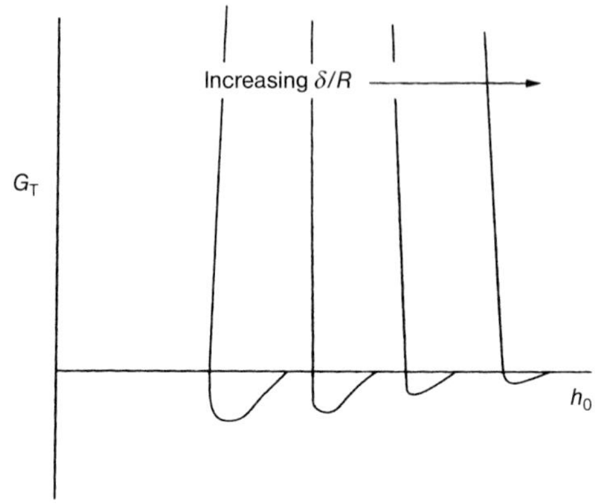

Fig. 10.16 Schematic representation of the total energy–distance curves for a sterically stabilized system at increasing values of δ/R.

diffusion of the droplets, which is of the order of kT, is sufficient to prevent any gravitational creaming or sedimentation. The transparent or translucent nano-emulsion will show no separation on storage.

The above sterically stabilized nano-emulsions will also show no flocculation (weak or strong) and the system remains fluid on storage. The small droplets also prevent their coalescence, since these droplets are non-deformable and hence surface fluctuations are prevented. In addition, the significant film thickness (compared with the particle radius) prevents any thinning or disruption of the liquid film between the droplets.

The only instability of nano-emulsions is Ostwald ripening, which arises from the small droplet size and its distribution. The solubility of the smaller droplets with radius r_1 is higher than that of the larger droplets with radius r_2. This is due to the higher radius of curvature of the smaller droplets giving them higher solubility (according to the Laplace equation, the solubility is inversely proportional to the droplet radius) when compared with that of the larger droplets.

The increase in solubility with decreasing droplet radius was analyzed by Lord Kelvin [25], who considered the difference in chemical potential of dispersed phase droplets with different sizes. The solubility $S(r)$ of a droplet with radius r is related to that of a droplet with infinite radius $S(\infty)$ (i.e. the bulk phase solubility) by

$$S(r) = S(\infty)\exp\left(\frac{2\gamma V_m}{rRT}\right) \tag{17}$$

where γ is the interfacial tension, V_m is the molar volume of the disperse phase, R is the gas constant and T is the absolute temperature.

For two droplets with radii r_1 and r_2 (where $r_1 < r_2$) with solubility S_1 and S_2 (where $S_1 > S_2$), the following equation (sometimes referred to as the Ostwald equation) can be derived using Eq. (17):

$$\left(\frac{RT}{V_m}\right)\ln\left(\frac{S_1}{S_2}\right) = 2\gamma\left(\frac{1}{r_1} - \frac{1}{r_2}\right) \tag{18}$$

Thus, on storage, oil molecules will diffuse from the smaller to the larger droplets and this results in a shift of the droplet size distribution to larger sizes. This process may continue until the droplets reach sufficient size for creaming or sedimentation to occur. In other words, the nano-emulsion will eventually becomes a macro-emulsion.

The Ostwald ripening rate can be obtained by plotting the cube of the radius r^3 versus time, whereby straight lines are produced according to the Lifshitz-Slesov-Wagner (LSW) equation [26]:

$$r^3 = \frac{8}{9}\left[\frac{S(\infty)\gamma V_{\mathrm{m}} D}{\rho RT}\right]t \tag{19}$$

where D is the diffusion coefficient of the disperse phase in the continuous phase and ρ is the oil density.

Several methods may be applied to reduce Ostwald ripening. Addition of a second disperse component with much lower solubility in the continuous phase compared with the oil used for preparing the nano-emulsions requires an oil with a higher hydrocarbon chain length such as squalane. In this case, partitioning between different droplets occurs, with the component having the lower solubility in the continuous phase expected to be concentrated in the smaller droplets. During Ostwald ripening in a two-component disperse system, equilibrium is established when the difference in chemical potential between different sized droplets (which results from differences in curvature) is balanced by the difference in chemical potential resulting from partitioning of the two components. This method may not be desirable in practice since addition of a second component may affect the preparation and steric stability of the nano-emulsion. An alternative and more useful method is to use a polymeric surfactant that strongly adsorbs at the O/W interface (enhancing the Gibbs dilational elasticity) and with limited solubility in the continuous phase. This is the basis of using hydrophobically modified inulin (INUTEC SP1), as discussed below.

Recently, we carried out a systematic study of the effect of INUTEC SP1 on the Ostwald ripening rate of nano-emulsions [27]. Several oils were used and the INUTEC SP1 concentration was varied between 8 and 12% based on the oil phase. For example, for a 20:80 (v/v) nano-emulsion the INUTEC SP1 concentration was 1.6 and 2.4% (based on the total composition). The nano-emulsions were prepared using a microfluidizer (Microfluidics, USA) at 700 bar for 1 min. The z-average droplet size was determined using PCS (HPPS Instruments, supplied by Malvern, UK).

As an illustration, Fig. 10.17 shows plots of the cube of the radius (R^3) versus time for 20:80 (v/v) dimethicone (50 cSt)–water nano-emulsions that were stored at 50 °C at two INUTEC SP1 concentrations.

The Ostwald ripening rate is obtained from the slope of the linear curves in Fig. 10.17 and this was found to be 1.1×10^{-29} and 2.4×10^{-30} m^3 s^{-1} for the 1.6 and 2.4% INUTEC SP1 concentrations, respectively. These rates are about three orders of magnitude lower than those obtained using conventional non-ionic

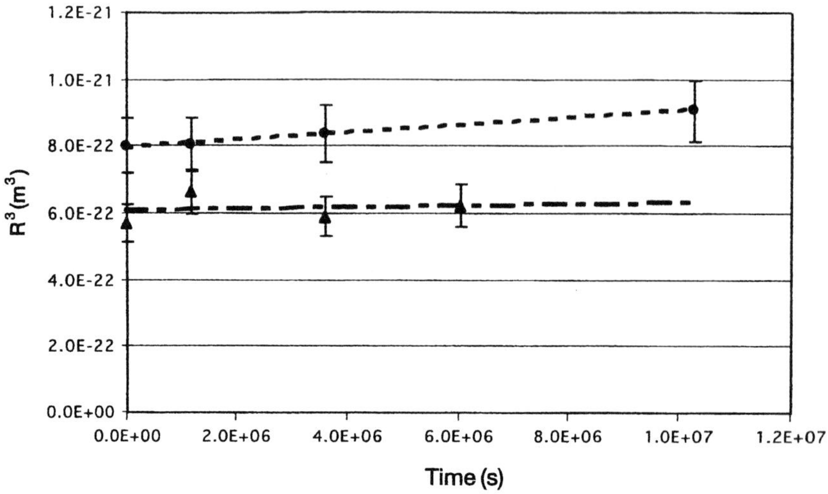

Fig. 10.17 Plots of R^3 versus time for nano-emulsions of dimethicone containing 1.6% (top line) and 2.4% (bottom line) INUTEC SP1.

surfactants such as alcohol ethoxylates [28]. As mentioned above, this reduction in Ostwald ripening rate using INUTEC SP1 is due to the strong adsorption of the polymeric surfactant with multi-point attachment with several alkyl groups and also enhancement of the Gibbs dilational elasticity, thus reducing the diffusion of the oil molecules from the smaller to the larger droplets.

10.9 Stabilization of Multiple Emulsions Using Polymeric Surfactants

Multiple emulsions are complex systems of emulsions. Two types may be distinguished, namely water-in-oil-in-water (W/O/W) and oil-in-water-in-oil (O/W/O). The first type may be considered as water-in-water emulsions separated by an oily membrane, whereas the second type may be considered as oil-in-oil emulsions separated by an aqueous membrane [29]. The internal droplets could also consist of a polar solvent such as glycol or glycerol.

In the early development of multiple emulsions, conventional surfactants with low HLB (for preparation of the primary W/O emulsion) and high HLB (for emulsification of the W/O emulsion into an aqueous solution) were used, but these systems had limited stability [29] with several breakdown processes as described before [29]. The stability of multiple emulsions was greatly improved by using polymeric surfactants, as will be discussed below.

For the preparation of a stable W/O emulsion, Arlacel P135 (an A–B–A block copolymer of PHS–PEO–PHS) described above is the most suitable. This polymeric surfactant can be used for the preparation of the primary W/O emulsion

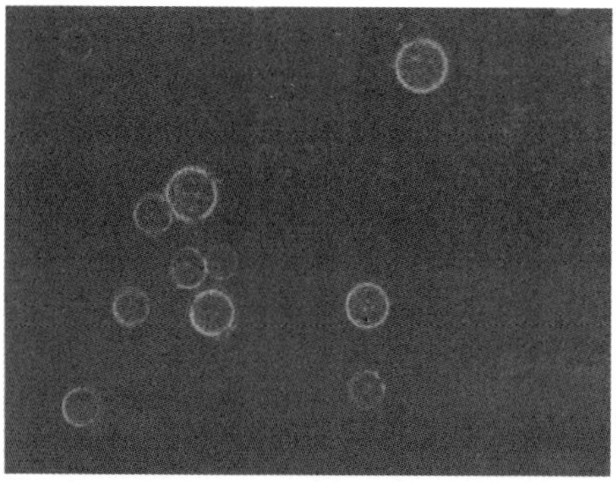

Fig. 10.18 Optical micrograph of a W/O/W multiple emulsion.

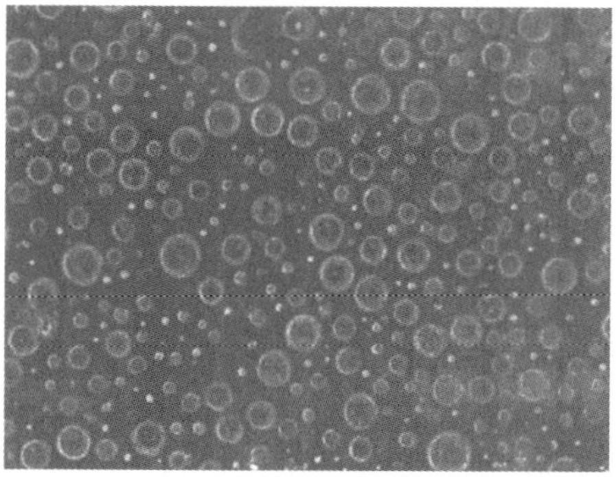

Fig. 10.19 Optical micrograph of an O/W/O multiple emulsion.

using a high-speed stirrer to produce droplets usually $<1\ \mu m$ in diameter. A high volume fraction of water (>0.7) can also be achieved. This primary emulsion can be emulsified using INUTEC SP1 (hydrophobically modified inulin) under conditions of relatively low agitation to produce multiple emulsion covering the range 5–50 μm radius. The osmotic pressure of the internal water droplets and the external aqueous medium can be balanced (to reduce water diffusion from the internal water droplets to the external medium and vice versa) using electrolytes or non-electrolytes. For O/W/O multiple emulsions one can start with an O/W emulsion or nano-emulsion that is stabilized by INUTEC SP1 and this is further emulsified into an oil solution of Arlacel P135.

As an illustration, both types of multiple emulsions were prepared using the above polymeric surfactants [30]. A two step emulsification process was used as described above. For the W/O/W multiple emulsion the osmotic pressure was balanced using 0.1 mol dm^{-3} $MgCl_2$ both in the internal water droplets and external continuous phase. Figure 10.18 shows optical micrographs of a W/O/W multiple emulsion that was stored for several months at 50 °C, and Fig. 10.19 shows the corresponding O/W/O multiple emulsion. The latter was prepared starting from an O/W nano-emulsion. In both cases, no change in droplet size of the internal primary emulsion and the final multiple emulsion was observed over a long period, indicating the high stability of these multiple emulsions on storage.

References

1 D.H. Napper, *Polymeric Stabilization of Colloidal Dispersions*, Academic Press, London (**1983**).

2 T.F. Tadros, *Applied Surfactants Principles and Applications*, Wiley-VCH, Weinheim (**2005**).

3 T.F. Tadros, in *Principles of Polymer Science and Technology in Cosmetics and Personal Care*, E.D. Goddard and J.V. Gruber (Eds.), Marcel Dekker, New York, 73–113 (**1999**).

4 C. Stevens, A. Meriggi, M. Peristeropoulou, P.P. Christov, K. Booten, B. Levecke, A. Vandamme, N. Pittivils, T.F. Tadros, *Biomacromolecules, 2*, 1256 (**2001**).

5 T.F. Tadros, A. Vandamme, B. Levecke, K. Booten, C.V. Stevens, *Advances in Colloid and Interface Science, 108/109*, 207 (**2004**).

6 P.J. Flory, *Principles of Polymer Chemistry*, Cornell University Press, Ithaca, NY (**1953**).

7 C.C. Han, B. Mozer, *Macromolecules, 12*, 146 (**1979**).

8 I. Piirma, *Polymeric Surfactants*, Marcel Dekker, New York (**1992**).

9 G.J. Fleer, M.A. Cohen-Stuart, J.M.H.M. Scheutjens, T. Cosgrove, B. Vincent, *Polymers at Interfaces*, Chapman and Hall, London (**1993**).

10 J. Lyklema, *Fundamentals of Interface and Colloid Science*, Vol. II, Academic Press, New York (**1995**).

11 T. Cosgrove, T.L. Crowley, T. Rayan, *Macromolecules, 20*, 2879 (**1987**).

12 J.M.H.M. Scheutjens, G.J. Fleer, *Advances in Colloid and Interface Science, 16*, 341 (**1982**).

13 P.G. de Gennes, *Scaling Concepts in Polymer Physics*, Cornell University Press, Ithaca, NY (**1979**).

14 W. Liang, G. Bognolo, T.F. Tadros, *Langmuir, 11*, 2899 (**1995**).

15 I.M. Krieger, *Advances in Colloid and Interface Science, 3*, 111 (**1972**).

16 J. Nestor, J. Esquena, C. Solans, B. Levecke, K. Booten, T.F. Tadros, *Langmuir, 21*, 4837 (**2005**).

17 J. Nestor, J. Esquena, C. Solans, P.F. Luckham, B. Levecke, K. Booten, T.F. Tadros, *Langmuir*, submitted.

18 T.F. Tadros, A. Vandamme, K. Booten, B. Levecke, C.V. Stevens, *Colloids and Surfaces A, 250*, 133 (**2004**).

19 D. Exerowa, P.M. Kruglyakov, *Foam and Foam Films*, Elsevier, Amsterdam (**1998**).

20 D. Exerowa, T.F. Tadros, B. Levecke, to be published.

21 M.S. Aston, T.M. Herrington, T.F. Tadros, *Colloids and Surfaces, 40*, 49 (**1989**).

22 T.F. Tadros, in *Emulsions – a Fundamental and Practical Approach*, J. Sjoblom (Ed.), NATO ASI Series: Series C, Vol. 363, Kluwer, Dordrecht (**1992**).

23 H. Nakajima, in *Industrial Applications of Microemulsions*, C. Solans (Ed.), Marcel Dekker, New York (**1997**).

24 T.F. Tadros, P. Izquierdo, J. Esquena, C. Solans, *Advances in Colloid and Interface Science, 108/109*, 303 (**2004**).

25 W. Thompson (Lord Kelvin), *Philosophical Magazine, 42*, 448 (1871).

26 I.M. Lifshitz, V.V. Slesov, *Soviet Physics JETP, 35*, 331 (**1959**); C. Wagner, *Zeitschrift für Electrochemie, 35*, 581 (**1961**).

27 T.F. Tadros, E. Vandekerchhove, A. Vandamme, B. Levecke, K. Booten, *Cosmetics and Toiletries, 120*, 45 (**2005**).

28 P. Izquierdo, J. Esquena, T.F. Tadros, C. Federen, M.J. Gracia, N. Azemar, C. Solans, *Langmuir, 18*, 26 (**2002**).

29 D. Attwood, A.T. Florence, *Surfactant Systems*, Chapman and Hall, London (**1983**).

30 T.F. Tadros, B. Levecke, K. Booten, to be published.

11
Foam Films, Foams and Surface Rheology of Non-ionic Surfactants: Amphiphilic Block Copolymers Compared with Low Molecular Weight Surfactants

Cosima Stubenrauch and Brita Rippner Blomqvist

11.1
Introduction

Generating a foam first leads to the formation of a so-called wet foam, which consists of spherical air bubbles surrounded by a liquid. While draining, the structure gradually changes and the bubbles are transformed into polyhedral air cells separated by thin liquid films which are stabilized by surfactants that have adsorbed at the water/air interface (Fig. 11.1). Although the application of liquid foams is widespread (in cleaning agents, beverages, fire-fighting, flotation and oil recovery, to mention just a few), too little is yet understood about the parameters with which their stability can be controlled. Hence the development of new products is often based on "trial and error".

In order to learn more about the properties of well-drained, so-called dry foams, the investigation of the foam's building blocks, i.e. foam films, is generally regarded as promising [1–4]. With respect to low molecular weight (LMW) surfactants, we usually distinguish between sterically and electrostatically stabilized films. The former are called *Newton black films* (NBFs), whereas the latter are referred to as *common black films* (CBFs) or *common thin films* (CTFs). For the sake of clarity, we will just use the abbreviation CBFs (the difference between CBFs and CTFs is explained in [5]). The formation of a CBF or an NBF is determined by the type of surfactant, the surfactant concentration, the electrolyte concentration, surface-active additives and the pH. The situation changes, however, when we consider foam films stabilized by amphiphilic polymers. In contrast to the LMW surfactants, the range of the steric and the electrostatic interactions can be of the same order of magnitude. Hence we can no longer clearly distinguish between long- and short-range repulsion: the transition from electrostatic to steric stabilization is now continuous. These long-range steric interactions can be described with the de Gennes scaling theory for interacting polymer brushes [6]. According to this theory, the steric interactions are a function of the applied pressure, which is in contrast to the behavior of NBFs. In other words, while a NBF does not change its thickness when the pressure is

Colloids and Interface Science Series, Vol. 1
Colloid Stability: The Role of Surface Forces, Part I. Edited by Tharwat F. Tadros

ISBN: 978-3-527-31462-1

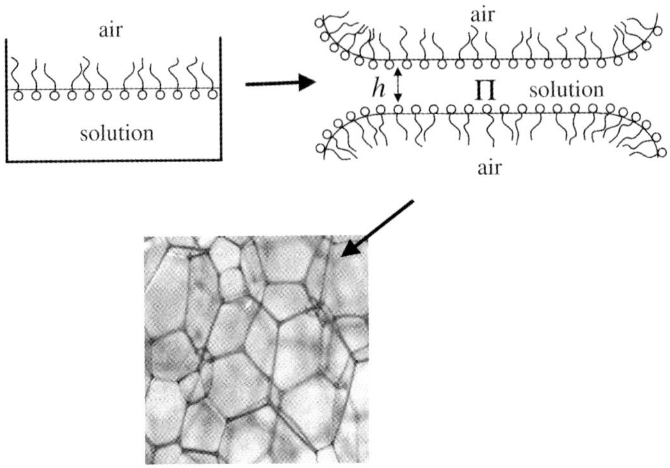

Fig. 11.1 (a) Schematic drawings of a surfactant-coated air/water surface and a free-standing foam film of thickness h. (b) Photograph of a dry foam consisting of foam films connected via plateau borders. For the formation of foam a large number of single foam films of different size and orientation are needed. Each single film consists of two surfaces separated by an aqueous core.

increased, the respective polymer film thickness decreases further [2, 7], because the polymeric surfactants are compressible. These films will be called *bilayer films* (BFs) to distinguish them from the NBFs.

By measuring the disjoining pressure Π as a function of the film thickness h, one obtains two important pieces of information. First, the mechanism stabilizing the film can be determined and quantified. Supposing that the nature of the surfactant is not known, a film of 20 nm thickness, for instance, can be stabilized (a) by steric forces in the case of an amphiphilic polymer, (b) by structural forces in the case of surfactant–polyelectrolyte mixtures (reviewed in [8, 9]) or (c) simply by long-range electrostatic forces. A distinction is easily possible when the Π–h curve is known. It is only with this knowledge that the properties can be tuned in a controlled way. For example, the thickness and stability of an electrostatically stabilized foam film can easily be tuned by adding an electrolyte, whereas the electrolyte will not influence the properties of a sterically stabilized film. The second important piece of information one obtains from Π–h curves is the maximum pressure sustainable by the film. This pressure is a measure for the film's stability and what one expects is increased stability with increasing repulsive interaction forces. This is, however, only part of the truth. It has been found experimentally for both LMW ionic [10] and LMW non-ionic surfactants [11–13] that equal electrostatic repulsion does not automatically result in equal foam film stability. Furthermore, long-range steric repulsion between amphiphilic polymers does not guarantee the formation of stable foam films [14]. Hence the stability of thin foam films cannot be explained solely by the magnitude of the repulsive interactions operating normal to the film sur-

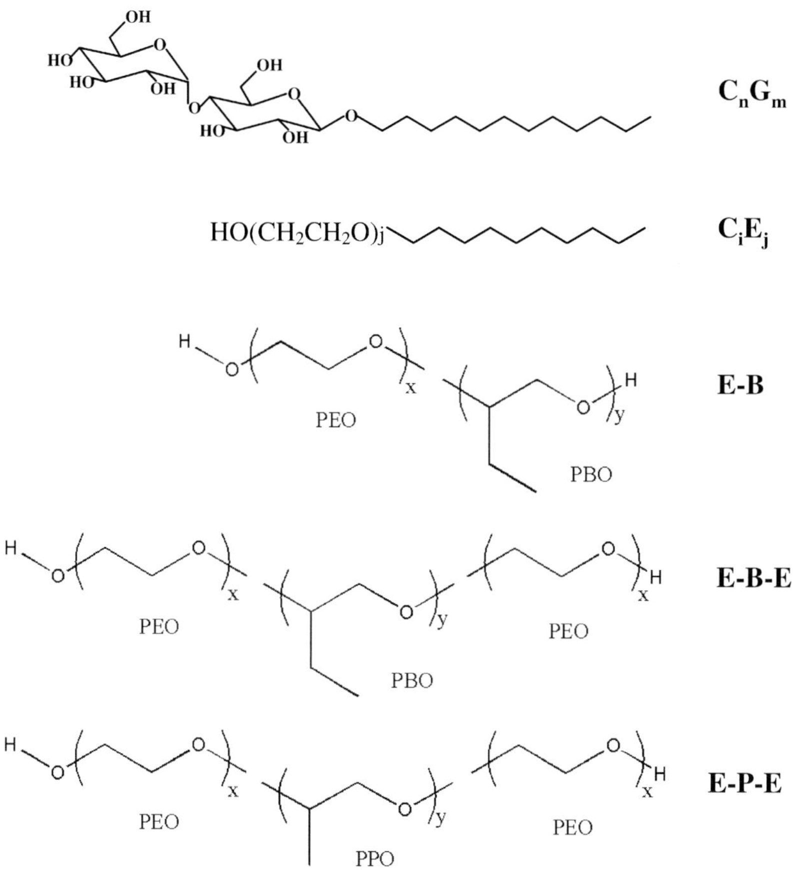

Fig. 11.2 Molecular structures of two low molecular weight non-ionic surfactants and three non-ionic amphiphilic block copolymers. C_nG_m, alkyl polyglucoside (the structure of *n*-alkyl-*β*-D-maltoside is shown); C_iE_j, alkyl polyglycol ether; E–B, diblock copolymer of poly(ethylene oxide) and poly(butylene oxide); E–B–E, triblock copolymer of two poly(ethylene oxide) blocks and one poly(butylene oxide) block; E–P–E, triblock copolymer of two poly(ethylene oxide) blocks and one poly(propylene oxide) block.

faces. What is needed is a surface which is able to dampen external disturbances and therefore prevents the film from rupturing. This ability is believed to be mirrored in the surface viscoelasticity of the monolayer [10, 12, 15–18].

The next question to be addressed is whether or not the properties of single surfaces and/or isolated foam films can be compared with those of foams. The mechanical strength of interfacial layers and their response to dilatational and shear deformations in the lateral direction are definitely important for the foam's stability. However, the exact correlation between surface rheological parameters and foam stability is far from clear [17, 19]. Moreover, specifying the correlation between isolated foam films and foams (see, for example, [1, 2, 20,

21] and references cited therein) also turned out to be a challenge. Isolated films used in model studies are either horizontal or vertical and of fixed sizes, whereas foams consist of a three-dimensional network of interconnected films, which are of various orientations and sizes. Moreover, foam films are usually studied under defined pressures whereas foams are studied under gravity, i.e. that the pressure is not constant but a function of the height of the foam column.

This chapter is organized as follows. In Section 11.2, the origin of the disjoining pressure in thin foam films will be discussed and foam films stabilized by LMW surfactants will be compared with those stabilized by amphiphilic block copolymers. On the basis of experimental results, the role the structure of the surfactant and the amphiphilic polymer will be analyzed. Section 11.3 deals with the drainage and stability of foams stabilized by either LMW surfactants or amphiphilic polymers. Whenever possible and reasonable, the question of how the properties of isolated foam films correlate with those of the respective foams is addressed. Section 11.4 is devoted to the role that surface rheology plays with respect to the stability of both foam films and foams. Surface elasticities measured under different experimental conditions will be presented and compared with (a) the Π–h curves of the respective foam films and (b) the drainage curves of the respective foams. The chapter concludes with an outlook and a discussion of the future of the field. The molecular structures of the non-ionic LMW surfactants and the non-ionic amphiphilic block copolymers referred to in the present review are shown in Fig. 11.2.

11.2 Disjoining Pressure in Foam Films

11.2.1 DLVO and Non-DLVO Contributions

The origin of the disjoining pressure is interaction between the two interfaces of the thin liquid film (reviewed in [2, 3, 22]). One usually defines the disjoining pressure as the sum of long-range repulsive electrostatic (Π_{elec}), short-range attractive van der Waals (Π_{vdW}) and repulsive steric (Π_{steric}) interactions, the range of which depends on the molecular structure of the film-stabilizing surfactant. In addition, under certain conditions, structural interactions ($\Pi_{\text{structural}}$) are present so that the total disjoining pressure is

$$\Pi(h) = \Pi_{\text{elec}}(h) + \Pi_{\text{vdW}}(h) + \Pi_{\text{steric}}(h) + \Pi_{\text{structural}}(h) \qquad (1)$$

where h is the film thickness. Note that the first two terms on the right-hand side account for the classical DLVO theory [23, 24]. Structural interactions caused by the confinement of a fluid between two walls have been reviewed only recently [8] and will not be dealt with here.

11.2.1.1 DLVO Interactions

Electrostatic Double-layer Interactions

On the assumption that the two interfaces of the thin liquid film are charged, the two electric double layers interact when the separation between the interfaces approaches twice the decay length of their ionic atmosphere, the so-called Debye length. As the ionic strength of a solution decreases, the Debye length increases, which, in turn, increases the range over which the electrostatic double layers interact. The same holds true for an increase in the surface charge density at a constant Debye length. To calculate the electrostatic component Π_{elec} of the disjoining pressure, the non-linear Poisson-Boltzmann equation has to be solved using appropriate boundary conditions [22]. Under certain conditions, a simplified form of Π_{elec} can be used. If one assumes small potentials in the middle of the film and very little overlap between the two double layers, the electrostatic repulsion between two charged interfaces in a 1:1 electrolyte can be expressed as

$$\Pi_{\mathrm{elec}}(h) = 64RTc\tanh^2\frac{F\psi_0}{4RT}\exp(-\kappa h) \tag{2}$$

where T is the temperature, R the gas constant, c the electrolyte concentration, F the Faraday constant, ψ_0 the surface potential and κ^{-1} the Debye length. Comparing the measured with the calculated Π–h curves, one can deduce the surface potential, from which the corresponding surface charge density q_0 is calculated using the Graham equation [22]:

$$q_0 = \sqrt{8\varepsilon\varepsilon_0 RTc}\,\sinh\left(\frac{F\psi_0}{2RT}\right) \tag{3}$$

where ε and ε_0 are the dielectric constants of the medium and of the vacuum, respectively. Note that the relevant thickness for electrostatic repulsions is defined by the location of the charges. In the case of ionic surfactants, the location of the charges corresponds to the location of the head groups, whereas in the case of non-ionic surfactants the charges are located at the air/water interface. The origin of the latter is mentioned in Section 11.2.2 and discussed in detail in [8, 35].

Van der Waals Interactions

The van der Waals component Π_{vdW} of the disjoining pressure is defined as

$$\Pi_{\mathrm{vdW}}(h) = -\frac{A}{6\pi h^3} \tag{4}$$

where A is the Hamaker constant [22, 25]. In the case of symmetric films such as those treated in the present review, the Hamaker constant is always positive. Hence the van der Waals interactions acting in foam films are always attractive.

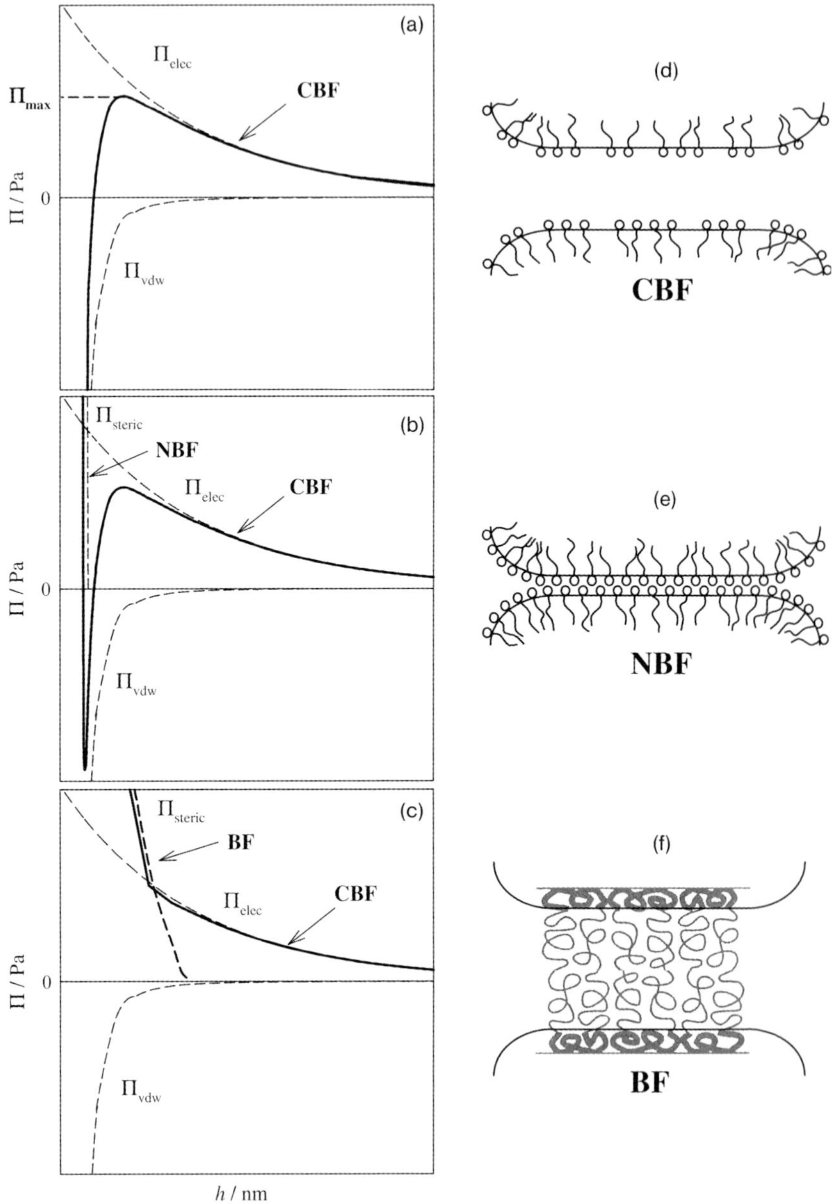

Fig. 11.3 (a–c) Schematic representation of the disjoining pressure Π as a function of the film thickness *h* (solid lines). The three main contributions to Π are repulsive electrostatic (Π_{elec}), attractive van der Waals (Π_{vdW}) and repulsive steric (Π_{steric}) interactions, all given by dashed lines. In a *common black film* (CBF) the stability is determined by Π_{elec}, whereas the *Newton black film* (NBF) and the *bilayer film* (BF) are stabilized by Π_{steric}. (a) DLVO interactions; (b) DLVO and short-range steric interactions; (c) DLVO and long-range steric interactions. Also shown are schematic drawings of a CBF (d), an NBF (e) and a BF (f).

In the presence of surfactant, the van der Waals component of the interaction should ideally be calculated using (at least) a five-layer model (e.g. air–surfactant–water–surfactant–air). However, in most cases it is sufficient to use the simple non-retarded Hamaker constant for the surfactant-free film, which is 3.7×10^{-20} J for an air–water–air film.

According to Eq. (1), the addition of Π_{vdW} and Π_{elec} leads to the theoretical Π–h curve, which is drawn schematically in Fig. 11.3 a. What we see is that the electrostatic double-layer interactions create an energy barrier, i.e. a maximum of the disjoining pressure $\Pi_{\max}$ is observed. Thus, when the applied external pressure P_c is lower than $\Pi_{\max}$, the film thins continuously, whereas the film ruptures when P_c exceeds $\Pi_{\max}$. However, film rupture is often observed at pressures far below $\Pi_{\max}$. Moreover, film rupture is found to take place not at a certain pressure but rather in a pressure range. A new approach to explain these experimental observations is made in [26–28].

11.2.1.2 Steric Interactions

In foam films stabilized by LMW surfactants, instead of a film rupture a stepwise transition to a *Newton black film* (NBF) can take place when the applied pressure P_c exceeds $\Pi_{\max}$. In that case two stable regions, i.e. regions of positive disjoining pressure, can be distinguished, as illustrated in Fig. 11.3 b. As mentioned in the Introduction, the thicker *common black film* (CBF) is stabilized electrostatically whereas the stability of the thinner NBF is determined by a repulsive steric force Π_{steric}. As undulations of the interface, peristaltic fluctuations, protrusion of surfactants at the interface and head group overlap have to be taken into account, no simple expression for Π_{steric} is yet known [3, 29]. Hence the NBF is usually presented as an incompressible surfactant bilayer, i.e. as a film the thickness of which does not depend on the applied pressure. Note that a stepwise CBF–NBF transition is observed when the derivative of Π with respect to the film thickness h, i.e. $\mathrm{d}\Pi/\mathrm{d}h$, becomes positive. This is the case when the van der Waals interactions become larger than the electrostatic interactions. The thickness at which this occurs is a function only of the surface charge density q_0 and the Hamaker constant A. However, if this thickness is approximately the same as the thickness where steric interactions start to stabilize the film, no discrete transition is seen and Π is a monotonically decreasing function of h, as can be seen in Fig. 11.3 c.

A continuous transition from electrostatic to steric stabilization has been observed experimentally for foam films stabilized by surfactants with large head groups. Examples will be given in Section 11.2.3. What is important in the present context is the fact that in the case of large non-ionic head groups, Π_{steric} is dominated by the head group overlap which, in turn, can be calculated by theories used for polymer brushes. Thus, in contrast to the LMW surfactants referred to above, an expression for Π_{steric} can now be given. According to the theory of de Gennes [6], the head group is considered to be compressible, which results in

$$\Pi_{\text{steric}}(h) = kT\Gamma^{\frac{3}{2}}\left[\left(\frac{2h_{\text{head}}}{h_{\text{core}}}\right)^{\frac{9}{4}} - \left(\frac{h_{\text{core}}}{2h_{\text{head}}}\right)^{\frac{3}{4}}\right] \quad (5)$$

for $h_{\text{core}} < 2h_{\text{head}}$, where k is Boltzmann's constant, T is the absolute temperature, Γ is the surface concentration of the surfactant, h_{head} is the length of the head group at infinite separation and h_{core} is the thickness of the film core consisting of the head groups and water [2, 7, 30]. (Note that the total film thickness equals $h = h_{\text{core}} + 2\ h_{\text{tail}}$, where the length of the hydrophobic tail h_{tail} is considered to be constant.) Hence the steric interaction is a function of the applied pressure since h_{core} changes with pressure, which is not the case for the typical NBFs. In other words, while an NBF does not – or at least not significantly – change its thickness when the pressure is increased, the thickness of the compressible film decreases further. Stubenrauch et al. proposed to call these films *bilayer films* (BFs) to distinguish them from the NBFs [5]. Accordingly, we will speak of NBFs and BFs in the following.

11.2.2
Foam Films Stabilized by Low Molecular Weight Surfactants

In the following, some selected Π–h curves of foam films stabilized by three different non-ionic surfactants, namely the sugar surfactant *n*-dodecyl-β-D-maltoside (β-$C_{12}G_2$) and the two alkyl polyglycol ethers hexaethylene glycol monododecyl ether ($C_{12}E_6$) and tetraethylene glycol monodecyl ether ($C_{10}E_4$), will be presented and briefly discussed (the molecular structures are shown in Fig. 11.2). Further, results and detailed discussions can be found in [2, 8]. All Π–h curves were measured with a *thin film pressure balance* (TFPB), which is described in detail in [2, 8, 31–33]. The solutions contained 10^{-4} M NaCl to adjust a definite Debye length for the fits that were done on the basis of the DLVO theory. The measurements were carried out at the "natural" pH of 5.5 ± 0.3 (due to the dissolution of CO_2 in water). The reason for choosing the Π–h curves shown in Fig. 11.4 was to discuss the influence of the surfactant concentration and also of the surfactant structure on both the magnitude of the disjoining pressure and the film stability. The adsorption isotherms and the critical micellar constant (CMC) values which are needed for this discussion were determined by evaluating the respective surface tension isotherms (data not shown).

11.2.2.1 Influence of the Surfactant Concentration

In Fig. 11.4, the Π–h curves of the three non-ionic surfactants are shown for different concentrations. The concentrations were chosen such that they are below, around and above the CMC. In all cases, CBFs and NBFs are formed. The thickness of the CBFs decreases monotonically as the disjoining pressure increases. The slope remains constant with changing surfactant concentrations but the curves shift towards lower disjoining pressures when the surfactant concentration

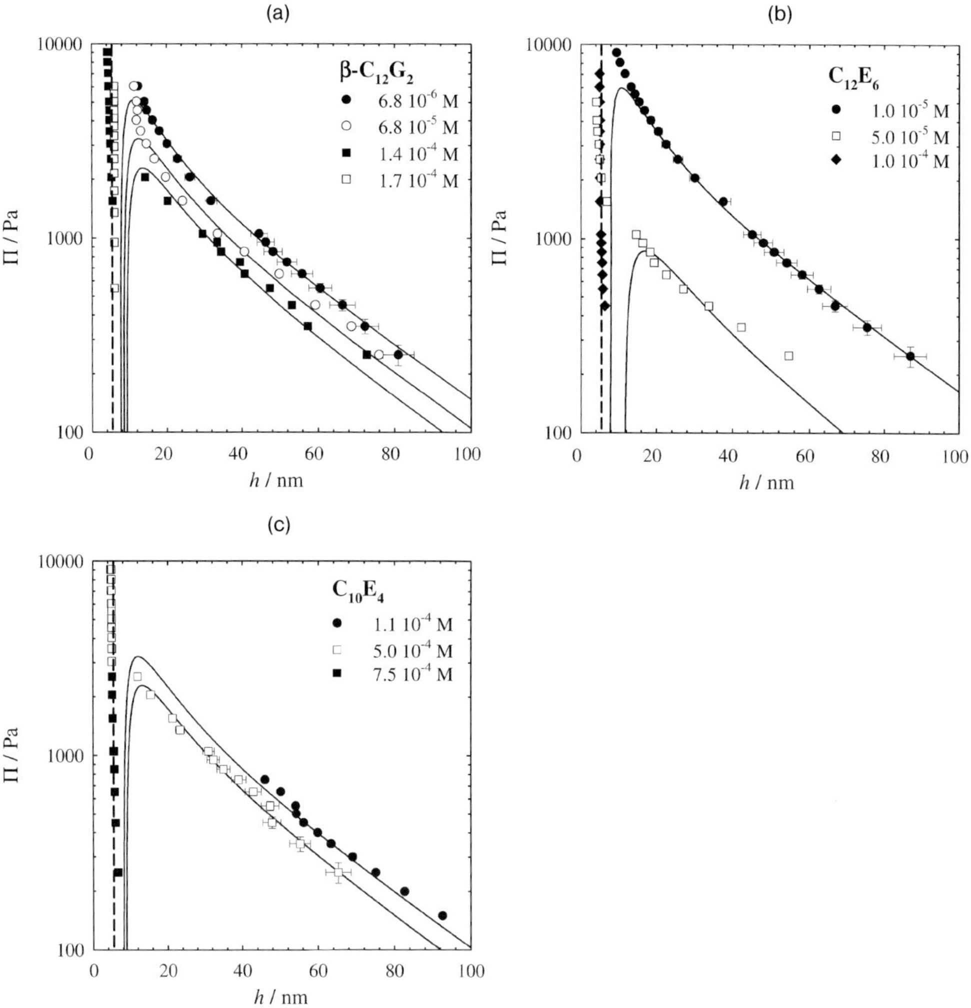

Fig. 11.4 *Π–h* curves for different concentrations of β-$C_{12}G_2$ (a), $C_{12}E_6$ (b) and $C_{10}E_4$ (c). The concentrations are chosen such that they are below and around the CMC, which are 1.5×10^{-4} M for β-$C_{12}G_2$, 8.0×10^{-5} M for $C_{12}E_6$ and 8.6×10^{-4} M for $C_{10}E_4$. The solid lines are calculated according to the DLVO theory, from which the surface charge densities listed in Table 11.1 were calculated. The data for β-$C_{12}G_2$ are taken from [12], for $C_{12}E_6$ from [35] and for $C_{10}E_4$ from [34].

is increased. This shift is accompanied by an increasing tendency to form an NBF, which is illustrated by the fact that the transition from a CBF to an NBF does not appear for the lowest concentrations, whereas it is observed at intermediate concentrations via black spot formation in the CBF. Moreover, at the highest concentrations investigated, no CBF is formed at all, which means that only an NBF with a constant thickness is observed. Once formed, the NBFs are stable over the whole pressure range investigated. As the thickness and the stability of the NBF do not change significantly within the investigated pressure range, the NBF is represented as a dashed vertical line.

Both the Π–h curves shown in Fig. 11.4 and the fact that the surfactants dealt with are non-ionic raise the question where the electrostatic repulsion, i.e. the charges at the water/air surface, comes from. At present, the widely accepted explanation for the origin of the charges in thin foam films stabilized by non-ionic surfactants is the presence of excess OH^- ions at the water/air surface [2, 8, 36–38]. However, it has not yet been clarified yet whether the OH^- ions are specifically adsorbed or if other mechanisms are responsible for the excess of OH^- ions at the water/air surface. A detailed review of the current discussion and the results obtained so far is given in [8]. Moreover, a new mechanism explaining the source of the OH^- ions at the water/air surface has been proposed [35]. Despite the fact that the origin of the OH^- ions is not yet clear, the experimental Π–h curves seen

Table 11.1 Surfactant concentration c, corresponding surface concentration Γ and area per molecule $A_{molecule}$ at the air/water surface [a].

$c \times$ (M) [b]	Γ (10^{-6} mol m^{-2})	$A_{molecule}$ (nm^2)	ψ_0 (mV)	q_0 (mC m^{-2})	A_{charge} (nm^2)
β-$C_{12}G_2$					
6.8×10^{-6}	2.3	0.72	56	1.55	100
3.4×10^{-5}	4.0	0.42	56	1.55	100
6.8×10^{-5}	4.3	0.39	45	1.17	130
1.4×10^{-4}	4.5	0.37	38	0.95	170
1.7×10^{-4}	4.7	0.35	– [c]	– [c]	– [c]
$C_{12}E_6$					
1.0×10^{-5}	3.0	0.55	60	1.71	90
5.0×10^{-5}	3.2	0.52	28	0.67	240
1.0×10^{-4}	3.2	0.52	– [c]	– [c]	– [c]
$C_{10}E_4$					
3.5×10^{-5}	2.9	0.57	45	1.17	130
1.1×10^{-4}	3.2	0.52	45	1.17	130
1.6×10^{-4}	3.2	0.52	43	1.10	150
2.5×10^{-4}	3.3	0.50	40	1.01	160
5.0×10^{-4}	3.3	0.50	38	0.95	170
7.5×10^{-4}	3.3	0.50	– [c]	– [c]	– [c]

a) Γ was obtained from fitting the surface tension isotherm to the Langmuir-Szyszkowski equation. Surface potentials ψ_0, surface charges q_0 and area per charge A_{charge} are from DLVO calculations. The calculated Debye length is $\kappa^{-1} = 30$ nm at the given electrolyte concentration of 10^{-4} M NaCl. Data for β-$C_{12}G_2$ are taken from [12], for $C_{12}E_6$ from [35] and for $C_{10}E_4$ from [34].

b) CMC (β-$C_{12}G_2$) = 1.5×10^{-4} M, CMC ($C_{12}E_6$) = 8.0×10^{-5} M and CMC ($C_{10}E_4$) = 8.6×10^{-4} M.

c) No measurable electrostatic repulsion under the given experimental conditions.

in Fig. 11.4 were fitted with the DLVO theory to obtain the surface charge density q_0. The results of these calculations are given in Table 11.1.

Looking at Fig. 11.4 and Table 11.1, one sees that the value of q_0 decreases with increasing surfactant concentration until it is so low that a CBF can no longer be stabilized and an NBF is directly formed under the chosen experimental conditions. These results are in absolute agreement with those reported by other workers (see, for example, [11, 39–41]) and are therefore a general phenomenon occurring in foam films stabilized by LMW non-ionic surfactants. Experimentally it has been observed repeatedly that the surface charge density is constant at low surfactant concentrations and decreases significantly above a certain concentration, which was found to be connected with the total number of ethylene oxide units [40]. What is important to realize is that the decrease in the surface charge contrasts sharply with the adsorption of the surfactant, which changes significantly at low concentrations and stays close to constant already far below the CMC (see Table 11.1). Although this observation is not yet understood, a typical competitive adsorption between surfactant and hydroxide can be excluded.

In conclusion, the stabilizing mechanism in aqueous foam films stabilized by LMW non-ionic surfactants can easily be tuned by varying the surfactant concentration. With increasing surfactant concentration the surface charge density decreases, which leads to a transition from a CBF to an NBF. Obviously the non-ionic head groups of the investigated surfactants are large enough to create a significant short-range repulsion which stabilizes the NBF over a broad pressure range. Whether or not this ability is a general feature of LMW non-ionic surfactants will now be discussed.

11.2.2.2 **Influence of the Surfactant Structure**

Head Group Size Looking at Fig. 11.4, one sees that films stabilized by β-$C_{12}G_2$ and $C_{12}E_6$ can hardly be distinguished. In other words, for these particular surfactants the influence of the head group structure on the Π–h curve is not very strong. This observation is in accordance with results obtained for the homologs $C_{10}E_4$ and $C_{10}E_8$ [38]. At a surfactant concentration of two-thirds of the CMC and a salt concentration of 3.0×10^{-4} M, both systems form a CBF the Π–h curves of which lie on top of each other. Obviously the size of the head group has no influence on the electrostatic interactions in the film. However, there is a lower limit of the head group size below which the stability of the film and its ability to form an NBF are affected. Whereas $C_{10}E_8$ and $C_{10}E_4$ form stable CBFs, the CBF of $C_{10}E_2$ ruptures between 2000 and 3000 Pa. Moreover, in contrast to $C_{10}E_8$ and $C_{10}E_4$, an NBF of $C_{10}E_2$ is not stable [38]. The differences in the CBF stabilities are believed to be due to different surface elasticities, which, however, has not yet been investigated in detail. We will return to this point in Section 11.3. In addition to a sufficient surface elasticity, a large enough steric repulsion is required to stabilize an NBF. Obviously, for non-ionic surfactants the size of the head group is the parameter which determines the ability of the surfactant to stabilize an NBF. It has been found for C_iE_j surfactants that an NBF is stabilized when the head group consists of more than two ethylene oxide units. A comparison with the sugar surfactants reveals that the smallest possible head group, i.e. glucoside, is already large enough to stabilize an NBF [41, 42]. Hence it is only for the C_iE_j surfactants – at least in a limited range – that the formation of an NBF can be tuned by adjusting the head group size.

Alkyl Chain Length Having shown that the head group has no significant influence on the long-range repulsion but mainly on the ability of the system to form an NBF, we can now discuss the influence of the alkyl chain length by comparing $C_{12}E_6$ and β-$C_{12}G_2$ with $C_{10}E_4$. As already mentioned, the same general trends as those found for $C_{12}E_6$ and β-$C_{12}G_2$ can be seen (Fig. 11.4): a CBF is formed at the lowest concentration, at the intermediate concentration a transition from a CBF to an NBF at about 3000 Pa is observed and at the highest concentration only an NBF is stable over the whole pressure range investigated. Moreover, fitting the Π–h curves with the DLVO theory, one obtains a decrease in the surface charge density with increasing surfactant concentration (see Table 11.1), which is analogous to what was found for β-$C_{12}G_2$ and $C_{12}E_6$. At first sight, no significant difference can be detected between the three surfactants investigated in the present chapter. However, a closer look reveals that the properties of the $C_{10}E_4$ films are completely different – not with respect to the magnitude of the disjoining pressure but with respect to the film stability. As can be seen in Fig. 11.4, the stability of the CBF increases with increasing surfactant concentration, whereas for $C_{12}E_6$ and β-$C_{12}G_2$ the reverse is observed. It is important to realize that the increase in the film stability in the case of $C_{10}E_4$ is accompanied by a decrease in the surface charge density, which, in turn, is expected to destabilize the film. In other words, the increasing stability cannot be explained in terms of disjoining pressures. We can go one step further and compare the surface charge densities of the $C_{10}E_4$ films with those of the films stabilized by $C_{12}E_6$ and β-$C_{12}G_2$. It is seen in Table 11.1 that – accidentally – films with equal surface charge densities q_0 have been investigated, for example the films of the 1.1×10^{-4} M $C_{10}E_4$ and that of the 6.8×10^{-5} M β-$C_{12}G_2$ solution. For both systems $q_0 = 1.17$ mC m^{-2} has been determined. However, it is seen in Fig. 11.4 that equal repulsive interactions do not automatically result in equal film stabilities. Although both films have the same q_0 values, the film of β-$C_{12}G_2$ is much more stable than that of $C_{10}E_4$. As such a significant effect of the head group can be excluded, it must be the alkyl chain length which is responsible for the difference in the stability. The observation that the film stability decreases with decreasing chain length has also been made for the two homolog series which have been investigated so far [10, 38]. From the data available we can now estimate the minimum chain length at which foam films of C_iE_j and C_nG_m surfactants, respectively, are stable. For the non-ionic alkyl tetraethylene glycol ethers C_nE_4 n has to be ≥ 10 [38] and for the alkyl glucosides C_nG_1 a hydrophobic octyl chain is sufficient for the formation of thin liquid films [41]. Unfortunately, of the C_nG_2 series only two surfactants have been investigated so far, namely β-$C_{12}G_2$ [12] and β-$C_{10}G_2$ [43]. Nevertheless, it is very reasonable to assume a minimum chain length of $n = 10$ because Persson et al. [43] reported that the β-$C_{10}G_2$ films were very unstable; hence stable β-C_8G_2 films are unlikely. Thus, the C_nG_2 and the C_nE_4 series seem to be very similar with respect to their foam film properties, both requiring at least a decyl chain to stabilize a film. In addition, both head groups are large enough to allow the formation of an NBF.

In conclusion, the structure of LMW surfactants plays a dominant role with regard to the stability of the foam film and also its ability to form an NBF, whereas its influence on the long-range electrostatic interactions is insignificant. We still have to answer the question of what is a suitable parameter to describe the film's stability. In other words, apart from repulsive interactions, what else is needed to stabilize a foam film? Before we come back to this question in Section 11.3, we will discuss the properties of foam films stabilized by amphiphilic block copolymers and compare them with those of the LMW surfactants.

11.2.3
Foam Films Stabilized by Amphiphilic Block Copolymers

As has already been discussed, for most LMW surfactants a stepwise transition from the thick CBF to the thin NBF was observed (see Figs. 11.3b and 11.4). The first continuous CBF–NBF transition was observed for aqueous solutions of eicosaoxyethylene nonylphenol ether (NP20) by Kolarov et al. [44]. The head group contains 20 EO units and is therefore considerably larger than the head groups discussed in connection with Fig. 11.4. Ten years later it was shown in an extensive disjoining pressure study [11] that the CBF–NBF transition is discrete for $C_{10}E_4$ and continuous for $C_{10}E_8$. Hence a crossover from discrete to continuous NBF formation is expected between E_4 and E_8, which indeed can be deduced from Fig. 11.4. A closer look at the Π–h curves for $C_{12}E_6$ reveals that at 1.0×10^{-5} M $C_{12}E_6$ the CBF thins continuously down to very low thicknesses. A stepwise transition occurring at pressures higher than 9000 Pa is very unlikely (but not impossible). What we will focus on in the following are Π–h curves of surfactants the head groups of which are significantly larger than E_8 so that we deal with polymeric surfactants rather than with LMW surfactants. For that purpose, Π–h curves of $E_{41}B_8$, $E_{106}B_{16}$, $E_{21}B_8E_{21}$ and $E_{122}P_{56}E_{122}$ that were measured with the TFPB technique are shown in Fig. 11.5. Note that the solutions from which the foam films were formed contained no salt (a), 10^{-4} M NaCl (b) and 10^{-2} M KBr (c). All measurements were carried out at the "natural" pH of 5.5–6.0. The concentrations were below or in the vicinity of the CMC, which is 2.0×10^{-5} M for $E_{106}B_{16}$ (24 °C) [14], 4.0×10^{-4} M (26 °C) for $E_{41}B_8$ [45], 1.2×10^{-3} M (40 °C) for $E_{21}B_8E_{21}$ [46] and 3.0×10^{-5} M (23 °C) for $E_{122}P_{56}E_{122}$ [47]. Note that these values can only be considered as rough estimates since every polymer preparation is strictly unique with respect to the exact composition. In addition, precise CMC determinations have often been made difficult owing to the pres-

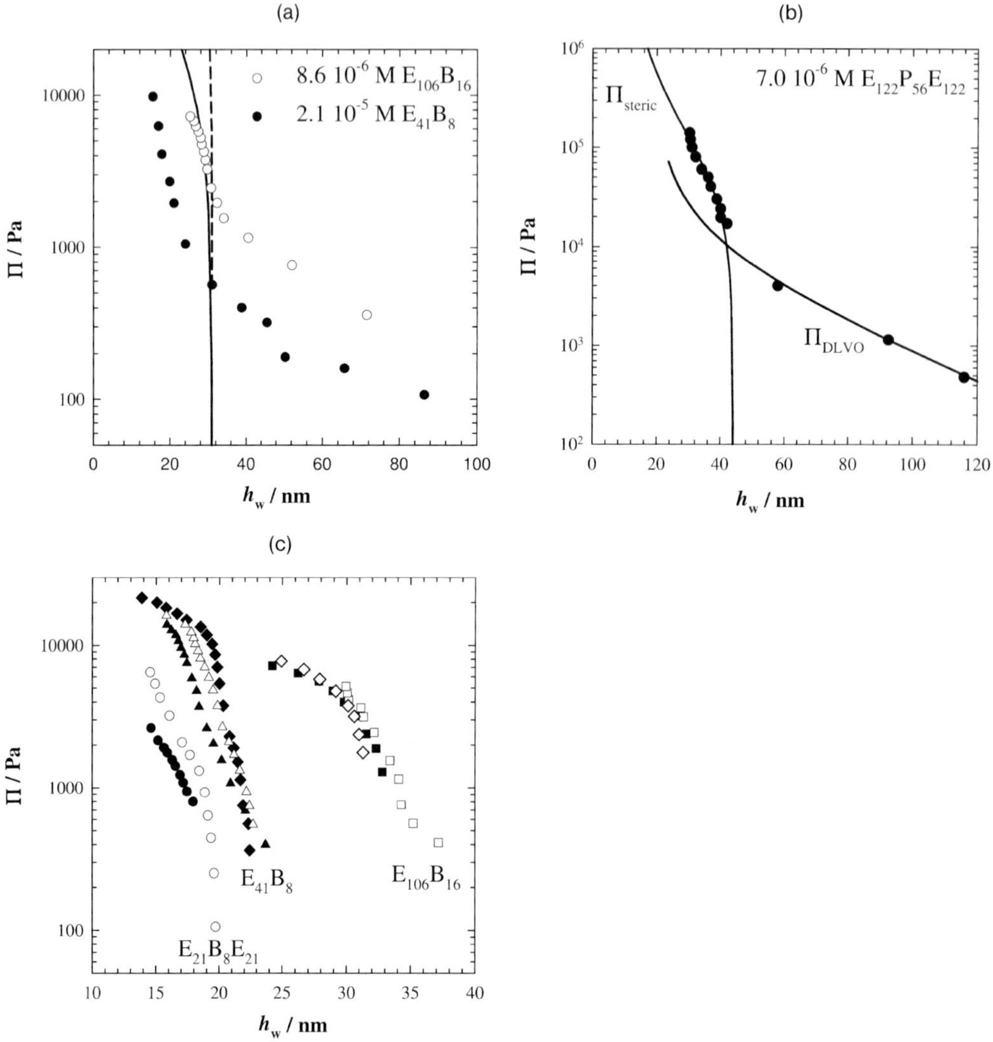

Fig. 11.5 Π–h_w curves for different diblock and triblock copolymers. The concentrations are all below or near the CMC (see text for details). (a) No electrolyte was added. The data are taken from [14, 33]. The lines represent fits to the de Gennes brush theory. The data were fitted using adsorbed amounts of 0.255 mg m^{-2} (solid line) and 2.55 mg m^{-2} (dashed line), respectively, and an h_{head} of 15.5 nm. (b) The solution contained 10^{-4} M NaCl. The solid lines were calculated according to the DLVO (assuming constant surface charge boundary conditions) and to the de Gennes brush theory (Eq. 5). The data are taken from [7]. (c) The solutions contained 10^{-2} M KBr. The concentrations were 1.7×10^{-6} (■), 3.4×10^{-6} (□) and 8.6×10^{-6} M (◇) for $E_{106}B_{16}$, 8.4×10^{-6} (▲), 4.2×10^{-5} (△) and 1.3×10^{-4} (◆) for $E_{41}B_8$ and 5.0×10^{-4} (●) and 1.3×10^{-3} (○) for $E_{21}B_8E_{21}$. The data are taken from [14]. Note that the scales of the *x*- and *y*-axes are different.

ence of several temperature-dependent inflection points in the surface tension isotherms. Moreover, these inflection points are often "unsharp", i.e. that it is rather a concentration range than a particular concentration that has to be considered as the CMC [14, 48, 49]. Another important experimental difference compared with LMW surfactant systems is that equilibration times are significantly longer. The time needed for a CBF to adjust its new thickness (in response to a change of the applied pressure) was up to 2 h [14]. Note that the equilibration time of the respective BFs was shorter, namely 15–20 min. The film thicknesses reported in Fig. 11.5 are so-called equivalent water layer thicknesses, h_w, which are calculated by assuming that the film's refractive index is homogeneous and equals that of water. The reason for this rather simplistic assumption is the difficulty in calculating or measure the actual refractive index of the polymer film which consists of flexible and compressible polymer chains. Moreover, the hydrophilic and hydrophobic segments of the polymer mix within the layer which leads to less well-defined sublayers. We will return to this topic in Section 11.4.3.

Figure 11.5 illustrates that all studied amphiphilic block copolymers form both CBFs and BFs with a continuous CBF–BF transition. The thickness of the foam film at which the steric interaction between opposing interfacial polymer layers starts to play a role is considerably larger than for the respective NBFs of the LMW surfactants (see Fig. 11.4). Moreover, the thickness of the BFs was found to decrease further with increasing pressure, which was not the case for the LMW surfactants. As is seen in Fig. 11.5 b, the data obtained for amphiphilic block copolymers can be fitted fairly well. The long-range interaction can be described with the DLVO theory under constant surface charge boundary conditions, i.e. that an electrostatic repulsion determines the film thickness at large separations (the origin of charged interfaces in systems of non-ionic surfactants has already been addressed in Section 11.2.2). In the steric regime, the Π–h curve is adequately described with de Gennes's equation (Eq. 5) by using h_{head} as the fitting parameter. Details are given in [7]. The respective Π–h curves of the salt-free E–B block copolymer solutions in Fig. 11.5 a were not fitted according to the DLVO theory to avoid speculations about the electrolyte concentration. However, as is shown for $E_{106}B_{16}$, the non-DLVO part of the Π–h curves can also be fitted reasonably well to the brush theory [50]. The fitted value of h_{head} is 15.5 nm. De Gennes's description (Eq. 5) is sensitive to the area per molecule at the interface. The adsorbed amount of $E_{106}B_{16}$ and hence the area per molecule was determined by ellipsometry [51] and the surface tension isotherm [14]. At the given concentration a value of 2.55 mg m^{-2} was obtained, which corresponds to an area per molecule of 400 Å^2. However, the fit corresponding to an adsorbed amount of 2.55 mg m^{-2} is in poor agreement with the experimental data, whereas arbitrarily using a 10 times smaller value, i.e. 0.255 mg m^{-2}, results in a more satisfactory description (see Fig. 11.5 a). Note in this context that determinations of the area per molecule based on surface tension measurements and Gibbs adsorption equation are difficult because of the complexity of the surface tension isotherms of block copolymers. Hence the area per molecule at the air/water interface has often been overestimated in the past, which is discussed in [14, 49, 51].

11.2.3.1 **Influence of the Electrolyte Concentration**

As is the case for films stabilized by LMW surfactants, the electrostatic contribution to the disjoining pressure can be suppressed by increasing the electrolyte concentration. In other words, at high electrolyte concentrations the steric interaction dominates the repulsive pressure that stabilizes the foam film, which is illustrated in Fig. 11.5. Looking at the $E_{106}B_{16}$ and $E_{41}B_8$ foam films, one sees that the electrostatic repulsive force (Fig. 11.5 a) vanishes if 10^{-2} M KBr is added to the solutions (Fig. 11.5 c). Note that the slope of the Π–h curves measured at high electrolyte concentrations decreases with pressure, which indicates the compressibility of the polymer layers in the steric regime. In conclusion, the Π–h curves of E–B, E–B–E and E–P–E polymers reported so far reveal that an increase in the monovalent *electrolyte concentration* (NaCl or KBr) influences the thickness of the CBF, whereas it does not change the thickness of the BFs [14, 33, 52], at least not in the presence of up to 0.05 M added salt. (The addition of electrolyte lowers the CMC [53]. At salt concentrations higher than 0.05 M, it has been suggested that trapped block copolymer aggregates increase the thickness of thin liquid films [54].) Further evidence was provided by measurements with the Scheludko cell with which thicknesses of $E_{122}P_{56}E_{122}$ and $E_{27}P_{39}E_{27}$ foam films were measured as a function of the electrolyte concentration at a constant disjoining pressure. It was shown that the film thickness decreases continuously with increasing electrolyte concentration until the film reaches a constant thickness at a so-called critical electrolyte concentration (c_{el}) [2, 7, 47]. Raising the salt content further does not affect the film thickness (at constant block copolymer concentration). Hence c_{el} marks the transition between electrostatic and steric dominance of Π, which is in analogy to LMW surfactants.

11.2.3.2 **Influence of the Block Copolymer Concentration**

In contrast to the effect the electrolyte concentration has on the Π–h curves, a change in the *block copolymer concentration* affects the thickness of the sterically stabilized BFs, whereas no significant effect has been observed on the electrostatic interactions of the film [7, 47]. In Fig. 11.5 c the Π–h curves of $E_{21}B_8E_{21}$, $E_{41}B_8$ and $E_{106}B_{16}$ block copolymers measured at concentrations of 4.6×10^{-4}–1.1×10^{-3}, 7.6×10^{-6}–1.1×10^{-4} and 1.7×10^{-6}–8.3×10^{-6} M are shown. As already mentioned, all these concentrations are below or in the vicinity of the respective CMC values. Figure 11.5 reveals that an increase in the block copolymer concentration results in a small (1–2 nm), although significant, increase in the film thickness for all block copolymers. Similarly, the thickness of BFs stabilized by $E_{122}P_{56}E_{122}$ increases by ca. 5 nm when the $E_{122}P_{56}E_{122}$ concentration increases from 7×10^{-7} to 7×10^{-6} M [7, 47]. Even larger changes of the film thickness have been observed, namely from $h_w=20$ to 47 nm on increasing the

$E_{122}P_{56}E_{122}$ concentration from 4×10^{-7} to 3×10^{-5} M [7, 30]. The thickness finally reaches a plateau value at concentrations close to the CMC. Thickness changes of a few nanometers can be explained by an increasing amount of polymer at the interface which causes the poly(ethylene oxide) head groups to adopt a more extended conformation towards the central part of the film. This, in turn, results in a slightly enhanced long-range steric repulsion at elevated concentrations [14]. The large thickness changes, however, are bulk rather than surface phenomena. It is argued that, with increasing bulk concentration, more and more molecules are confined between the two polymer brushes, which increases the confinement energy per unit area of the film and hence the thickness of the film core if the capillary pressure is kept constant. The plateau thickness corresponds to two polymer layers plus a film core the thickness of which equals the radius of gyration [7].

Looking carefully at the concentration dependence of $E_{106}B_{16}$ foam films in Fig. 11.5 c, one finds an exception to the general trend, namely a decreasing thickness with increasing block copolymer concentration. A somewhat thinner film forms at the highest concentration, which can be explained by the polydispersity of the polymer samples. It is reasonable to argue that at this concentration, which is close to the CMC, the most hydrophobic molecules of the "polymer mixture" already form micelles and are therefore no longer available for adsorption at the surface. The effect that the polydispersity of polymers has on the film thickness is described in detail in [14]. Similar effects of the polydispersity have been observed for the adsorption of block copolymers on solid surfaces [55].

11.2.3.3 Influence of the Block Copolymer Structure

Figure 11.5 c contains several important pieces of information about the impact of the block copolymer structure on BFs. First, films of $E_{21}B_8E_{21}$ ruptured already at pressures between 2000 and 3000 Pa for $c=5.0\times10^{-4}$ M and between 5000 and 6000 Pa for $c=1.1\times10^{-3}$ M, whereas the $E_{41}B_8$ and the $E_{106}B_{16}$ films remained stable under the given experimental conditions. Therefore, we have no information about the influence that the block size of the two diblock copolymers has on the film stability. Second, the concentrations needed to stabilize foam films with the triblock copolymer were 2–3 orders of magnitude higher than those of the diblock copolymers. Note that the overall composition and the length of the hydrophobic block (B) of $E_{21}B_8E_{21}$ and $E_{41}B_8$ are identical. Hence the polymer architecture, i.e. the number of head groups per molecule in the present case, is a crucial factor with regard to film stability. There are (at least) two possible explanations for the higher stability of the diblock copolymer film. First, the triblock copolymer may pack less efficiently at the air/water interface because of the two junctions between the hydrophobic (B) and the hydrophilic (E) parts of the molecule, which leads to lower surface concentrations compared with the diblock. Smaller adsorbed amounts of triblock copolymers have indeed been observed at hydrophobic solid surfaces [46, 55]. The second reason could be that the E blocks are simply too short to provide sufficient steric repulsion.

Note that the onset of steric forces occurs at an h_w of about 19–20 nm for $E_{21}B_8E_{21}$, of 22–23 nm for $E_{41}B_8$ and of 34–35 nm for $E_{106}B_{16}$. It is obvious that the largest polymer forms the thickest film. However, the film thickness is not a simple function of the E block size. The films of $E_{41}B_8$ are, for example, only slightly thicker than but not twice as thick as those of $E_{21}B_8E_{21}$. Indeed, h_w of the diblock copolymer films was found to scale with the radius of gyration (R_g) of the respective polymer [14]. The onset of steric repulsion in $E_{122}P_{56}E_{122}$ films was found to occur at 42–43 nm, which confirms the general trend, namely that a larger polymer forms a thicker bilayer film. The comparison between $E_{41}B_8$ and $E_{21}B_8E_{21}$ indicates that the length of the poly(ethylene oxide) chain determines the range of steric repulsions in foam films. It is important to bear in mind that, unlike LMW surfactants, amphiphilic block copolymers change their conformation with increasing surface concentration, as is described theoretically in [6, 56]. Experimental evidence for concentration-dependent structural transitions in block copolymer surface layers has also been found [57, 58]. The structure changes with increasing surface concentration from a nearly flat conformation to a three-dimensional extended structure. Therefore, the thickness of the polymer layer always depends on the surface coverage [7]. Ideally, block copolymers should be compared at an equal area per molecule rather than at equal bulk solution concentration; however, this is difficult and would require control of the surface concentration by an independent method such as ellipsometry or neutron reflectivity. A systematic study, at controlled surface concentration, of a block copolymer series with equal sizes of the hydrophobic B or P blocks and varying lengths of the E blocks and vice versa remains to be performed. These TFPB measurements are expected to clarify the influence the hydrophilic and hydrophobic blocks of block copolymers have on the foam film properties.

11.3 Drainage and Stability of Foams

This section deals with the drainage and stability of foams stabilized by non-ionic surfactants. However, it is not intended to be a review about foam drainage and foam stability in general. What we shall focus on are those studies that were carried out to find direct correlations between foams and the respective foam films. Whenever possible and reasonable, the properties of isolated foam films will be compared with those of the respective foams. The question of how foam and foam film properties can be correlated quantitatively is still to be answered.

11.3.1 Correlation Between Foams and Foam Films

Pure liquids do not foam. Surfactants, polymers, proteins or particles are needed for the formation and stabilization of foams. As foams are thermodynamically unstable, they will eventually collapse. The main mechanisms leading to this collapse are drainage, film rupture (coalescence) and Ostwald ripening (coarsening). On the other hand, there are conditions under which the foam collapse is slowed, namely a high bulk liquid viscosity, a high surface viscoelasticity and repulsive interactions across the foam lamellae. Hence the lifetime of the foam depends on the interplay between stabilizing and destabilizing mechanisms. In general, the properties of a foam are described by the foam volume formed during the foam generation process (foamability), its lifetime (foam stability) and its liquid content. The liquid content can, in addition to the absolute values of liquid volume, be expressed in terms of the foam density (total liquid volume of the foam V_{liquid}/total foam volume V_{foam}) or its inverse, the so-called foam number ($V_{\text{foam}}/V_{\text{liquid}}$). Speaking of liquid and foam volumes, one always has to keep in mind that a time-scale is involved. Once the foam is generated, drainage due to gravity comes into play, which reduces both V_{foam} and V_{liquid} as a function of time (see Fig. 11.6a). It is during this process that a wet foam turns into a dry foam. Whereas the former consists of spherical air bubbles surrounded by the liquid phase (Kugelschaum), in the latter polyhedral air cells are separated by thin liquid foam films. For a detailed description of foam structures, the book by Weaire and Hutzler is recommended [59].

Returning to the relation between foams and foam films, it should be mentioned that in a dry foam the liquid is mainly to be found in the plateau bor-

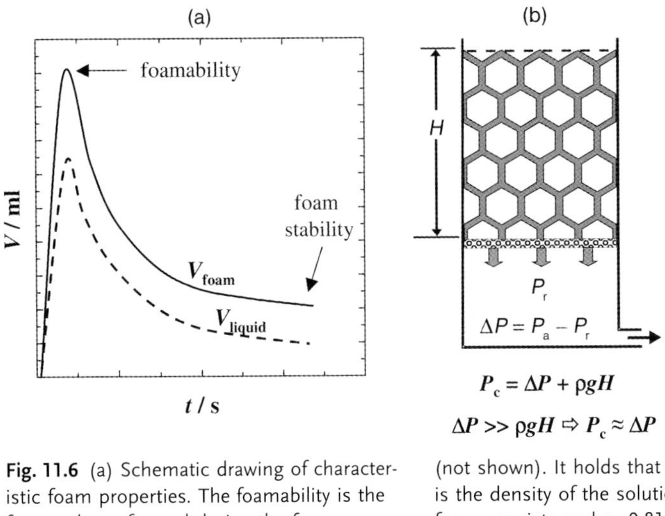

Fig. 11.6 (a) Schematic drawing of characteristic foam properties. The foamability is the foam volume formed during the foam generation process, the foam stability the time until the foam collapses (the so-called lifetime), V_{foam} represents the total foam volume and V_{liquid} the liquid volume of the foam. (b) Under gravity the capillary pressure P_c and hence the properties of a foam are a function of the foam height H (not shown). It holds that $P_c = \rho g H$, where is the density of the solution of which the foam consists and $g = 9.81\ \text{m s}^{-2}$. Applying a reduced pressure P_r so that $\Delta P = P_a - P_r \gg \rho g H$ (P_a is the atmospheric pressure), one obtains $P_c \cong \Delta P$. In that case, P_c no longer depends on gravitational forces and the properties of the foam are expected to be constant along the foam column as illustrated.

ders. However, it is the thin foam films separating the air bubbles which are the stabilizing building blocks of a dry foam. To be more precise, the stability of single foam films is believed to be directly correlated with that of the respective foam. The first investigations into the correlation between foams and foam films were carried out by Exerowa and Scheludko [61, 62]. They demonstrated that the surfactant concentration c_{bl} at which black spots are formed in single foam films are correlated with the stability of the respective foam. Since then, a lot of work has been done on both foam films and foams to specify their correlation (see, for example, [1, 2, 20, 21] and references cited therein). However, quantitative correlations between single foam films and foams are not trivial and far from being clarified, mainly for the following reasons. (1) The macroscopic foam is a three-dimensional system of interrelated films and plateau borders. If one film ruptures, the whole system has to rearrange, which in turn has an effect on the properties of the remaining films ("collective" effect). In contrast to the foam, the foam film is to be regarded as an isolated system. (2) Whereas each film in the macroscopic foam is oriented differently, the respective single film is horizontal and well aligned. (3) Investigations with single foam films are performed under defined capillary pressures P_c. However, in foams P_c is not constant but a function of the height of the foam column. That is why a direct comparison between single films subjected to a defined pressure and foams subjected to a pressure gradient is unreasonable and cannot yield quantitative correlations.

What requirements have to be met in order to compare the properties of films and foams? A foam is defined as a three-dimensional network of interrelated and differently oriented films, so that the first two points mentioned above cannot be changed or influenced. However, it is possible to investigate films and foams at the same constant capillary pressure. The former are investigated with the *thin film pressure balance* (TFPB), which is described in detail in [2, 8, 31–33]. Briefly, with this technique a constant pressure can be applied to the film and the corresponding film thickness h is measured interferometrically. Calculating the disjoining pressure Π from the applied pressure, one obtains Π–h curves with which the nature of the interaction forces stabilizing the foam film can be determined. The corresponding technique for foams dates back to ideas published about 20 years ago [63]. Recently it was further improved and is now used as the *foam pressure drop technique* (FPDT) [64], which is commercially available as FA1 (Sinterface Technologies). As in the TFPB, a constant pressure is applied and the rate of drainage and the lifetime of the foam at this particular pressure can be determined. The main difference between the FPDT and traditional techniques is the fact that with the latter foam properties are measured in a gravitational field where the capillary pressure P_c is a function of the foam column height H, i.e. $P_c = \rho g H$ with ρ being the density of the solution of which the foam consists and g the acceleration due to gravity. Therefore, all foam properties (e.g. film thickness, border and film radii, foam stability) change along the foam column. However, if we apply a pressure drop ΔP, which is larger than the hydrostatic pressure, i.e. $\Delta P >> \rho g H$, the capillary pressure in the plateau borders will be $P_c \cong \Delta P$. In that case, the above-mentioned foam properties can be considered equal along the whole foam column (see Fig. 11.6 b). Moreover, foam properties studied at constant capillary pressures can be directly compared with single film properties. Examples will be given in Section 11.3.2.

The development of the FPDT was not only motivated by the wish to find quantitative correlations between films and foams, it was also designed to provide a technique with which the mechanisms leading to foam collapse can be studied separately. Under gravity, it is always the superposition of drainage, film rupture and Ostwald ripening that is measured. As these processes depend on different parameters, their superposition makes a quantitative analysis very difficult, if not even impossible. However, with the FPDT the time the drainage process takes is controllable via the magnitude of ΔP – the larger is ΔP, the shorter is the drainage time. Let us assume for the sake of clarity that the films are very stable and that Ostwald ripening can be neglected. In this case, the effect of film rupture can be neglected during the short drainage time at high ΔP so that it is the pure drainage process that we look at. On the other hand, after drainage is completed, the stability of the foam depends solely on the stability of the films. Note that there is no other technique available with which the drainage on the one hand and coalescence and coarsening on the other can be investigated independently of each other. It is mainly for this reason that the FPDT has a high potential to achieve a "standard technique" for the characterization of stable foams.

Despite the above-mentioned advantages of the FPDT, it goes without saying that studying foams only with the FPDT is not sufficient as most of the processes that we need to understand take place under gravity. Hence the FPDT mainly has to be seen as a technique that allows us to learn more about how drainage on the one hand and film rupture and Ostwald ripening on the other contribute to the destabilization of foams. To understand foam properties not only under these particular conditions but also in general, an additional technique that works under gravity is needed for comparison. A suitable technique for complementary measurements under gravity is the foam column [65], which is commercially available as FoamScan (I.T. Concept). In brief, a foam is generated by sparging gas through a porous frit placed beneath the liquid. The drainage of the foam generated is then followed by monitoring both V_{liquid} and V_{foam} as a function of time (see Fig. 11.6 a). In both the FPDT and the FoamScan the amount of water is determined via conductivity measurements. However, with the FPDT a relative water content W of the foam is determined [64], whereas the FoamScan measures the absolute liquid volume of the foam (V_{liquid}) and the corresponding total foam volume (V_{foam}) is extracted from images recorded by a CCD camera. Examples will be given in Section 11.3.3. Unfortunately, no study has been published in which the same solutions were investigated with the two complementary techniques. We believe that such a study would improve our understanding considerably. What we will focus on in the following are results obtained for aqueous solutions of amphiphilic block copolymers that were studied with the FPDT (Section 11.3.2) and the FoamScan (Section 11.3.3).

Whenever reasonable and feasible, the results obtained by these two techniques will be compared and discussed with regard to the respective film properties. Differences between amphiphilic block copolymers and LMW surfactants with regard to their foaming properties will be addressed.

11.3.2
Drainage and Stability of Foams Under Reduced Pressure

To date there have been only a few studies of measurements with the *foam pressure drop technique* (FPDT) [Khr79, 5, 52, 60, 63, 66–71]. Moreover, most of these studies deal with the stability of well-drained foams so that drainage data under reduced pressure are available only for four systems, namely for the anionic surfactant sodium dodecyl sulfate [67], the non-ionic surfactant eicosaoxyethylene nonylphenol ether and for two non-ionic amphiphilic block copolymers [5]. The results obtained for the latter will be discussed below. To begin with, Fig. 11.7 presents results for the drainage (top) and the lifetime (bottom) of foams measured with the FPDT under reduced pressure.

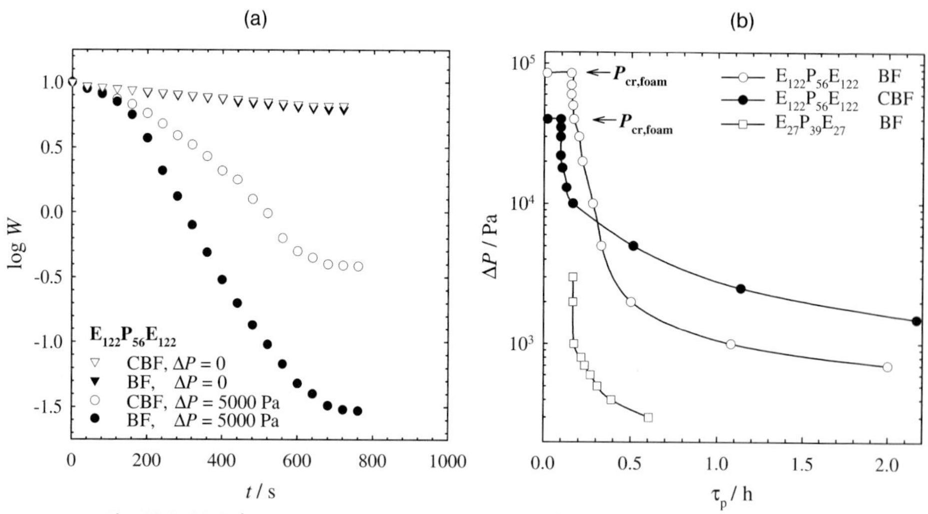

Fig. 11.7 (a) Relative water content W as a function of time t for foams stabilized by aqueous solutions of $E_{122}P_{56}E_{122}$ at a concentration of $c=10^{-5}$ M. To the solutions salt was added, namely 10^{-4} M NaCl to obtain CBFs and 10^{-1} M NaCl to obtain BFs. Drainage was investigated under gravity, i.e. $\Delta P=0$ Pa (triangles) and after applying a pressure drop of $\Delta P=5000$ Pa (circles). (b) Foam lifetimes τ_p measured at different applied pressure drops ΔP for foams stabilized by aqueous solutions of the amphiphilic block copolymers $E_{122}P_{56}E_{122}$ and $E_{27}P_{39}E_{27}$. For $E_{122}P_{56}E_{122}$ the same concentrations as above were used. In the case of $E_{27}P_{39}E_{27}$, a polymer concentration of $c=7\times10^{-5}$ M in 10^{-1} M NaCl solution was studied. Under these conditions a BF is formed. $P_{cr,foam}$ is the critical capillary pressure of foam destruction as defined in [63]. (All data are taken from [5, 52]).

11.3.2.1 **Foam Drainage**

The drainage of aqueous foams stabilized by the non-ionic amphiphilic block copolymer $E_{122}P_{56}E_{122}$ (see Fig. 11.2) is shown in Fig. 11.7a. The experiments were carried out at a polymer concentration of $c=10^{-5}$ M and an NaCl content of 10^{-4} M (CBF) and 10^{-1} M (BF), respectively. The type of foam film formed under the given experimental conditions was determined from Π–h curves of the same solutions [52]. As can be seen in Fig. 11.7a, the drainage under gravity ($\Delta P=0$) is very slow and hardly dependent on the type of foam film during the measuring time of about 700 s. However, at an applied pressure drop of $\Delta P=5000$ Pa significant differences are observed. The drainage of the foam consisting of BFs (called *BF foam* in the following) is much faster than that of the foam consisting of CBFs (called *CBF foam* in the following). Moreover, the water content which is finally reached is one order of magnitude lower in the BF foam. As already mentioned above, in a dry foam the liquid is mainly to be found in the plateau borders. Hence the water content of the BF foam is lower not because BFs are thinner than CBFs but because the plateau borders contain less liquid. The same general trend, namely a faster drainage and a lower final water content of BF foams, has been observed for foams stabilized by a 7×10^{-5} M $E_{27}P_{39}E_{27}$ solution [5]. As was the case for $E_{122}P_{56}E_{122}$, 10^{-4} M and 10^{-1} M NaCl were added to adjust the type of foam film, which was confirmed by the respective Π–h curves [52]. The main difference between these two polymers is their ability to stabilize foams and foam films [52]: The foams generated from $E_{27}P_{39}E_{27}$ solutions were so unstable that the drainage experiments could not be carried out at a pressure drop of $\Delta P=5000$ Pa. Hence the data reported in [5] for $E_{27}P_{39}E_{27}$ were measured at $\Delta P=1500$ Pa. We will return to the different stabilities below.

In order to understand the effect that the foam film type has on the drainage, it is important to realize that the thin films (CBF, NBF or BF) do form immediately after applying the pressure. Thus at the beginning of the drainage, foam films which are orders of magnitudes thicker than a CBF or a BF, coexist with the respective thin films. During the drainage, more and more CBFs and BFs are formed until finally the whole foam consists of one type of foam film. Speculative as it may be, a different hydrodynamic behavior of these films could be the reason for the difference in the drainage curves seen in Fig. 11.7a. This might not be the only reason, which becomes clear if we look at the results for foams stabilized by SDS. (Note that respective data for non-ionic LMW surfactants are not available.) In contrast to the results obtained for the two polymers discussed above, faster drainage was observed for CBF foams (i.e. foam formed from solutions of low electrolyte concentration) [67]. The explanation given by the authors is a different contact angle between the plateau borders and the thin liquid film. The smaller contact angle in the CBF results in a larger plateau region and thus in a lower resistance with regard to the draining liquid. A question that still need to be answered is why we observe opposite behavior for foams stabilized by SDS and E–P–E polymers regarding the effect of added salt on the drainage behavior. It is very likely that additional mechanisms need to be taken into account. For example, the addition of electrolyte not only screens the electrostatic forces but also changes the aqueous solubility of the amphiphilic block copolymers [53, 72–74]. This change in solubility mainly depends on the anion of the salt [53, 72]. For example, NaCl decreases the solubility of E–P–E block copolymers, which leads to a decrease in the CMC [53, 72–74]. Consequently, the number of aggregates might increase on addition of salt. In this respect, addition of NaCl has the same effect as an increase in temperature [48, 75]. Moreover, the adsorption (rate and amount) of block copolymers is also influenced by salt [72]. How these "salt-induced" changes affect the drainage rate of foams and foam films has not yet been studied in detail.

11.3.2.2 **Foam Stability**

In Fig. 11.7, the lifetimes τ_p of aqueous foams stabilized by the polymeric surfactants $E_{122}P_{56}E_{122}$ and $E_{27}P_{39}E_{27}$ are shown at different applied pressure drops ΔP. In addition, the critical capillary pressures $P_{cr,foam}$ at which the foam is destroyed very fast ("avalanche-like") are indicated. We will return to this value below. As was the case for the drainage studies, the experiments were carried out at an $E_{122}P_{56}E_{122}$ concentration of $c=10^{-5}$ M and an NaCl content of 10^{-4} M (CBF) and 10^{-1} M (BF), respectively. For $E_{27}P_{39}E_{27}$ a polymer concentration of 7×10^{-5} M and a salt content of 10^{-1} M were chosen, which led to the formation of a BF foam. Note that results obtained with the FPDT are usually presented in terms of τ_p–ΔP curves. However, we will speak of ΔP–τ_p curves in the following in order to compare the data with the respective Π–h curves of foam films. In this context, it is important to remember that the thickness h of a single foam film stays constant when the capillary pressure P_c in the plateau borders equals the disjoining pressure Π between the two interfaces. In other words, what is the disjoining pressure Π in single films is the pressure drop ΔP in foams. Hence we are able to investigate and compare films and foams at equal capillary pressure.

As can be seen in Fig. 11.7b, the lifetime τ_p of the $E_{122}P_{56}E_{122}$ foams decreases significantly with increasing pressure drop ΔP. Moreover, the type of film the foam consists of has a significant influence on the lifetime as was the case for the drainage process. At low ΔP values it is the CBF foam that is more stable, whereas it is the BF foam at high ΔP values. The "crossover" in stability appears around 10^4 Pa for this particular system. Another point worth mentioning is that the lifetime reaches a plateau value for both the BF and the CBF foams. The plateau of the latter, however, is located at shorter lifetimes. Let us try to understand these differences by looking at the Π–h curves of the corresponding single foam films (see Fig. 3 in [52]). The higher stability of the CBF foam at low ΔP can be explained by the fact that its films are thicker, which, in turn, reduces the probability of film rupture. The observation that the lifetimes reach a plateau value can also be explained with the Π–h curves of the respective foam films. At high pressures both systems form a BF the thickness of which changes only slightly compared with that of a CBF. It is known from

SDS that the lifetime of NBF foams does not depend on the pressure drop (the initial steep decrease of the τ_p–ΔP curve seen in Fig. 2 of [63] is mainly due to drainage which is not completed at these low ΔP). The same pressure insensitivity is seen with respect to the thickness of a single NBF. Hence if the film thickness does not change, no change is observed in the stability of either the foam or the film. The high stability of both the NBF and the NBF foam can be explained with the molecular structure of the NBF: The NBF consists of a densely packed surfactant bilayer which resembles a liquid crystalline phase rather than a liquid phase. The fact that the same overall trend, namely a constant foam stability at constant film thickness, is also observed for the BF foams is most likely due to the similar structures of NBFs and BFs, namely the rather densely packed adsorbed surface layers that are capable to stabilize foam films by steric repulsion acting normal to the film surface.

However, what cannot be explained yet are the different $P_{cr,foam}$ values and the different plateau values for τ_p that were observed for high and low electrolyte concentrations, respectively. The Π–h curves (Fig. 3 in [52]) indicate that at these high pressures BFs of equal thickness are formed in both systems. [The type of foam film and its thickness depend not only on the composition (surfactant and/or salt concentration) but also on the applied pressure. Most of the foams studied with the FPDT so far consisted of one type of foam film over the whole experimental pressure range. However, in the case of block copolymers, a continuous transition from a CBF to a BF takes place with increasing pressure, as was discussed in connection with Fig. 11.5. Hence we have to distinguish between pressures under which the films are CBFs and BFs, respectively, which was ignored in the past. The pressure at which the steric forces start to dominate over the electrostatic forces, i.e. at which the CBF becomes a BF, can be determined by fitting the respective Π–h curves as is seen in Figs. 11.3 c and 11.5.] Thus with regard to the film thickness the systems no longer differ and equal $P_{cr,foam}$ and equal τ_p would have been expected if the thickness were the relevant parameter. It is important to mention that it is not only $P_{cr,foam}$ but also the critical capillary pressure for film rupture $P_{cr,film}$ that increases with increasing electrolyte concentration [52]. Moreover, the P_{cr} values of the foams and films nearly match quantitatively. For the foams, $P_{cr,foam}$ values of 9×10^4 and 4×10^4 Pa are reported (see Fig. 11.7 b), whereas $P_{cr,film} \geq 10^5$ Pa and $P_{cr,film} = (4–7) \times 10^4$ Pa are found for the respective foam films [52]. Although the observation of different critical pressures and lifetimes at equal film thicknesses still needs to be explained, the fact that $P_{cr,foam} \sim P_{cr,film}$ demonstrates the high potential of the new concept, which directly compares ΔP–τ_p and Π–h curves.

A last point worth mentioning in connection with Fig. 11.7 is the difference between BF foams stabilized by $E_{122}P_{56}E_{122}$ and $E_{27}P_{39}E_{27}$, respectively. Foams stabilized by the latter are stable only up to 3000 Pa. Moreover, the lifetime is not long enough to separate drainage from other foam destroying processes such as film rupture and Ostwald ripening, as was discussed above. This example demonstrates the dilemma we are faced with if the foams are not very stable, i.e. the limitations of the *foam pressure drop technique*. ΔP has to be high enough to overcompensate the hydrostatic pressure ρgH and to induce a drainage which is fast compared to the lifetime – these conditions are not fulfilled in the case of $E_{27}P_{39}E_{27}$. However, if these conditions are met we obtain reproducible lifetimes with which the stability of well-drained foams can be described quantitatively. Comparing the $\Delta P \tau_p$ curves of $E_{122}P_{56}E_{122}$ and $E_{27}P_{39}E_{27}$ with the corresponding Π–h curves of the foam films, one sees that the different foam stabilities are reflected in the film stabilities. Whereas the $E_{122}P_{56}E_{122}$ BFs rupture at pressures $\geq 10^5$ Pa, rupture of $E_{27}P_{39}E_{27}$ BFs was observed at 3000 Pa, which is in agreement with the observations made for the corresponding foams. What is still unclear, however, is the reason for the different stabilities. In this context, the surface elasticity is often mentioned as a parameter that determines the film and thus the foam stability. We return to this point in Section 11.4.

In conclusion, the processes leading to the collapse of the foam can – at least partly – be separated with the FPDT. Applying a very high pressure drop, ΔP reduces the drainage time significantly so that film rupture and Ostwald ripening can be neglected during the short drainage time in a sufficiently stable foam. In other words, it is indeed the pure drainage process that we are looking at. On the other hand, after drainage is completed, we can study the destruction of a well-drained foam which should be due mainly to film rupture and Ostwald ripening. This is of significant importance because in traditional experimental setups such as the Ross-Miles foam test drainage, film rupture and Ostwald ripening are always superposed, which makes a quantitative analysis of the single processes difficult, if not impossible. Another important result is the fact that the type of film the foam consists of, i.e. CBF or BF in the case of amphiphilic block copolymers, has a significant influence on its drainage and lifetime, an observation that has been widely ignored. As single foam films can be studied with the TFPB, it is beyond any doubt that a combination of the TFPB and the FPDT is a very promising way to find quantitative correlations between films and foams. The main drawback of the FPDT is the fact that it is restricted to the study of stable foams. Note that foams stabilized by non-ionic LMW surfactants are not usually stable enough to resist the applied pressure drops, which explains why no FPDT data are available for typical LMW surfactants. With the results obtained so far, many questions can be answered. However, just as many questions remain open. Why do we observe different lifetimes for $E_{27}P_{39}E_{27}$ foams at different electrolyte concentrations although the film thicknesses are equal under the experimental conditions? Are $P_{cr,foam}$ and $P_{cr,film}$ the appropriate parameters to correlate quantitatively the stability of foams and foam films? How can we quantify the influence that the molecular structure has on both the film and the foam stability? How can we describe theoretically the influence that the type of foam film has on the properties of the respective foam?

11.3.3
Drainage and Stability of Foams Under Gravity

The processes of drainage, Ostwald ripening and film rupture are superposed in foams subjected to gravity. All three processes lead to a decrease in the foam volume, i.e. to foam destruction. Let us compare two systems whose foam stabilities are different. If we only measure the decrease in the foam volume as a function of time, we cannot specify which of the three processes is responsible for the different stabilities. Disregarding Ostwald ripening, one can still distinguish three cases that would explain why the evolution of the foam volume with time is different for two systems: (a) the drainage of the two systems is the same, whereas the film stabilities are different; (b) the drainage of the two systems is different, whereas the film stabilities are the same; (c) both drainage and film stabilities are different. However, there is an indirect way to distinguish between drainage and film rupture, namely to measure simultaneously the liquid volume of the foam (V_{liquid}) and the total foam volume (V_{foam}). In Fig. 11.8a the evolution of V_{liquid} and V_{foam} over a period of 1 h is shown for two different amphiphilic block copolymers, namely $E_{103}P_{40}E_{103}$ and $E_{26}P_{39}E_{26}$ [76]. The data were measured with the FoamScan with which a predetermined volume of foam (150 mL) was generated by sparging of air (rate 30 mL min^{-1}) through the aqueous polymer solution. The polymer concentrations were equal to half the CMC and the solutions contained 10^{-2} M NaCl. Note that very similar results were obtained for these block copolymers at a polymer concentration of 1.6 mM [76].

As can be seen in Fig. 11.8a, the generation of 150 mL of foam took about 220 s for both polymers so that their foamabilities can be regarded as equal. Once generated, the foam volume decreases owing to the processes mentioned above. In the following we will neglect Ostwald ripening and discuss the foam destruction in terms of drainage and film rupture only. What we learn from Fig. 11.8a is that there are indeed two processes that we can distinguish because their time-scales are different. Up to $t \approx 800$ s, i.e. ~ 580 s after the foam generation is completed, V_{foam} changes only slightly and is nearly equal for the two different block copolymers. On the other hand, V_{liquid} changes enormously for both systems. In this region the foam decay is clearly dominated by the drainage of bulk solution and the drainage process seems to be very similar for the two polymers. However, at $t > 800$ s, V_{foam} decreases significantly, whereas V_{liquid} (which is already very low) changes only slightly. Hence changes in the liquid content are not expected to affect the total foam volume significantly. In other words, at $t > 800$ s the foam decay can only be caused by film rupture because drainage no longer plays a role. The films we are talking about are supposed to be BFs as electrostatic repulsion across block copolymer foam films are screened in the presence of 10^{-2} M NaCl. This example demonstrates that it is also under gravity that a distinction between drainage and film rupture can be made as was the case for the FPDT (see Section 11.3.2). However, the slower the drainage under gravity, the greater is the superposition of different foam destruction processes. The main problem is the fact that the drainage rate under gravity is given by the solution properties (and gravity),

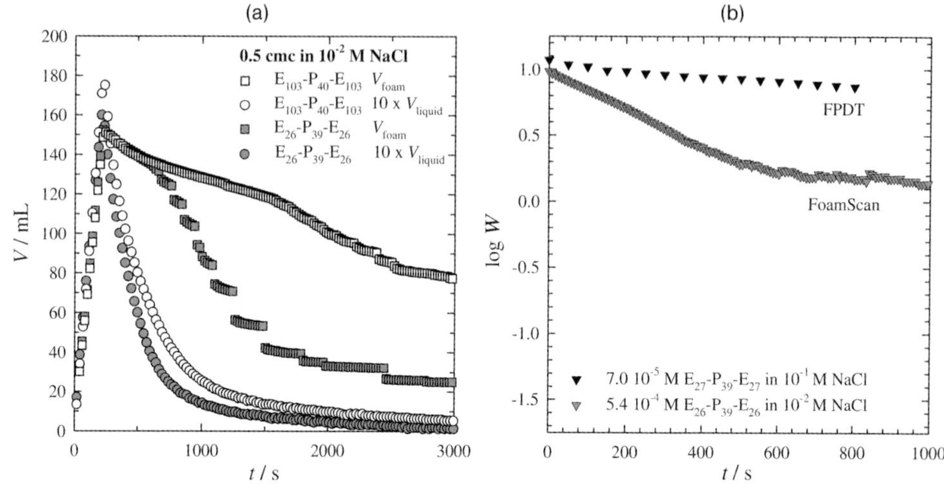

Fig. 11.8 (a) Total foam volume V_{foam} (squares) and liquid volume of the foam V_{liquid} (circles) as a function of time t for foams generated from aqueous solutions of $E_{103}P_{40}E_{103}$ and $E_{26}P_{39}E_{26}$, respectively. The solution concentrations were 4.4×10^{-4} M $E_{103}P_{40}E_{103}$ and 5.4×10^{-4} M $E_{26}P_{39}E_{26}$, respectively, corresponding to half of the CMC of the polymers. 10^{-2} M NaCl was added to the solutions to obtain BFs. Note that V_{foam} and V_{liquid} are shown both during and after foam generation. The data were taken from [76] and rearranged for this chapter. (b) Relative water content W as a function of time t for foams stabilized by aqueous solutions of $E_{27}P_{39}E_{27}$ and $E_{26}P_{39}E_{26}$. To both solutions salt was added to obtain BFs. The $E_{27}P_{39}E_{27}$ solution was studied with the FPDT and the $E_{26}P_{39}E_{26}$ solution with the FoamScan. Note that the water content obtained with the FPDT was measured under gravity, i.e. at $\Delta P=0$ Pa. Data for $E_{27}P_{39}E_{27}$ are taken from [5]. The $\log W$ data for $E_{26}P_{39}E_{26}$ were calculated from the data shown in (a). Note that the block copolymers $E_{27}P_{39}E_{27}$ and $E_{26}P_{39}E_{26}$ are to be regarded as the same polymer despite the small difference in the E block length.

whereas it can be controlled via the applied pressure ΔP in the case of the FPDT. (Note that film rupturing processes at a certain applied pressure drop ΔP and under gravity are expected to be different. It is very likely that ΔP enhances not only the drainage but also the rupturing of films, which makes it difficult to compare quantitatively the FoamScan with the FPDT results.)

As already mentioned, the foamability and the drainage of the two polymers do not differ very much. However, clear differences between the two polymers are found once the foam has drained, i.e. at $t > 800$ s. The $E_{26}P_{39}E_{26}$ foam collapses considerably faster than the $E_{103}P_{40}E_{103}$ foam. If we argue that it is the film rupture that determines the foam destruction in this region, we can conclude that films stabilized by $E_{103}P_{40}E_{103}$ are more stable than those of $E_{26}P_{39}E_{26}$. This is in agreement with the observations made for the free-standing foam films, as was discussed in connection with Fig. 11.5. In the absence of electrostatic repulsion, the following rule of thumb holds true: "The larger the block copolymer, the thicker is the BF; and the thicker the BF, the more stable are the film and the foam." Hence we can correlate the thickness of single BFs with the stability of foams. Note that $E_{103}P_{40}E_{103}$ and $E_{26}P_{39}E_{26}$ differ only with regard to the length of the PEO chains, which demonstrates very well the influence they have on the foam stability.

Further, it is worth mentioning that the total foam volume decreased stepwise at $t>800$ s (apparently from the top of the column), which was very pronounced for the smaller of the two polymers, $E_{26}P_{39}E_{26}$. The stepwise decrease of V_{foam} is most probably due to collective effects such as collective film rupture. This observation gives further evidence for the view that foam stability is controlled via the stability of the single foam films. Note that Ostwald ripening adds to the overall properties since it occurs continuously and cannot be separated from the coalescence process, i.e. film rupturing, as was the case for the FPDT.

With the data published so far, we are able to compare – at least qualitatively – the log W data obtained with the FPDT under gravity (i.e. at $\Delta P=0$) and those obtained with the FoamScan. What we will focus on is the log W data of $E_{27}P_{39}E_{27}$ and $E_{26}P_{39}E_{26}$. The former data are published in [5], whereas the latter were calculated for this chapter according to W (vol.%) = 100 (V_{liquid}/V_{foam}) from the data shown in Fig. 11.8 a. {Note that the results obtained for $E_{27}P_{39}E_{27}$ with the FPDT are qualitatively the same as those obtained for $E_{122}P_{56}E_{122}$ (see Fig. 11.7 a). The same general trend, namely a faster drainage and a lower final water content of BF foams compared with CBF foams, has been observed for foams stabilized by $E_{27}P_{39}E_{27}$ [5]. Moreover, under gravity ($\Delta P=0$) the water content decreased from 10% to ~ 8% during the measuring time of 800 s, which is similar to the decrease from 10 to 7% observed for $E_{122}P_{56}E_{122}$. The main difference between these two polymers is their ability to stabilize foams and foam films [52], which was discussed in Section 11.3.2.} The respective log W–t curves are presented in Fig. 11.8 b. Note that the FPDT measures the water content only after the foam is generated, whereas the FoamScan starts to collect data when foam generation is started. This difference is taken into account in Fig. 11.8 b, where the time-scale of the FoamScan data is shifted such that at $t=0$ s foam generation is finished. The most striking difference between the FPDT and the FoamScan results is the drainage rate. In the first case the water content is reduced from ~ 12% (log $W=1.1$) to ~ 7% (log $W=0.82$) during the first 800 s, whereas in the latter case a reduction from 10% (log $W=1.0$) to ~ 2% (log $W=0.25$) is observed. There are three possible reasons for this difference. First, equal drainage behavior is expected only for foams whose properties are very similar. For example, comparable bubble sizes and comparable amounts of liquid at the beginning of the drainage process are important. Whereas the latter happens to be the same (W is ~ 10% for both systems), we do not have information about the former. Parameters which influence the bubble size are the porosity of the filter through which air is sparged and the gas flow rate during foam generation. Second, the total amount of foam investigated with the FoamScan and the FPDT is different. The total foam volume in the FoamScan study (150 mL) was approximately twice the foam volume that was studied with the FPDT. As was discussed in Section 11.3.1, the capillary pressure P_c is a function of the foam column height H, i.e. $\Delta P_c=\rho g H$. Hence the larger foam volume in the FoamScan study results in a higher P_c, which, in turn, leads to faster drainage compared with the drainage measured with the FPDT. Further, the compositions of the solutions are different regarding polymer and electrolyte concentration (and the polymers are not the same although their overall composition is nearly equal). Hence in order to compare FPDT and FoamScan data quantitatively, we need to study exactly the same systems under experimental conditions which are as similar as possible. It is not until we obtain the same data with the two techniques under gravity that we can proceed and compare data obtained under gravity with those obtained under reduced pressure.

In conclusion, measuring simultaneously V_{liquid} and V_{foam} is the most convincing way to study foams under gravity. If drainage happens to be very fast, these measurements allow us to distinguish between drainage and film rupture, the two major foam destruction processes if we neglect Ostwald ripening. So far only V_{liquid} and V_{foam} data for three different block copolymers are available [76]; corresponding data for non-ionic LMW surfactants have not yet been published. (For the sake of completeness, we mention a FoamScan study carried out with foams stabilized by the non-ionic LMW surfactant $C_{12}E_6$ [77]. However, the focus of that study was on the question of how the drainage depends on the surfactant concentration; what is more, the study provided no data evaluation of the kind presented in Fig. 11.8.) Despite the lack of data, we believe that simultaneous measurements of V_{liquid} and V_{foam} in combination with the FPDT will considerably enhance our knowledge about foams. To achieve this aim, it is imperative to agree on a standard procedure when it comes to the evaluation of experimental data.

11.4 Surface Rheology of Surfactant Monolayers

11.4.1 Surface Rheology and Film Stability

"Equal surface forces do not automatically result in equal foam film stabilities". This statement cannot be called into question because it has been found experimentally for both LMW ionic [10] and LMW non-ionic surfactants [11–13]. An example will be given in Section 11.4.2. Furthermore, long-range steric repulsion between amphiphilic polymers does not guarantee the formation of stable foam films [14], which will be discussed in Section 11.4.3. Thus the stability of thin foam films cannot be explained solely by the magnitude of the repulsive interactions operating normal to the film surfaces. What we need is a surface which is able to dampen external disturbances and thus prevents the film from rupturing. This ability is believed to be mirrored in the surface dilational elasticity of the monolayer [10, 12, 15–17, 43]. (In the area of surface rheology, the response to two types of surface deformations are generally distinguished, namely the change of the interfacial area by expansion and compression and the change of the surface shape at a constant area, which are treated in terms of dilational rheology and shear rheology, respectively. What we will focus on in this chapter is the surface dilational elasticity and its relationship to foam and foam film stability.) In contrast to the surface forces which operate normal to the surface, the surface dilational elasticity is related to processes which operate tangential to the surface. However, the exact relationship between surface dilational elasticity and film stability is still unclear and certainly there is no direct proportionality between the two. Moreover, experimental data are rare. The dilemma we are faced with is that of the systems investigated so far either the surface rheology or the film properties have been investigated in detail.

The surface elasticity ε and viscosity η are the real and the imaginary parts, respectively, of the surface dilational modulus $\bar{\varepsilon}$, which describes the linear response of a surfactant monolayer to a sinusoidal deformation of frequency ν, i.e.

$$\bar{\varepsilon} = \varepsilon + i\omega\eta \qquad (6)$$

with $\omega = 2\pi\nu$. In simplified terms, the surface elasticity ε reflects the ability of the monolayer to adjust its surface tension in an instant of stress, whereas the surface viscosity η is a measure for the speed of relaxation processes which restore the equilibrium after the disturbance. The surface viscosity represents the energy loss due to relaxation processes that might origin from the exchange of molecules between the bulk and the surface phase or from rearrangements of molecular segments within a surface region. What is important is the fact that the surface rheological parameters depend on both the frequency of disturbance ν and the surfactant concentration c. We will return to this in connection with Fig. 11.9. Equations for $\varepsilon(\nu,c)$ and $\eta(\nu,c)$ were first derived by Lucassen and van den Tempel [78]. According to their model, there are no adsorption/desorption barriers at the interface and exchanges between bulk and surface are controlled by simple diffusion. With these assumptions, one obtains

$$\varepsilon(\nu, c) = \varepsilon_0 \frac{1 + \xi}{1 + 2\xi + 2\xi^2} \qquad (7)$$

and

$$\eta(\nu, c) = \frac{\varepsilon_0}{2\pi\nu} \frac{\xi}{1 + 2\xi + 2\xi^2} \qquad (8)$$

with

$$\xi = \sqrt{\frac{\omega_0}{4\pi\nu}} \qquad (9)$$

Hence the parameters ε_0 and ω_0 determine the $\varepsilon(\nu,c)$ and the $\eta(\nu,c)$ curves. The former is the high-frequency limit of the surface elasticity [according to Eq. (7), one obtains $\varepsilon = \varepsilon_0$ for $\nu \to \infty$] and the latter is the molecular exchange parameter

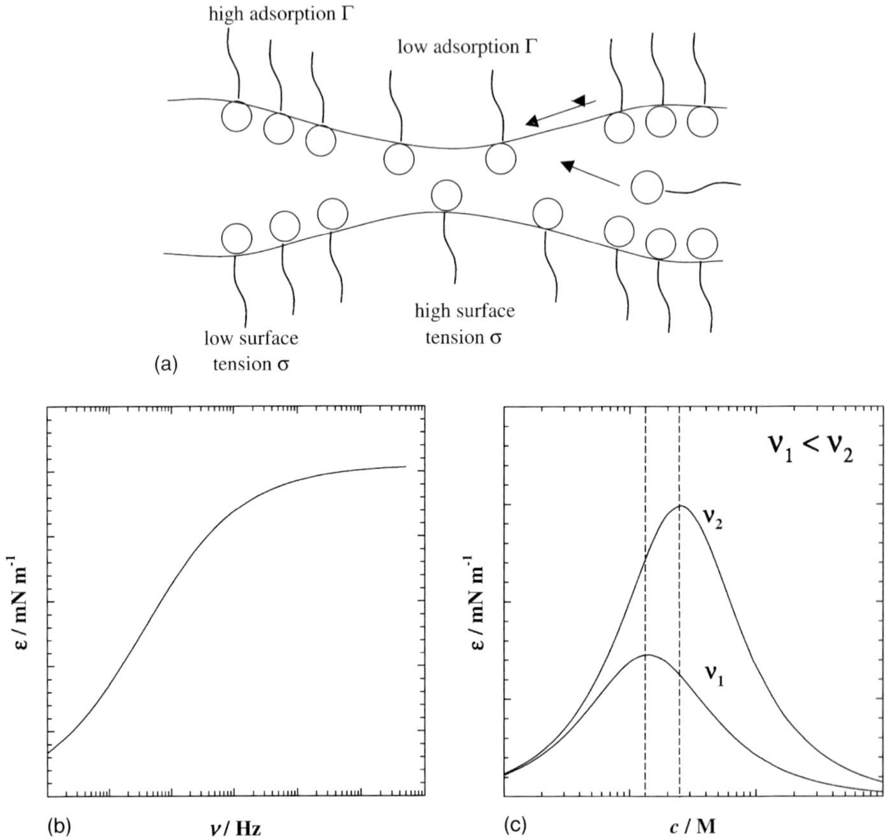

Fig. 11.9 (a) Schematic diagram of a thin foam film. Spatial fluctuations lead to changes in the surfactant's surface concentration, which, in turn, results in local surface tension differences. These differences can be evened out by surfactant molecules diffusing along the surface and/or from the bulk to the surface. (b, c) Schematic graphs of the dilational surface elasticities ε as a function of the frequency ν (b) and the surfactant concentration c (c).

ω_0. Note that both ε_0 and ω_0 are functions of c but independent of ν. In principle, it is possible to calculate ε_0 and ω_0 and thus the frequency and concentration dependencies of ε and η, respectively, from surface tension isotherms. However, if the respective equation of state is not known (which is often the case), ε_0 and ω_0 can only be treated as fitting parameters. Further details can be found in [19, 79–81] and references cited therein.

If not quantitatively, we can at least explain qualitatively that correlations between surface rheology and film stability are expected. For that purpose we have to recall the mechanism leading to film rupture. Rupture of thin liquid films is always preceded by local thinning during which part of the film's surface area is extended (see Fig. 11.9 a). This leads to surface tension differences and thus to a surface dilational modulus $\bar{\varepsilon}$, which tends to restore the original shape of the film. It holds that

$$\bar{\varepsilon} = \frac{\mathrm{d}\sigma}{\mathrm{d}\ln A} \tag{10}$$

where σ is the surface tension and A the surface area ([82] and references cited therein). Note that Eq. (10) coincides with half of the Gibbs elasticity, i.e. $2\bar{\varepsilon} = E_{\mathrm{Gibbs}}$ [81] (in Gibbs' original work, the elasticity of the whole film, i.e. of two film surfaces and not that of a single monolayer, was considered, which is mirrored in $E_{\mathrm{Gibbs}} = 2\bar{\varepsilon}$ [83]). Let us consider the real part of the modulus $\bar{\varepsilon}$, i.e. the surface elasticity ε, of LMW surfactants, in more detail. As the monolayer is directly connected with the bulk phase, the surface elasticity depends strongly on adsorption and desorption processes. Compressing (expanding) the monolayer leads to desorption (adsorption) of the surfactant molecules into (from) the bulk to restore the equilibrium surface concentration. Two extreme cases are easy to understand: when the frequency of the compression is low, the monolayer has time to reach equilibrium and there is no resistance to the compression ($\varepsilon = 0$ for $\nu \rightarrow 0$). When the frequency is high, the monolayer has no time to respond and behaves as if it were insoluble ($\varepsilon = \varepsilon_0$ for $\nu \rightarrow \infty$). Thus ε increases with increasing ν until a plateau value is reached, which is shown schematically in Fig. 11.9b. The situation is even more complex because the time to reach equilibrium depends strongly on the surfactant concentration. The higher the surfactant concentration, the faster is the molecular exchange between bulk and surface. It is because of this fast exchange that at high concentrations any surface tension difference $\mathrm{d}\sigma$ is evened out immediately, which results in $\varepsilon = 0$. The experimental observation related to this exchange is a maximum in the ε–c curve which shifts towards higher concentrations with increasing frequency. The interrelation between frequency and concentration is illustrated by the two vertical lines drawn in Fig. 11.9c. These two lines determine a concentration range in which the elasticity decreases with increasing c at ν_1, whereas ε increases with c at the higher frequency ν_2.

The main difference between LMW surfactants and amphiphilic block copolymers with regard to surface rheological properties is their diffusion–relaxation time. This time is usually very short in the case of LMW surfactants and mainly given by the diffusion-controlled exchange between bulk and surface. In the case of amphiphilic block copolymers, however, any diffusional exchange with the bulk generally occurs on time-scales that are much longer than those used in surface rheological experiments. Hence it is not the diffusion to the air/water interface but the rearrangement of polymer segments within the surface layer that is the main relaxation mechanism. The surface rheological behavior differs accordingly.

The surface rheology of non-ionic homopolymers is reviewed in [84] and will not be discussed in the following. What we will focus on are the few studies that deal with the surface rheology of non-ionic amphiphilic block copolymers

[57, 85–87]. Munoz et al. [85] measured the surface elasticity ε of $E_{76}P_{29}E_{76}$ at frequencies ranging from 20 Hz to 2 kHz (see Fig. 6 in [85]) with the capillary wave technique. It was found that ε is independent of frequency at extremely low concentrations, i.e. at $c \leq 4 \times 10^{-9}$ M. However, at $c \geq 10^{-8}$ M, the elasticity was found to pass through a maximum with increasing frequency, which clearly differs from the ε–ν curve for LMW surfactants shown in Fig. 11.9b. It was concluded that the maximum of the ε–ν curve is not in agreement with the Lucassen and van den Tempel model, which means that it is not due to adsorption/desorption processes. That diffusional adsorption/desorption processes are usually unimportant for the surface rheology of polymers was also concluded from studies with other E–P–E block copolymers. Plots of the surface dilational modulus $\bar{\varepsilon}$ as a function of the surface pressure π measured at different frequencies between 0.1 and 800 Hz (capillary wave [85], oscillating bubble [86], Langmuir trough [87] and oscillating ring trough [57]) show remarkable similarity, at least up to about $\pi = 10$ mN m^{-1}. We will return to this point in Section 11.4.3. A possible explanation for the maximum of the ε–ν curve observed by Munoz et al. might be the incomplete hydrodynamic description that is needed to transfer capillary wave data to dilational parameters [88]. Limitations of the hydrodynamic theory are also believed to account for negative values of the dilational surface viscosity, which has been observed by several groups [85, 89].

As was the case for the ε–ν curves, ε–c curves of amphiphilic block copolymers are also expected to be different from those of LMW surfactants. At low polymer concentrations, the surface elasticity increases with the polymer concentration (as is the case for LMW surfactants) simply because the amount adsorbed increases. However, assuming that the molecular exchange between the bulk and the surface is irrelevant, one expects no maximum in the respective ε–c curves. The main problem we are faced with is that plotting ε–c curves is not reasonable since the adsorption equilibrium is reached very slowly, i.e. the surface dilational modulus $\bar{\varepsilon}$ and hence the surface elasticity ε continuously change with time [57]. On the other hand, it has been found that the dilational parameters of a block copolymer monolayer are unique functions of the surface pressure π and the surface concentration Γ, respectively. Therefore, as will be discussed in Section 11.4.3, the most informative way of expressing the rheology of block copolymers is in terms of ε–π or ε–Γ curves.

11.4.2
Surface Rheology of Low Molecular Weight Surfactants

There have been numerous studies of the surface rheology of LMW surfactants. As most of these studies were performed at only one or two frequencies, they are not very informative and might lead to wrong conclusions, as will be discussed below. More information can be extracted from extensive studies where both the concentration and the frequency dependence of the surface rheological parameters were investigated [15, 16, 20, 90–97]. However, it is only in two studies where the surface rheology was investigated of the same solutions with which the respective foam films were stabilized [18, 98]. It is on the latter that we will focus.

An example of the independence of surface forces and film stability is given in Fig. 11.10 a, where two of the *Π–h* curves already presented in Fig. 11.4 are shown. Note that the stability we are referring to in this context is not the lifetime of the film but a measure for the maximum pressure that can be applied to the film. Moreover, the following discussion will be restricted to the stability of the CBF. In Fig. 11.10 a, the *Π–h* curves of 6.8×10^{-5} M β-$C_{12}G_2$- and 1.1×10^{-4} M $C_{10}E_4$-solutions are shown. With respect to the surface forces, no significant difference can be detected between the two *Π–h* curves; in both cases the surface charge density q_0 is calculated to be 1.17 mC m^{-2}. However, the corresponding foam film stabilities are far from being comparable. Whereas the β-$C_{12}G_2$ film is stable up to 9000 Pa, the $C_{10}E_4$ film already ruptures at pressures around 800 Pa.

Moreover, it was found that the stability of the CBF increases with increasing $C_{10}E_4$ concentration, whereas for β-$C_{12}G_2$ the reverse is observed (see Fig. 11.4). It is important to realize that the increase of the film stability in the case of $C_{10}E_4$ is accompanied by a decrease in the surface charge density (see Table 11.1), which, in turn, is expected to destabilize the CBF. To sum up, one can conclude that neither the low stability of the $C_{10}E_4$ film at $q_0=1.17$ mC m^{-2} nor the increasing stability of the $C_{10}E_4$ film with decreasing surface charge can be explained in terms of surface forces.

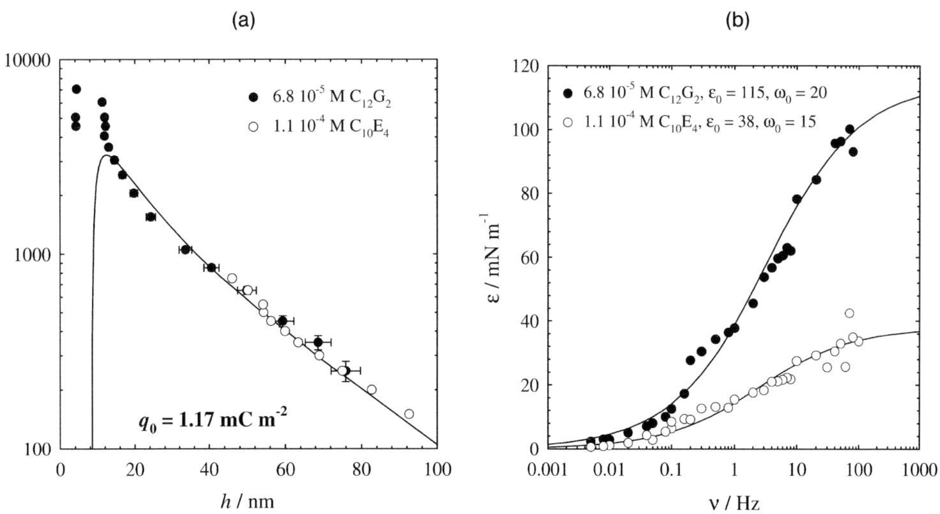

Fig. 11.10 (a) *Π–h* curves of thin foam films stabilized by aqueous solutions of β-$C_{12}G_2$ and $C_{10}E_4$. The solutions contain 10^{-4} M NaCl. The data are fitted with the DLVO theory from which the surface charge density is calculated to be $q_0=1.17$ mC m^{-2} (see Table 11.1). Data are the same as shown in Fig. 11.4. (b) Dilational surface elasticities ε of solutions containing 6.8×10^{-5} M β-$C_{12}G_2$ and 1.1×10^{-4} M $C_{10}E_4$ as a function of the frequency ν (data are taken from [98]).

To gain a deeper insight into a possible correlation between film stability and surface rheology, oscillating bubble studies were carried out [18, 98]. The dilational surface elasticities ε and viscosities η were measured as a function of the frequency ν and the surfactant concentration c for the two particular surfactants mentioned above, namely β-$C_{12}G_2$ and $C_{10}E_4$. The investigated frequency range was 0.005–100 Hz and the concentrations investigated were equal to those for which the respective Π–h curves have already been measured. In the following we will restrict ourselves to the influence the frequency has on the surface elasticities of the two solutions with which the Π–h curves seen in Fig. 11.10 a were measured. In Fig. 11.1 b the respective ε–ν curves for β-$C_{12}G_2$ and $C_{10}E_4$ are seen. The observation that ε increases continuously with increasing frequency is in agreement with Eq. (7). Moreover the experimental data can be described very well with the model of Lucassen and van den Tempel. Fitting the results according to Eqs. (7) and (8) with ε_0 and ω_0 being the fitting parameters one obtains $\varepsilon_0 = 114$ mN m^{-1} for β-$C_{12}G_2$ and $\varepsilon_0 = 36$ mN m^{-1} for $C_{10}E_4$. The corresponding ω_0 values are 22 and 12 s^{-1}. Further details are given in [98].

The results in Fig. 11.10 clearly demonstrate a correlation between foam film stability and surface elasticity, while the different film stabilities are not reflected in the surface forces. More precisely, the surface elasticity of β-$C_{12}G_2$ is larger than that of $C_{10}E_4$ over the whole frequency range investigated. Hence the higher stability of the β-$C_{12}G_2$ film is mirrored in the higher ε values. Anticipating that the calculated ε–ν curves describe correctly both the low- and high-frequency ranges, we can even argue that for this particular concentration the elasticity of β-$C_{12}G_2$ is higher than that of $C_{10}E_4$ at any frequency. This is a very important result because it is not yet clear which is the relevant frequency with respect to the stability of free-standing foam films. The second observation that we cannot explain with surface forces, namely that a decrease in the surface charge density is accompanied by an increase in film stability in the case of $C_{10}E_4$, can also be explained with surface elasticities: the higher the surfactant concentration c, the larger is the high-frequency limit of the surface elasticity ε_0 [98]. The results presented in [98] clearly illustrate that studying the surface elasticity at one frequency only instead of over a frequency range may lead to wrong conclusions. In the studied concentration range of 5.0×10^{-6}–2.5×10^{-4} M, the elasticity decreases with increasing c at low ν, while the opposite is observed at high ν. This change is simply due to the fact that at low ν the frequency of disturbance is slower than the frequency of the molecular exchange ω_0, whereas it is faster than ω_0 at high ν. In the former case the elasticity is mainly given by the molecular exchange: the higher the concentration, the faster is the exchange and the lower is the elasticity, as explained in Section 11.4.1. In the latter case the elasticity is determined by the surface concentration Γ: the higher the bulk concentration, the higher is Γ and the higher is the elasticity. From the ε–c curves in Fig. 11.9 c we can conclude that the studied concentration range lies between the two vertical lines, i.e. on the right hand side of the maximum at low ν and on the left-hand side of the maximum at high ν. A detailed discussion can be found in [18, 98].

So far, a correlation between the surface elasticity and the corresponding Π–h curves has only been discussed for the cationic alkyl trimethylammonium bromides (C_nTAB). For this particular homologous series, a strong increase in ε is observed when the chain length is increased from $n=12$ to 14 [15, 91]. A further increase of the chain length does not have any significant influence. Comparing these results with the fact that the respective films are only stable for $n>12$ [10], one can conclude that foam film stability and surface elasticity are directly correlated. However, the surface rheology study of the C_nTAB series was restricted to two different frequencies and the respective foam film stabilities were only investigated for one particular concentration. It is only a complete set of ε–c and ε–ν curves which will enable us to quantify the correlation between the surface elasticity and the film stability, i.e. the pressure threshold for film rupture. Moreover, these curves are needed to clarify whether or not thermally induced thickness and concentration fluctuations affect the DLVO forces. A promising concept to describe the correlation between fluctuations and DLVO forces was published by Bergeron [3, 10]. It is proposed that the energy needed to create these fluctuations could be a function of both the surface elasticity ε and the disjoining pressure Π. The lower the surface elasticity the higher the fluctuations and the higher the probability to exceed the energy barrier, i.e. Π_{max}. Therefore, the surface elasticity and the pressure threshold for film rupture are expected to be correlated, first experimental evidence of which is provided by the results presented above.

11.4.3
Surface Rheology of Amphiphilic Block Copolymers

The surface dilational rheology of the block copolymers $E_{99}P_{65}E_{99}$, $E_{103}P_{40}E_{103}$ and $E_{26}P_{39}E_{26}$ has been studied as a function of the bulk solution concentration at a fixed frequency of 0.13 Hz [57] with an oscillating ring trough method [99]. As already mentioned in Section 11.4.1, the surface rheology of E–P–E block copolymers has also been investigated with the capillary wave technique [85], the Langmuir trough (the surface dilational modulus can be extracted from the π–Γ isotherms) [87] and with the oscillating bubble technique [86]. In the following, the results obtained by Rippner Blomqvist et al. [57] will be discussed. It will be shown that the surface rheology of block copolymers in a selective solvent, i.e. in a solvent (water in this case) in which one of the blocks is more soluble (the E block) than the other (the P block), depends only on the surface concentration Γ of the polymer. As the polymer conformation at the interface changes with Γ, dilational rheology reflects structural changes within the surface layer that occur by rearrangements of the polymer segments. Note that neither the results in [85] nor those in [57] were compatible with relaxation processes due to diffusional bulk–surface exchange (only at very high bulk concentrations such effects cannot be excluded). Unfortunately, foam film studies of the systems whose surface rheology is known do not exist. However, in [76] the foam stability of foams stabilized by the same block copolymer solutions that were used in the surface rheology study [57] was investigated. The relationship between foam stability and the surface dilational moduli of $E_{103}P_{40}E_{103}$ and $E_{26}P_{39}E_{26}$ will be dealt with at the end of the section.

In [57] the surface dilational modulus $\bar{\varepsilon}$, the surface elasticity ε and the surface viscosity η were measured as functions of time (for 1 h) during adsorption at solution concentrations between 2×10^{-8} and 1.6×10^{-3} M. The layers were mainly elastic, i.e. the value of ε was much higher than that of η over the whole range of surface concentrations so that $\bar{\varepsilon}$ can be considered to be equal to ε in this case. The surface dilational moduli for $E_{103}P_{40}E_{103}$ and $E_{26}P_{39}E_{26}$ that were obtained after 1 h are shown as a function of the surface pressure π in Fig. 11.11 a. Note that in Fig. 11.11 $\bar{\varepsilon}$ is given, whereas Fig. 11.10 shows the real part of the modulus (namely the elasticity). As can be seen in Fig. 11.11, $\bar{\varepsilon}$ was found to be a function only of the actual surface pressure π, whereas it is independent of the bulk concentration c. This is one of the strongest experimental supports for the argument that any concentration-dependence of $\bar{\varepsilon}$ is not due to the exchange of molecules, but to the rearrangement of polymer segments. Further evidence for the concentration independence of $\bar{\varepsilon}$ was found by comparing the $\bar{\varepsilon}$–π curves of adsorbed layers (Fig. 11.11 a) with those of spread layers (Fig. 11.11 b). The spread layers were subjected to a continuous compression in a Langmuir trough and the $\bar{\varepsilon}$ values were calculated from the resulting Π–A curves [57]. The investigation of spread block copolymer layers for three polymers whose number of segments is known made it possible to relate structural transitions to the surface concentration Γ (mg m^{-2}) and the area per molecule A [57]. To conclude, one can say that the surface dilational modulus of block copolymers is a measure for the *resistance to compression* of the layer, i.e. for the repulsion between polymers or polymer segments.

Although knowing that $\bar{\varepsilon}$ is only a function of π, we still have to explain the shape of the $\bar{\varepsilon}$–π curves. As already mentioned, the values of $\bar{\varepsilon}$ depend on the conformation of the polymer chains at the interface [57, 85]. It is proposed in [57] that an almost flat conformation is adopted at π values of 0–5 mN m^{-1} where $\bar{\varepsilon}$ increases with π. At $\pi = 5$ mN m^{-1} the surface concentration is found to be 0.4 mg m^{-2} irrespective of the block copolymer size. Between $\pi = 5$ and 10 mN m^{-1}, where $\bar{\varepsilon}$ decreases with π, the more hydrophilic E segments of the polymer start to form loops and tails.

With further increase in the surface concentration Γ and thus of π, the E chains adopt a more and more extended conformation leading to increasing repulsive interactions, whereas most of the P segments still adopt a flat conformation at the surface ($\bar{\varepsilon}$ increases once again with π). Finally, at surface pressures beyond the second maximum of the $\bar{\varepsilon}$–π curve, $\bar{\varepsilon}$ decreases once more, which indicates that now some of the P segments start to form loops and tails [57]. Theoretical considerations [49] and experimental neutron reflectivity data [58] suggest that the more hydrophobic P segments are not only located in the top layer but also penetrate into the E–water sublayer. Furthermore, it was found that the simple brush model does not fully fit the neutron reflectivity data for some E–P–E block copolymers [100]. These results are consistent with the expla-

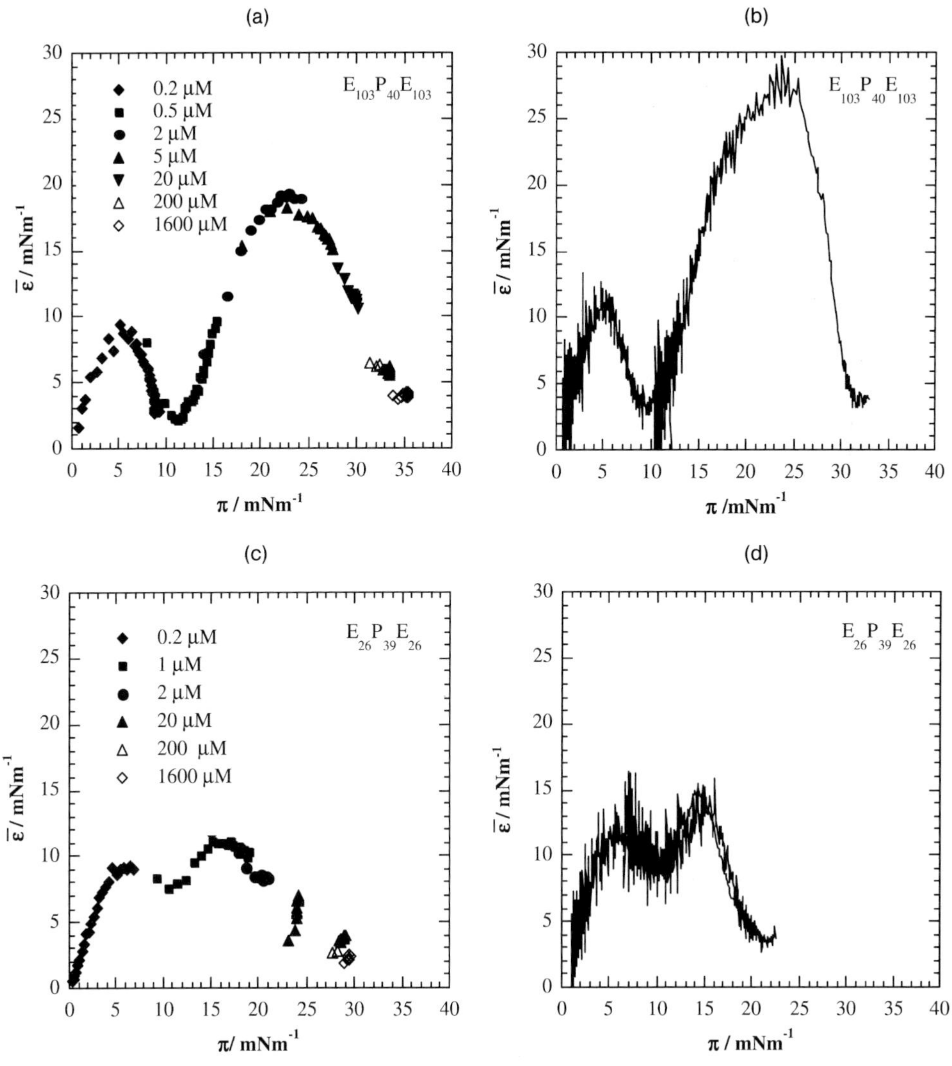

Fig. 11.11 Surface dilational moduli $\bar{\varepsilon}$ of the block copolymers $E_{103}P_{40}E_{103}$ (a, b) and $E_{26}P_{39}E_{26}$ (c, d) as functions of the surface pressure π of adsorbed layers (a, c) and spread layers (b, d). The adsorption layers were formed from solutions of various concentrations as indicated. The frequency of dilation was 0.13 Hz. The spread layers were continuously compressed in a Langmuir trough.

nation given previously by Munoz et al. [85] except for the behavior at the highest surface pressures. Based on ellipsometric measurements of the layer thickness, Munoz et al. argue that at the second maximum of the $\bar{\varepsilon}$–π curve the mushroom to brush transition of the E chains occurs, whereas the P segments remain in the top layer.

We will conclude this section by discussing the relationship between the surface rheology on the one hand and the stability of foam films and foams on the other. The rupture pressure of the $E_{21}B_8E_{21}$ foam films (Fig. 11.5 c) was found to be much lower than that of the diblock copolymers $E_{41}B_8$ and $E_{106}B_{16}$. Based on the results presented in Section 11.4.2, it is reasonable to assume that the high-frequency limit of the surface elasticity of $E_{21}B_8E_{21}$ surfaces is lower than that of $E_{41}B_8$ and $E_{106}B_{16}$ surfaces. Unfortunately, the surface rheology has not been studied either for these block copolymers or for other polymers for which Π–h curves of the respective foam films exist. Hence experimental evidence for a correlation between foam film stability (in the sense of rupture pressure) and surface rheology is not available. Let us try to answer the question of whether there is a correlation between surface rheology and foam stability. The surface dilational modulus $\bar{\varepsilon}$ of $E_{103}P_{40}E_{103}$ and $E_{26}P_{39}E_{26}$ shown in Fig. 11.11 and the foam data in Fig. 11.8 a were obtained with exactly the same block copolymer samples. Searching for a correlation between the stability of a three-dimensional foam and $\bar{\varepsilon}$, one faces several problems. First, the fact that the actual surface concentration of the foam bubbles is not known makes a quantitative comparison impossible. The polymer will start to adsorb as soon as each bubble emerges from the frit and adsorption may continue during the experiment (note that the foam properties presented in Fig. 11.8 were studied as a function of time). Second, adsorption from the confined environment of well-drained foams is not likely to be the same as adsorption from a bulk phase which is studied in surface rheology experiments. Third, the frequencies of the various disturbances the foam encounters, e.g. ambient vibrations, are unknown and impossible to measure. Hence we do not know the relevant frequency that we have to study in surface rheological experiments in order to obtain more information about the corresponding foam. Despite these problems, we will briefly compare the foam stability with the corresponding $\bar{\varepsilon}$ values for the two polymers $E_{103}P_{40}E_{103}$ and $E_{26}P_{39}E_{26}$. First, the time dependence of V_{foam} and $\bar{\varepsilon}$, respectively, is different. Whereas $\bar{\varepsilon}$ is constant for both polymers at concentrations around the CMC over a period of 1 h [57], V_{foam} decreases considerably, as can be seen in Fig. 11.8 a. Another observation worth mentioning is the fact that the lower stability of the $E_{26}P_{39}E_{26}$ foam is reflected in slightly lower $\bar{\varepsilon}$ values. The difference between the $\bar{\varepsilon}$ values of $E_{103}P_{40}E_{103}$ and $E_{26}P_{39}E_{26}$, however, is small (about 2 mN m^{-1}) and its relevance may be questioned (the foam stabilities and surface dilational moduli we are referring to were both measured for 1.6×10^{-3} M solutions of $E_{103}P_{40}E_{103}$ and $E_{26}P_{39}E_{26}$, respectively). On the other hand, the fact that the $E_{103}P_{40}E_{103}$ layer reaches a maximum $\bar{\varepsilon}$ of 19 mN m^{-1} whereas the maximum of $E_{26}P_{39}E_{26}$ is only 11 mN m^{-1} might be an indication of a correlation between $\bar{\varepsilon}$ values and foam stability. In order to prove this statement, foams need to be studied at the surface pressures where the maximum of $\bar{\varepsilon}$ was observed, which is not possible so far. Hence we have to conclude that a quantitative relationship between the surface dilational modulus and the foam stability of block copolymers has not yet been found.

11.5 Conclusions

This chapter is centered around disjoining pressure Π versus thickness h curves of foam films stabilized by non-ionic amphiphilic block copolymers and non-ionic low molecular weight (LMW) surfactants, respectively (Section 11.2). The respective Π–h curves are presented as a function of various parameters. From these curves, we obtain two important pieces of information. First, they allow us to quantify DLVO and non-DLVO contributions to the disjoining pressure. Second, we can clearly distinguish between LMW and polymeric surfactants via the range of the non-DLVO contribution, i.e. the steric contribution.

In Section 11.3, the drainage and stability of foams stabilized by non-ionic amphiphilic block copolymers are presented. Studies carried out under reduced pressure are compared with studies carried out under gravity. On the basis of the results presented in Section 11.2 the correlation between the stability of isolated foam films and foams is discussed. The most important result is the observation that "the larger the block copolymer the thicker is the BF (sterically stabilized bilayer film), and the thicker the BF, the more stable are the film and the foam, respectively". Unfortunately, corresponding data for LMW surfactants are not available.

The Π–h curves allow us not only to establish correlations between foam film and foam properties but also to deduce that equal repulsive interactions (normal to the film's surfaces) do not automatically result in equal foam film stabilities. As a large number of workers consider the surface elasticities ε of the corresponding monolayers to be a suitable parameter to describe the film stability, respective data are presented in Section 11.4 and are discussed in connection with the film stability. Indeed, a clear correlation between film stability and surface elasticity was found in the case of LMW surfactants: higher film stabilities are reflected in higher ε values. No such correlation has been found for amphiphilic polymers so far, which we believe is simply due to the lack of appropriate data.

In our opinion, the next two steps to take in order to achieve major breakthroughs are as follows:

- Although all data presented in this chapter are very important and informative, most of the time we were not able to compare directly the results obtained for foam films, foams and surfaces. What we need to agree on are model systems which are then studied under experimental conditions that are as similar as possible. We believe that a complete set of surface, film and foam data for both one non-ionic LMW surfactant and one non-ionic amphiphilic polymer will answer a number of open questions.

- Having a complete set of data for pure LMW and polymeric surfactants, we have to address surfactant mixtures as a first approach to technically relevant systems. For example, common laundry detergents are mixtures of different ionic and non-ionic surfactants. The problem we are faced with is that the existing knowledge about the influence different components have on foam properties is far from satisfactory so that it is often impossible to predict how changes that need to be made – for ecological, economic, cosmetic or commercial reasons – will influence the foaming properties of the product. In order to reach this goal, in fundamental studies of foam films and foams attention to technically relevant systems has to be paid. Not only will various surfactant mixtures have to be studied, but it is also the influence of additives (perfume oils, alcohols, salts) that needs to be investigated thoroughly to optimize products and processes where foam films and foams are involved.

Acknowledgments

C. S. is indebted to the DFG, the Fond der Chemischen Industrie and the European Commission (Marie Curie RTN SOCON, Contract No. MRTN-CT-2004-512331) for financial support. B.R. wants to thank the Swedish foundation for Strategic Research (the Colloid and Interface Technology Program) and the European Commission (Marie Curie Training Site fellowship at the Institute for Food Research in Norwich, Contract No. QLK1-CT-2000-60030) for financial support. We gratefully acknowledge the support of Dr. Rossen Sedev, who sent us the data for Fig. 11.5 (b).

References

1 Pugh, R., *Adv. Colloid Interface Sci.*, **1996**, *64*, 67.
2 Exerowa, D., Kruglyakov, P., *Foam and Foam Films – Theory, Experiment, Application*, Elsevier, Amsterdam, **1998**.
3 Bergeron, V., *J. Phys.: Condens. Matter*, **1999**, *11*, R215.
4 Stubenrauch, C., *Tenside Surf. Deterg.*, **2001**, *38*, 350.
5 Stubenrauch, C., Makievski, A., Khristov, K., Exerowa, D., Miller, R., *Tenside Surf. Deterg.*, **2003**, *40*, 196.
6 de Gennes, P., *Adv. Colloid Interface Sci.*, **1987**, *27*, 189.
7 Sedev, R., Exerowa, D., *Adv. Colloid Interface Sci.*, **1999**, *83*, 111.
8 Stubenrauch, C., v. Klitzing, R., *J. Phys. Condens. Matter*, **2003**, *15*, R1197.
9 Bergeron, V., Claesson, P., *Adv. Colloid Interface Sci.*, **2002**, *96*, 1.
10 Bergeron, V., *Langmuir*, **1997**, *13*, 3474.
11 Karraker, K., PhD Thesis, Berkeley, **1999**.
12 Stubenrauch, C., Schlarmann, J., Strey, R., *Phys. Chem. Chem. Phys.*, **2002**, *4*, 4504; **2003**, *5*, 2736 (erratum).
13 Schlarmann, J., Stubenrauch, C., *Tenside Surf. Deterg.*, **2003**, *40*, 190.
14 Rippner, B., Boschkova, K., Claesson, P., Arnebrant, T., *Langmuir*, **2002**, *18*, 5213.
15 Monroy, F., Giermanska-Khan, J., Langevin, D., *Colloids Surf. A*, **1998**, *143*, 251.
16 Fruhner, H., Wantke, K.-D., Lunkenheimer, K., *Colloids Surf. A*, **2000**, *162*, 193.
17 Langevin, D., *Adv. Colloid Interface Sci.*, **2000**, *88*, 209.
18 Stubenrauch, C., Miller, R., *J. Phys. Chem. B*, **2004**, *108*, 6412.
19 Bos, M., van Vliet, T., *Adv. Colloid Interface Sci.*, **2001**, *91*, 437.
20 Lucassen, J., in *Anionic Surfactants: Physical Chemistry of Surfactant Action*, Lucassen-Reynders, E. (Ed.), Marcel Dekker, New York, **1981**, p. 217.
21 Kruglyakov, P., in *Thin Liquid Films*, Ivanov, I. (Ed.), Marcel Dekker, New York, **1988**, 767–827.
22 Israelachvili, J., *Intermolecular and Surface Forces*, Academic Press, San Diego, 2nd edn, **1998**.

23 Derjaguin, B., Landau, L., *Acta Physicochim. USSR*, **1941**, *14*, 633.
24 Verwey, E., Overbeek, J., *Theory of the Stability of Lyophobic Colloids*, Elsevier, Amsterdam, **1948**.
25 Hamaker, H., *Physica*, **1937**, *4*, 1058.
26 Stubenrauch, C., Kashchiev, D., Strey, R., *J. Colloid Interface Sci.*, **2004**, *280*, 244.
27 Stubenrauch, C., Strey, R., *Langmuir*, **2004**, *20*, 5185.
28 Stubenrauch, C., *ChemPhysChem*, **2005**, *6*, 35.
29 Israelachvili, J., Wennerström, H., *J. Phys. Chem.*, **1992**, *96*, 520.
30 Sedev, R., Exerowa, D., Findenegg, G., *Colloid Polym. Sci.*, **2000**, *278*, 119.
31 Mysels, K., Jones, M., *Discuss. Faraday Soc.*, **1966**, *42*, 42.
32 Exerowa, D., Scheludko, A., *Chim. Phys.*, **1971**, *24*, 47.
33 Claesson, P., Ederth, T., Bergeron, V., Rutland, M., *Adv. Colloid Interface Sci.*, **1996**, *67*, 119.
34 Schlarmann, J., Stubenrauch, C., Strey, R., *Phys. Chem. Chem. Phys.*, **2003**, *5*, 184.
35 Stubenrauch, C., Rojas, O., Schlarmann, J., Claesson, P., *Langmuir*, **2004**, *20*, 4977.
36 Exerowa, D., *Kolloid.-Z.*, **1969**, *232*, 703.
37 Marinova, K., Alargova, R., Denkov, N., Velev, O., Petsev, D., Ivanov, I., Borwankar, R., *Langmuir*, **1996**, *12*, 2045.
38 Karraker, K., Radke, C., *Adv. Colloid Interface Sci.*, **2002**, *96*, 231.
39 Manev, E., Pugh, R., *Langmuir*, **1991**, *7*, 2253.
40 Waltermo, Å., Manev, E., Pugh, R., Claesson, P., *J. Dispers. Sci. Technol.*, **1994**, *15*, 273.
41 Bergeron, V., Waltermo, A., Claesson, P., *Langmuir*, **1996**, *12*, 1336.
42 Persson, C., Claesson P., Johansson, I., *Langmuir*, **2000**, *16*, 10227.
43 Persson, C., Claesson, P., Lunkenheimer, K., *J. Colloid Interface Sci.*, **2002**, *251*, 182.
44 Kolarov, T., Cohen, R., Exerowa, D., *Colloids Surf.*, **1989**, *42*, 49.
45 Yu, G.-E., Yang, Z., Ameri, M., Attwood, D., Collett, J.H., Price, C., Booth, C., *J. Phys. Chem. B*, **1997**, *101*, 4394.
46 Yang, Z., Pickard, S., Deng, N.-J., Barlow, R.J., Attwood, D., Booth, C., *Macromolecules*, **1994**, *27*, 2371.
47 Exerowa, D., Sedev, R., Ivanova, R., Kolarov, T., Tadros, Th., *Colloids Surf. A*, **1997**, *123–124*, 277.
48 Alexandridis, P., Holzwarth, J.F., Hatton, T.A., *Macromolecules*, **1994**, *10*, 2414.
49 Linse, P., Hatton, T.A., *Langmuir*, **1997**, *13*, 4066.
50 Rippner Blomqvist, B., **2005**.
51 Rippner Blomqvist, B., Benjamins, J.-W., Nylander, T., Arnebrant, T., *Langmuir*, **2005**, *21*, 5061.
52 Khristov, K., Jachimska, B., Malysa, K., Exerowa, D., *Colloids Surf. A*, **2001**, *186*, 93.
53 Booth, C., Attwood, D., *Macromol. Rapid Commun.*, **2000**, *21*, 501.
54 Elisseeva, O., Besseling, N., Koopal, L., Cohen Stuart, M., *Langmuir*, **2005**, *21*, 4954.
55 Schillén, K., Claesson, P.M., Malmsten, M., Linse, P., Booth, C., *J. Phys. Chem.*, **1997**, *101*, 4238.
56 Cohen Stuart, M.A., Cosgrove, T., Vincent, B., *Adv. Colloid Interface Sci.*, **1986**, *24*, 143.
57 Rippner Blomqvist, B., Wärnheim, T., Claesson, P.M., *Langmuir*, **2005**, *21*, 6373.
58 Vieira, J.B., Li, Z.X., Thomas, R.K., Penfold, J., *J. Phys. Chem. B*, **2002**, *106*, 10641.
59 Weaire, D., Hutzler, S., *The Physics of Foams*, Clarendon Press, Oxford, **1999**.
60 Stubenrauch, C., Khristov, K., *J. Colloid Interface Sci.*, **2005**, *286*, 710.
61 Exerowa, D., Scheludko, A., in *Chemistry, Physics and Application of Surface Active Substances*, Overbeek, J. (Ed.), Gordon and Breach, London, **1964**, p. 1097.
62 Scheludko, A., *Colloid Chemistry*, Elsevier, Amsterdam, **1966**.
63 Khristov, K., Exerowa, D., Kruglyakov, P., *Colloid Polym. Sci.*, **1983**, *261*, 265.
64 Khristov, K., Exerowa, D., Christov, L., Makievski, A., Miller, R., *Rev. Sci. Instrum.*, **2004**, *75*, 4797.
65 Guillerme, C., Loisel, W., Bertrand, D., Popineau, Y., *J. Texture Stud.*, **1993**, *24*, 287.

66 Kruglyakov, P., Exerowa, D., Khristov, K., *Adv. Colloid Interface Sci.*, **1992**, *40*, 257.

67 Khristov, K., Exerowa, D., *Colloids Surf. A*, **1995**, *94*, 303.

68 Khristov, K., Exerowa, D., Yankov, R., *Colloids Surf. A*, **1997**, *129*, 257.

69 Khristov, K., Exerowa, D., *J. Dispers. Sci. Technol.*, **1997**, *18*, 561.

70 Khristov, K., Exerowa, D., Malysa, K., in *Foams, Emulsions and Their Applications*, Zitha, P., Banhart, J., Verbist, G. (Eds.), Verlag Metall Innovation Technologie, Bremen, **2000**, 21–31.

71 Khristov, K., Exerowa, D., Minkov, G., *Colloids Surf. A*, **2002**, *210*, 159.

72 Malmsten, M., Lindman, B., *Langmuir*, **1990**, *6*, 357.

73 Su, Y.-L., Liu, H.-Z., Wang, J., Chen, J.-Y, *Langmuir*, **2002**, *18*, 865.

74 Desai, P., Jain, N., Sharma, R., Bahadur, P., *Colloids Surf. A*, **2001**, *178*, 57.

75 Kjellander, R., Florin, E.J., *J. Chem. Soc., Faraday Trans.*, **1981**, *1*, 2053.

76 Rippner Blomqvist, B., Folke, S., Claesson, P.M., *J. Dispers. Sci. Technol.*, **2006**, *27*, 469.

77 Beneventi, D., Pugh, R.J., Carré, B., Gandini, A., *J. Colloid Interface Sci.*, **2003**, *268*, 221.

78 (a) Lucassen, J., van den Tempel, M., *Chem. Eng. Sci.*, **1972**, *27*, 1283;
(b) Lucassen, J., van den Tempel, M., *J. Colloid Interface Sci.*, **1972**, *41*, 491.

79 Gaydos, J., in *Drops and Bubbles in Interfacial Research*, Möbius, D., Miller, R. (Eds.), Elsevier, Amsterdam, **1998**, 1–60.

80 Loglio, G., Pandolfini, P., Miller, R., Makievski, A.V., Ravera, F., Ferrari, M., Liggieri, L., in *Novel Methods to Study Interfacial Layers*, Möbius, D., Miller, R. (Eds.), Elsevier, Amsterdam, **2001**, 439–483.

81 Lucassen-Reynders, E.H., Cagna, A., Lucassen, J., *Colloids Surf. A*, **2001**, *186*, 63.

82 Lucassen-Reynders, E.H., *Food Struct.*, **1993**, *12*, 1.

83 Gibbs, J.W., *The Scientific Papers*, Vol. 1, Dover, New York, **1961**.

84 Noskov, B.A., Akentiev, A.V., Bilibin, A.Y., Zorin, I.M., Miller, R., *Adv. Colloid Interface Sci.*, **2003**, *104*, 245.

85 Munoz, M.G., Monroy, F., Hernandez, P., Ortega, F., Rubio, R.G., Langevin, D., *Langmuir*, **2003**, *19*, 2147.

86 Hambardzumyan, A., Aguié-Béghin, V., Daoud, M., Douillard, R., *Langmuir*, **2004**, *20*, 756.

87 Kim, C., Yu, H., *Langmuir*, **2003**, *19*, 4460.

88 Rubio, R.G., **2005**.

89 Peace, S.K., Richards, R.W., Williams, N., *Langmuir*, **1998**, *14*, 667.

90 (a) Lucassen, J. Giles, D., *J. Chem. Soc., Faraday Trans. 1*, **1975**, *71*, 217;
(b) Lucassen, J., *Faraday Discuss. Chem. Soc.*, **1975**, *59*, 76.

91 Stenvot, C., Langevin, D., *Langmuir*, **1988**, *4*, 1179.

92 (a) Jiang, Q., Chiew, Y., Valentini, J., *J. Colloid Interface Sci.*, **1993**, *155*, 8; **1993**, *159*, 477.

93 Jayalakshmi, Y., Ozanne, L., Langevin, D., *J. Colloid Interface Sci.*, **1995**, *170*, 358.

94 Wantke, K.-D., Fruhner, H., Fang, J., Lunkenheimer, K., *J. Colloid Interface Sci.*, **1998**, *208*, 34.

95 Wantke, K.-D., Fruhner, H., Lunkenheimer, K., *Prog. Colloid Polym. Sci.*, **2000**, *115*, 227.

96 Wantke, K.-D., Fruhner, H., *J. Colloid Interface Sci.*, **2001**, *237*, 185.

97 Liggieri, L., Attolini, V., Ferrari, M., Ravera, F., *J. Colloid Interface Sci.*, **2002**, *255*, 225.

98 Santini, E., Ravera, F., Ferrari, M., Stubenrauch, C., Makievski, A., Krägel, J., *Colloids Surf. A*, in press.

99 Kokelaar, J.J., Prins, A., Degee, M., *J. Colloid Interface Sci.*, **1991**, *146*, 507.

100 Sedev, R., Steitz, R., Findenegg, G., *Physica B*, **2002**, *315*, 267.

12
Effect of the Intrinsic Compressibility on the Dilational Rheology of Adsorption Layers of Surfactants, Proteins and Their Mixtures

V. B. Fainerman, V. I. Kovalchuk, M. E. Leser and R. Miller

Abstract

Measurements of dilatational rheology provide important information on the properties of adsorption layers. Although simple models, such as the adsorption isotherms of Langmuir and Frumkin, describe the adsorption equilibrium fairly well, the dynamic and mechanical properties often deviate from these models significantly, sometimes even by orders of magnitude. Therefore, accurate experiments of the dilatational elasticity and viscosity provide much more accurate information about the interaction between molecules in interfacial layers. It is shown here that the consideration of the intrinsic compressibility of molecules in adsorbed layers allow a quantitative understanding of the dilatational rheology even at higher surface coverages, where simple models fail. The thermodynamic models including interfacial compressibility were successfully applied to surfactant, protein and also mixed surfactant and protein/surfactant systems.

12.1 Introduction

Many technical and natural systems with fluid interfaces are to a great extent characterized by surface rheological properties. Examples of such systems are foams and emulsions in general, food colloids, coating flows, the air/sea interface and biological liquids. Studies of the dilatational viscoelastic behavior of liquid/fluid interfaces are also of fundamental importance for understanding relaxation mechanisms. Moreover, such studies give an insight into the equilibrium thermodynamic characteristics of adsorption layers.

The mechanical response of an interface to small area deformations is defined by the complex dilatational modulus, which is a constitutive property of the system and independent of the functional form of the imposed perturbation [1–3]. Accordingly, the complex dilatational elasticity as a function of frequency, $E(i\omega)$, can be obtained from measurements of the interfacial tension response

Colloids and Interface Science Series, Vol. 1
Colloid Stability: The Role of Surface Forces, Part I. Edited by Tharwat F. Tadros

ISBN: 978-3-527-31462-1

to respective area perturbations. A variety of methods are suitable for these studies, such as the oscillating barrier [3–5], drop and bubble shape [6–10], elastic ring [1], capillary wave damping [11–15] and oscillating drop and bubble [16–20] techniques. The frequency ranges in which all these methods work are different and complement each other, so that a wide range of frequencies between ≤0.001 Hz and several kHz is covered. Such a wide frequency range is important because the large number of possible relaxation processes at fluid interfaces have very different characteristic times.

Any changes in the interfacial area immediately initiate processes that tend to restore the equilibrium state of the system. These relaxation processes can be monitored easily via measurements of the dynamic surface tension. A variety of relaxation processes, such as diffusional exchange and adsorption–desorption of surfactant molecules, interfacial reorientations, conformational changes or aggregation of adsorbed molecules and micellization kinetics, can contribute to the dynamic surface tension behavior. Under conditions not far from equilibrium, the link between surface perturbation and surface tension response can be expressed through the surface dilatational modulus, which accounts for all interfacial relaxation processes [1, 2]. $E(\omega)$ is a complex quantity, the real $E_r(\omega)$ and imaginary $E_i(\omega)$ parts of which reflect the elastic and viscous response of the interfacial system. Accordingly, the values $E_r(\omega)$ and $E_i(\omega)/\omega$ are usually referred to as the surface dilatational elasticity and viscosity, respectively [21–23].

The surface dilatational modulus is a constitutive property of the system, independent of the type of excitation. As a result, the surface dilatational modulus appears to be the most suitable characteristic of the interface to compare results of different dynamic experiments. Moreover, studies of the dilatational modulus also provide information about the equilibrium thermodynamic characteristics of the interfacial layer and the kinetic coefficients of processes occurring at the interface. These two circumstances explain the recently increased interest in surface rheological studies.

For a surfactant adsorption layer with a pure diffusional relaxation mechanism, the real and imaginary parts of the surface dilatational modulus are described by the equations [3, 21, 22]

$$E_r(\omega) = E_0 \frac{1+\zeta}{1+2\zeta+2\zeta^2}, \quad E_i(\omega) = E_0 \frac{\zeta}{1+2\zeta+2\zeta^2} \tag{1}$$

where $\zeta = \sqrt{\omega_D/2\omega}$, $E_0(c) = -\mathrm{d}\gamma/\mathrm{d}\ln\Gamma$ is the limiting (high-frequency) elasticity, $\omega_D(c) = D(\mathrm{d}\Gamma/\mathrm{d}c)^{-2}$ is the diffusion relaxation frequency, γ is the interfacial tension and Γ, c and D are the surfactant adsorption, bulk concentration and bulk diffusion coefficient, respectively. When other relaxation processes occur at or near the surface (micellar breakdown, adsorption-barrier processes, electrical double layer (DL) relaxation, etc.), more complicated models have to be developed, some of which are reviewed in [23]. The effect of electrical DL relaxations is considered in [24, 25]. These models consider not only the dependence on kinetic constants of the processes involved but also on the equilibrium thermodynamic characteristics of the interfacial layer, such as $d\gamma/d\ln\Gamma$ and $d\Gamma/dc$.

Hence the equilibrium characteristics of the adsorption layer can be obtained from the frequency dependence of the complex dilatational elasticity. The same characteristics are accessible via the equation of state and adsorption isotherm of the interfacial layer. The comparison shows, however, that a large discrepancy exists between the characteristics E_0 and $d\Gamma/dc$ obtained from surface rheological studies and those predicted by the Frumkin adsorption model at high surface coverage. As has been shown in a number of experimental studies with different surfactant systems, the limiting elasticity E_0 obtained from rheological studies levels off or even passes through a maximum with increasing concentration, whereas that predicted by the Frumkin isotherm increases continuously and can exceed the experimental values by several orders of magnitude [11, 12, 22, 26–29]. In addition, at high concentrations the Frumkin isotherm gives strongly underestimated values for $d\Gamma/dc$ [27, 28, 30]. Using a polynomial fit of the surface pressure vs. concentration data instead of the Frumkin isotherm can improve the situation, as was shown in [12]; such an approach, however, does not provide physical insight into the interfacial behavior. A new interpretation was proposed recently [18, 19, 30–32], which explains the described effect on the basis of a finite compressibility of the adsorbed surfactant molecules. It is the main aim of this chapter to summarize the physical picture, the basics of the theoretical model and experimental data to demonstrate the capacity of this new means of interpretation of the interfacial behavior at higher surface coverage.

12.2 Dilational Elasticity of Surfactant Adsorption Layers

12.2.1 Two-dimensional Molecular Compressibility

The saturation adsorption, Γ_∞, and the molar area of a surfactant, $\Omega = 1/\Gamma_\infty$, are usually assumed to be independent of surface pressure. Such an assumption, based on a "hard core" model of the molecules, can be used only when the surface layer is in a state not very close to saturation. A number of synchrotron GIXD (grazing incidence X-ray diffraction) data for liquid-condensed surface layers show that the molar area of a surfactant in the condensed state can be approximately represented by a linear dependence on surface pressure Π [18, 19]:

$$\Omega = \Omega_0(1 - \varepsilon\Pi) \tag{2}$$

where Ω_0 is the molar area at zero surface pressure and ε is the two-dimensional relative surface layer compressibility coefficient, which characterizes the intrinsic compressibility of molecules in the surface layer. As an example, the dependences of the area per molecule on surface pressure in a condensed

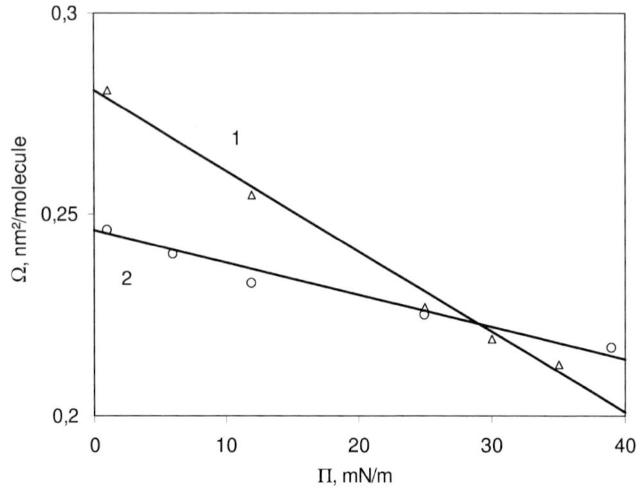

Fig. 12.1 Dependences of the area per molecule as a function of surface pressure in a condensed monolayer. (1) (*S*)-AMD-18 [34], (2) *rac*-ESD-18 [33].

monolayer state are shown for 1-stearylamine monoglycerol in the *S*-(–)-chiral form [(*S*)-AMD-18] [33] and 1-monosteroyl-rac-glycerol molecule in racemic *R*-(+)-chiral form (*rac*-ESD-18) [34] in Fig. 12.1. One can see that with increasing surface pressure, the area per molecule decreases. This decrease is almost linear and can be described by Eq. (2) with $\varepsilon = 0.009$ mN m^{-1} for (*S*)-AMD-18 and 0.0032 mN m^{-1} for *rac*-ESD-18. The two-dimensional relative surface layer compressibility coefficient ε is usually in the range 0.009–0.0025 mN m^{-1} [18]. This value depends on the type of insoluble surfactant, whereas in the same homologous series there is a significant decrease with increasing hydrocarbon chain length. The intrinsic compressibility possibly reflects changes in the tilt angle of adsorbed molecules upon surface layer compression, accompanied by an increase in surface layer thickness [18]. Note that the upper limit for ε for soluble surfactants can be larger than the maximum value found experimentally for insoluble condensed monolayers. This conclusion is justified by the fact that molecules in adsorption layers do not exist in a condensed state (i.e. less closely packed). Moreover, the length of the hydrocarbon chain in soluble surfactants is shorter than that characteristic of insoluble surfactants.

12.2.2
Models for Non-ionic Surfactants

The adsorption isotherm and the corresponding equation of state for the Frumkin model can be written as

$$bc = \frac{\theta}{(1-\theta)} \exp(-2a\theta) \qquad (3)$$

$$\Pi = -\frac{RT}{\Omega}[\ln(1-\theta)+a\theta^2] \tag{4}$$

where $\theta = \Gamma\Omega$ is the surface layer coverage, R is the gas constant, T is the absolute temperature, b is the adsorption constant and a is the Frumkin interaction constant.

In [18, 19, 30] the molecular compressibility was discussed thermodynamically in the framework of some approximation models. In [31], based on the two-dimensional solution theory and assuming adsorption also of solvent molecules, more rigorous Frumkin-type equations were presented, which take into account the 2D compressibility of the surface layer, the non-ideality of entropy (resulting from the difference between the molar areas of solvent and surfactant) and the enthalpy of mixing (resulting from intermolecular interactions). The corresponding model (Model 1) is described by the following set of equations:

$$\Pi = -\frac{RT}{\Omega_0}\ln(1-\theta)+(1-1/n)\theta+a\theta^2] \tag{5}$$

$$bc = \frac{\theta}{n(1-\theta)^n}\exp(-2an\theta) \tag{6}$$

where $n=\Omega/\Omega_0$ and Ω_0 is the molar area of a solvent molecules. Here, $\theta=\Gamma\Omega$ is again the surface layer coverage, as in case of the classical Frumkin isotherm, but the molar area of the surfactant molecules Ω is given by Eq. (2). Accordingly, one can express θ and n as

$$\theta = \Gamma\Omega_0(1-\varepsilon\Pi) \tag{7}$$

$$n = 1-\varepsilon\Pi \tag{8}$$

This rigorous model (Eqs. 5–8) can be simplified further by neglecting the contribution of the non-ideality of entropy [31]. In this case (simplified Model 2) $n=1$ and, therefore, the equations of state and adsorption isotherm in Eqs. (5) and (6) turn into the ordinary Frumkin model, given by Eqs. (3) and (4) with Ω_0 instead of Ω, but with Ω and θ now described by Eqs. (2) and (7).

A semi-empirical model (Model 3) was also proposed recently [19, 30] based on the simultaneous solution of Eqs. (2) and (3) with the Gibbs equation. It neglects solvent adsorption and, therefore, differs from the previous models by another localization of the Gibbs dividing surface:

$$\Pi = -\frac{RT}{\Omega_0(1-\varepsilon\Pi/2)}[\ln(1-\theta)+a\theta^2] \tag{9}$$

The last two models describe the surface layer behavior similarly to Eqs. (2) and (5) to (8) [31].

The relationship (2) in fact implies that the molar area of a surfactant is a variable quantity which becomes lower with increasing surface pressure or adsorption. This fact could be taken into account also in a different way by assuming that there are two or more orientations of the adsorbed surfactant molecule at the interface, with different molar areas Ω_1 and Ω_2 (for definiteness we assume $\Omega_1 > \Omega_2$). This reorientation model (Model R), for the case of ideal enthalpy and entropy of mixing of the surface layer, obeys the equation of state [35]:

$$\Pi = -\frac{RT}{\Omega}\ln(1-\theta) \tag{10}$$

and the adsorption isotherm

$$b_2 c = \frac{\Gamma_2 \Omega}{(1-\theta)^{\Omega_2/\Omega}} \tag{11}$$

Here the total adsorption Γ and mean molar area Ω are defined by

$$\Gamma = \Gamma_1 + \Gamma_2 \tag{12}$$

$$\Omega\Gamma = \theta = \Omega_1\Gamma_1 + \Omega_2\Gamma_2 \tag{13}$$

The ratio of adsorption values in the two possible states of the molecule can be expressed by

$$\frac{\Gamma_1}{\Gamma_2} = \exp\left(\frac{\Omega_1 - \Omega_2}{\Omega}\right)\left(\frac{\Omega_1}{\Omega_2}\right)^a \exp\left[-\frac{\Pi(\Omega_1 - \Omega_2)}{RT}\right] \tag{14}$$

where b_2 is the adsorption equilibrium constant in state 2 and a is a constant. The parameter a accounts for the fact that the adsorption equilibrium constant for surfactant molecules in state 1 (with larger area) could exceed that in state 2, which results in an additional (in comparison with $a=0$) increase in the fraction of molecules adsorbed in states of larger area. For $a=0$ the adsorption activity b_i of all states is the same. Other reorientation models (e.g. those which account for non-ideal entropy of mixing) were discussed in [35]. The reorientation model provides, for example, a perfect description of the adsorption behavior of oxethylated surfactants [35, 36], because the oxyethylene chain is capable of unfolding at the interface at low coverages. Note that the reorientation model differs essentially from the Frumkin-type 2D compressibility Models 1–3 in respect of the underlying principle employed to account for the variation of the molar area of the surfactant. In Models 1–3 this is based on the phenomenological Eq. (2), which was verified experimentally for the condensed state of surfactant molecules in a monolayer [18]. In contrast, the reorientation model, given by Eqs. (10) to (14), is based solely on the Butler equation [37], which states that the chemical potential of adsorbed molecules is determined by their molar area. In this case, as the chemical potential depends on the surface pressure, an increase in surface pressure leads to an increase in the fraction of adsorbed molecules in states characterized by a minimum area, cf. Eq. (14), which results in a decrease in the mean molar area.

The mechanism responsible for the variation of the molar area in the Frumkin-type intrinsic compressibility monolayer Models 1–3 is different from that characteristic of the reorientation model. Hence it seem possible to arrange a combined model (RC model), which assumes both the reorientation of molecules and the variation of the molar area in the state with lower area 2 caused by a two-dimensional compressibility given by Eq. (2), i.e. $\Omega_2=\Omega_0(1-\varepsilon\Pi)$. For the sake of simplicity, we avoid the increase in the number of parameters in the RC model by assuming $a=0$. In Table 12.1 the peculiarities of all proposed models are summarized.

The limiting (high-frequency) elasticity $E_0=\mathrm{d}\Pi/\mathrm{d}\ln\Gamma$ can be calculated by differentiation of the equation of state taking into account the dependence of Γ on Π. For the Frumkin model (Eqs. (3) and (4) with $\Omega=\Omega_0=$constant), one obtains

$$E_0=\frac{RT}{\Omega_0}\left(\frac{\theta}{1-\theta}-2a\theta^2\right) \tag{15}$$

whereas, for example, the consideration of intrinsic compressibility (Model 2) yields [35]

$$E_0=\frac{\dfrac{\theta}{1-\theta}-2a\theta^2}{\dfrac{\Omega_0}{RT}+\dfrac{\varepsilon\theta}{1-\varepsilon\Pi}\left(\dfrac{1}{1-\theta}-2a\theta\right)} \tag{16}$$

It is interesting to compare the limiting elasticity for the Frumkin model with the values obtained from Models 1–3 at $\theta\to 1$. For $\theta\to 1$, Eq. (15) yields $E_0\to\infty$, in contradiction with all experimental results available. In contrast, all three Models 1–3 (see Eq. 16) yield the same finite value of E_0 at $\theta\to 1$:

$$E_0=\frac{1-\varepsilon\Pi}{\varepsilon} \tag{17}$$

For $\varepsilon=0.005$–$0.01\ \mathrm{mN\ m^{-1}}$, which is usual for common surfactants, and $\Pi=40\ \mathrm{mN\ m^{-1}}$ we obtain from Eq. (17) that $E_0=160$–$60\ \mathrm{mN\ m^{-1}}$, which is in qualitative agreement with experimental data.

Table 12.1 Characteristics of the theoretical models.

Model	Characteristics of the interfacial layer
Frumkin	Non-ideal enthalpy ($a\neq 0$), no intrinsic compressibility ($\varepsilon=0$)
1	$a\neq 0$, $\varepsilon\neq 0$, non-ideal entropy
2	$a\neq 0$, $\varepsilon\neq 0$, ideal entropy
3	$a\neq 0$, $\varepsilon\neq 0$, Gibbs dividing surface
R	$a=0$, $\varepsilon=0$, ideal entropy, two states of the surfactant with different molar area
RC	$a=0$, $a=0$, ideal entropy, two states, $\varepsilon\neq 0$ at minimum molar area

Rusanov [38], analyzing the effect of molecular orientation on the interfacial layer characteristics, derived equations for the molecular area and the thickness of the adsorption layer similar to those discussed above, which also predict a finite value of the derivative $d\Pi/d\Gamma$ in the limit of high concentrations.

12.2.3
Selected Experimental Results

From the measured real and imaginary parts of the complex dilatational elasticity, the characteristics of the adsorption layer $E_0 = -d\gamma/d\ln\Gamma$ and $d\Gamma/dc$ can be obtained as discussed above. It is easy to find from Eqs. (1) that for a pure diffusion relaxation

$$E_0 = \frac{E_r^2 + E_i^2}{E_r - E_i} \tag{18}$$

$$\frac{d\Gamma}{dc} = \frac{E_r - E_i}{E_i}\sqrt{\frac{D}{4\pi f}} \tag{19}$$

where $f=\omega/2\pi$ is the frequency of oscillations [30]. Eqs. (18) and (19) give the equilibrium characteristics of the adsorption layer at a given concentration and should therefore not depend on frequency, although the right-hand sides of these equations are frequency-dependent. The frequency dependence disappears when the diffusional model [represented by Eqs. (1)] adequately describes the behavior of the studied system. When the characteristics obtained by Eqs. (18) and (19) are frequency dependent, this means that a pure diffusional relaxation model is not applicable and a more general model is required or the experimental error is too large. It can be seen from these equations that the experimental error can be especially significant when the real and imaginary parts are close to each other or when the imaginary part is very small [30].

The values of E_0 and $d\Gamma/dc$ or dodecyldimethylphosphine oxide (C_{12}DMPO) solutions, obtained via Eqs. (18) and (19), for the oscillating bubble experiments are shown in Figs. 12.2 and 12.3 (taken from [28] and [39]). The experimental dependences are well described by the models accounting for the intrinsic 2D compressibility. At the same time, the Frumkin model predicts a strong increase in E_0 with concentration, in strong contrast with experimental data (Fig. 12.2). The models which assume an internal compressibility of the molecules provide a much better fit to the experimental dependence of $d\Gamma/dc$ than the Frumkin model does (Fig. 12.3). The compressibility coefficient ε obtained from this fit is of the order of 0.01 mN m^{-1}. Such a value is in very good agreement with data obtained from independent experiments (from GIXD and Π–A isotherms in the liquid-condensed state) on the dependence of the molecular area of insoluble amphiphilic molecules in the condensed state on the surface pressure [18, 19].

The surface pressure vs. concentration dependences obtained for the same theoretical models and the same sets of parameters are presented in Fig. 12.4.

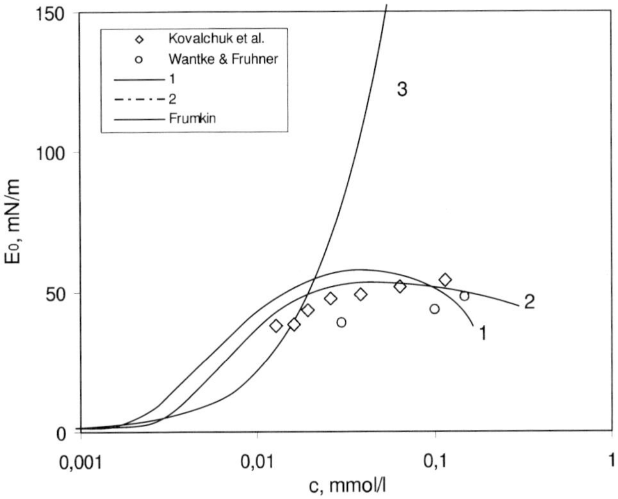

Fig. 12.2 Experimental limiting elasticity E_0 vs. concentration for C_{12}DMPO solutions obtained by the oscillating bubble method [28] (○) and [39] (◇); theoretical dependences calculated for Frumkin model (3), rigorous model 1 (1) and simplified model 2 (2) (after [31]).

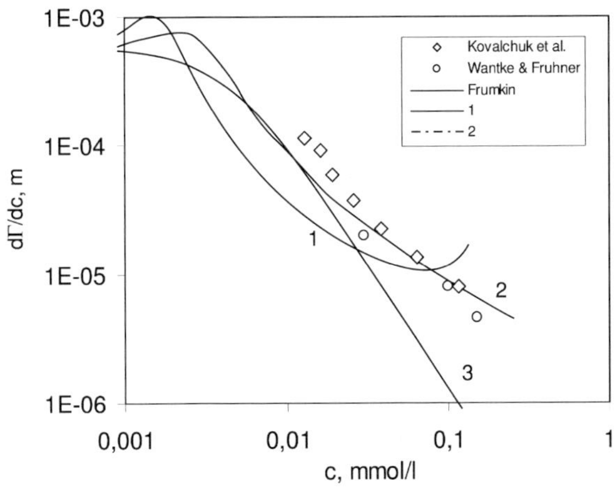

Fig. 12.3 Experimental values $d\Gamma/dc$ vs. concentration of C_{12}DMPO obtained by the oscillating bubble method [28] (○) and [39] (◇). Notation of the theoretical curves as in Fig. 12.2 (after [31]).

One can see that the influence of the compressibility coefficient on the shape of the $\Pi(c)$ isotherm is almost negligible. Hence this coefficient can be determined only from dynamic experiments, such as small-amplitude harmonic perturbations. This is in line with the conclusions drawn by Lucassen-Reynders that the rheological surface layer characteristics are far more sensitive to varia-

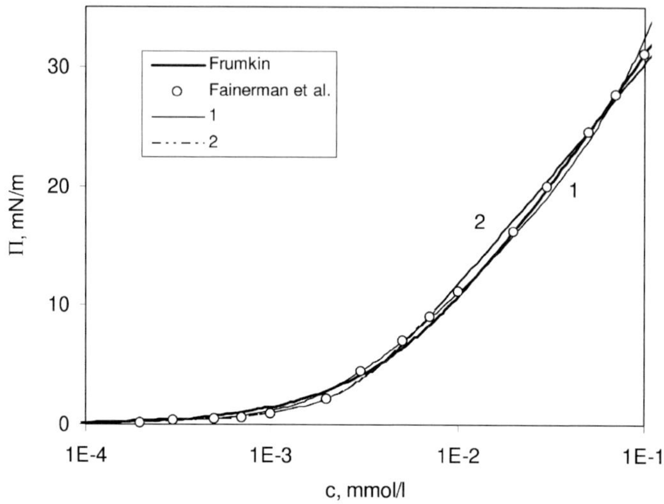

Fig. 12.4 Experimental dependence of surface pressure Π vs. concentration for C_{12}DMPO from [36] (○). Notation of the theoretical curves as in Fig. 12.2 (after [31]).

tions in the equation of state parameters than are the equilibrium $\Pi(c)$ curves and that the information obtained from rheological data can be more accurate than those obtained from equilibrium surface pressure data [21]. Results obtained for undecyldimethylphosphine oxide solutions are very similar to those for C_{12}DMPO solutions shown here [30].

The surface dilatational modulus for solutions of some oxethylated alcohols ($C_{12}EO_3$, $C_{12}EO_6$ and $C_{14}EO_6$) was studied by Lucassen and Giles by using the oscillating barrier method [4]. In Fig. 12.5, the experimental dependence $E_0(\Pi)$ for $C_{12}EO_6$ solutions is compared with the predictions calculated from several models [31]. At high surface coverage the Frumkin isotherm predicts much higher elasticities than those obtained experimentally. The simple reorientation model (Model R) neglecting 2D compressibility shows similar problems [31]. However, the combined reorientation model (Model RC accounting for reorientation and 2D compressibility) and the approximate Model 2 (neglecting the non-ideality of entropy) agree fairly well with the experimental data, indicating that a limited elasticity modulus exists at high surface coverage. Similar results were discussed for $C_{12}EO_3$ and $C_{14}EO_6$ in [31].

As for the alkyldimethylphosphine oxides, all models predict practically the same behavior of the equilibrium $\Pi(c)$ curves also for oxethylated alcohols [31]. It should be noted, however, that for these surfactants the Frumkin model yields large negative values of the interaction parameter a, which could not be ascribed to van der Waals forces. Hence, from a physical point of view, the Frumkin model is not adequate for the description of non-ionic oxethylated surfactants.

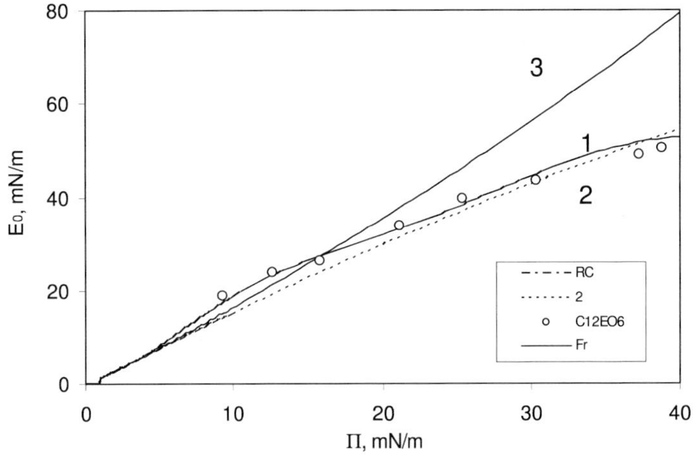

Fig. 12.5 Dependence of the limiting elasticity E_0 of $C_{12}EO_6$ solutions on surface pressure obtained by the oscillating barrier method [4] (○); theoretical dependences calculated for the Frumkin model (3), the combined reorientation model RC (reorientation with 2D compressibility) (1) and the simplified model 2 (2) (after [31]).

It was also shown in [31] that at higher surface coverage the RC model yields molar areas which agree much better with the molar areas at the CMC provided by neutron reflection experiments [40]. At the same time, the corresponding molar areas obtained from models accounting only for 2D compressibility are much lower than the experimental values. Hence the available experimental data for surface layers of oxethylated surfactants (especially those with large numbers of EO groups) at the solution/air interface are described more adequately by the combined reorientation model [31].

12.2.4
Ionic Surfactants

It has been shown that the consideration of 2D compressibility can be successfully adapted for the description of the rheological behavior also of ionic surfactants, for example for the cationic alkyltrimethylammonium bromides, C_nTAB [41]. In the framework of an electroneutral surface layer model, arranged by a certain choice of the Gibbs dividing surface, the equations of state of the surface layer and the adsorption isotherm for the Frumkin model for ionic surfactants are [42]

$$\Pi = -\frac{2RT}{\Omega}[\ln(1-\theta) + a\theta^2] \tag{20}$$

$$b[c(c+c_2)]^{\frac{1}{2}} f^{\pm} = \frac{\theta}{1-\theta}\exp(-2a\theta) \tag{21}$$

where Ω is the partial molar area of the ionic surfactant (about two times larger than the molar area of the solvent), $f^{\pm}$ is the average activity coefficient of ions in the bulk solution, c is the ionic surfactant concentration and c_2 is the inorganic (1:1) salt concentration.

The Debye-Hückel equation corrected for short-range interactions accurately represents the values of the average activity coefficient $f^{\pm}$:

$$\log f^{\pm} = -\frac{0.5115\sqrt{I}}{1+1.316\sqrt{I}} + 0.055I \tag{22}$$

where $I = c + c_2$ is the ionic strength expressed in mol L^{-1} and the numerical constants correspond to 25 °C [43].

Taking into account that the surfactants dissociate into two ions, introducing the expressions for the mean activity of ions in the solution and involving rigorous expressions for the surface layer non-ideality of enthalpy and entropy, instead of Eqs. (5) and (6), we obtain for 2D compressible surface layers the following equation of state and adsorption isotherm:

$$\Pi = -\frac{RT}{\Omega_0}\left[\ln(1-\theta) + (1-1/n)\theta + a\theta^2\right] \tag{23}$$

$$b[c(c+c_2)]^{\frac{1}{2}} f^{\pm} = \frac{\theta}{n(1-\theta)^n}\exp[-2an\theta] \tag{24}$$

where Ω is the molar area of the ionic surfactant, Ω_0 is the molar area of a solvent molecule, $n = 1 - \varepsilon\Pi = \Omega/2\Omega_0$, $\Omega = 2\Omega_0(1-\varepsilon\Pi)$ and $\theta = \Gamma\Omega = 2\Gamma\Omega_0(1-\varepsilon\Pi)$.

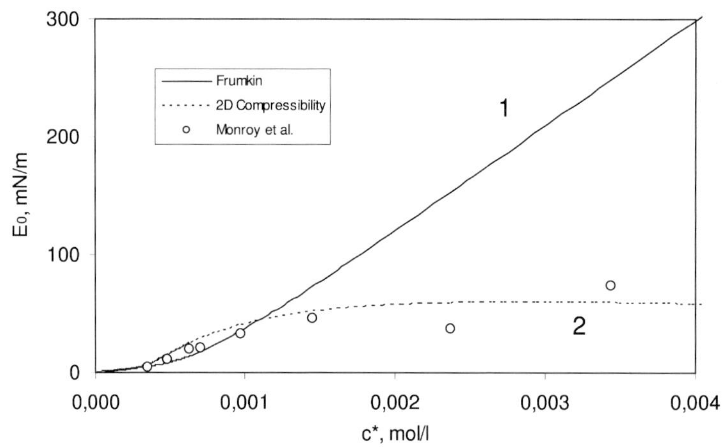

Fig. 12.6 Dependence of the limiting elasticity E_0 of C_{14}TAB solutions on the activity $c^* = [c(c+c_2)]^{\frac{1}{2}} f^{\pm}$ (○) [44]; theoretical dependences calculated for the Frumkin model (1) and 2D compressibility model (2) (after [41]).

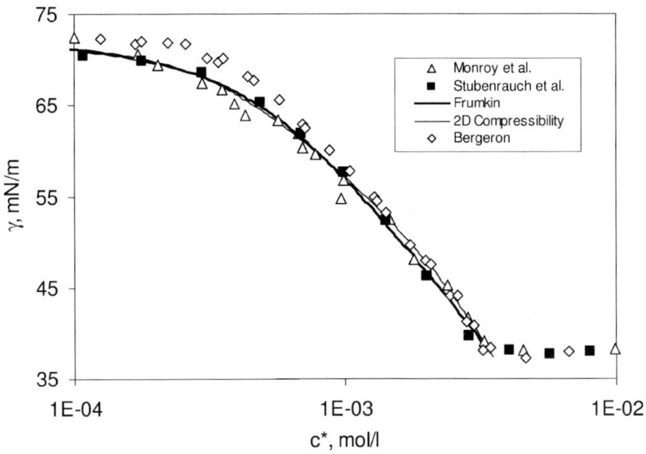

Fig. 12.7 Dependence of equilibrium surface tension for C_{14}TAB solutions on the activity c^*: (◇) data [45], (△) data [44], (■) data [41]; theoretical curves calculated from the Frumkin model (thick line) and 2D compressibility model (thin line) (after [41]).

The rheological behavior of some cationic surfactants (C_{12}TAB, C_{14}TAB and C_{16}TAB) was analyzed in [41] based on the experimental data obtained in literature by the capillary wave methods [11, 44]. In Fig. 12.6, the experimental values of the limiting elasticity E_0 for C_{14}TAB surface layers are compared with results calculated for the Frumkin model given by Eqs. (20) and (21) and for the model assuming intrinsic compressibility given by Eqs. (23) and (24) with $\varepsilon = 0.01$ mN m^{-1}. One can clearly see that only by assuming an intrinsic compressibility can the experimental data be described satisfactorily. Similar results were obtained also for C_{12}TAB and C_{16}TAB [41].

The experimental results published in [44, 45] and theoretical dependences of surface tension on activity $c^* = [c(c + c_2)]^{\frac{1}{2}} f^{\pm}$ calculated for C_{14}TAB using the same theoretical models and the same sets of parameters as in Fig. 12.6 are presented in Fig. 12.7. Both models describe the experimental data very well, while the influence of the compressibility coefficient on the surface pressure isotherm is negligible.

The experimental adsorption values of C_{12}TAB and C_{14}TAB measured directly using the neutron reflection method [46] were compared in [41] with the results calculated from the Frumkin and intrinsic compressibility models (data for C_{14}TAB are presented in Fig. 12.8). It is shown that the intrinsic compressibility model predicts an almost linear increase in the adsorption at high concentrations, in complete agreement with the neutron reflection experiments, whereas the Frumkin model predicts an asymptotic behavior of the $\Gamma(c^*)$ dependences, with practically constant Γ at concentrations near the CMC.

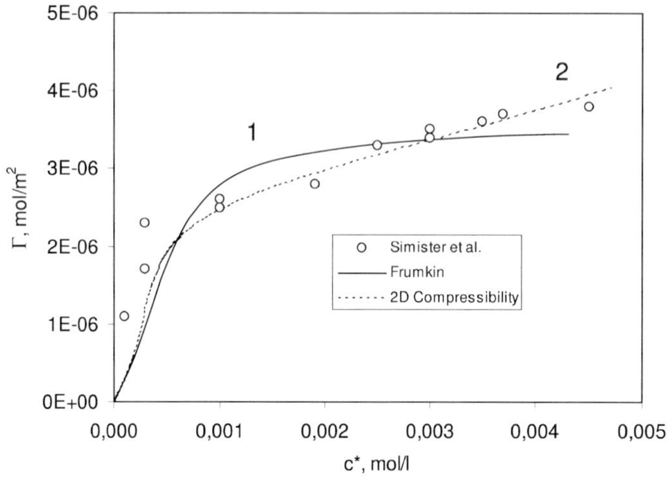

Fig. 12.8 Dependence of $C_{14}TAB$ adsorption on the activity c^* (○) [46]; theoretical dependences calculated for the Frumkin model (1) and 2D compressibility model (2) (after [41]).

12.3 Elasticity of Protein Adsorption Layers

The rheological characteristics of protein adsorption layers exhibit an unusual dependence of the surface elasticity on bulk concentration c or adsorption Γ of a protein [47–54]. For globular proteins, a monotonic increase in the elasticity with increasing adsorption or surface pressure up to a maximum value is observed and this maximum is considerably higher than that for flexible proteins. Many previous theoretical models have failed to explain this behavior, possibly because these models do not account for the known experimental fact that the conformational distribution of protein molecules depends on the interfacial coverage and hence the average molecular area becomes smaller with increasing adsorption [55, 56]. A theoretical model has been proposed to explain this behavior for several common proteins [57]. It was demonstrated that a single set of model parameters allows all experimental dependences to be reproduced: surface pressure, amount adsorbed and adsorption layer thickness as functions of the concentration.

This model assumes that protein molecules can exist in a number of states with different molar areas, varying from a maximum value (Ω_{max}) at very low surface coverage to a minimum value (Ω_{min}) at high surface coverage [57]. Assuming that the molar areas of two 'neighboring' conformations differ from each other by the value Ω_0 (molar area increment, chosen equal to the molar area of the solvent or the area occupied by one segment of the protein molecule) and that the total number of possible states of the protein molecule is n, one obtains the molar area in the ith state $\Omega_i = \Omega_1 + (i-1)\Omega_0$ and the maximum area $\Omega_{max} = \Omega_1 + (n-1)\Omega_0$, where $1 \le i \le n$ and $\Omega_1 = \Omega_{min} \gg \Omega_0$.

The following surface equation of state based on a first-order model for both the non-ideal entropy and the heat of mixing was also derived in [57]:

$$-\frac{\Pi\Omega_0}{RT} = \ln(1-\theta) + \theta(1-\Omega_0/\Omega) + a\theta^2 \tag{25}$$

where Ω is the average molar area of the protein, Γ_i is the protein adsorption in the ith state, $\theta = \Omega\Gamma = \sum_{i=1}^{n}\Omega_i\Gamma_i$ is the total surface coverage by protein and $\Gamma = \sum_{i=1}^{n}\Gamma_i$ is the total protein adsorption.

The equation for the adsorption isotherm for each state (j) of the adsorbed protein is

$$b_j c = \frac{\Omega\Gamma_j}{(1-\theta)^{\Omega_j/\Omega}}\exp[-2a(\Omega_j/\Omega)\theta] \tag{26}$$

where c is the protein bulk concentration and b_j is the adsorption equilibrium constant for the protein in the jth state. It is assumed that the values of the constants b_j for all states j from $i=1$ to $i=n$ are the same and therefore the adsorption constant for the protein molecule as a whole is $\Sigma b_j = nb_j$ [57]. This assumption together with Eq. (26) allows the calculation of the distribution function of adsorptions over the adsorption states of the protein molecule:

$$\Gamma_j = \Gamma\frac{(1-\theta)^{\frac{\Omega_j-\Omega_1}{\Omega}}\exp\left(2a\theta\frac{\Omega_j-\Omega_1}{\Omega}\right)}{\sum_{i=1}^{n}(1-\theta)^{\frac{\Omega_i-\Omega_1}{\Omega}}\exp\left(2a\theta\frac{\Omega_i-\Omega_1}{\Omega}\right)} \tag{27}$$

It is a general feature of the equation-of-state theory leading to Eq. (25) that with increasing surface coverage the adsorption of molecules requiring a smaller area increases at the expense of those with a larger area. This is a special type of the intrinsic compressibility of protein adsorption layers, reflected by the exponential factor in Eq. (27). The factor implies that the probability of the existence of a protein molecule in the states characterized by larger area becomes essentially lower with increasing surface coverage θ, because $(\Omega_j-\Omega_1)/\Omega>0$. Therefore, at $\theta\to 0$ all states are equally probable and $\Omega=(\Omega_{min}+\Omega_{max})/2$, but with increase in θ the value of Ω becomes smaller and approaches Ω_{min} at $\theta\to 1$. Hence, from Eq. (27), we obtain $d\Omega/d\Gamma<0$.

This dependence of Ω on θ (or Γ) drastically affects the limiting elasticity E_0 of the protein adsorption layer. Differentiation of the equation of state Eq. (25) with respect to $\ln\Gamma$ (assuming $\Omega_0\ll\Omega$, which is generally fulfilled for proteins) and taking into account the dependence $\Omega(\Gamma)$ results in [58]

$$E_0 = \frac{d\Pi}{d\ln\Gamma} = \frac{RT}{\Omega_0}\left(\frac{\theta}{1-\theta} - \theta - 2a\theta^2\right)\left(1+\frac{d\ln\Omega}{d\ln\Gamma}\right) \tag{28}$$

It was shown above that $d \ln \Omega / d \ln \Gamma < 0$. Considering the factor in last parentheses on the right-hand side of Eq. (28), it becomes clear that the limiting elasticity E_0 for the adsorption layers of proteins should be essentially lower than that for molecules with constant Ω, for which Eq. (28) has the simple form [48, 54]

$$E_0 = \frac{RT}{\Omega_0}\left(\frac{\theta}{1-\theta} - \theta - 2a\theta^2\right) \tag{29}$$

Let us now compare experimental and theoretical dependences of the surface dilatational elasticity. Figure 12.9 illustrates the experimental and theoretical dependences of surface pressure on the adsorption of BSA and β-casein. The experimental results shown were reported in [48, 59, 60] and the theoretical curves (solid lines) were calculated from Eqs. (25) to (27) [57] using the following parameter values: for BSA, $\Omega_0 = 2.5 \times 10^5$ m^2 mol^{-1}, $\Omega_{min} = 3.0 \times 10^7$ m^2 mol^{-1}, $\Omega_{max} = 7.5 \times 10^7$ m^2 mol^{-1}, $a = 1$, $b_j = 3 \times 10^5$ L mol^{-1} (or $\Sigma b_j = 5.43 \times 10^7$ L mol^{-1} for the molecule as a whole), and for β-casein, $\Omega_0 = 2.5 \times 10^5$ m^2 mol^{-1}, $\Omega_{min} = 4.5 \times 10^6$ m^2 mol^{-1}, $\Omega_{max} = 4.5 \times 10^7$ m^2 mol^{-1}, $a = 1$, $b_j = 2.5 \times 10^6$ L mol^{-1} (or $\Sigma b_j = 4.075 \times 10^8$ L mol^{-1} for the molecule as a whole).

It can be seen that the theoretical curves in Fig. 12.9 agree well with the experimental data. It should be noted that the parameter sets given above simultaneously provide a good description of the surface pressure dependence on concentration for these proteins [57].

Comparing the parameters for BSA and β-casein, one can see that the main difference between the two proteins is in the values for Ω_{min} and Ω_{max} and in

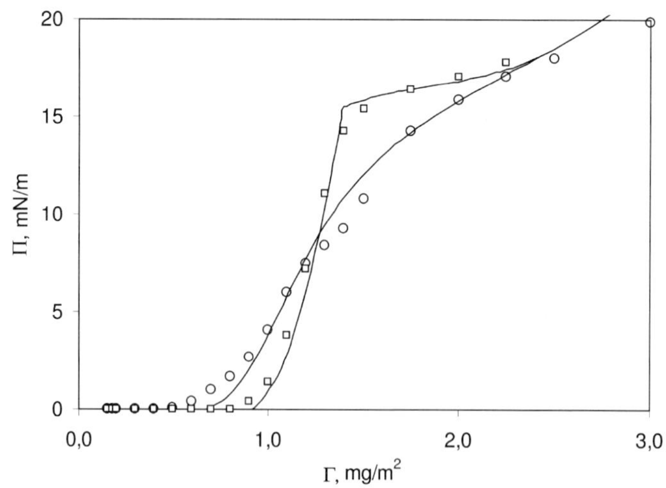

Fig. 12.9 Dependence of the surface pressure for BSA (□) and β-casein (○) as a function of protein adsorption: experimental data [48, 59, 60]; theoretical lines calculated from Eqs. (25) to (27) (after [57]).

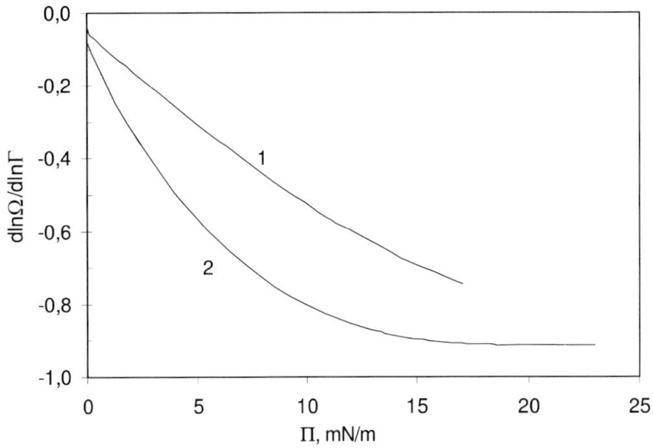

Fig. 12.10 Dependence of the derivative d ln Ω/d ln Γ on surface pressure for BSA (1) and β-casein (2).

the ratio between these values: for BSA the ratio $\Omega_{max}/\Omega_{min}=2.5$, whereas for the flexible β-casein this ratio is much larger, $\Omega_{max}/\Omega_{min}=10$. Therefore, the absolute value of the derivative $d\ln\Omega/d\ln\Gamma$ (or intrinsic compressibility) for β-casein should exceed that for the BSA. The dependence of the derivative $d\ln\Omega/d\ln\Gamma$ on the surface pressure for these proteins is shown in Fig. 12.10. One can see that the values are indeed negative, as expected. For β-casein the absolute value of this derivative is higher than that for BSA, i.e. at $\Pi > 15$ mN m^{-1} we obtain $d\ln\Omega/d\ln\Gamma=-0.9$ for β-casein.

The theoretical values of the limiting elasticity modulus E_0, as calculated numerically from Eqs. (25) to (27) and from Eq. (29), are shown in Fig. 12.11. The limiting elasticity calculated from Eq. (29) (or from Eqs. (25) to (27) with $\Omega_{max}=\Omega_{min}$) show a significant increase in E_0 with increasing surface pressure. In contrast, the calculations using the actual Ω_{min} and Ω_{max} lead to significantly lower E_0 values. For β-casein the values are not only considerably lower than those for BSA, but they also exhibit two local extrema, i.e. a maximum followed by a shallow minimum, corresponding to the kink in the experimental $\Pi(\Gamma)$ curve at $\Gamma \approx 1.2$ mg m^{-2} (cf. Fig. 12.9). This phenomenon was first noted by Graham and Phillips [49], who qualitatively ascribed it to a transition in the configuration from all-trains to trains- and loops. To our knowledge, the present theory offers the first quantitative interpretation.

Also shown in Fig. 12.11 are the experimental values of the elasticity modulus E measured for these proteins using the oscillating drop method [9, 48, 54]. The data obtained with this method at two values of drop surface oscillation frequency (0.033 and 0.84 rad s^{-1} for β-casein and 0.084 and 0.84 rad s^{-1} for BSA) are fairly similar, which indicates that at these frequencies there is no significant exchange of the protein between the drop surface and the bulk (conservation of the adsorbed mass of protein, $A\Gamma \cong$ constant), which means $E_0 \cong E$.

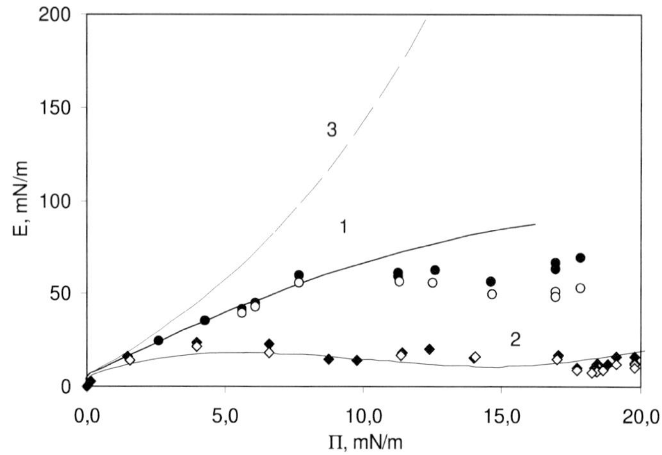

Fig. 12.11 Dependence of the limiting surface elasticity E_0 on surface pressure calculated from the model given by Eqs. (25) to (27) for BSA (1) and β-casein (2), curve 3 calculated from Eq. (29); experimental values for BSA at frequencies of 0.084 (○) and 0.84 rad s^{-1} (●) and for β-casein at frequencies of 0.033 (◇) and 0.84 rad s^{-1} (◆) (data taken from [9, 48, 54]).

Moreover, viscous phase angles measured simultaneously [48] were negligible at the highest frequency, implying that the measured values at this frequency were indeed pure elasticities. It can be seen that the agreement between the theoretical and experimental dependences presented in Fig. 12.11 is fairly good. Not only does the theory correctly predict the limiting elasticity values for the two proteins, but it also reproduces in many details the shape of the $E_0(\Pi)$ curve: for BSA this dependence is monotonic and approaches a limiting value, and for β-casein the curve exhibits a maximum and a minimum of E_0 located fairly close to those found experimentally. Note that the dependence of E_0 on Π as determined in [61] for β-casein at a frequency of 0.2 Hz is also in good agreement with the theoretical curve in Fig. 12.11.

12.4
Rheology of Mixed Protein/Surfactant Layers

Addition of surfactants can modify adsorbed protein layers at liquid/fluid interfaces, which leads to changes in the adsorption and rheological characteristics [62–67]. Although the dilatational rheology of proteins or protein/surfactant mixtures is extremely important from a practical point of view, no corresponding theory was available until recently. For much simpler systems, e.g. surfactant mixtures, it appears possible to predict the rheological behavior of a mixture using data for the individual components. One of the first attempts to analyze theoretically the rheology of surfactant mixtures was made by Lucassen-

Reynders [68]. A theoretical analysis of the dilatational rheology of surfactant mixtures was later performed by Garrett and Joos [69], who generalized the theory by Lucassen and van den Tempel [3]. The following expression for the complex dilatational modulus was obtained:

$$E = E_r + iE_i = E_0 \sum \frac{\Gamma_j}{\Gamma} \left(\frac{1 + \zeta_j + i\zeta_j}{1 + 2\zeta_j + 2\zeta_j^2} \right) \tag{30}$$

where $\zeta_j = \sqrt{\omega_{Dj}/2\omega}$, $\omega_{Dj} = D_j(\partial c_j/\partial \Gamma_j)^2$ is the characteristic frequency of diffusional relaxation, $\Gamma = \sum \Gamma_j$ and Γ_j and D_j are the adsorption and the diffusion coefficient for the jth component of the solution, respectively.

To determine the derivatives $\partial c_j/\partial \Gamma_j$, the equation

$$\frac{\partial c_j}{\partial \Gamma_j} = \left[\sum_i \left(\frac{\partial \Gamma_j}{\partial c_i} \right)_{c \neq c_i} \left(\frac{\partial c_i}{\partial t} \right) \bigg/ \left(\frac{\partial c_j}{\partial t} \right) \right]^{-1} \tag{31}$$

was proposed in [69]. The derivatives $\partial c/\partial t$ (with respect to time t) refer to the dynamic subsurface layer. In some cases it becomes possible to neglect the cross terms $i \neq j$ and in Eq. (31) the derivatives $\partial c_j/\partial \Gamma_j$ can be expressed by $\partial c_j/\partial \Gamma_j = (\partial c_j/\partial \Gamma_j)_{c_i}$ as proposed in [68]. A similar result for two-component systems (with the components distinguished by subscripts 1 and 2) follows, for example, for component 1 if the condition $(\partial \Gamma_1/\partial c_1)_{c_2} \gg (\partial \Gamma_1/\partial c_2)_{c_1}$ is valid. If the values ζ_2 for component 2 are fairly high ($\zeta_2 \gg 1$, viscous surface layer behavior), then Eq. (30) can be transformed into

$$|E| = (E_r^2 + E_i^2)^{\frac{1}{2}} = \frac{E_0(\Gamma_1/\Gamma)}{(1 + 2\zeta_1 + 2\zeta_1^2)^{\frac{1}{2}}} \tag{32}$$

An expression for the limiting (high-frequency) elasticity E_0 for a mixture of two surfactants was given by Joos (cf. Eq. (2.179) in [55]):

$$E_0 = \frac{(\Gamma_2 + \Gamma_1 \Gamma_2^0/\Gamma_1^0) E_1^0 E_2^0}{\Gamma_2 E_1^0 + \Gamma_1 \Gamma_2^0/\Gamma_1^0 E_2^0} \tag{33}$$

where the superscript 0 refers to the individual solutions with the same concentration as in the surfactant mixture, i.e. $E_j^0 = -(d\gamma/d\ln \Gamma_j^0)$ is the limiting elasticity of the jth surfactant individual solution and Γ_j^0 is the corresponding adsorption.

The theory of Garrett and Joos [69] was further developed in [70, 71]. In [70], an irreversible thermodynamic approach was applied and the general method for the derivation of equations for the surface elasticity of the mixture of an arbitrary number of surfactants and mixed adsorption kinetics was proposed. For mixtures of two surfactants, an analytical expression for the surface dilatational modulus was derived in [71]. This expression involves four partial derivatives of the adsorptions with respect to concentrations and two partial derivatives of the surface tension with respect to adsorptions:

$$E = -\frac{1}{B}\left(\frac{\partial\gamma}{\partial \ln \Gamma_1}\right)_{\Gamma_2}\left[\sqrt{\frac{i\omega}{D_1}}a_1 + \sqrt{\frac{i\omega}{D_2}}a_2\frac{\Gamma_2}{\Gamma_1} + \frac{i\omega}{\sqrt{D_1 D_2}}(a_1a_4 - a_2a_3)\right]$$
$$-\frac{1}{B}\left(\frac{\partial\gamma}{\partial \ln \Gamma_2}\right)_{\Gamma_1}\left[\sqrt{\frac{i\omega}{D_2}}a_4 + \sqrt{\frac{i\omega}{D_1}}a_3\frac{\Gamma_1}{\Gamma_2} + \frac{i\omega}{\sqrt{D_1 D_2}}(a_1a_4 - a_2a_3)\right] \quad (34)$$

where $a_1 = (\partial\Gamma_1/\partial c_1)_{c_2}$, $a_2 = (\partial\Gamma_1/\partial c_2)_{c_1}$, $a_3 = (\partial\Gamma_2/\partial c_1)_{c_2}$, $a_4 = (\partial\Gamma_2/\partial c_2)_{c_1}$ and $B = 1 + \sqrt{i\omega/D_1}a_1 + \sqrt{i\omega/D_2}a_4 + (i\omega/\sqrt{D_1D_2})(a_1a_4 - a_2a_3)$.

It can be expected that the results obtained with the theory for surfactant mixtures are applicable to protein/surfactant mixtures. Clearly, such application should account for the peculiarities of the adsorption and rheological behavior of proteins.

As Eqs. (31) to (34) involve the adsorptions of the components both in the mixture Γ_j and in the individual solutions Γ_j^0, we consider the results obtained for the mixed protein/surfactant solutions. With the approximation $\Omega_0 \cong \Omega_S$ (where Ω_S is the molar area of surfactant), the following equation of state for a protein/non-ionic surfactant mixture was derived recently [72]:

$$-\frac{\Pi\Omega_0}{RT} = \ln(1 - \theta_P - \theta_S) + \theta_P(1 - \Omega_0/\Omega) + a_P\theta_P^2 + a_S\theta_S^2 + 2a_{PS}\theta_P\theta_S \quad (35)$$

where a_{PS} is a parameter which describes the interaction between the protein and surfactant molecules; the subscripts P and S refer to protein and surfactant, respectively. A small difference between Ω_0 and Ω_S can be accounted for by introducing

$$\Omega_0^* = \frac{\Omega_0\theta_P + \Omega_S\theta_S}{\theta_P + \theta_S} \quad (36)$$

into Eq. (35) instead of Ω_0. For the protein in state $j=1$ and the surfactant, the adsorption isotherms read [72]

$$b_{P1}c_P = \frac{\Omega\Gamma_{P1}}{(1 - \theta_P - \theta_S)^{\Omega_1/\Omega}}\exp[-2a_P(\Omega_1/\Omega)\theta_P - 2a_{PS}\theta_S] \quad (37)$$

$$b_Sc_S = \frac{\theta_S}{(1 - \theta_P - \theta_S)}\exp(-2a_S\theta_S - 2a_{PS}\theta_P) \quad (38)$$

and the distribution of protein adsorptions over the states j is given by the expression

$$\Gamma_{Pj} = \Gamma_P \frac{(1-\theta_P-\theta_S)^{\frac{\Omega_j-\Omega_1}{\Omega}} \exp[2a_P\theta_P(\Omega_j-\Omega_1)/\Omega]}{\sum_{i=1}^{n}(1-\theta_P-\theta_S)^{\frac{\Omega_i-\Omega_1}{\Omega}} \exp[2a_P\theta_P(\Omega_i-\Omega_1)/\Omega]} \tag{39}$$

The compressibility of the surfactant molecules in the mixed adsorption layer can be treated in the framework of the approximate Frumkin-type Model 2 (neglecting the contribution of non-ideality of entropy in Eqs. (5) and (6)). In this case Ω_S and θ_S in Eqs. (35) to (39) should be described by equations that take into account the 2D compressibility of surfactant adsorption layer:

$$\Omega_S = \Omega_{S0}(1-\varepsilon\Pi) \tag{40}$$

$$\theta_S = \Gamma_S\Omega_{S0}(1-\varepsilon\Pi) \tag{41}$$

Note that the use of Eqs. (40) and (41) for surfactant in protein/surfactant mixtures and the assumption $a_{PS}=0$ are probably more preferable when strong inhomogeneities in the mixed surface layers have to be expected. In fact, protein and surfactant molecules practically do not mix in surface layers but form domains containing essentially one of the components [73, 74]. Therefore, the problem of the theoretical description of a mixture can be formulated as follows. Given the known values of Ω_0, Ω_{S0}, Ω_{min}, Ω_{max}, a_S, a_P, b_{P1} and b_S for the individual protein and surfactant solutions, the parameters Γ_P, Γ_S, θ_P, θ_S, Π and partial derivatives of Eqs. (31) and (34) for mixture can be calculated.

Figure 12.12 illustrates the experimental surface tension isotherm for decyldimethylphosphine oxide and β-lactoglobulin (β-LG)/C_{10}DMPO mixtures for different C_{10}DMPO concentrations at a fixed β-LG concentration of 10^{-6} mol L^{-1} (taken from [75]). The two experimental series of data, referring to solutions with and without the addition of sodium azide, are in good agreement. The theoretical curve for C_{10}DMPO shown in Fig. 12.12 was calculated from the Frumkin model, while the theoretical curve for the mixtures was calculated assuming $a_{PS}=0$. Fairly satisfactory agreement between the experimental results and the theory was established.

To calculate the limiting elasticity E_0 and the frequency-dependent elasticity module $|E|$, the adsorption values of β-LG and C_{10}DMPO in the mixture and in the individual solutions should also be known. Figure 12.13 exhibits the dependence of the calculated adsorption values for C_{10}DMPO in the individual solution and the adsorptions of β-LG and C_{10}DMPO in the mixture as a function of the C_{10}DMPO concentration at a fixed β-LG concentration of 10^{-6} mol L^{-1}. For C_{10}DMPO concentrations above 10^{-4} mol L^{-1}, the protein is significantly displaced from the adsorption layer by the surfactant, whereas at C_{10}DMPO concentrations below 10^{-5} mol L^{-1} the adsorptions of β-LG are equal to that in absence of surfactant.

The dependences of the dilatational elasticity modulus on the oscillation frequency at various C_{10}DMPO concentrations in the β-LG/C_{10}DMPO mixtures are

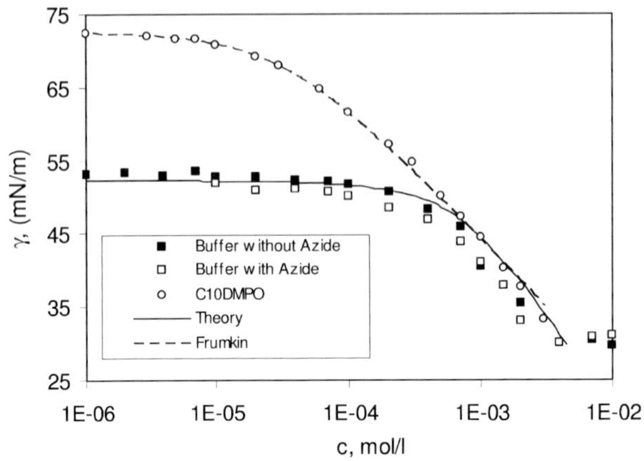

Fig. 12.12 Surface tension isotherms for individual C_{10}DMPO (○) and β-LG/C_{10}DMPO mixtures (■ without sodium azide; □ with sodium azide) vs. C_{10}DMPO concentration; ○ isotherms calculated from Eqs. (3), (4) and (35) to (39) (after [75]).

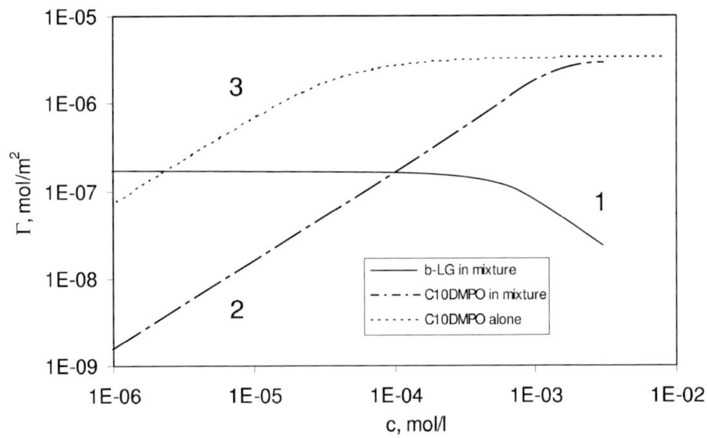

Fig. 12.13 Adsorption of β-LG (1) and C_{10}DMPO (2) in β-LG/C_{10}DMPO mixtures vs. C_{10}DMPO concentration; (3) C_{10}DMPO adsorption in absence of protein (after [75]).

shown in Fig. 12.14. Here the values of the phase angle φ determined as $\cos\varphi = E_r/|E|$ were in the range 3–12°, i.e. the mixed surface layers behave almost ideally elastic. With increasing oscillation frequency, the φ values decrease. At low concentrations the increase in C_{10}DMPO resulted in a decrease of the angle φ compared with the pure β-LG solution and subsequently in an increase. This last increase is attributable to the increasing area covered by C_{10}DMPO alone. Note that for pure C_{10}DMPO solutions in the frequency range studied, viscoelastic behavior was observed. It follows from Fig. 12.14 that, with increas-

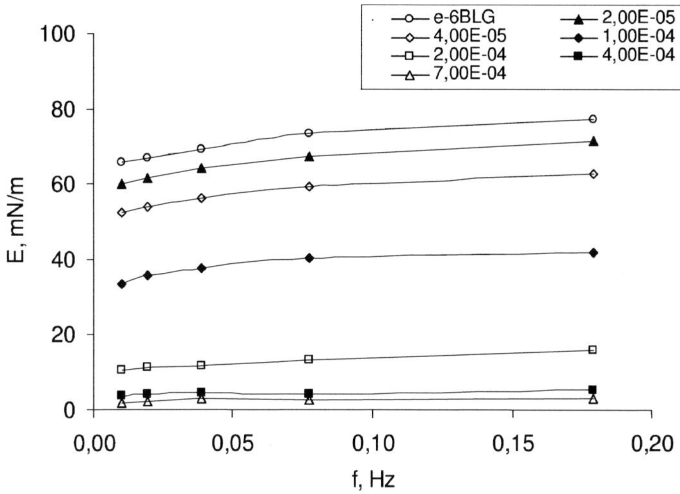

Fig. 12.14 Experimental dependence of surface dilatational modulus on oscillations frequency for β-LG/C_{10}DMPO mixtures at various C_{10}DMPO concentrations: ○ 10^{-6} mol L^{-1} β-LG without C_{10}DMPO; ▲ with addition of 2×10^{-5} mol L^{-1} C_{10}DMPO; ◇ 4×10^{-5} mol L^{-1}; ◆ 10^{-4} mol L^{-1}; □ 2×10^{-4} mol L^{-1}; ■ 4×10^{-4} mol L^{-1}; △ 7×10^{-4} mol L^{-1} (after [75]).

ing C_{10}DMPO concentration, the elasticity modulus of the β-LG/C_{10}DMPO mixture decreases significantly. For example, the modulus for β-LG mixed with 7×10^{-4} mol L^{-1} C_{10}DMPO is 20 times lower than that for pure β-LG.

The limiting elasticity E_0 and the diffusion relaxation frequency ω_D for the β-LG solution of 10^{-6} mol L^{-1} were calculated from the corresponding frequency dependence shown in Fig. 12.14 using the extrapolation procedure proposed in [4]. The estimated values of $E_0=82$ mN m^{-1} and $\omega_D=0.02$ rad s^{-1} agree well with the data obtained in [54]. The corresponding values of E_0 and ω_D for the C_{10}DMPO solutions were obtained from oscillating bubble experiments in [28]. For the C_{10}DMPO concentrations studied in the present work, the E_0 values vary in a narrow range, i.e. between 30 and 35 mN m^{-1} [28]. At the same time, it was found that the experimental dependence of ω_D on c_S is almost linear and can be described fairly well by the linear relationship $\omega_D=120c_S$, where ω_D is in rad s^{-1} and c_S in mmol L^{-1}.

Figure 12.15 illustrates calculations of the elasticity modulus for β-LG/C_{10}DMPO mixtures according to Eq. (32). Here E_0 for the mixtures were calculated from Eq. (33). The values ω_{Dj} for the β-LG/C_{10}DMPO mixture components are determined from the partial derivatives $\partial c_j/\partial\Gamma_j[\omega_{Dj} = D_j(\partial c_j/\partial\Gamma_j)^2]$, which were approximated for β-LG by $(\partial c_j/\partial\Gamma_j)_{c_i}$. This substitution is correct (cf. Eq. 31) because the dependence of β-LG adsorption on C_{10}DMPO concentration in the studied concentration range is very weak (cf. Fig. 12.13). The values ζ for the individual C_{10}DMPO adsorption layers in the given frequency range are fairly high ($\zeta \gg 1$) and become still higher for mixtures with β-LG. Therefore,

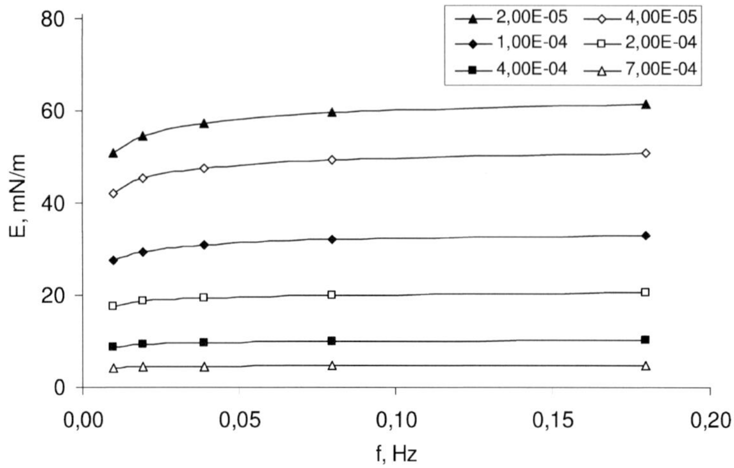

Fig. 12.15 Theoretical dependence of surface dilatational modulus on oscillation frequency for β-LG/C_{10}DMPO mixtures at various C_{10}DMPO concentrations calculated from Eqs. (32) and (33); notations as in Fig. 12.14 (after [75]).

the influence of all the derivatives in Eq. (31) on the value $\partial c_j/\partial \Gamma_j$ for C_{10}DMPO is not significant and Eq. (30) for the system studied can be reduced to Eq. (32), where the subscript 1 refers to the protein.

Comparing the curves in Fig. 12.14 with those in Fig. 12.15, we find fairly satisfactory agreement between the experimental elasticity values and those predicted by the theoretical model. Probably, using Eq. (34) together with Eqs. (35) and (41) would give better agreement between the experiment and the theoretical model.

12.5 Conclusions

The analysis of experimental data for surfactants of different types leads to the conclusion that the rheological characteristics of surface layers are far more sensitive to the interactions between the adsorbed molecules than the equilibrium surface pressure/surface tension isotherms. The well-known Frumkin model describes equilibrium tensiometry data very well; however, the resulting surface dilatational rheological behavior deviates drastically from experimental findings. At the same time, the theoretical models assuming intrinsic compressibility of adsorbed molecules [18, 19, 31, 41] are capable of describing not only equilibrium tensiometric data but also the surface rheological behavior. It is also worth noting that the parameters of these models (e.g. compressibility coefficient, molar areas) are in good agreement with the results of independent experiments and that these models allow one to avoid the assumption of a negative interaction parameter in the Frumkin model. Hence the analysis performed confirms the physical adequacy of the theoretical models accounting for an intrinsic molecular compressibility. In addition, it emphasizes the importance of surface rheological studies which are much more sensitive than the molecular peculiarities of surfactants in adsorption layers.

Measured values of the surface dilatational elasticity of protein adsorption layers can be described satisfactorily by means of a theoretical model [57, 58] which assumes that adsorbed protein molecules can exist in multiple conformations characterized by different molecular areas. The main factor that affects the dependence of the limiting elasticity $E_0(\Pi)$ is the intrinsic compressibility of the protein adsorption layer: the change of the mean molar area with the protein adsorption. The theoretical value of E_0 for proteins is lower than that characteristic of adsorbed molecules with constant area by the factor $(1 + \mathrm{d}\ln\Omega/\mathrm{d}\ln\Gamma)$. This results in lower values of E_0, since configurations with lower molecular areas are favored at the expense of configurations requiring larger area, especially at high surface pressures. For β-casein as a flexible protein, the dependence of the average molar area on surface pressure is much stronger than that for the globular protein BSA. This stronger dependence accounts for two characteristic differences between the two proteins: (1) surface elasticities are much lower for the flexible β-casein; (2) for β-casein, the elasticity E_0 passes through a maximum followed by a minimum at intermediate values of the surface pressure, whereas for BSA the elasticity exhibits a monotonic increase with increasing surface pressure.

The theoretical model discussed here for qualitative analysis of the dilatational elasticity of mixed surfactant solutions demonstrates fairly satisfactory agreement with experimental results obtained from oscillating bubble experiments for mixtures of the non-ionic surfactant C_{10}DMPO with the protein β-LG. Note that the parameters of the model used to describe the adsorption behavior of protein in the mixture account for the specific features characteristic of the protein molecule in the solution, namely the capability of folding and unfolding of the molecule in the surface layer. In this context, the ability of the protein to decrease its molar area in a saturated monolayer was an important fact to be considered in the presence of increasing amounts of surfactant.

Acknowledgments

This work was financially supported by projects of the ESA (FAST and FASES) and the German Science Foundation (Mi 418/14-1).

References

1 G. Loglio, R. Miller, A. M. Stortini, U. Tesei, N. Degli Innocenti, R. Cini, *Colloids Surf. A, 90* (**1994**) 251; *95* (**1995**) 63.

2 E. K. Zholkovskij, V. I. Kovalchuk, V. B. Fainerman, G. Loglio, J. Krägel, R. Miller, S. A. Zholob, S. S. Dukhin, *J. Colloid Interface Sci., 224* (**2000**) 47.

3 J. Lucassen, M. van den Tempel, *Chem. Eng. Sci., 27* (**1972**) 1283.

4 J. Lucassen, D. Giles, *J. Chem. Soc., Faraday Trans. 1, 71* (**1975**) 217.

5 B. A. Noskov, *Adv. Colloid Interface Sci., 95* (**2002**) 236.

6 G. Loglio, P. Pandolfini, R. Miller, A. V. Makievski, F. Ravera, M. Ferrari, L. Liggieri, in *Novel Methods to Study Interfacial Layers. Studies in Interface Science,* Vol. 11, D. Möbius, R. Miller (eds.). Elsevier, Amsterdam, **2001**, p. 439.

7 R. Miller, R. Sedev, K.-H. Schano, Ch. Ng, A. W. Neumann, *Colloids Surf. A, 69* (**1993**) 209.

8 R. Miller, Z. Policova, R. Sedev, A. W. Neumann, *Colloids Surf. A, 76* (**1993**) 179.

9 J. Benjamins, A. Cagna, E. H. Lucassen-Reynders, *Colloids Surf. A, 114* (**1996**) 245.

10 E. H. Lucassen-Reynders, A. Cagna, J. Lucassen, *Colloids Surf. A, 186* (**2001**) 63.

11 C. Stenvot, D. Langevin, *Langmuir, 4* (**1988**) 1179.

12 Y. Jayalakshmi, L. Ozanne, D. Langevin, *J. Colloid Interface Sci., 170* (**1995**) 358.

13 B. A. Noskov, D. O. Grigoriev, R. Miller, *J. Colloid Interface Sci., 188* (**1997**) 9.

14 B. A. Noskov, D. A. Alexandrov, R. Miller, *J. Colloid Interface Sci., 219* (**1999**) 250.

15 B. A. Noskov, A. V. Akentiev, A. Yu. Bilibin, I. M. Zorin, R. Miller, *Adv. Colloid Interface Sci., 104* (**2003**) 245.

16 K.-D. Wantke, H. Fruhner, in *Drops, Bubbles in Interfacial Research. Studies in Interface Science,* Vol. 6, D. Möbius, R. Miller (eds.). Elsevier Science, Amsterdam, **1998**, p. 327.

17 V. I. Kovalchuk, J. Krägel, E. V. Aksenenko, G. Loglio, L. Liggieri, in *Novel Methods to Study Interfacial Layers. Studies in Interface Science,* Vol. 11, D. Möbius, R. Miller (eds.). Elsevier, Amsterdam, **2001**, p. 485.

18 V. B. Fainerman, R. Miller, V. I. Kovalchuk, *Langmuir, 18* (**2002**) 7748.

19 V. B. Fainerman, R. Miller, V. I. Kovalchuk, *J. Phys. Chem. B, 107* (**2003**) 6119.

20 L. Liggieri, V. Attolini, M. Ferrari, F. Ravera, *J. Colloid Interface Sci., 255* (**2002**) 225.

21 E. H. Lucassen-Reynders, in Anionic Surfactants, *Physical Chemistry of Surfactant Action. Surfactant Science Series,* Vol. 11, E. H. Lucassen-Reynders (ed.). Marcel Dekker, New York, **1981**, p. 173.

22 J. Lucassen, R. S. Hansen, *J. Colloid Interface Sci.,* 23 (**1967**) 319; J. Lucassen, M. van den Tempel, *J. Colloid Interface Sci., 41* (**1972**) 491.

23 M. van den Tempel, E. H. Lucassen-Reynders, *Adv. Colloid Interface Sci., 18* (**1983**) 281.

24 A. Bonfillon, D. Langevin, *Langmuir, 10* (**1994**) 2965.

25 P. Warszynski, K.-D. Wantke, H. Fruhner, *Colloids Surf. A, 189* (**2001**) 29.

26 A. Bonfillon, D. Langevin, *Langmuir, 9* (**1993**) 2172.

27 K-D. Wantke, H. Fruhner, J. Fang, K. Lunkenheimer, *J. Colloid Interface Sci., 208* (**1998**) 34.

28 K.-D. Wantke, H. Fruhner, *J. Colloid Interface Sci., 237* (**2001**) 185.

29 V. I. Kovalchuk, J. Krägel, A. V. Makievski, G. Loglio, F. Ravera, L. Liggieri, R. Miller, *J. Colloid Interface Sci., 252* (**2002**) 433.

30 V. I. Kovalchuk, G. Loglio, V. B. Fainerman, R. Miller, *J. Colloid Interface Sci., 270* (**2004**) 475.

31 V. B. Fainerman, V. I. Kovalchuk, E. V. Aksenenko, M. Michel, M. E. Leser, R. Miller, *J. Phys. Chem. B, 108* (**2004**) 13700.

32 V. I. Kovalchuk, R. Miller, V. B. Fainerman, G. Loglio, *Adv. Colloid Interface Sci. 114–115* (**2005**) 303–313.

33 U. Gehlert, D. Vollhardt, G. Brezesinski, H. Möhwald, *Langmuir, 12* (**1996**) 4892.

34 U. Gehlert, D. Vollhardt, *Langmuir, 18* (**2002**) 688.

35 V. B. Fainerman, S. A. Zholob, E. H. Lucassen-Reynders, R. Miller, *J. Colloid Interface Sci., 261* (**2003**) 180.

36 V. B. Fainerman, R. Miller, E. V. Aksenenko, A. V. Makievski, In *Surfactants – Chemistry, Interfacial Properties, Applica-*

tion. *Studies in Interface Science*, Vol. 13, V. B. Fainerman, D. Möbius, R. Miller (eds.). Elsevier, Amsterdam, **2001**, pp. 189–286.
37 J. A. V. Butler, *Proc. R. Soc. London, Ser. A, 138* (**1932**) 348.
38 A. I. Rusanov, *Mendeleev Commun.* (**2002**) 218–220.
39 V. I. Kovalchuk, J. Krägel, A. V. Makievski, F. Ravera, L. Liggieri, G. Loglio, V. B. Fainerman, R. Miller, *J. Colloid Interface Sci., 280* (**2004**) 498.
40 J. R. Lu, R. K. Thomas, J. Penfold, *Adv. Colloid Interface Sci., 84* (**2000**) 143.
41 C. Stubenrauch, V. B. Fainerman, E. V. Aksenenko, R. Miller, *J. Phys. Chem. B, 109* (**2005**) 1505.
42 V. B. Fainerman, E. H. Lucassen-Reynders, *Adv. Colloid Interface Sci., 96* (**2002**) 295.
43 R. A. Robinson, R. H. Stokes, *Electrolyte Solutions.* Butterworths, London, **1965**.
44 F. Monroy, J. Giermanska Khan, D. Langevin, *Colloids Surf. A, 143* (**1998**) 251.
45 V. Bergeron, *Langmuir, 13* (**1997**) 3474.
46 E. A. Simister, E. M. Lee, R. K. Thomas, J. Penfold, *J. Phys. Chem., 96* (**1992**) 1373.
47 G. Serrien, G. Geeraerts, L. Ghosh, P. Joos, *Colloids Surf., 68* (**1992**) 219.
48 J. Benjamins, E. H. Lucassen-Reynders, in *Proteins at Liquid Interfaces,* D. Möbius, R. Miller (eds.). Elsevier, Amsterdam, **1998**, p. 341.
49 D. E. Graham, M. C. Phillips, *J. Colloid Interface Sci., 76* (**1980**) 227.
50 N. Puff, A. Cagna, V. Aguie-Beghin, R. Douillard, *J. Colloid Interface Sci., 208* (**1998**) 405.
51 A. Williams, A. Prins, *Colloids Surf. A, 114* (**1996**) 267.
52 M. Mellema, D. C. Clark, F. A. Husband, A. R. Mackie, *Langmuir, 14* (**1998**) 1753.
53 M. Blank, J. Lucassen, M. van den Tempel, *J. Colloid Interface Sci., 33* (**1970**) 94.
54 J. Benjamins, Static, Dynamic Properties of Proteins Adsorbed at Liquid Interfaces. Thesis, Wageningen University, **2000**.
55 P. Joos, *Dynamic Surface Phenomena.* VSP, Dordrecht, **1999**.
56 P. Joos, G. Serrien, *J. Colloid Interface Sci., 145* (**1991**) 291.
57 V. B. Fainerman, E. H. Lucassen-Reynders, R. Miller, *Adv. Colloid Interface Sci., 106* (**2003**) 237.
58 E. H. Lucassen-Reynders, V. B. Fainerman, R. Miller, *J. Phys. Chem. B, 108* (**2004**) 9173.
59 J. Benjamins, J. A. de Feijter, M. T. A. Evans, D. E. Graham, M. C. Phillips, *Discuss. Faraday Soc., 59* (**1975**) 218.
60 J. A. de Feijter, J. Benjamins, F. A. Veer, *Biopolymers, 17* (**1978**) 1759.
61 A. Aschi, A. Charbi, L. Bitri, P. Calmettes, M. Daoud, V. Aguie-Beghin, R. Douillard, *Langmuir, 17* (**2001**) 1896.
62 R. Miller, V. B. Fainerman, A. V. Makievski, J. Krägel, D. O. Grigoriev, V. N. Kazakov, O. V. Sinyachenko, *Adv. Colloid Interface Sci., 86* (**2000**) 39.
63 A. Dussaud, G. B. Han, L. Ter Minassian-Saraga, M. Vignes-Adler, *J. Colloid Interface Sci., 167* (**1994**) 247.
64 N. J. Turro, X.-G. Lei, K. P. Ananthapadmanabhan, M. Aronson, *Langmuir, 11* (**1995**) 2525.
65 R. Wüstneck, J. Krägel, R. Miller, P. J. Wilde, D. C. Clark, *Colloids Surf. A, 114* (**1996**) 255.
66 J. Krägel, M. O'Neill, A. V. Makievski, M. Michel, M. E. Leser, R. Miller, *Colloids Surf. B, 31* (**2003**) 107.
67 E. Dickinson, *Colloids Surf. B, 15* (**1999**) 161.
68 E. H. Lucassen-Reynders, *J. Colloid Interface Sci., 42* (**1973**) 573.
69 P. R. Garrett, P. Joos, *J. Chem. Soc., Faraday Trans. 1, 72* (**1976**) 2161.
70 B. A. Noskov, G. Loglio, *Colloids Surf. A, 141* (**1998**) 167.
71 Q. Jiang, J. E. Valentini, Y. C. Chiew, *J. Colloid Interface Sci., 174* (**1995**) 268.
72 V. B. Fainerman, S. A. Zholob, M. Leser, M. Michel, R. Miller, *J. Colloid Interface Sci., 274* (**2004**) 496.
73 A. R. Mackie, A. P. Gunning, M. J. Ridout, P. J. Wilde, V. I. Morris, *Langmuir, 17* (**2001**) 6593.
74 A. R. Mackie, A. P. Gunning, M. J. Ridout, P. J. Wilde, J. R. Patino, *Biomacromolecules, 2* (**2001**) 1001.
75 R. Miller, M. E. Leser, M. Michel, V. B. Fainerman, *J. Phys. Chem., 109* (**2005**) 13 327–13 331.

13
Metastability and Lability in Surface Phase Transitions: Surface Forces and Line Tension Effects

Borislav V. Toshev

Abstract

The transition of a homogeneous into a heterogeneous system may occur spontaneously in a labile region of supersaturations; in the metastable region of supersaturations it is realized as an energy barrier-determined process. This chapter examines the factors that affect the boundary separating these two zones. One of these factors is the line tension of a three-phase contact line. The origin of this quantity is in the long-range surface forces in the zone where three bulk phases meet each other. The analysis of the equilibrium and stability conditions of such systems is carried out by the omega potential thermodynamic formalism. It is shown that the core of this method can be found in Gibbs' theory of capillarity.

13.1
Introduction

This chapter is organized into four main sections. In Section 13.2, the special position of the omega potential in the theory of capillarity is evidenced: no other extensive thermodynamic functions possess its merits, especially when the mechanical equilibrium conditions are studied.

The advantage of using the thermodynamic approach with the omega potential, even as a teaching strategy for the theory of capillarity, is demonstrated in Section 13.3 by considering the case of condensation on ions and deriving the equation of Tohmfor and Volmer.

The core point of this chapter is the boundary that separates the labile region of supersaturations from the metastable region of supersaturations and the factors that affect that boundary [1] (Sections 13.3 and 13.4). One of these factors, unknown earlier, is the line tension. The line tension of the three-phase contact line is a product of long-range surface forces in the zone where three bulk and surface phases meet each other. This thermodynamic quantity can have positive or negative values, as Gibbs mentioned [2] (Section 13.5).

Colloids and Interface Science Series, Vol. 1
Colloid Stability: The Role of Surface Forces, Part I. Edited by Tharwat F. Tadros

ISBN: 978-3-527-31462-1

As a definition, we adopt that the nucleus of the new phase is that formation which is in *unstable* equilibrium with its surroundings. Therefore, no nuclei exist in the labile region of supersaturations.

13.2
Omega Potential Thermodynamic Formalism

Despite of the statement, e.g. [3, 4], that the Ω potential – expressed as

$$\Omega = F - G \tag{1}$$

where F is the Helmholtz free energy, $G \equiv \sum_i \mu_i N_i$ is the Gibbs free energy (μ_i and N_i are the chemical potential and mass of the ith component, respectively) – is irrelevant for thermodynamics, this extensive thermodynamic function, rather than F and G, seems to be the most convenient for presenting the results of Gibbs's theory of capillarity [2] (cf. [5]).

Linford's well-known interpretation [6] of the Δ quantities of the basic extensive thermodynamic functions is as follows:

$$\begin{aligned} (\Delta F)_{\mathrm{T}} &= \text{reversible work (mechanical + useful)} \\ (\Delta G)_{\mathrm{T,P}} &= \text{reversible work (useful)} \\ (\Delta \Omega)_{\mathrm{T,V},\mu_i} &= \text{reversible work (mechanical)} \end{aligned} \tag{2}$$

where T=absolute temperature, P=pressure and V=volume. Even these simple declarations imply that the Ω potential may be of advantage in comparison with the other thermodynamic functions when the condition for mechanical equilibrium together with its stability is searched for.

An observation shows expressively the special position of the Ω potential among the other thermodynamic functions in Gibbs' surface thermodynamics [7].

The basic idea of the Gibbs surface thermodynamics is in the substitution of the *real* heterogeneous system of bulk phases and surfaces of discontinuity with an *idealized* Gibbs' system that consists of bulk homogeneous phases only. The correspondence between the two systems – the real system and the idealized one – is realized by introducing Gibbs' excesses of the extensive thermodynamic quantities. These *excess* quantities depend on the position of the Gibbs *dividing surface.*

Let us consider a real system of two bulk phases, say g and l, and a surface of discontinuity between them. The idealized system contains only the bulk phases with volumes V^g and V^l determined at a given position of the dividing surface $V^g+V^l=V$, where V is the volume of the real system. The definition of the excess of any extensive thermodynamic function E then will be

$$\tilde{E} = E - e^g V - \Delta e V^l \, ; \quad \Delta e = e^l - e^g \tag{3}$$

where $\Delta e > 0$, $e \equiv$ volume density of E. Obviously the quantities $\tilde{E}$ are not physical properties of the system under consideration because their values are not invariant with respect to the position of the dividing surface (this position determines the volume V^{l}).

It is convenient to rearrange Eq. (3) as

$$\tilde{e} = \tilde{E}/O = \tilde{e}^0 - \Delta e \frac{V^{\mathrm{l}}}{O}\,; \quad \tilde{e}^0 = \frac{E - e^{\mathrm{g}} V}{O} \tag{4}$$

where O stands for the area of the surface phase, and, for a plane interface, this transforms into

$$\tilde{e} = \tilde{e}^0 - \Delta e H \tag{5}$$

In the latter case, the real system is supposed to be in a parallelepiped box with volume V, the two phases are taken to be homogeneous right to the dividing surface (plane), the phase l is in the "under" part of the box, the phase g is in the "upper" part of the box and H is the distance between the "under" side of the box and the dividing surface. The quantity $\tilde{e}^0$ does not depend on the position of the dividing surface; its value corresponds formally to $\tilde{e}$ at H (or V^{l}) $= 0$.

When applying Eq. (5), for instance, to F, G and N_{i}, equations for $\tilde{f}$ (specific surface Helmholtz free energy), $\tilde{g}$ (specific surface Gibbs energy) and Γ_{i} (Gibbs' adsorption) are readily obtained:

$$\begin{aligned} \tilde{f} &= \tilde{f}^0 - \Delta f H \\ \tilde{g} &= \tilde{g}^0 - \Delta g H \\ \Gamma_{\mathrm{i}} &= \Gamma_{\mathrm{i}}^0 - \Delta n_{\mathrm{i}} H \end{aligned} \tag{6}$$

However, when applying Eq. (5) to the Ω potential, then

$$\tilde{\omega} = \tilde{f}^0 - \tilde{g}^0 \tag{7}$$

i.e. $\tilde{\omega}$, defined as an excess quantity, is not in fact an excess quantity and, therefore, it should be system's property that would be determined by some experiment.

In fact, Eq. (7) coincides with Gibbs' equation (659) [2]. The latter equation is a definition of Gibbs' σ quantity, i.e., $\tilde{\omega} \equiv \sigma$. In [2, p. 315], instructive remarks can be found: "We may regard σ as expressing the work spent in forming a unit of the surface of discontinuity ... but it cannot properly be regarded as expression the tension of the surface. The latter quantity depends upon the work spent in stretching the surface $[(\partial\tilde{\Omega}/\partial O)_{\mathrm{T},\mu_{\mathrm{i}}}]$, while the quantity σ depends upon the work spent in forming the surface $[\sigma \equiv \tilde{\omega} = \tilde{\Omega}/O]$. With respect to perfectly fluid masses, these processes [and quantities] are not distinguishable".

Many years later, Eriksson followed such a line of consideration [8] but he did not recognize that the quantity g^{s}_{ex}, interpreted as a work of reversible cleavage work, is the quantity $\tilde{\omega}$, as defined by Eq. (7).

13.3
Metastability and Lability in Homogeneous Condensation

This section reveals our explanatory strategy for using of the Ω potential thermodynamic formalism in the theory of capillarity. The example considered is condensation on ions [9]. It seems to be worthwhile to follow such a narrative because this case demonstrates explicitly the existence of two zones of supersaturation and the boundary that separates the metastable zone from the labile zone.

The Ω potential of the homogeneous gas phase g containing an ion with charge e and radius a is

$$\Omega_{\mathrm{I}} = -p^{\mathrm{g}}V + \frac{1}{2}\frac{e^2}{a} \tag{8}$$

The second term on the right-hand side is the electrostatic part of Ω_{I}, i.e. the free energy Ω_{e}, calculated by $\Omega_{\mathrm{e}} = \int_{\mathrm{V}}(\varepsilon/8\pi)E^2\mathrm{d}V$ with $E = e/r^2$, where r is the radial distance from the ion and ε (=1 for vapors) is the dielectric constant.

When a liquid drop l is formed around the ion

$$\Omega_{\mathrm{II}} = -p^{\mathrm{g}}(V - v) - p^{\mathrm{l}}v + \sigma o + \frac{1}{2}e^2\left(1 - \frac{1}{\varepsilon}\right)\frac{1}{R} + \frac{1}{2}\frac{e^2}{\varepsilon a} \tag{9}$$

with $v = (4/3)\pi R^3$ and $o = 4\pi R^2$, where R is the radius of the spherical liquid drop and σ is the surface tension of the gas/liquid interface. The free energy Ω_{e} is calculated again by $\Omega_{\mathrm{e}} = \int_{\mathrm{V}}(\varepsilon/8\pi)E^2\mathrm{d}V$ with $E = e/r^2$ for $r > R$ and $E = e/\varepsilon r^2$ for $r < R$.

Hence the work of formation of a drop will be

$$\Delta\Omega = -\Delta P v + \sigma o + \frac{1}{2}e^2\left(1 - \frac{1}{\varepsilon}\right)\frac{1}{R} - \frac{1}{2}e^2\left(1 - \frac{1}{\varepsilon}\right)\frac{1}{a} \tag{10}$$

with $\Delta P = p^{\mathrm{l}} - p^{\mathrm{g}}$.

By making use of $\mathrm{d}\Delta\Omega/\mathrm{d}R = 0$, one obtains for the *capillary pressure* of the drop

$$\Delta P = \frac{2\sigma}{R} - \frac{1}{8\pi}\left(1 - \frac{1}{\varepsilon}\right)\frac{1}{R^4} \tag{11}$$

This equation is a generalization of the Laplace equation for the capillary pressure (at $e = 0$):

$$\Delta P = 2\sigma/R\,. \tag{12}$$

Equation (11) exhibits an extremum at

$$R_{\rm m}^3 = \frac{e^2}{4\pi\sigma}\left(1 - \frac{1}{\varepsilon}\right) \tag{13}$$

The capillary pressure is the pressure difference between two bulk phases in equilibrium and the Laplace formula (12) holds for the *surface of tension* of the surface of discontinuity between those phases [2]. If Eq. (11) is inserted in Eq. (10), an expression for the work of formation W of an equilibrium drop from vapors is readily obtained:

$$W = \frac{1}{3}\sigma o + \frac{2}{3}e^2\left(1 - \frac{1}{\varepsilon}\right)\frac{1}{R} + \text{constant}. \tag{14}$$

At $e=0$, Eq. (14) gives the well-known Gibbs' result [2]:

$$W = \frac{1}{3}\sigma o = \frac{1}{2}\Delta P v \tag{15}$$

Equation (11) determines the difference in pressure within and outside the equilibrium drop, $p^{\rm l}$ and $p^{\rm g}$, respectively. Obviously, at constant temperature and chemical potentials, both pressures are also constant. These pressures can be calculated separately provided that the starting equation is [5]

$$\mu^{\rm l}(T, p^{\rm l}) = \mu^{\rm g}(T, p^{\rm g}) \tag{16}$$

Therefore,

$$v^{\rm l}{\rm d}p^{\rm l} = v^{\rm g}{\rm d}p^{\rm g} \tag{17}$$

[$v^{\rm j}$ ($j=l$, g) are molar volumes]. As usual, $v^{\rm l} \ll v^{\rm g}$ and $v^{\rm g}=kN_{\rm a}T/p^{\rm g}$; k and $N_{\rm a}$ are the Boltzmann constant and Avogadro's number, respectively. Then, by integrating Eq. (17) with some algebra and neglecting $v_0^{\rm l}(p^{\rm g} - p_\infty^{\rm g})$ ($v_0^{\rm l}$ is the molecular volume; $p_\infty^{\rm g}$ is the drop vapor pressure at drop radius $R \to \infty$), one obtains

$$kT \ln p^{\rm g}/p_\infty^{\rm g} = kT \ln s = v_0^{\rm l}\left[\frac{2\sigma}{R} - \frac{1}{8\pi}e^2\left(1 - \frac{1}{\varepsilon}\right)\frac{1}{R^4}\right] \tag{18}$$

where s is the supersaturation, defined as $s \equiv p^{\rm g}/p_\infty^{\rm g}$. This is a generalization of the Gibbs-Thomson equation, which in the theory of homogeneous nucleation from vapors is usually written as

$$kT\ln\left(\frac{p^{g}}{p^{g}_{\infty}}\right) = kT\ln s = \frac{2\sigma v^{l}_{0}}{R} \tag{19}$$

In contrast with the case described by Eq. (19), Eq. (18) exhibits an extremum at $R=R_m$ when $s=s_m$. At $s=1$, $R=R_o$ and then

$$(R_m/R_0)^3 = 4 \tag{20}$$

The vapor pressure dependence $\ln s$ vs. R is shown in Fig. 13.1. The analysis below shows that s_m is just the boundary that separates the labile region of supersaturations ($s>s_m$) from the metastable region of supersaturations ($s<s_m$). Let us mention again that in the metastable (fluctuational) region the formation of nuclei of the new phase is a barrier-determined process at $s>s_{cr}$ where the critical supersaturation s_{cr} is defined by the well-known Volmer's procedure [1]. It is expected that $s_{cr}<s_m$.

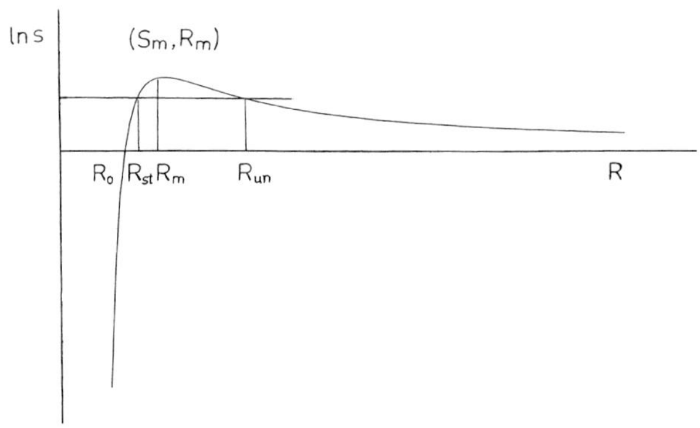

Fig. 13.1 Dependence of the vapor pressure of a charged drop on its size (notations according to the text).

At $s<s_m$, two equilibrium drops of different size could coexist: a smaller one with radius R_{st} and a larger one with radius R_{un}; $R_{st}<R_{un}$ (Fig. 13.1). By using Eqs. (10) and (11), one readily obtains

$$\left(\frac{\partial^2\Omega}{\partial R^2}\right)_{eq} = -8\pi\sigma + 2e^2\left(1-\frac{1}{\varepsilon}\right)\frac{1}{R^3} = 8\pi\sigma\left[\left(\frac{R_m}{R}\right)^3 - 1\right] \tag{21}$$

This implies $(\partial^2\Omega/\partial R^2)_{eq}>0$ when $R<R_m$. Hence, at $s<s_m$ the equilibrium drop with radius R_{st} is in stable equilibrium with its surroundings (minimum of the Ω potential).

For the other equilibrium drop with $R_{un}>R_m$, $\partial^2\Omega/\partial R^2)_{eq}<0$. Therefore, this drop is in unstable equilibrium with its surroundings (maximum of the Ω potential).

At $s \geq s_m$, no condensation occurs by a fluctuational mechanism and, consequently, no nuclei of the new phase exist. Let us consider the special case of $s = s_m$. It is easily established that the first and second derivatives of Ω at $R = R_m$ are both zero. The third derivative is

$$\left(\frac{\partial^3 \Omega}{\partial R^3}\right)_{R=R_m} = -24\pi\sigma / R_m < 0 \tag{22}$$

On the other hand

$$\Omega(R) - \Omega(R_m) = \frac{1}{3!}\left(\frac{\partial^3 \Omega}{\partial R^3}\right)_{R=R_m} (R - R_m)^3 \tag{23}$$

i.e. the non-equilibrium drops with $R < R_m$ will grow spontaneously, because $\Omega(R) > \Omega(R_m)$. The same is valid for non-equilibrium drops with $R > R_m$, since in this case $\Omega(R) < \Omega(R_m)$.

The drop that corresponds to the minimum of the Ω potential has a radius R_{st}. Any other non-equilibrium drop, smaller or larger than the equilibrium drop, will spontaneously change its size until the equilibrium radius R_{st} is reached. Then the work of nucleus formation W_{TV} is obtained by applying Eq. (14):

$$W_{TV} = W_{un} - W_{st} = \frac{1}{3}\sigma(o_{un} - o_{st}) - \frac{2}{3}e^2\left(1 - \frac{1}{\varepsilon}\right)\left(\frac{1}{R_{st}} - \frac{1}{R_{un}}\right) \tag{24}$$

This is the equation of Tohmfor and Volmer obtained in 1938 in the framework of the method of Becker and Döring by considering nucleus formation in a medium containing ions [10] (cf. also [1, 11, 12]).

The above analysis shows that at a given ion charge e the boundary between the metastable and labile regions of supersaturation s_m depends on temperature only (σ is also a temperature-dependent quantity). Thus the metastable region expands and the labile region shrinks when the temperature decreases. However, at $e = 0$ (classical theory of homogeneous nucleation), no labile region can be detected.

13.4
Metastability and Lability in Heterogeneous Condensation

The labile regions of supersaturation cannot be recorded if the theory of homogeneous nucleation is based on Eq. (19). This is the case of supersaturated vapors in the absence of ions within them. We have a similar situation in the case of heterogeneous nucleation with cap-shaped nuclei formed on a substrate. Equation (19) again determines the vapor pressure of such small drops. In both cases, homogeneous and heterogeneous nucleation, Gibbs' equation (15) is valid and, therefore, heterogeneous condensation is more favorable than the homogeneous condensation, because, at a given supersaturation, the volume of the cap-shaped nucleus (spherical segment) is smaller than the volume of the homogeneous nucleus (sphere).

However, the theory of heterogeneous condensation even in its capillary approximation exhibits new features if the line tension κ of the nucleus contact perimeter is taken into account. In this case, the capillary pressure of the drop-nucleus is not affected by the line energy of the three-phase contact line; it remains as $2\sigma/R$. However, in contrast to the case of homogeneous nucleation, where all the values between 0 and ∞ of R, the radius curvature of the drop, are allowed, now the values of R are restricted by the condition of tangential mechanical equilibrium. This requires, for the case of a drop on a plane surface (Fig. 13.2):

$$\frac{\kappa/\sigma}{R} = (\cos\theta_\infty - \cos\theta)\sin\theta \tag{25}$$

where $\sin\theta = r/R$ (r is the radius of the contact perimeter; $\theta \to \theta_\infty$ when $r \to \infty$). Therefore, instead of Eq. (19), one should write

$$kT\ln\frac{p^{g}}{p^{g}_{\infty}} = kT\ln s = \frac{2\sigma^2 v_0^{l}}{\kappa}(\cos\theta_\infty - \cos\theta)\sin\theta \tag{26}$$

Consequently, the supersaturation s_m, as a boundary between a labile region of supersaturations and a metastable region of supersaturations, appears again. To s_m corresponds θ_m, determined by the condition

$$\cos\theta_{m1,2} = \frac{1}{4}\cos\theta_\infty\left(1 \pm \sqrt{1 + 8/\cos^2\theta_\infty}\right) \tag{27}$$

with two roots: $\theta_{m1} < \theta_\infty$ at $\kappa < 0$; $\theta_{m2} > \theta_\infty$ at $\kappa > 0$ [13, 14].

Figure 13.3 shows the dependence of $\ln(p^g/p^g_\infty)$ on the drop size n (= the number of the molecules in a water droplet formed on a solid substrate) [14]. This dependence is different for the both cases, for positive line tension $\kappa = +10^{-5}$ dyn (curve 3) and for negative line tension $\kappa = -10^{-5}$ dyn (curve 4). The calculations of these curves are carried out with the following data: $v_0^{l} = 2.93\times10^{-23}$ cm^3, $\sigma = 72$ dyn cm^{-1}, $kT = 4\times10^{-14}$ erg, $\theta_\infty = 45^\circ$, i.e., $\theta_{m1} = 25^\circ$ and $\theta_{m2} = 124^\circ$ (Eq. 27). For comparison we show also the κ-unaffected curves 1 ($\theta_\infty = 180^\circ$) and 2 ($\theta_\infty = 45^\circ$). It should be emphasized that the κ-affected curve of $\ln(p^g/g^g_\infty)$ vs. n lies

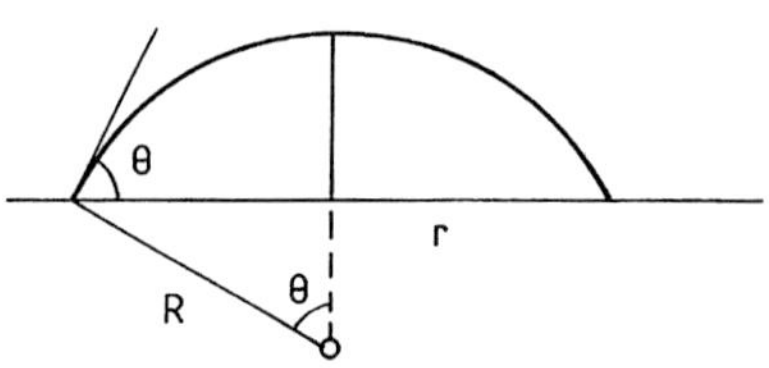

Fig. 13.2 A drop on a plane surface. R = radius of curvature of the drop, r = radius of the drop contact perimeter, θ = contact angle.

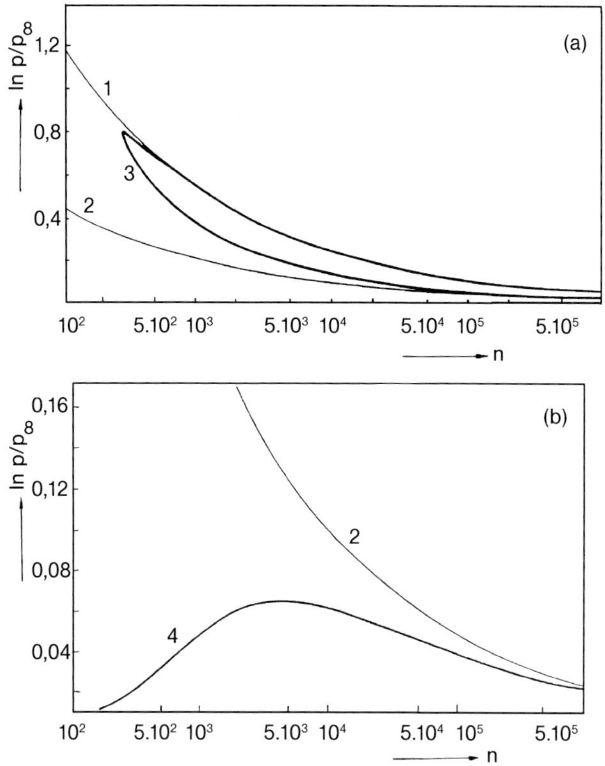

Fig. 13.3 Dependence of ln p^g/p^g_∞ on n, the number of the molecules in a water drop formed on a flat surface. Curve 1, $\theta_\infty = 180°$ and $\kappa = 0$; curve 2, $\theta_\infty = 45°$ and $\kappa = 0$; curve 3, $\theta_\infty = 45°$ and $\kappa = 10^{-5}$ dyn; curve 4, $\theta_\infty = 45°$ and $\kappa = -10^{-5}$ dyn. Note the change in the vertical scale between (a) and (b).

between the uncorrected curves for homogeneous and heterogeneous phase formation when $\kappa > 0$, whereas for $\kappa < 0$ it lies below those curves.

It is important that, owing the line tension effect, two regions of supersaturations are now exhibited. The metastable region of supersaturations for $\kappa > 0$ is larger than that for $\kappa < 0$ and, correspondingly, the labile region of supersaturations for $\kappa < 0$ is larger than that for $\kappa > 0$.

Nuclei exist in the metastable region of supersaturation only. In this region, at a given supersaturation, two equilibrium drops of different size could coexist. For $\kappa > 0$ both equilibrium drops are in unstable equilibrium with their surroundings. Therefore, both drops could be considered as nuclei of the new phase. In fact, the nucleus is the smaller drop because it is formed more easily than the larger drop. For $\kappa < 0$, two equilibrium drops can coexist again. However, the smaller drop cannot be considered as a nucleus of the new phase, because it is in stable equilibrium with its surroundings. Then the work of nucleus formation has to be expressed by

$$W = W_{un} - W_{st} = \frac{1}{2}[\Delta P(v_{un} - v_{st}) + \kappa(l_{un} - l_{st})] \tag{28}$$

where v and l are the volume and contact perimeter length of the unstable (un) and stable (*st*) drops, respectively. Hence the latter case seems to be similar to that of condensation upon electrically charged free droplets (cf. Eqs. 24 and 28).

It is well known that in many cases the critical supersaturation s_{cr}, determined experimentally, is lower than that predicted by the classical theory of heterogeneous nucleation. This may be due to the negative line tension effect. Additionally, one may assume that the both quantities, s_{cr} and s_m, are close each to other. Then $s_{cr} \approx s_m$ can be interpreted as an onset of barrierless condensation. This is an advantage of such an interpretation of the heterogeneous condensation experimental data; in this way, the difficulties with determining the pre-exponential factor of the kinetic Volmer equation [1] can be avoided. For verification of these theoretical predictions, the reverse Wilson chamber (RWC) method was developed [15–19]. In these studies, water vapor condenses on a hexadecane substrate. For this case, the above interpretation of the experimental data obtained gives $\kappa = -1.9 \times 10^{-5}$ dyn. This value of κ is in a good agreement with that obtained earlier for the case of a gas bubble attached to the liquid surface through a Newton black film [20, 21]. The temperature dependence of κ for this case (water drops on a hexadecane substrate) has also been established [22]. Similar studies have been carried out with different immiscible liquid substrates [23].

13.5 Origin and Properties of Line Tension

For the sake of simplicity, we confine our considerations to the simplest case of three flat surfaces of equal surface tensions that meet each other at a straight line. The pressures of the bulk phases are equal and the angles between the plane surfaces are also equal (each one is 120°). Gibbs treats this case [2, p. 292] with the assumption that the line is substituted by a filament of a different phase. The bulk phases in equilibrium are A, B and C and the introduction of the new phase D into the contact line region, in accordance with Stevin's principle, does not disturb the equilibrium of the system. According to Gibbs, in the transition from the volume fluid phase D to a filament in stable equilibrium in the contact line, the system changes "in size, remaining always similar to itself in form, and … the tensions diminish in the same ratio as lines, while the pressures remain constant" [2, p. 292].

According to Gibbs, the theory of equilibrium and stability of such systems may alternatively be developed on the idea of line tension [2, p. 288], which could take a negative value [2, p. 296]. If this method is used, the introduction of an additional volume phase D is no longer necessary since the state of the substances in the line of discontinuity where the molecular fields of all phases are overlapped cannot be identified with the states of the substances in any of the A, B and C phases.

Now, in accordance with Gibbs' prescription [2, p. 288], the line tension κ (for a unit of the contact line length) is defined by

$$3\overline{\sigma}\delta = 3\sigma\delta + \kappa \tag{29}$$

where δ is the conventionally chosen boundary of the contact region in which $\sigma(x) \neq \sigma$, while the axis x is in any of the phase surfaces with an origin at the contact dividing line where $\sigma(x=0) = \sigma_0$; $3\overline{\sigma}\delta = 3\int_0^\delta \sigma(x)\mathrm{d}x$. The left-hand side of Eq. (29) is the Ω potential of the real system under consideration with variable surface tensions in a region close to the contact line. The right-hand side is the Ω potential of the idealized system with undisturbed surface tensions. The excess quantity κ provides a correspondence of both systems – real and idealized. If one puts in Eq. (29) $\overline{\sigma} = 1/2\,(\sigma + \sigma_0)$ with $3\sigma_0 = \sigma$:

$$\kappa = -\sigma\delta \tag{30}$$

This result is similar to the well-known estimation of the edge energy of a crystal, expressed by the specific surface energy σ: $\kappa = \sigma\delta$ (δ is the atomic size). If we define κ as the work spent forming a unit of the contact line length and σ as the work spent forming a unit of surface area, then the contact line with a unit length has an "area" of 1δ, so that $\kappa/1\delta = \sigma$. However, owing to the formality, such a consideration is obviously unsatisfactory.

As, according to Gibbs, the surface tensions and the respective surface areas vary in the same ratio, the calculation of κ could be carried out without introducing $\overline{\sigma}$ and using σ_0. Gibbs' equation (631) [2] must be written here in the form

$$x\mathrm{d}\sigma = \frac{1}{2}\mathrm{d}(x\sigma) \tag{31}$$

and after integration

$$\int_0^\delta \sigma \mathrm{d}x = \frac{1}{2}\sigma\delta$$

so that

$$\kappa = -\frac{3}{2}\sigma \tag{32}$$

Equations (30) and (32), which determine κ with a negative sign, are obtained in [24]. They are similar to the simple equation $\kappa = -\sqrt{3/4}\sigma\delta$, proposed by Kerins and Widom [25].

As is well known, the surface tension can only be positive, as follows from the mechanical equilibrium stability conditions. The above equations give a negative value for the line tension. What is the reason for this?

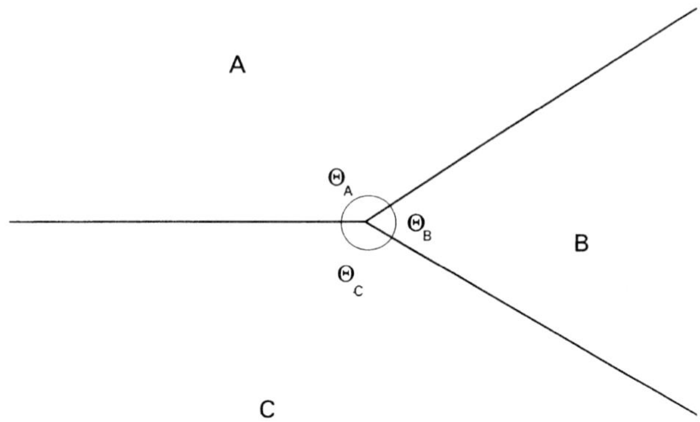

Fig. 13.4 Three bulk phases A, B and C, separated by three plane surfaces, AB, BC and AC, with contact angles θ_A, θ_B and θ_C.

The system under consideration is constructed as follows. Three bulk phases A, B and C are separated by plane surfaces AB, BC and AC, which meet each other in a straight line. Then each of the surfaces tensions σ^{AB}, σ^{BC} and σ^{AC} is less than the sum of the other two [2]. This important conclusion follows directly from the condition of mechanical equilibrium (Fig. 13.4):

$$\frac{\sigma^{AB}}{\sin\theta_C} = \frac{\sigma^{BC}}{\sin\sigma_A} = \frac{\sigma^{AC}}{\sin\theta_B} \tag{33}$$

which implies that the three surface tensions would be considered as sides of a triangle (Neumann's triangle) [26].

In fact, the Gibbs surface tension inequalities follow from the stability conditions of the capillary system studied. These conditions can be deduced by means of the system Ω potential that, as has already been emphasized, is the proper thermodynamic function for such an aim.

The thermodynamic state of the system considered is described by the Ω potential for which the equilibrium value Ω_e ought to be

$$\Omega_e = -PV + \sigma^{AB}o^{AB} + \sigma^{AC}o^{AC} + \sigma^{BC}o^{BC} + \kappa l \tag{34}$$

where P is the system's pressure (the pressures of the bulk phases are equal because all the surfaces are plane). The contact line of length l is considered as an axis of a cylinder of volume $V=\pi L^2 l$ and the surfaces AB, BC, AC are rectangles of areas $\sigma^{AB}=\sigma^{AC}=\sigma\text{BC}=Ll$ (Fig. 13.5).

Obviously, the parameters L and l must be large enough in comparison with the molecular dimensions. Then the line of discontinuity is entirely located within the core of the vessel containing the system studied and vessel's cylindrical surface is positioned somewhere in the bulk of the A, B and C phases where the three surfaces of discontinuity are not disturbed by the three-phase contact

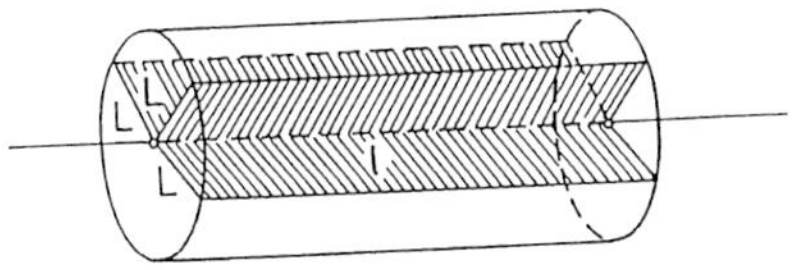

Fig. 13.5 The system is in a cylindrical vessel of volume $v=\pi L^2 l$.

zone. Hence for the system studied the non-physical parameter L can be increased by a suitable choice without any restriction.

We shall examine the stability of the mechanical equilibrium of the system studied. At constant temperature T, volume V and chemical potentials μ_i, after a random increase in the contact line length, $\Delta l > 0$, the system will deviate from its equilibrium state and the equilibrium of the system will be *stable* when

$$(\Delta\Omega)_{T,V,\mu_i} = \Omega - \Omega_e > 0 \tag{35}$$

For instance, the increase in the length of the contact line could result in a decrease in the area o^{BC} and, at the same time, in an increase in the areas o^{AB} and o^{AC}. Then the new areas of the surfaces AB, AC and BC will be $L(l+\Delta/2)$, $L(l+\Delta/2)$ and $L'(l+\Delta/2)$, respectively; if $\Delta l > 0$ is small, $L' < L$ is close to L.

There are three possibilities for such contact line length fluctuations and, therefore, it is easy to obtain three alternative expressions for $(\Delta\Omega)_{T,V,\mu_i}$ [27]:

$$\begin{aligned}
&(\sigma^{AB} + \sigma^{AC} - \sigma^{BC} + 2\kappa/L)\Delta l = \\
&(\sigma^{AC} + \sigma^{BC} - \sigma^{AB} + 2\kappa/L)\Delta l = \\
&(\sigma^{AB} + \sigma^{BC} - \sigma^{AC} + 2\kappa/L)\Delta l
\end{aligned} \tag{36}$$

This result shows that Gibbs' surface tensions inequalities

$$\begin{aligned}
&\sigma^{BC} < \sigma^{AB} + \sigma^{AC} \\
&\sigma^{AB} < \sigma^{AC} + \sigma^{BC} \\
&\sigma^{AC} < \sigma^{AB} + \sigma^{BC}
\end{aligned} \tag{37}$$

follow directly from the equilibrium stability condition (35). This condition also permits the values of κ to be either positive or negative and, for the system studied, the inequalities (37) are not affected by the quantity κ because of the parameter L present in the expressions (36).

For the special case of three equal surface tensions $\sigma^{AB} = \sigma^{AC} = \sigma^{BC} = \sigma$, expressions (36) reduce to

$$(\Delta\Omega)_{T,V,\mu_i} = \sigma\Delta l > 0 \tag{38}$$

Hence the well-known statement for surface tension being only positive follows immediately from Eq. (38).

The quantity κ of the boundary of two two-dimensional phases is entirely analogous to the quantity σ appearing at the surface of discontinuity between two bulk phases. Indeed, for the case of two two-dimensional equilibrium phases α and β with $\sigma^{\alpha}=\sigma^{\beta}=\sigma$ and a straight contact line, a random increase in the length of the contact line $\Delta l>0$ leads to a change $\Delta\Omega$ in the system Ω potential from its equilibrium value [28]:

$$\Omega_{\mathrm{e}} = \sigma^{\alpha}o^{\alpha} + \sigma^{\beta}o^{\beta} + \kappa l = \sigma o + \kappa l \tag{39}$$

$(o^{\alpha}+o^{\beta}=o=\text{constant})$

$$(\Delta\Omega)_{\mathrm{T},o,\mu_{\mathrm{i}}} = \kappa\Delta l \tag{40}$$

This change can be only positive, otherwise (a negative κ) the system will spontaneously deviate from its equilibrium state and destruction of the system will then occur.

The considerations above lead to plain results. However, to some it might look as if the model systems used were constructed in an artificial manner. To go back to reality, we consider the case of a thin liquid film in contact with a meniscus with a straight contact line (Fig. 13.6). Then [29],

$$\Delta P\mathrm{d}v + 2\sigma\mathrm{d}o = \Delta P\mathrm{d}v^{(\mathrm{id})} + 2\sigma^{(\mathrm{id})}\mathrm{d}o^{(\mathrm{id})} + \mathrm{d}\kappa \tag{41}$$

The left-hand side of Eq. (41) is the Ω potential of the real system (ΔP is the capillary pressure of the meniscus). The terms with index (*id*) on the right-hand side refer to the Ω potential of the same system after a suitable idealization. With $\mathrm{d}v=Z\mathrm{d}x$, $\mathrm{d}v^{(\mathrm{id})}=Z^{(\mathrm{id})}\mathrm{d}x$, $\mathrm{d}o=\mathrm{d}x/\cos\theta$ and $\mathrm{d}o^{(\mathrm{id})}=\mathrm{d}x/\cos\theta^{(\mathrm{id})}$, an expression for the correction term $\mathrm{d}\kappa$ follows immediately:

$$\kappa = \int_0^{\infty}\left[\left(\frac{2\sigma}{\cos\theta}+\Delta PZ\right) - \left(\frac{2\sigma}{\cos\theta}+\Delta PZ\right)^{(\mathrm{id})}\right]\mathrm{d}x \tag{42}$$

One should distinguish between σ, θ and $Z(x)$ for the real and for the idealized system. Whereas in the real system σ is variable in the transition from the meniscus (surface tension σ^{l}) to the film (surface tension σ^{f}) [30], the surface tensions in the idealized system, σ^{l} and σ^{f}, are constant. Hence the profiles of the surfaces in the real and idealized systems are different (θ is the local surface slope angle; Z is the distance between the two symmetrically positioned surfaces; Fig. 13.6) (both films with curved contact line and asymmetric films are analyzed elsewhere [31]).

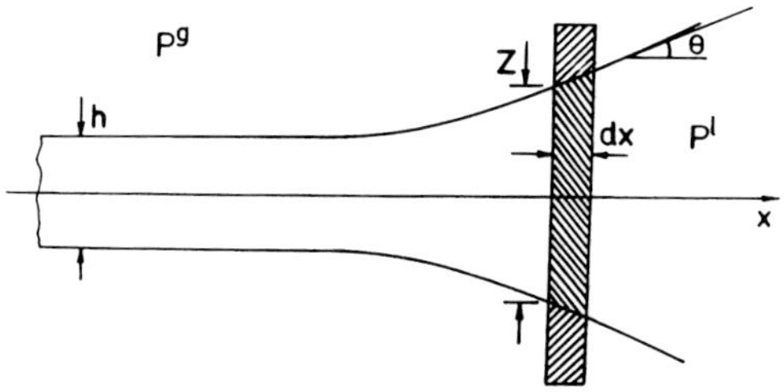

Fig. 13.6 A liquid film in contact with a cylindrical meniscus: the subsystem with volume $dv+dv^g$=constant (shaded) is in the form of rectangular parallelepiped with side dx (in the plane of the drawing) and side unit (in the direction perpendicular to the drawing and parallel to the straight contact line).

The condition of tangential mechanical equilibrium with the tension of the film γ [30, 32, 33] is

$$\begin{aligned}
&\gamma = 2\sigma^{\mathrm{f}} + \Delta Ph = \\
&2\sigma\cos\theta + \Delta PZ = \\
&2\sigma^{\mathrm{l}}\cos\theta^{(\mathrm{id})} + \Delta PZ^{(\mathrm{id})} = \\
&2\sigma^{\mathrm{f}}\cos\theta_{\mathrm{h}} + \Delta Ph = 2\sigma^{\mathrm{l}}\cos\theta_0
\end{aligned} \tag{43}$$

where θ_{h} and θ_0 are the film contact angles: θ_{h} is defined by extrapolation of the surface of the meniscus at constant capillary pressure until the surface of the film at $Z=h$ is intersected; to define the contact angle θ_0 this extrapolation goes to the middle plane of the film where the film *surface of tension* [32] is situated. From Eq. (43) with $\mathrm{d}x = \frac{1}{2}\mathrm{ctg}\theta\mathrm{d}Z$,

$$\mathrm{d}x = \frac{1}{2}\mathrm{ctg}\theta^{(\mathrm{id})}\mathrm{d}Z^{(\mathrm{id})}$$

it follows from Eq. (42) that

$$\kappa = \int_0^{\mathrm{h}} [\sigma\sin\theta - \sigma^{\mathrm{l}}\sin\theta^{(\mathrm{id})}]\mathrm{d}Z \tag{44}$$

Equations (42) and (44) were first derived by de Feijter and Vrij [34].

The long-range surface forces (disjoining pressure) in a liquid film with constant or variable thickness produce a variable surface tension $\sigma(x)$ and surface tension of the film σ^{f} [35]. Then, for $\sigma^{\mathrm{f}}<\sigma^{\mathrm{l}}$, in accordance with Eq. (44), the typical sign of κ will be negative. However, positive values of κ also cannot be excluded. Figure 13.7 and Eq. (44) imply such a possibility. As is known (e.g. [12]), the maximum in Fig. 13.7a decreases with increasing electrolyte concen-

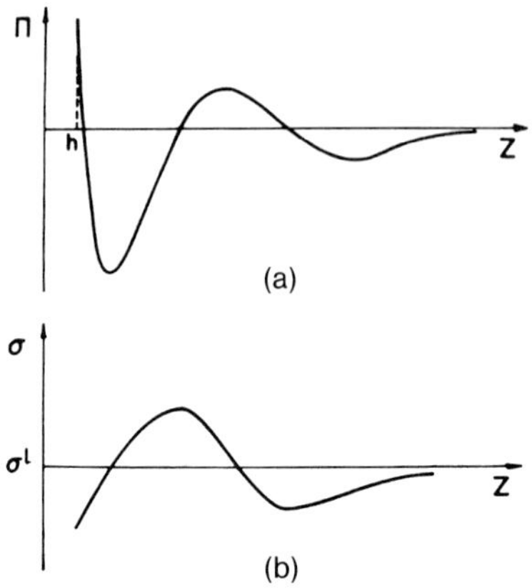

Fig. 13.7 (a) Typical isotherm for the disjoining pressure $\Pi(Z)$; (b) isotherm $\sigma(Z)$ obtained from $\partial 2\sigma/\partial Z = -\Pi$.

tration of the solution from which the film is formed. This implies the possibility of changing the sign of κ (Eq. 44) from the less typical positive to more expected (according to the theory) negative. Experimental evidence of such behavior exists [36].

Taking a backward glance at the theory of heterogeneous condensation (Section 13.4), we may conclude that temperature and the sign of the line tension govern the boundary between the metastable and labile regions of supersaturation. With a change in the sign of the line tension (by temperature or composition of the system studied), the ratio of these two zones will also be changed, i.e. one of these zones can be expanded at the expense of the other.

13.6
Historical Context and Conclusion

The Department of Physical Chemistry at the University of Sofia was established at the end of 1925. During the Professorships of Ivan N. Stranski (1896–1979) and Rostislaw Kaischew (1908–2002), research was concentrated mainly in the field of crystal growth and nucleation phenomena [37]. During the Professorship of Alexei Scheludko (1920–1995), new research areas were developed: thin liquid films, capillary phenomena, wetting, flotation, surface layers, etc., with biophysical and nanoscience applications [38, 39].

With the present chapter a double objective is pursued: (1) to bridge the gap between the older results of Stranski and Kaischew in the field of the molecular physics of crystals and the newer achievements in the field of colloid science after Scheludko – by outlining their common solid scientific basis – and (2) in the modern development of the latter problems, to acknowledge the crucial participation of Dochi Exerowa (Institute of Physical Chemistry, Bulgarian Academy of Sciences) and Dimo Platikanov. Thus with the present chapter I pay tribute to them for a fruitful and happy 70th anniversary.

References

1 M. Volmer, *Kinetik der Phasebildung.* Theodor Steinkopff, Dresden, **1939**.

2 J. W. Gibbs, *The Scientific Papers of J. Willard Gibbs. In Two Volumes. Vol. I: Thermodynamics.* Dover, New York, **1961**.

3 A. Münster, *Chemische Thermodynamik.* Akademie Verlag, Berlin, **1969**.

4 D. Ter Haar, H. Wergeland, *Elements of Thermodynamics.* Addison-Wesley, Publ. Co., Reading, Mass., USA, **1966**.

5 L. D. Landau, E. M. Lifshitz, *Statistical Physics. Part I.* Pergamon Press, Oxford, **1980**.

6 R. G. Linford, The derivation of thermodynamic equations for solid surfaces. *Chem. Rev., 78,* 81 (**1978**).

7 B. V. Toshev, Remarks on the classical theory of capillarity. *J. Disp. Sci. Technol., 18,* 801 (**1997**).

8 J. C. Eriksson, Thermodynamics of surface phase systems. V. Contribution to the thermodynamics of the solid–gas interface. *Surf. Sci., 14,* 221 (**1969**).

9 B. V. Toshev, Condensation on ions. *J. Math. Phys. Sci., 1,* 120 (**2002**).

10 G. Tohmfor, M. Volmer, Germ formation under the influence of electrical charging. *Ann. Phys., 33,* 109 (**1938**).

11 J. P. Hirth, G. M. Pound, *Condensation and Evaporation. Nucleation and Growth Kinetics.* Pergamon Press, Oxford, **1963**.

12 A. Scheludko, *Kolloidnaya Khimiya.* Mir, Moscow, **1984**.

13 B. V. Toshev, A. Scheludko, Line tension and its application to the theory of heterogeneous phase formation. *Lect. Notes Phys., 386,* 138 (**1991**).

14 B. V. Toshev, D. Platikanov, A. Scheludko, Line tension in three-phase equilibrium systems. *Langmuir, 4,* 488 (**1988**).

15 A. Scheludko, V. Chakarov, On the barrier-limited condensation of water on hexadecane. *Colloid Polym. Sci., 261,* 776 (**1983**).

16 V. Chakarov, On the condensation of water on hexadecane. *Colloid Polym. Sci., 261,* 452 (**1983**).

17 V. Chakarov, A. Scheludko, M. Zembala, The effect of initial humidity on water condensation on hexadecane. *J. Colloid Interface Sci., 92,* 35 (**1983**).

18 V. Chakarov, M. Zembala, O. Novozhilova, A. Scheludko, Determination of the regime in the reverse Wilson chamber at critical supersaturation measurement. *Colloid Polym. Sci., 265,* 347 (**1987**).

19 A. D. Alexandrov, On the temperature distribution in the reverse Wilson chamber (RWC). *Colloid Polym. Sci., 274,* 384 (**1996**).

20 D. Platikanov, M. Nedyalkov, A. Scheludko, Line tension of Newton black film. I. Determination by critical bubble method. *J. Colloid Interface Sci., 75,* 612 (**1980**).

21 D. Platikanov, M. Nedyalkov, V. Nasteva, Line tension of Newton black film. II. Determination by diminishing bubble method. *J. Colloid Interface Sci., 75,* 620 (**1980**).

22 A. D. Alexandrov, B. V. Toshev, A. Scheludko, Nucleation from supersaturated water vapors on *n*-hexadecane: temperature dependence of critical supersaturation and line tension. *Langmuir, 7,* 3211 (**1991**).

23 A. D. Alexandrov, B. V. Toshev, A. Scheludko, Nucleation from supersaturated water vapour on immiscible liquid substrates: effect of the macroscopic geometry of the three-phase system on the critical supersaturation and line tension. *Colloids Surf. A, 79,* 43 (**1993**).

24 A. Scheludko, B. V. Toshev, On Gibbs' negative line tension. *C. R. Acad. Bulg. Sci., 40*(1), 75 (**1987**).

25 J. Kerins, B. Widom, The line of contact of three fluid phases. *J. Chem. Phys., 77,* 2061 (**1982**).

26 J. S. Rowlinson, B. Widom, *Molecular Theory of Capillarity.* Oxford University Press, Oxford, **1982**.

27 B. V. Toshev, On Gibbs' surface tension inequalities. *C. R. Acad. Bulg. Sci., 43*(1), 69 (**1990**).

28 B. V. Toshev, On Gibbs' phase rule. *Langmuir, 7,* 569 (**1991**).

29 B. V. Toshev, Simple derivation of de Feijter and Vrij's formula for line tension. *C. R. Acad. Bulg. Sci., 39*(10), 87 (**1986**).

30 B. V. Toshev, Some problems of Gibbs' two dividing surfaces: thermodynamics of multicomponent foam films. *Colloids Surf.*, *2*, 243 (**1981**).
31 B. V. Toshev, Return to de Feijter and Vrij's formula for line tension at thin liquid films. *Colloid Polym. Sci.*, *273*, 807 (**1995**).
32 J. C. Eriksson, B. V. Toshev, Disjoining pressure in soap films thermodynamics. *Colloids Surf.*, *5*, 241 (**1982**).
33 J. C. Eriksson, B. V. Toshev, On the mechanics of thin liquid films. *Ann. Univ. Sofia Fac. Chim.*, *76*, 65 (**1982**).
34 J. A. de Feijter, A. Vrij, Transition region, contact angles and line tension in free liquid films. *J. Electroanal. Chem.*, *37*, 9 (**1972**).
35 B. V. Toshev, D. Platikanov, Disjoining pressure, contact angles and line tension in free thin liquid films. *Adv. Colloid Interface Sci.*, *40*, 157 (**1992**).
36 D. Exerowa, D. Kashchiev, D. Platikanov, B. V. Toshev, Linear energy with positive and negative sign. *Adv. Colloid Interface Sci.*, *49*, 303 (**1994**).
37 B. V. Toshev, *University of Sofia. Department of Physical Chemistry. Bibliography 1925–1961.* St. Kliment Ohridski University Press, Sofia, **1997**; *Ambix*, *48*, 212 (**2001**).
38 B. V. Toshev, *Colloid and Interface Science Group. Bibliography.* Sofia University Press, Sofia, **1977**.
39 B. V. Toshev, A. Fabrikant, *Colloid and Interface Science. Reference List. Part II (1976–1987).* Bulgarian Academy of Sciences, Sofia, **1988**.

14
Structure and Stability of Black Foam Films from Phospholipids

Mickael Nedyalkov

Abstract

We applied the X-ray reflectivity technique to determine the molecular structure of black foam films of various phospholipids and different mixtures of phospholipids and proteins. We studied the relations between the chemical nature of the lipids and the stability and structural properties of the film. In particular, for zwitterionic lipids, we show that the presence of water is controlled mainly by the chemical nature of the lipid headgroup and not by the electrostatic interactions. Other experiments were performed with a mixture of one non-charged lipid, dimyristoylphosphatidylcholine (DMPC), and with a negatively charged lipid, dimyristoylphosphatidylglycerol (DMPG). Bovine serum albumin (BSA) was added to the above mixture. The experiments showed a very strong influence of DMPG on the structure and stability of the phospholipid films and strong lipid–protein interactions. This result can be explained by the presence of protein molecules inserted between the two film layers. This process is completely controlled by adjusting the protein chemical potential in the solution. We observed two different behaviors of the film: at the highest lipid concentration of a DMPC film there was the usual protein (lysozyme) diffusion into the film, and at the lowest lipid concentration we observed the spontaneous formation of a sandwich structure immediately after the drainage. We show that this process of "swelling" is reversible. We also studied the action of a new type of amphiphilic cyclodextrins on DMPC phospholipid films. The stability of these films depends strongly on the molar ratio of phospholipid to cyclodextrins in the mixture. A molar ratio of 3 provides a highly stable film the molecular structure of which was investigated in detail. The cyclodextrins are anchored in the membrane through the cholesterol arm, making the cavity available for the inclusion of the guest molecule dosulepine, which makes them of interest for drug delivery. Additional surface tension measurements of all the solutions were performed in order to obtain complementary information about the state of the monolayers.

Colloids and Interface Science Series, Vol. 1
Colloid Stability: The Role of Surface Forces, Part I. Edited by Tharwat F. Tadros

ISBN: 978-3-527-31462-1

14.1 Black Films

Black films of surfactants are simple free-standing bilayer systems involving most of the basic physical interactions existing in more complex structures, such as biological membranes [1]. These black films, common and Newton, represent the final stage of thinning of the thicker soap films due to draining of water from them. They are generally formed from solutions of common surfactants or other surface-active substances. The films are called "black" because they do not reflect natural light as a result of their smaller thickness compared with visible wavelengths.

Depending on the salt concentration, two types of black films can be observed. For the common black films (CBFs), a balance between van der Waals attraction and double-layer repulsion forces determines the equilibrium thickness [2, 3]. The CBFs are known to be thicker than the Newton black films (NBFs), which involves subtler short-range forces. Free amphiphilic black films have generally been studied by classical techniques such as electrical and contact-angle measurements, optical reflectance, infrared (IR) absorption [4–9] measurements etc. It has been shown [10] that X-ray reflectivity is the most powerful technique because it is sensitive to electron density gradients, which are strong at the two air/film interfaces. With this technique a few years ago it was demonstrated that the NBF was thinner than usually expected and very well organized [11]. It consists of two opposite walls of molecules, without any liquid water between them. The aqueous core is reduced to an ultimate hydration layer of the polar heads and the roughness is limited to the capillary waves. Such a feature is general even for very different surfactants. The study of the structure and stability of NBF is therefore a very good means for direct investigations of interfacial phenomena such as interactions between surfactant bilayers.

The last step of our investigations in this area was devoted to the use of these films in studying the interfacial properties of biological systems.

14.2 Phospholipid Films

Recently, many studies have focused on the measurement of interaction forces within the film, for both insoluble [12, 13] and soluble [14–16] phospholipidic compounds playing the main role in the structure and organization of a real cell membrane.

Almost all phospholipids share the same central function but can differ in their aliphatic chains (R_1 and R_2) and their hydrophilic group R_3 (Fig. 14.1). We have used in our studies [17] three chains of different lengths: myristoyl (M) with 14 carbons, palmitoyl (P) with 16 carbons and oleoyl (O) with 18 carbons and a C=C double bond. We also used three different functions for R_3: choline

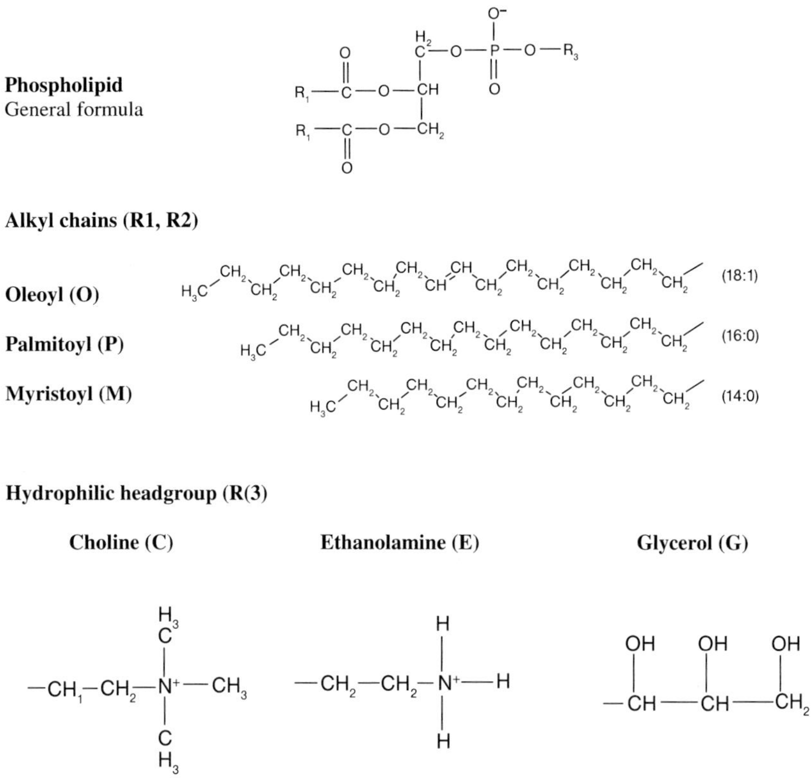

Fig. 14.1 Chemical formulae of the lipids. All the phospholipids (except DOTAB) share the same central function but could differ in their aliphatic chains (R_1 and R_2) and their hydrophilic group (R_3).

(C), ethanolamine (E) and glycerol (G). As one can see, the PC lipids are zwitterionic and share the same headgroup (R_3=C), but the aliphatic chains differ: R_1=R_2=M for DMPC (dimyristoylphosphatidylcholine), R_1=R_2=P for DPPC (dipalmitoylphosphatidylcholine) and R_1=R_2=O for DOPC (dioleoylphosphatidylcholine). The PE lipids are also zwitterionic but have a smaller headgroup (R_3=E). For these lipids we selected the same type of chains as in the case of PC lipids: R_1=R_2=M for DMPE (dimyristoylphosphatidylethanolamine), R_1=R_2=P for DPPE (dipalmitoylphosphatidylethanolamine) and R_1=R_2=O for DOPE (dioleoylphosphatidylethanolamine).

DMPG (dipalmitoylphosphatidylglycerol), characterized by R_1=R_2=M and R_3=G, is negatively charged, whereas DOTAB {*N*-[1-(2,3-dioleoyloxy)propyl]trimethylammonium salt} is positively charged. In the last case, the phosphate group is not present and the headgroup (R_3=E) is directly connected to the aliphatic chains via the glycerol group. The chains are R_1=R_2=O.

The preparation of solutions containing small unilamellar vesicles of phospholipids was carried out following the procedure described in [18].

In the first part of this chapter, we present the results on the roles of the aliphatic chains and the headgroup in the stability and structure of the phospholipid films. The effect of the electrical charges that sometimes occurred is also discussed.

In the second part, we report the first results of the confinement of a single layer of proteins in a surfactant NBF and in a mixed phospholipid NBF.

The third part presents the results obtained on mixed phospholipid/cyclodextrin derivative films in presence of some guest molecules of high interest for drug delivery.

14.3 Methods

For studying the structure and the forces acting within the film, measurements of the film thickness were carried out for the first time by Scheludko [19] and Exerowa et al. [20]. In this technique, the intensity of the monochromatic light reflected by the film is used to calculate its thickness. Unfortunately, with this optical method, it is not possible to obtain direct information on the internal structure of the film, as the incident wavelength is too large. Even when a more complex model was used for the film structures (three-layer model) [21, 22], most of the model parameters were assessed from other experiments [23], which were made on different systems.

Recently, Fourier transform infrared (FTIR) spectroscopy was also used to measure the absorbance of thin lipid films [21, 24]. By measuring the absorption corresponding to the surfactant stretching bands, this method provided new information on the molecule orientation [25], but there was a 25% discrepancy in the values of the thickness for the central water core of the film.

Up to now, most studies have examined the response of the film to the variation of a control parameter (e.g. ionic force of the solution, presence of different ions, etc.). The study of the relation between the film structure and stability and the chemical nature of the surfactants has been neglected. To solve this problem, we used X-ray reflectivity to determine the internal structure of lipid films. This technique, which has been used successfully in the case of soluble surfactants and described in detail in [26], measures the electronic density profile perpendicular to the surface of the film.

14.4 Experimental

To make a film, a metallic frame was immersed in phospholipid solution and then drawn up to form a film suspended on the central part of the frame. An X-ray beam then illuminated the film. The X-ray reflectivity measurement was performing using a high-resolution diffractometer (Micro-Control Optix) described in detail elsewhere [27, 28]. The measurement procedure was described in [17].

14.5
Modeling of the Phospholipidic Bilayers

In most cases, the films were described using a symmetrical five-box model (Fig. 14.2). This model, combining several assumptions with its validity, was described in detail in [17]. In that paper, on the structure of films drawn from liposomal suspensions of different phospholipids, we showed that the reflectivity curves of phospholipid films have very specific shapes with well-defined Kiessig fringes (Fig. 14.3).

The strong contrast between the different interference fringes also induced a strong modulation of the electron density profile. Therefore, we used a five-layer model [17] for the fitting procedure (Fig. 14.2). In the first layer, the aliphatic chains of the lipids have a relatively low electron density owing to the presence of C and H atoms. In the second layer, including the central part of the molecules (around the P atom), there is a high electron density owing to the presence of heavy atoms. In the third layer, including the headgroups of the lipids, a water layer is inserted between the two monolayers. Its thickness is a variable parameter; in the cases of choline, for example, the headgroup is bulky and hydrophobic. We expected the water to be expelled from this zone, according to the null hydration number given by Israelachvili [2]. For ethanolamine, the headgroup is hydrophilic and of small size. Therefore, we made the assumption that this layer contains many water molecules and that the lipid contribution to the electron density is insignificant.

The model permits the calculation of the lipid density on the surface of the film, and thus the mean area per lipid molecule.

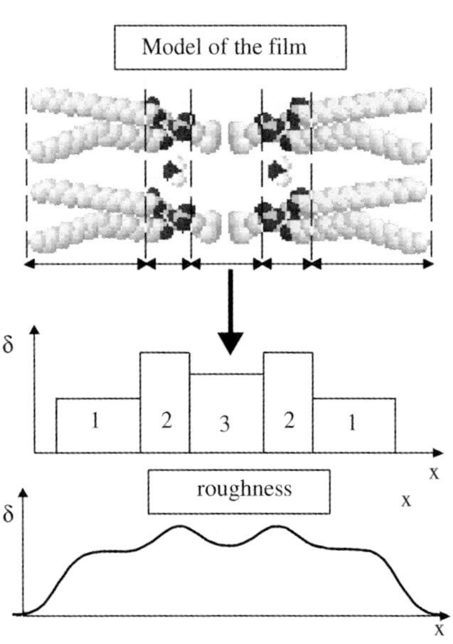

Fig. 14.2 Model of a phospholipid bilayer. Three regions could be distinguished in a lipid molecule. Taking into account the stricture of a foam film, this corresponds to a five-box model for the electron density.

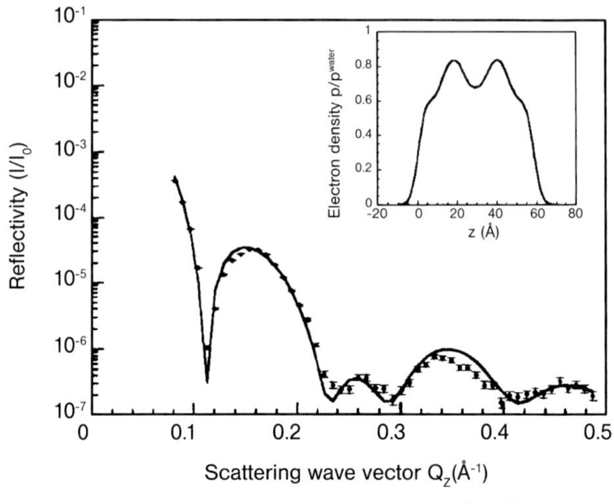

Fig. 14.3 X-ray reflectivity intensity collected for DMPC films. The inset shows the electron density profile normalized to the electron density of the water extracted from the data analyses versus the film thickness.

Finally, the density discontinuities between the different layers were smoothed using an interfacial roughness (variable but identical for all the interfaces). This roughness accounts both for the thermally excited collective motions and for the local disorder of the lipid molecules.

In addition to this detailed model, we also used simpler one- or three-layer models to analyze the reflectivity data. This was done when the experimental accuracy was lower (e.g. short lifetime of the film).

14.6 Results and Discussion

14.6.1 Phospholipid Films

14.6.1.1 DMPC, DOPC, DPPE, DOTAB and DMPG Films

Black films from DMPC, DPPC and DOPC were obtained at 27 °C from suspensions of these lipids without adding salt. The stability of the DPPC films was limited to a lifetime of few minutes and no X-ray reflectivity experiments could be performed. By contrast, the DMPC and DOPC films were stable for several days and we obtained highly accurate measurements of the reflectivity curves.

The results for these films are summarized in Table 14.1.

In addition to the quality of the fitting, the validity of our model could be verified by its internal and external coherency. The internal coherency was tested by comparing the results for different lipids, and the external coherency by

Table 14.1 X-ray reflectivity results.

Parameter	Lipid						
	DMPC	**DOPC**	**DOPE**	**DOPE+0.001 M NaCl**	**DOTAP+0.001 M NaCl**	**DOTAP+0.15 M NaCl**	**DMPC+0.15 M NaCl**
Overall thickness (Å)	55 ± 0.2	59 ± 0.3	69 ± 0.4	47 [c] ± 0.2	300 ± 2	52 ± 0.5	90 ± 0.3
Area per lipid (Å^2)	57 ± 0.2	64 ± 0.3	74 ± 0.4	ND [a]	70–80	76 ± 2	ND
Roughness (Å)	3.4 ± 0.06	3.6 ± 0.1	4.8 ± 0.15	ND	8 ± 0.2	2 ± 0.5	2 ± 0.7
Aliphatic chains:							
– Density [b]	0.8 ± 0.05	0.9 ± 0.05	0.9 ± 0.05	ND		ND	ND
– Thickness (Å)	12.9 ± 0.05	14.2 ± 0.07	11.6 ± 0.03	ND		ND	ND
Hydrophilic part:							
– Density [b]	1.2 ± 0.04	1.1 ± 0.06	1.1 ± 0.04	ND		ND	ND
– Thickness (Å)	11.4 ± 0.03	11.8 ± 0.04	10.1 ± 0.03	ND		ND	ND
– Bound water	11.5 ± 0.2	13 ± 0.2	12 ± 0.5	ND		ND	ND
Terminal part:							
– Density [b]	0.8 ± 0.05	0.7 ± 0.04	1	ND	1	ND	1
– Type [d]	Choline	Choline	Water [f]	ND	Water	ND	Water
– Thickness (Å)	3.3 ± 0.02	3.3 ± 0.03	(8+3) [e] ± 0.01	ND	130 ± 1	ND	~ 20 Å

a) Not determined.
b) The electronic density is normalized by the density of water ($\rho^{water} = 0.3\ e^-/Å^{-3}$).
c) Total lipid thickness: ~ 20 Å. Mean density: 0.8.
d) In the 5-box model, the central layer is symmetric and groups lipids of both sides of the film. In this table, the given value is for one part of the film only. Then the thickness of the central layer is obtained by multiplying the given value by 2.
e) 3 Å estimated for the ethanolamine group + 8 Å of liquid water.
f) Estimated.

comparison of our results with those already published for lipid layers, e.g. in Langmuir monolayers [29], absorbed layers [30–32] or lamellar phases [33, 34].

All the values summarized in Table 14.1 are consistent. Moreover, the parameters corresponding to the constant part of the different lipids (i.e. their intermediate parts) are almost constant: they have the same thickness and the same number of hydration water molecules. The value of 12 water molecules per lipid might appear high, but it is compatible with that observed in Langmuir films and with the high hydrophilicity of this region. On the other hand, the measured thickness for this layer is larger than the values obtained from a simple molecular model (8–9 Å) or those measured in other systems [29, 35]. This could indicate an unusual conformation of this lipidic portion (elongated and perpendicular to the plane of the film) or a problem with the definition of the different zones used in the model. For example, for some authors [36], the headgroup is oriented parallel to the surface place. In our case, this could slightly change the results of our model and especially the number of water molecules in the different parts of the film. However, thanks to the particular shape of the reflectivity curves our model remains globally valid even in this unfavorable case. To be more specific, the results concerning the presence or the absence of liquid water in the central layer will not be modified.

Up to now, we have focused our discussion on the case of a five-layer model. However, even when only simpler models are used for the data analysis, X-ray reflectivity has a crucial advantage over optical techniques: the film thickness can be determined in absolute terms, with no influence from the internal composition of the film.

14.6.1.2 **Stability of Phospholipid Films**

The stability of the films could be easily verified visually: DMPC, DOPC and DOPE films are stable for several hours and even up to several days. The lifetime of the DOTAB and DMPG films is limited to a few hours. In the other cases the films burst during drainage.

It is well known that macroscopic NBF can be formed only if the surfactant forms a dense monolayer at the interface [1]. In our case, the lipid surface concentration depends on the state of the lipid in the solution: below the gel/liquid crystal transition temperature, the surface film is in a gaseous state [37] and both the surface pressure and the surface concentration are near zero. It is therefore expected that stable films will only be obtained for lipids with a transition temperature lower than room temperature.

For example: in the case of the PC films, only the DMPC (T_c=23.5 °C) and the DOPC (T_c<7 °C) form stable films, in contrast to DPPC (T_c=41 °C). Moreover, the X-ray reflectivity can directly probe the state of the aliphatic chains in the film. In every case, an area per molecule was measured well above the limiting area corresponding to the close packing of the lipids (ca. 40 Å^2). In the PC case, the areas measured in the film (DMPC 57, DPPC 64 Å^2) largely correspond to those measured for DPPC in a liquid crystal phase: 60.8 Å^2 by Meuse et al. [30] in an absorbed bilayer and 62.9 Å^2 by Nagle et al. [34] and 59.4 Å^2 by Wiener and White [38] in multilayers. The small difference between DOPC and DMPC is due to the presence of a C=C double bond in the aliphatic chains of DOPC that interferes with the stacking of the lipid chains. There is, therefore, a decrease in the surface density and an increase in the interfacial roughness due to the disorder.

Therefore, our X-ray reflectivity results confirm, at the molecular scale, that only lipids in their liquid crystal state are able to form stable films.

14.6.1.3 **Influence of the Headgroup**

The force balance between the two lipid layers controls the thickness of the lipid films. In the case of thick films, the DLVO theory [3, 39] provides a good description of the variation of these forces with the modification of control parameters, e.g. disjoining pressure, ionic strength or pH of the water solution. On the other hand the relation between the chemical nature of the headgroup and the forces is not very well documented. The thickness of the lipid films for different headgroups can be seen in Table 14.1.

For charged lipids, the main forces in the film have an electrostatic origin [40, 41]. This is clearly visible in the case of DOTAB, a positively charged lipid. In the absence of salt, the electrostatic forces are not screened and the film remains thick (300 Å) with a large quantity of water. On addition of a large amount of NaCl (0.15 M), the film becomes extremely thin (52 Å) owing to the total screening of the electrostatic forces. The film structure has not been explicitly measured, but the thickness of the central water layer is most likely reduced to a few Ångstroms only: one has an NBF.

For negatively charged lipids (DMPG), only a film of limited stability was obtained (with 0.15 M NaCl). An overall thickness of 90 Å was measured. Although the internal structure has not been determined, the thickness of the water layer is estimated to be around 40 Å, corresponding to a CBF. The difference between the two charged lipids (DMPG vs. DOTAB) is due to the metastability of the CBF made with DMPG. In the latter case, the film lifetime is short and we had to record the reflectivity rapidly whereas the film was still in the CBF state and had not transformed to an NBF state.

In Table 14.1 are presented also the results for films of neutral lipids. These lipids are zwitterionic (DMPC, DMPE) with a negative charge on the phosphatidyl group and a positive charge on the nitrogen atom of the headgroup. In this situation, electrostatic interaction should be reduced to low dipole–dipole repulsion and should not play an important role in the structure of the film. This is mostly the case for the DMPC films, for which the thickness does not depend on the ionic strength. In this case, the water is only present as hydration water associated with the hydrophilic part of the lipids. Even in the middle of the film, no liquid water could be detected. We have an NBF.

In some films containing PC lipids (e.g. [14]), the presence of a liquid water core was detected, in contradiction with our results. Actually, the probability of the transition between a CBF and an NBF depends on the size of the film and this probability increases for films of large surface area. Such a film has a higher probability of transforming to the NBF state and this could be the reason why we do not observe a CBF.

A priori, one could expect the same behavior for PE and PC films if the structure is imposed mainly by the electrostatic interactions. However, this is not the case. We detected the presence of a 16-Å water core for a DOPE film without salt. This thickness corresponds to a layer of free water. This could be due mainly to the hydrophilicity of the ethanolamine group. This chemical group could largely favor the presence of a large quantity of water, by its ability to form an H-bond. Therefore, this film is in a CBF state in the absence of added salt. The electrostatic origin of the forces that prevent the formation of an NBF was verified by adding some salt to the water. A concentration of 10^{-3} M NaCl (to be compared with 0.15 M for DOTAB) is sufficient to screen totally the electrostatic repulsion and to obtain an NBF of thickness 47 Å.

These examples obviously indicate that, for films made from charged lipids, the electrostatic interactions regulate the state and thickness of the films. In the case of zwitterionic lipids, the situation is much subtler and the chemical nature of the headgroup has to be taken into account to understand the behavior of the film.

14.6.2 Surfactant–Protein Interaction in NBF

14.6.2.1 Introduction

The behavior of lipid films being established, we aimed to study the interaction of these films with proteins, dissolved in the subphase. This situation could be a suitable tool for the formation of large two-dimensional mixed films of proteins and lipids.

A few studies have reported attempts to form black films of proteins. Nevertheless, where microscopic CBF and NBF have been obtained, their structures were either bilayers of denatured proteins, more complex multilayer films or thick films [42–47].

We managed to obtain an alternatively stabilized freestanding film, confining a single protein layer in a surfactant bilayer. This first step of our experimental processes was necessary for a better understanding of the mechanism of involving a protein in a simple bilayer. To explain this process, we proposed a simple model [48] by comparing the protein chemical potential in the solution at the Gibbs interface (air/water) and in the NBF. Before pulling the film, the Gibbs interface is in thermodynamic equilibrium with the bulk reservoir. This sets the concentration of the protein in the Gibbs film as a function of the bulk concentration C_{BSA} (BSA = bovine serum albumin). At time $t=0$, when the NBF is formed, the surface concentration of each of the two Gibbs layers constituting the NBF does not have time to change. Hence the initial protein surface fraction Φ_{NBF} ($t=0$) in the film is just $2\Phi_G$ (where Φ_G is the protein surface fraction under the Gibbs film). This state is not in equilibrium and there is protein diffusion towards the NBF. At long enough times, a surface fraction Φ_{NBF}^{eq} is reached, for which the chemical potential of the protein in the NBF equals that in the bulk. If the protein molecules are attracted to the Gibbs interface and in the attraction within the two surfactant walls is weak, then the final concentration in the NBF will be considerably larger than the initial concentration after the NBF formation. Under the experimental conditions of this previous investigation, we estimated that, at equilibrium, the NBF surface fraction should by 0.6 and that the diffusion coefficient is $D \approx 10^{-7}$ $cm^2\ s^{-1}$. This is enough to allow protein–protein interactions in the NBF. Therefore, to permit the formation of this new sandwich structure made of two walls of surfactant and a single protein layer, the protein must be soluble and it must have only slight interactions with the surfactant, which cannot denature it.

14.6.2.2 **Films of Surfactant and BSA**

The nonionic surfactant $C_{12}E_6$ (hexaethylene glycol monododecyl ether) was chosen for its ability to form a very stable NBF. The respective surfactant and protein ranges of possible concentrations were found empirically after testing the film stability [49]. We found the minimum concentration of the surfactant (C_{C12E6}=0.075 mg mL^{-1} ≈ 2 cmc), which provides stable films only when mixed with BSA. This concentration, used for all the experiments, allows the formation of stable mixed films. This is the first experimental proof of the presence of BSA within the NBF. The surfactant concentration for which large stable films can be obtained without BSA is C_{C12E6}=0.5 mg mL^{-1} ≈ 15 cmc. The pH of the solutions was stable for roughly 1 h after the preparation and was found to be 7.4±0.1, above that of the BSA isoelectric point, 4.8 [46]. At this pH, BSA is known to be globular with an ellipsoidal shape of dimensions 4.16 and 14.09 nm.

The first reflectivity experiments were carried out on a film drawn from a solution at C_{BSA}=0.5 mg mL^{-1}. The reflectivity curve recorded immediately after film formation differs surprisingly little from that of the pure surfactant. At this stage, a first conclusion can be drawn: the protein interacts with the surfactant to stabilize the film, but its concentration within the NBF remains very low. This indicated only a slight increase in the thickness (~1 nm). Our next experiment, carried out at C_{C12E6}=2 mg mL^{-1}, was crucial because we observed a remarkable time-dependent "swelling" of the film, characterized by a continuous shift of the Kiessig fringes. After a few hours, a stable reflectivity profile was reached. In order to understand this phenomenon and to be more quantitative, another series of experiments were carried out at a higher concentration, C_{BSA}=4 mg mL^{-1} (Fig. 14.4). The experimental results are consistent with previous experiments because a "swelling" was again observed. We call the difference between the overall thickness of the NBF and that of the pure surfactant (6.3 nm) the extra thickness; it represents the matter swelling the NBF. The lowest value (0.3 nm) is observed just after the drainage (t=0) and corresponds to the initial thickness of a pure $C_{12}E_6$ NBF with a very small amount of protein. After 45 h, there is a plateau that indicates the end of thickness evolution.

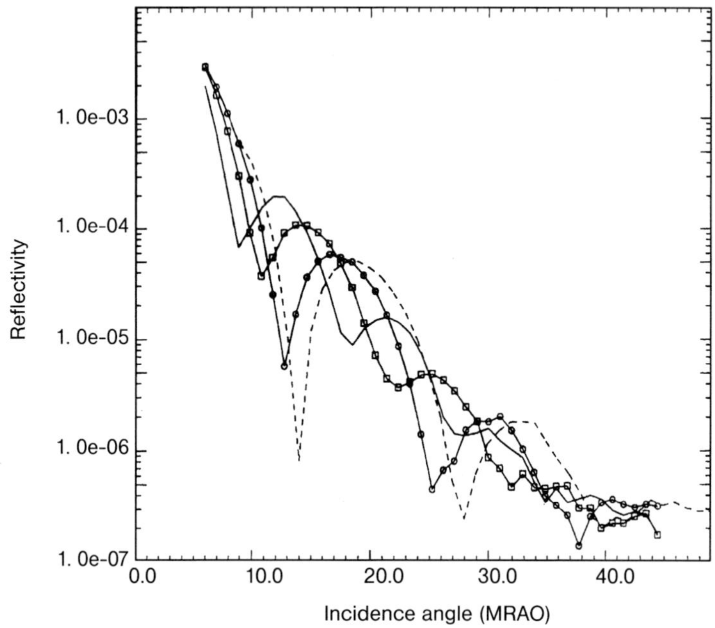

Fig. 14.4 Set of experimental reflectivity profiles. A time-function shift of the curves towards smaller angles is observed. The different profiles correspond to $C_{12}E_6$ NBF (dashed line) and a BSA–$C_{12}E_6$ NBF (C_{C12E6}=0.075 mg mL^{-1} and C_{BSA}=4 mg mL^{-1}), after 1 h (circles), after 10 h (squares) and after 18 h (solid line). The central core thicknesses between the surfactants walls are 0, 0.3, 1.5 and 3 nm, respectively.

The system then reaches an equilibrium state in which the overall thickness is 10.3 ± 0.2 nm, corresponding to a 4-nm final extra thickness. This extra thickness remains constant (30 h) until the film bursts. We attribute the swelling of the film to the protein insertion and not to water because the increase in the overall thickness is much smaller than that resulting from a transition to CBF due to the formation of an aqueous core (in general, > 10 nm). Hence the swelling cannot be accounted for by the absorption of water. The film remains an NBF whose overall thickness is smaller than that characteristic of a CBF. The swelling is therefore due to the sole insertion of protein.

The last crucial problem is to locate the protein with respect to the surfactant and to interpret the maximum extra thickness value (~4 nm). This value may correspond either to roughly twice the size of an unfolded molecule or to the width of a native molecule (4.1 nm) situated in the central core of the NBF. To obtain a definitive answer, we first formed a stable, pure $C_{12}E_6$ NBF and then injected the solution of pure BSA with a syringe. To make this experiment possible, we increased C_{C12E6} to 0.5 mg mL^{-1} and, subsequently, we also increased C_{BSA} to 6.6 mg mL^{-1}, thus reducing the film stability. We obtained a series of reflectivity profiles at regular time intervals and we again observed the swelling process. This is direct experimental proof of the protein insertion within the NBF. It is therefore clear that at equilibrium the new system is a "sandwich NBF" whose central core is a single layer of protein molecules. An important question concerns the state (unfolded or denatured) of the protein itself. As a result, the extra thickness can be attributed to the central core formed of a close-packed single layer of BSA molecules, probably in a native state. In the NBF, the protection against the denaturation could result from the "2D encapsulation" by the surfactant walls. A very simple model [50] may explain our observations by comparing the BSA chemical potentials in the solution, at the Langmuir (air/water) interface and in the NBF.

14.6.3
Interactions in Films of Phospholipids and Protein Mixtures

In this part of the chapter we report X-ray reflectivity investigations of the structure of mixed black films made of biological molecules: DMPC/DMPG mixture with and without protein (BSA) [51]. The two phospholipids have the same hydrophobic chain length of 14 C atoms but different headgroups. DMPC has a relatively hydrophobic choline headgroup and it is completely insoluble in water. The headgroup forms a zwitterion. DMPG has strong hydrophilic glycerol headgroup, it is soluble in water and it is negatively charged. The two phospholipids were mixed in the molar ratio DMPC:DMPG=8.5:1.5. This ratio was chosen for the following reasons:

1. According to the study of Lalchev et al. [52], this ratio gives a stable NBF.
2. The lateral diffusion coefficients in the interfacial layers for the mixed films are three times higher than the same coefficient for the DMPC films only [52].
3. This molar ratio is close to the natural ratio between PC and PG in a natural alveolar surfactant mixture [53], where PC:PG=6:1.

The DMPC concentration was 0.5 mg mL^{-1} and it allowed the formation of a stable, macroscopic NBF [17, 18].

For the preparation of solutions with and without NaCl, we used the following procedure: DMPC was dispersed in water and, after agitation for 30 min, the solution was sonicated for 10 min. DMPG was added to the liposomal DMPC dispersion and was completely dissolved. The films obtained from these solutions were stable [51]. All experiments were performed at 26 °C, well above the phase transition temperature of the DMPC/DMPG mixture (20 °C) [54].

14.6.3.1 Films of DMPC/DMPG Mixture Without Protein

In order to understand the influence of DMPG, we first studied the structure of films formed from the DMPC/DMPG phospholipid mixture without protein. A comparison of the DMPC and DMPC/DMPG reflectivity curves is presented in Fig. 14.5. For the fitting procedure we used a five-layer model. The results of

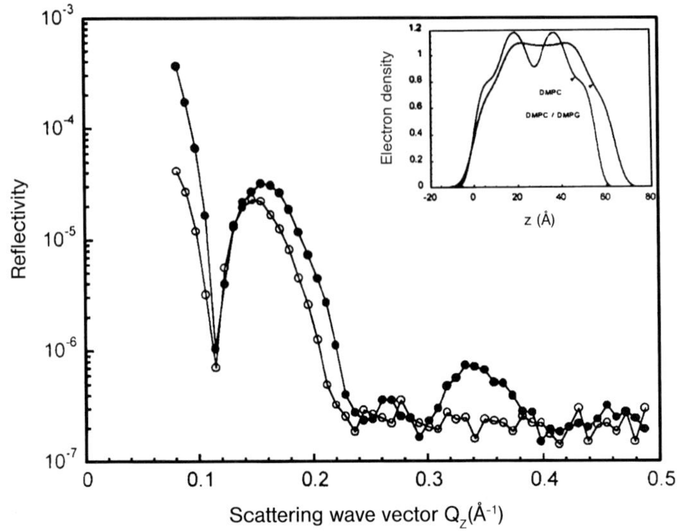

Fig. 14.5 Reflectivity curves and density profiles of DMPC and DMPC/DMPG films in the molar ratio PC:PG=8.5:1.5.

the fits are presented in Table 14.2. They show that the addition of DMPG dramatically changes the shape of the curve and also the total thickness of the film. The mixed film is thicker, which could be explained by the electrostatic repulsion between the film monolayers that occur due to the charges of the glycerol groups. The strong hydrophilicity of the PG headgroup induces an increase in the number of bonded water molecules and then increases the electron density of the central layer (ρ_3). The calculated area per molecule has a higher value (A=63.5 Å^2) than in DMPC films. The addition of NaCl (C=0.5 M) to the lipid

Table 14.2 Results of the fits for the films of DMPC and DMPC/DMPG [a].

Parameter	DMPC	DMPC/DMPG
h_1 (Å)	12.9	12.9
ρ_1	0.8	0.7
h_2 (Å)	11.4	10.0
ρ_2	1.2	1.1
h_3 (Å)	6.6	17.9
ρ_3	0.8	1.1
h_{total} (Å)	55	63.5
A (Å^2)	57	77.5

a) The fits were made using a five-layer model. The parameters h, ρ and A are the thickness, the electron density normalized to the density of the water and the area per molecule, respectively.

Table 14.3 Results of the fits for the films from DMPC/DMPG with and without NaCl [a)].

Parameter	DMPC + DMPG	DMPC + DMPG + NaCl
h_1 (Å)	12.9	15.5
ρ_1	0.7	0.7
h_2 (Å)	10.0	10.0
ρ_2	1.1	0.9
h_3 (Å)	17.9	13.5
ρ_3	1.1	0.7
h_{total} (Å)	63.5	64.5
A (Å^2)	77.5	64.3

a) The concentration of NaCl was 0.5 M.

mixture (Table 14.3) increases the film thickness and especially the thickness of the first layer (h_1) corresponding to the hydrophobic tails of the phospholipids. The presence of the salt decreases the area per molecule and also the tilt angle of the aliphatic chains. The addition of salt leads to screening of the electrostatic repulsion between the charged heads and condensation of the monolayers. The thickness (h_3) and the electron density (ρ_3) of the central layer are smaller, indicating that the quantity of the water is reduced. Hence the film is NBF.

Additional surface tension measurements of all the solution were performed in order to obtain complementary information about the state of the monolayers.

In Fig. 14.6, all the surface tension isotherms (DMPC, DMPG and DMPC/DMPG with and without NaCl solutions) are represented. The adsorption kinetics in the lipid solutions are very slow (more than 15 h) and it is in agreement with the model proposed by Vassilieff et al. [55] for the mechanism of vesicle disruption. The DMPC/DMPG mixture isotherms have two parts with different slopes. They could be considered as two-component adsorption isotherms with different adsorption rates for the two lipids. The surface saturation time (i.e. to reach a constant surface tension value) of the mixed solution is roughly two times longer than that for a one-component phospholipid solution; this exactly corresponds to the time necessary for the formation of a stable black film. In the presence of NaCl, the surface tension is always lower and the influence of the salt mainly acts in the first part of the curve. As mentioned above, DMPG was dissolved in a liposomal suspension of DMPC. Therefore, one can expect that the adsorption of this lipid will be faster than that of DMPC and that the salt plays a role mainly in the first part of the curve. The surface tension measurements show a strong influence of DMPG on the adsorption kinetics and formation of mixed DMPC/DMPG monolayers. These results are in agreement with the X-ray reflectivity results with and without NaCl.

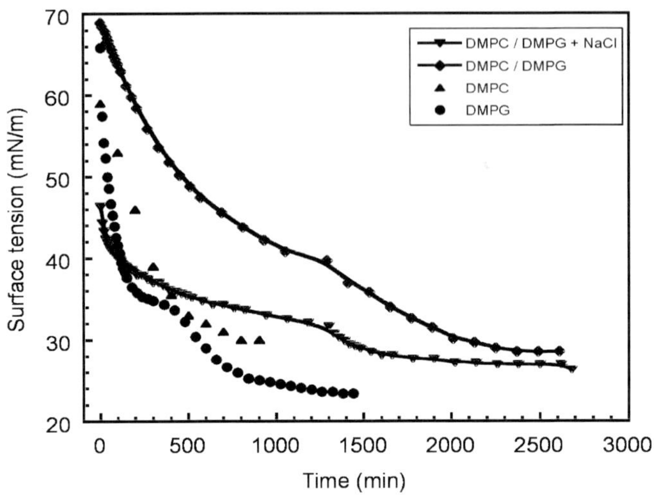

Fig. 14.6 Surface tension isotherms of DMPC 0.5 mg mL^{-1}, DMPG 0.089 mg mL^{-1} and DMPC/DMPG mixtures (with and without NaCl).

14.6.3.2 **Films of DMPC/DMPG Mixture With Protein BSA**

After the described preliminary experiments, we added BSA to the DMPC/DMPG mixture in order to form mixed lipid/protein films. The main idea was to realize the same diffusion process of a protein into the black film as those which we observed for the first time previously [48], replacing the simple surfactant $C_{12}E_6$ with phospholipids. First we tried to form mixed films with only DMPC and BSA. The only effect was a strong decrease in the time for film formation; the film structure remained identical with that of the DMPC film. The addition of BSA to the DMPC/DMPG mixture led to a thickening of the films.

In Fig. 14.7, the reflectivity curves of DMPC/DMPG films and those of mixed with BSA films are compared at two different protein concentrations. The increasing of BSA concentration tends to shift of the interference pattern to the small angle corresponding to an increase of the total film thickness. The films with BSA are thicker immediately after their formation and their thickness remains for fits (Table 14.4) show that the thickness of the hydrophobic layer (h_1) decreases which is related with an increase of the area per molecule. The electron density of the second (ρ_2) and the third (ρ_3) layer increases as well as the thickness of the central layer. These results show the presence of BSA into the films and the occurrence of interactions between the lipids and the protein.

The results of the measurements of the surface tension of the phospholipid solutions containing BSA (not shown here) show a strong reduction in the time for adsorption. The saturation value of the surface tension is lower than the value of the surface tension for a phospholipid solution without BSA. It seems that the adsorbed protein molecules are removed from the surface by the more surface-active lipid molecules. Several workers [56–58] have observed this phenomenon. How-

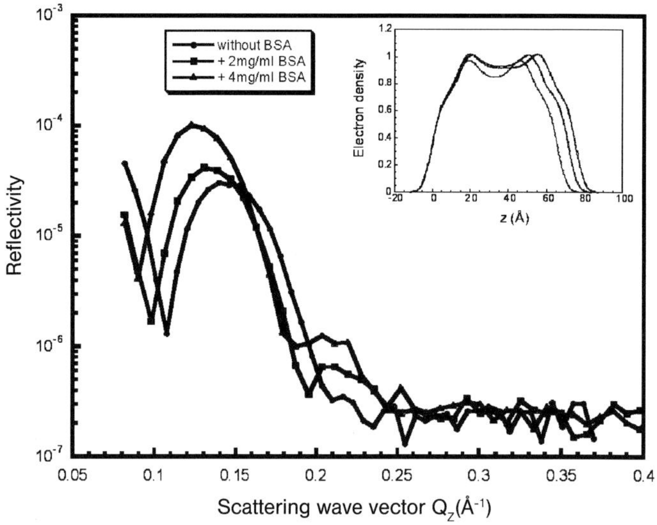

Fig. 14.7 Reflectivity curves and density profiles of DMPC/DMPG films with and without BSA at two BSA concentrations, 2 and 4 mg mL^{-1}, with 0.5 M NaCl.

Table 14.4 Results of the fits for DMPC/DMPG films with and without BSA[a)].

Parameter	DMPC + DMPG + NaCl	DMPC + DMPG + NaCl + 2 mg mL^{-1} BSA	DMPC + DMPG + NaCl + 4 mg mL^{-1} BSA
h_1 (Å)	15.5	14.4	13.6
ρ_1	0.7	0.7	0.7
h_2 (Å)	10.0	10.0	10.0
ρ_2	0.9	1.0	1.0
h_3 (Å)	13.5	22.5	28.0
ρ_3	0.7	0.9	0.9
h_{total} (Å)	64.5	71.4	75.2
A (Å^2)	64.3	69.3	73.4

a) The concentration of NaCl was 0.5 M.

ever, the presence of a protein in the film allows us to suppose the presence of protein molecules adsorbed in the subphase layer under the lipid molecules.

The reflectivity results and the surface tension isotherms show that BSA interacts with the lipids but it is very difficult to determine the position and conformation of the protein. The difference between the film structure and thickness with and without DMPG with BSA means that PG is at the origin of the lipid/protein interactions within the film. The headgroups of DMPG molecules are completely different from those of DMPC molecules, as mentioned above. The structure of DMPG headgroups offers the possibility of the creation of H-bonds and of electrostatic interactions. It is well known that albumin interacts very strongly with anionic surfactants; this process is accompanied by a change in the protein conformation [59, 60]. However, the quantity of salt in the solutions is sufficient to screen the negative net charge of the glycerol. The reduction in the surface saturation time and the results for the films from DMPC/BSA mixtures could be explained by involving the hydrophobic interactions, the process of disruption of the vesicles and a faster surface saturation. These processes take place in the bulk and, in the case of DMPC/BSA films, there are no proteins in the film.

Hence we expect two types of interactions, hydrophobic interactions between the aliphatic tails of the lipid molecules and BSA hydrophobic sites and creation of H-bonds between the PG headgroups and some residues of BSA.

14.6.3.3 Films of DMPC/DMPG Mixture With Protein Lysozyme

In this part of our investigations, we replaced BSA with a smaller and positively charged protein and we also used the non-charged phospholipid DMPC. Lysozyme is a globular soluble protein. It is obtained from chicken egg white and its isoelectric point is pH=11.4. It has a hard ellipsoidal structure whose dimensions are 30×30×45 Å and it is very stable against denaturation. For the experimental conditions (pH=5.5 and T=28 °C without addition of salt), the lysozyme is highly positively charged [61, 62].

In the absence of protein, the lipid concentration for which large, stable films can be obtained is $C_{DMPC}=0.5$ mg mL^{-1}. We also used $C_{DMPC}=0.12$ and 0.08 mg mL^{-1}, which do not allow the formation of a stable NBF without protein.

Our results show that, at a given protein concentration, there is a strong influence of the lipid concentration on the structure of the mixed films and on their behavior. Our results will therefore be separated into two parts according to the lipid concentration.

Effect of High Concentration of Lipid

At a high lipid concentration ($C_{DMPC}=0.5$ mg mL^{-1}) it is possible to obtain an NBF of DMPC in the absence of protein. The structure of such film is described in detail in [17] and at the beginning of this chapter. The mixed films are drawn from a solution prepared as described above with concentrations of 0.5 mg mL^{-1} DMPC and 0.7 mg mL^{-1} lysozyme.

The most important procedure of our purpose was to follow the time evolution of the total thickness of the film and the extra thickness. The central core thickness is increased owing to the presence of the lysozyme in the film. Table 14.5 gives the structural parameters corresponding to the fit. They are compared with the corresponding parameters of DMPC film. One can see that the roughness value (3 Å) remains as low as usually observed for DMPC film surfaces and that the insertion process does not significantly increase its value.

Table 14.5 Structural parameters for the films of DMPC and DMPC with lysozyme resulting from the fit using a five-layer model [a)].

Films of	h_1 (Å)	h_2 (Å)	h_3 (Å)	$\delta_1 \times 10^6$	$\delta_2 \times 10^6$	$\delta_3 \times 10^6$	σ (Å)	h_{total} (Å)
0.5 mg mL^{-1} DMPC	14.6	11.4	6.6	2.6	3.8	2.7	3.4	58.0
0.5 mg mL^{-1} DMPC + 0.7 mg mL^{-1} lysozyme	13.1	10.0	33.8	2.0	2.4	2.0	3.0	80.1
0.08 mg mL^{-1} DMPC + 0.7 mg mL^{-1} lysozyme	14.6	10.0	31.3	2.0	2.7	2.3	3.0	80.4

a) h_1, δ_1, h_2, δ_2 and h_3, δ_3 are the thickness and the real part of the refractive index of matter of X-rays, of the tails, the heads and the central core, respectively; σ is the roughness of the film.

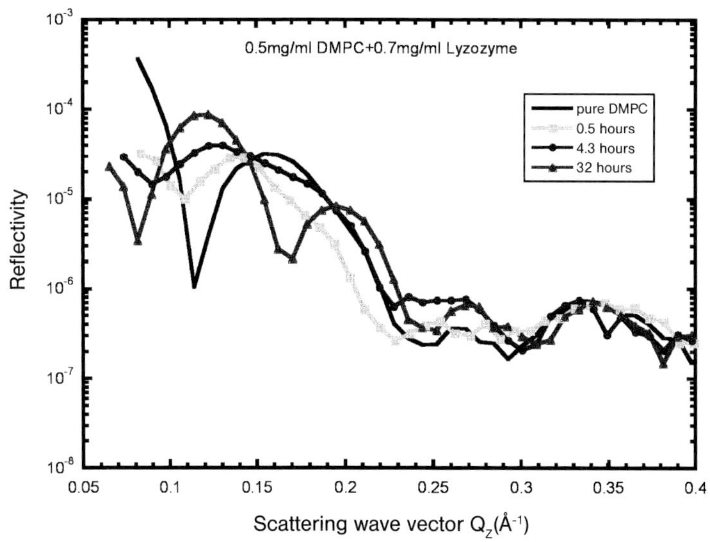

Fig. 14.8 Set of experimental reflectivity curves recorded at different times on an NBF. A time function shift of the fringes toward smaller angles is observed and shows that the initial film swells due to the protein insertion process.

The whole set of experimental reflectivity profiles recorded at different times is reported in Fig. 14.8. It displays a clear shift of the Kiessig fringes towards smaller angles, which evidences the swelling of the initial film due to the protein insertion. Figure 14.9 shows the time dependence of the central core thickness due to the lysozyme insertion within the DMPC NBF. The equilibration time for the formation of a complete protein single layer inserted in the surface

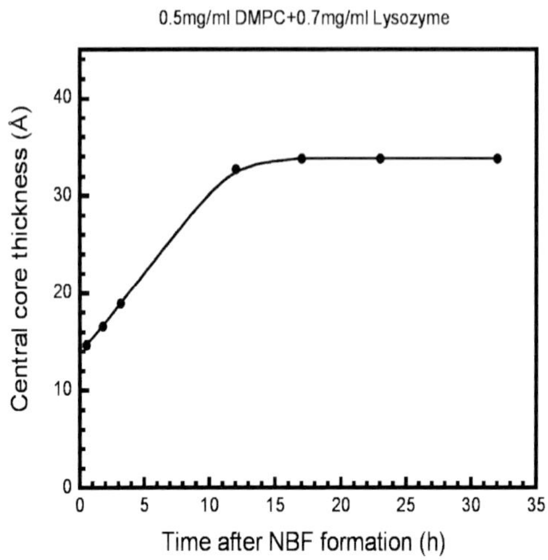

Fig. 14.9 Time dependence of the central core thickness due to the lysozyme insertion within the NBF. Time zero is chosen just after the drainage of the film.

bilayer by diffusion of BSA in the previous experiment was 48 h. In the present case with lysozyme, this time is reduced to 11 h. The equilibrium time for lysozyme is roughly four times shorter than for BSA. A very simple explanation of this value could by given by considering the difference of the contact area of the BSA and lysozyme with the surfactant walls (for an expected similar value of the interaction between the protein and the surfactants). For BSA, which has an ellipsoidal shape of dimensions 14.9×4.16 nm, the contact area is about four times higher than that of lysozyme (whose dimensions are 3×4.5 nm) and therefore the BSA diffusion rate should be four times slower than for lysozyme.

To understand the film architecture, we tried to relate the state of the Gibbs monolayer with the state of the black film by investigating the time evolution of the surface tension. Various investigations of protein monolayers [63–67] have revealed the complex behavior of a protein adsorbed at the Gibbs interface. The results in Fig. 14.10 show that the adsorption of the DMPS is a very slow process that requires more than 10 h to reach surface saturation [13]. The presence of the lysozyme in the solution increases the rate of the adsorption kinetics (about 2 h). The values of the surface tension corresponding to saturation are lower than that obtained for DMPC only. The balance between the protein adsorption and the lipid adsorption could explain this. From Fig. 14.10, one can observe that the adsorption of a protein is much faster than lipid adsorption. Thus, for the mixture at the beginning, there is mainly adsorption of protein, but the smaller lipid molecules progressively replace this protein. The same process has been described in several papers on mixed lipid/protein monolayers [58, 60, 68, 69]. Nevertheless, the value of the surface tension indicates the presence of a small quantity of protein molecules at the Gibbs interface. Such an organization at the surface before the film formation is in agreement with the model of the protein diffusion in the center of the black film after its formation, as reported above.

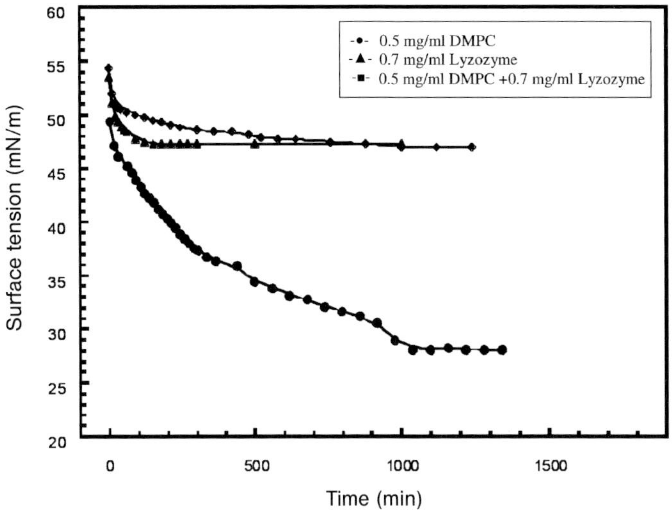

Fig. 14.10 Time dependence of the surface tension of solutions of lysozyme, DMPC and the DMPC/lysozyme mixture.

Effect of Low Concentration of Lipid

Some studies of lipid–protein interactions in the bulk have shown that the proteins adsorb on the vesicle surface [70–73]. To increase the number of free protein molecules capable of adsorbing at the air/water interface and to obtain better control of our mixed films, we reduced the lipid concentration. The minimum concentration that allows the formation of stable mixed films at the given protein concentration was found to be ~ 0.08 mg mL^{-1} for DMPC. It should be pointed out that, at this concentration, stable films could not be obtained without adding the lysozyme. This means that, at such a low phospholipid concentration during the film formation, there is no reservoir of single lipid molecules that are able to adsorb at the film surfaces and to stabilize the film. A mixed film was drawn from a solution, leaving it to equilibrate for 3 h. The films were very viscose and their drainage was much slower than in the previous case with high lipid concentration.

Reflectivity curves were recorded at different times after the complete drainage of the water. All the curves displayed fringes whose minimum stayed at an identical position. The fringes evolved with time and their intensity increased, leading to better fringe contrast. This is clearly indicative of molecule reorganization involving only a slight and irregular change of the total thickness of this "sandwich film". It was not possible to follow the time evolution of this film structure owing to the inhomogeneity of the film after its formation (i.e. only one fringe is visible). Three hours later, the second fringe appears, indicating an improvement in the film homogeneity. The last reflectivity profile is quantitatively similar to that obtained at equilibrium for higher concentrations of lipids after the swelling process. This means that the film structure should be very close to that described previously – mainly composed of two DMPC layers with a single layer of proteins between them. If most of the proteins are located in the center of the film, it may be possible that a small amount remains located within the surfactant layers. The surface tension isotherms of DMPC, lysozyme and mixture [74] (not shown here) demonstrate that the equilibrium value of the surface tension is reached more rapidly for the mixture (3 h) than for the lysozyme (15 h). This adsorption time coincides with the time necessary to draw and obtain stable black films. As we show further in the text, the initial Gibbs monolayer is mainly composed of proteins, including a small quantity of lipids. If we now compare the reflectivity profile recorded in equilibrium in the two cases, protein insertion by diffusion and spontaneous formation of a mixed film, we can see that the final thickness is the same, as is also possibly the structure. The formation of a sandwich structure that consists mainly of two walls of lipids inserting proteins thus requires exchange between lipids and proteins during the drainage to form the external surfactant walls.

14.6.4 Action of Amphiphilic Cyclodextrins in Phospholipid Films

14.6.4.1 Introduction

Recently, new cyclodextrin derivatives were synthesized and shown to exhibit strong amphiphilic properties. Here we studied the action of these new amphiphilic cyclodextrins on phospholipids [75]. The present experimental approach used NBFs stabilized by the phospholipid DMPC as an artificial target to simulate the action of modified amphiphilic cyclodextrin.

Our attention was drawn to a new hydrophobically modified cyclodextrin [76], namely 6′-(cholest-5-en-3α-ylamido)succinylamido-6′-deoxy-per(2,6-di-*O*-methyl)-cyclomaltoheptaose (chol-DIMEB). An analysis [77] of its behavior in aqueous solution using surface tension measurements and light, small-angle X-ray and neutron scattering techniques proved that it self-assembles into monodisperse spherical micelles with an average aggregation number of 24. These highly water-soluble micelles have been shown to be two-shell objects, the cyclodextrin moieties being exposed towards the aqueous medium, making them prone to include guest molecules in the cavities.

Using small-angle X-ray and neutron scattering (SAXS and SANS, respectively), the microstructure at the supramolecular scale of the modified cyclodextrin (chol-DIMEB) micelle, i.e. aggregation number, charge and volume, in the presence guest molecules could be defined. One of the guests, i.e. the neurotropic molecule dosulepine, has been shown to interact with chol-DIMEB micelles [78]. This molecule has been shown to form inclusion complexes with the native [79] and modified [80] β-cyclodextrin and is extensively used in laboratory studies as a guest model.

The selective interaction of dosulepine with cyclodextrin cavities of chol-DIMEB micelles has been clearly evidenced by diffusion-assisted NOE-pumping experiments. Moreover, on inclusion of this guest, the aggregation number of the micelle was shown not to change noticeably. These results indicated that the packing of the chol-DIMEB into spherical micelles was exceptionally robust.

These studies have clearly demonstrated that the inclusion properties of the cyclodextrin's cavities of chol-DIMEB aggregated into micellar objects are retained. The stability and specificity of the mixed micelle involving target molecules, such as dosulepine, make these hydrophobically modified cyclodextrins good candidates as molecular carriers. Chol-DIMEB can therefore be of great interest for targeting of biologically important molecules and especially for the delivery of drugs.

Our aim in this study initially was to see if it was possible to form a stable mixed cyclodextrin/DMPC film and to determine its structure at a molecular scale: how would the cyclodextrin be incorporated into the membrane model? What would be the prevalent interaction? Then we took advantage of the property of the cyclodextrin carrier molecule to follow *in situ* the behavior of an active guest molecule, dosulepine, trapped in the cyclodextrin cavity [75]. The preparations of the materials and of the solutions have been described elsewhere in detail [75, 77].

14.6.4.2 **Mixed DMPC/chol-DIMEB Films**

Mixed DMPC/chol-DIMEB films were studied with various DMPC:chol-DIMEB molar ratios R varying from $R=\infty$ to $R=0$ ($R=\infty$, pure DMPC film; $R=0$, pure chol-DIMEB film). The different thicknesses of the films extracted from the reflectivity curves collected for various ratios are shown in Fig. 14.11. This curve has a bell shape and exhibits a maximum of thickness (75 ± 1 Å) for a molar ratio of 3. The rise of the film until the maximum coincides also with high stability of the mixed film (more than 2 days). If we continue to add cyclodextrin, the thickness decreases and the film becomes more and more unstable. For the different ratios $R<3$ and $R>3$ (results not shown), the shapes of the curves are similar with the same level of intensity. By contrast, for $R=3$, the reflectivity curve exhibits a maximum intensity and a specific modulation of the Kiessig

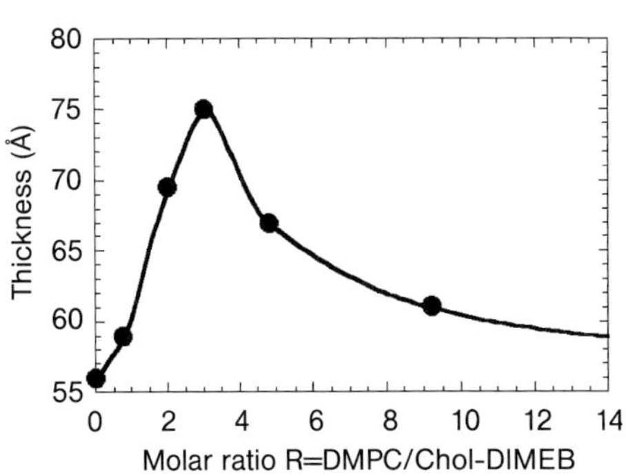

Fig. 14.11 Thickness extracted from experimental reflectivity curves of mixed DMPC/chol-DIMEB film at various molar ratios R (R=DMPC/chol-DIMEB) versus R.

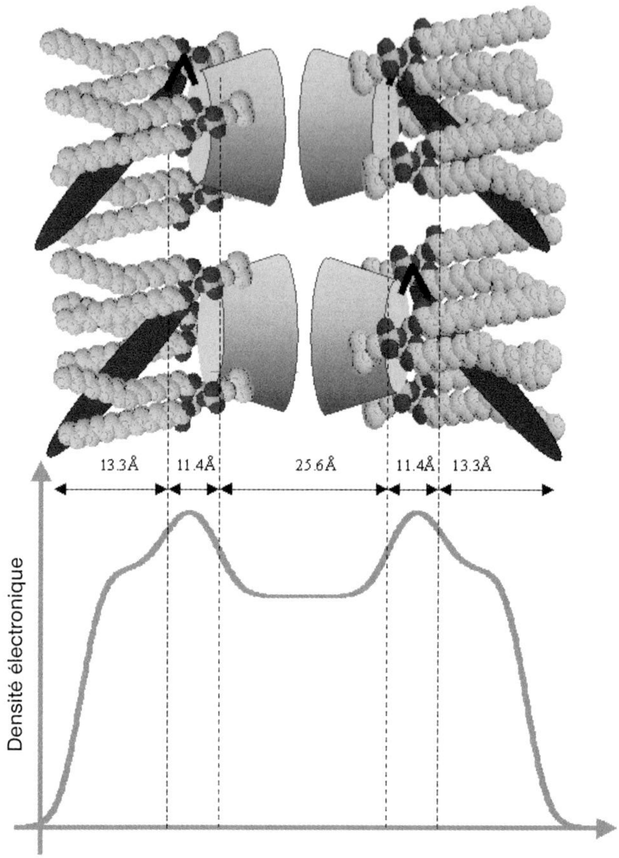

Fig. 14.12 Proposed structure of a mixed DMPC/chol-DIMEB film with a molar ratio of 3. The cyclodextrin cavities are embedded between two biological membrane-like walls composed of cholesterol and DMPC.

fringes. This clearly indicates that there are some molecular rearrangements within the film that yield a film structure specific to that particular ratio.

As for DMPC data, a five-layer model fitting procedure was necessary to describe the experimental data correctly [76]. We found an overall thickness of 75 Å. In comparison with a DMPC film, the 15 Å thickness increase observed is therefore due only to the specific structure of that film. The very low electronic density, 2×10^{-6}, in the central core, evidences the absence of any water molecules in this layer of the film. A thickness of 13.4 Å with an electronic density value of 1.8×10^{-6} was measured for the external layer. The intermediate layers display a high electron density of 2.45×10^{-6} and have an 11.4 Å extension. The central core thickness (25.5 Å) is comparable to the size of the cyclodextrin cavity (2×11 Å) [78]. Hence, in the mixed film, the central core should consist of the cyclodextrin cavities facing each other (see Fig. 14.12). The low value of the roughness (3 Å) supports the hypothesis of a homogeneous structure.

The ratio of the electronic density of the internal layer to that of the external layer is similar in both cases (1.35 for the mixed film and 1.43 for the pure DMPC). In the mixed case, the value is lower simply because we have to take into account the steric arrangement of three DMPC tails around the grafted cholesterol and the covalent succinyl bond. The surface [77] covered by one cyclodextrin cavity ring (295 $Å^2$) corresponds roughly to the additional surface of a cholesterol tail (120 $Å^2$) and three DMPC molecules (55 $Å^2$). Further, additional experiments (results not shown) indicated that the presence of a cholesterol molecule in a DMPC film, in the same molar ratio, did not lead to sensitive changes in the film thickness or in the density profile. Therefore, in the mixed film, such an arrangement results in an additional compaction that provides high stability to this molecular architecture. Indeed, owing to the film structure shown in Fig. 14.12, derived from the electronic density profile, it appears that the cyclodextrin insertion process in the DMPC bilayers can be reduced to a DMPC–cholesterol interaction. The condensing effects of cholesterol on the physical properties of lipid bilayers have been studied extensively by a variety of experimental methods [81–83]. Cholesterol is known as a regulator of membrane ordering [84] and to increase the dynamic rigidity of lipid bilayers [85]. However, up to now, what has remained unexplained is the specific molar ratio that leads to stability in the film. As observed previously [76], a mixed DMPC/chol-DIMEB lamellar phase was found to be stable at a molar ratio of 3.6, which is in the same range as that observed with our black films.

14.6.4.3 **Mixed DMPC/chol-DIMEB Films Including Dosulepine Guest Molecules**

Figure 14.13 shows the reflectivity intensity measured for this film versus the intensity reflected by the mixed film without dosulepine (dotted line). The large shift of the position of first minimum towards small-angle wavevectors indicates that dosulepine perturbs the film structure with a large swelling. It was difficult to interpret the reflectivity data in terms of a layer model, because of the film inhomogeneity and the complexity of the system. However, using a single-layer model, the precise overall thickness of the film could be measured as 100 ± 1 Å. To be sure the swelling observed was due only to dosulepine/chol-DIMEB complex formation, a test experiment was carried out. A mixed dosulepine/DMPC film was formed with respective concentrations of 0.3 and 0.5 mg L^{-1}, keeping the molar ratio constant ($R = 3$). The reflectivity curve is presented in the inset in Fig. 14.13. The position of its first minimum coincides with the first minimum of the pure DMPC curve. Consequently, the dosulepine molecule should not disturb the DMPC bilayer arrangement. This qualitatively confirms that the swelling of the DMPC/chol-DIMEB/dosulepine mixture film mainly originates in dosulepine/chol-DIMEB complex formation.

Assuming that the above-described mixed film structure (Fig. 14.12) is reliable, we can imagine this four-component film schematically as being formed by a cen-

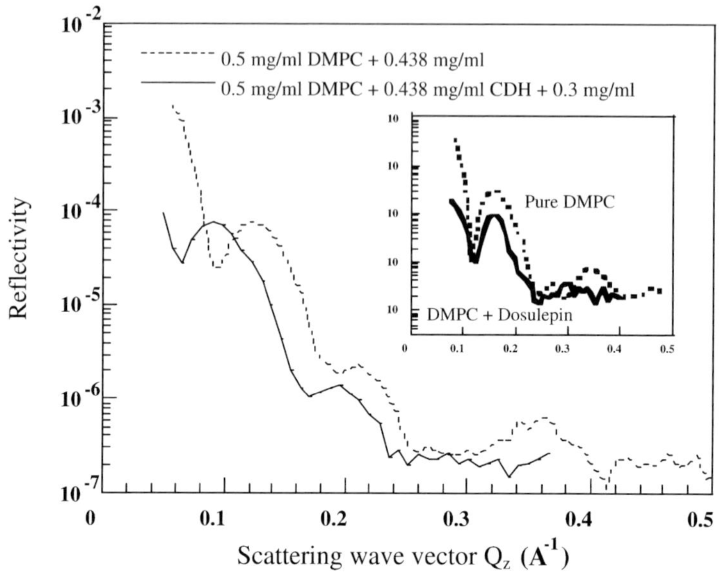

Fig. 14.13 Qualitative effect of dosulepine on the mixed DMPC/chol-DIMEB ($R=3$) film measured by X-ray reflectivity. Solid line, 0.5 mg mL^{-1} DMPC + 0.438 mg mL^{-1} chol-DIMEB + 0.3 mg mL^{-1} dosulepine; dashed line, 0.5 mg mL^{-1} DMPC +0.438 mg mL^{-1} chol-DIMEB. The swelling of the film is mainly ascribed to the occurrence of dosulepine/chol-DIMEB complexes as dosulepine alone does not perturb the DMPC bilayer structure (see the inset).

tral layer gathering the cyclodextrin cavities (each of them holding a dosulepine molecule), embedded between two external walls composed of cholesterol/DMPC mixtures, representing the biological membrane. Hence the overall thickness increase should be mostly related to partial dosulepine inclusion inside the cyclodextrin hydrophobic cavity. It has already been reported [79] that only one of the three rings of the dosulepine molecule is trapped in the cavity. The charged aliphatic chain remains outside the cavity. Nevertheless, this peculiar dosulepine/chol-DIMEB complexation is not able to explain the 25 Å thickness increase (the larger theoretical size of the molecule is no more than 9 Å). The difference may be accounted for by electrostatic repulsions (charged molecules of dosulepine) that generate a gap between the two layers of the film, which would imply a thickness increase. Very complex interactions (maybe constant exchange of dosulepine between the cyclodextrin receptors) should occur in the center of the film. Moreover, this electrostatic repulsion combined with steric interactions could be involved in the thickness increase. Besides, neither a five- nor a seven-box-model was able to reproduce correctly the experimental reflectivity profile since very complex interactions could occur in the center of the film.

14.7 Conclusion

Our results show that X-ray reflectivity is perfectly suited for the study of phospholipidic black films. In contrast with the other techniques already used in such investigations, the internal substructure can be determined with molecular-scale resolution. Therefore, we were able to study accurately the relation between the film structure and the molecular composition of the lipids. This permits fine tuning of the film properties and especially the selection of the hydrophil–hydrophobe balance for the central part of the film.

The behavior of lipid films being well established, we studied the interactions of these films with proteins dissolved in the subphase. Initially, we made the first NBF containing a single and close-packed layer of proteins. We found a time-dependent insertion of protein after the NBF formation, which we explained by the difference in chemical potential between the NBF and the solution after drawing.

The addition of DMPG made possible the formation of a stable mixed lipid/protein film. The film structure determination and the surface tension isotherms suggest two types of interactions between the lipids and protein, hydrophobic interactions and creation of H-bonds.

We evidenced two different behaviors depending on the lipid concentration. At high lipid concentration, we observed a swelling process, whereas at a lower concentration, we obtained spontaneously a mixed structure, with only molecular reorganization. If the two processes are clearly identified, the detailed structure and the process of reorganization require further investigations (such as surface diffraction at the interfaces). The difference between the two cases was also confirmed by surface tension experiments. It should be pointed out that this protein insertion under controlled conditions could be used to form model systems of biological interest for fundamental investigations of interactions between specific lipids and proteins.

The effect promoted by inclusion of cyclodextrins in a DMPC biological membrane model does not entail homogeneous mixed films at a macroscopic scale but results in a well-defined and organized structure at a microscopic scale, which remains a five-layer model. The cyclodextrins are anchored in the membrane through the cholesterol arm, making the cavity available for inclusion of dosulepine. Even if some effort is still needed to elucidate the structure of the analogous film with dosulepine, the molecular carrier property of cyclodextrin is clearly evidenced at the molecular scale. The strategy developed here should now be extended so as to be able to control the selective release of the active guest through the aqueous medium.

In summary, we believe that the use of NBFs for the study of specific interactions in membranes is very promising. NBFs combined with the X-ray reflectivity technique should provide a new tool for structural biology.

References

1 D. Exerowa, P.M. Kruglyakov, *Foam and Foam Films: Theory, Experiment, Application. Studies in Interface Science*, Vol. 5. Elsevier, Amsterdam, **1998**.

2 J.N. Israelachvili, *Intermolecular and Surface Forces with Applications to Colloid and Biological Systems*. Academic Press, Orlando, FL, **1985**.

3 E.J. Verwey, J. Overbeek, *Theory of the Stability of Lyophobic Colloids*. Elsevier, Amsterdam, **1948**.

4 M. N. Jones, K. Mysels, P.C. Scholten, *Trans. Faraday Soc.*, **1966**, *62*, 1336.

5 J.S. Clunie, J.F. Goodman, B.T. Ingram, in *Surface and Colloid Sci.*, Vol. 3, E. Matijevic (ed.). Wiley, New York, **1971**.

6 K. Maysels, *J. Phys. Chem.*, **1964**, *68*, 3441.

7 A. Scheludko, D. Exerowa, *Kolloid-Z.*, **1959**, *165*, 14.

8 J.M. Corkill, J.F. Goodman, D.R. Haisman, S.P. Harold, *Trans. Faraday Soc.*, **1961**, *57*, 821.

9 A. de Feijter, A. Vrij, *J. Colloid Interface Sci.*, **1979**, *70*, 456.

10 J.-J. Benattar, A. Schalchli, O. Belorgey, *J. Phys. I Fr.*, **1992**, *2*, 9.

11 O. Bélorgey, J.-J. Benattar, *Phys. Rev. Lett.*, **1991**, *66*, 313.

12 R. Cohen, R. Koyanova, B. Tenchov, D. Exerowa, *Eur. Biophys. J.*, **1991**, *20*, 203.

13 J. Toca-Herrera, H.-J. Müller, R. Krustev, D. Exerowa, H. Möhwald, *Colloids Surf. A*, **1998**, *144*, 319.

14 R. Cohen, D. Exerowa, T. Kolarov, T. Yamanaka, V. M. Müller, *Colloids Surf.*, **1992**, *65*, 201.

15 R. Cohen, D. Exerowa, *Colloids Surf. A*, **1994**, *85*, 271.

16 R. Cohen, D. Exerowa, T. Yamanaka, *Langmuir*, **1996**, *12*, 5419.

17 N. Cuvillier, F. Millet, V. Petkova, M. Nedyalkov, J.-J. Benattar, *Langmuir*, **2000**, *16*, 5029.

18 T. Yamanaka, M. Hayashi, R. Matuura, *J. Colloid Interface Sci.*, **1982**, *88*, 458.

19 A. Scheludko, *Adv. Colloid Interface Sci.*, **1967**, *1*, 391.

20 D. Exerowa, D. Kashchiev, D. Platikanov, B. Toshev, *Adv. Colloid Interface Sci.*, **1994**, *49*, 303.

21 R. Cohen, D. Exerowa, T. Kolarov, T. Yamanaka, T. Tano, *Langmuir*, **1997**, *13*, 3172.

22 S.P. Frankel, K.J. Maysels, *J. Appl. Phys.*, **1966**, *37*, 3725.

23 W.A.B. Donners, J.B. Rijnbout, A. Vrij, *J. Colloid Interface Sci.*, **1977**, *61*, 249.

24 C. Smart, W. Senior, *Trans. Faraday Soc.*, **1966**, *62*, 3253.

25 T. Tano, T. Umemura, J., *Langmuir*, **1997**, *13*, 5718.

26 O. Bélorgey, J.-J. Benattar, *Phys. Rev. Lett.*, **1991**, *66*, 313.

27 A. Schalchli, D. Sentenac, J.-J. Benattar, *J. Chem. Soc., Faraday Trans.*, **1996**, *92*, 553.

28 J.-J. Benattar, A. Schalchli, D. Sentenac, F. Rieutord, *Prog. Colloid Polym. Sci.*, **1997**, *105*, 11113.

29 H. Möhwald, *Annu. Rev. Phys. Chem.*, **1990**, *41*, 441.

30 C.W. Meuse, S. Krueger, C.F. Majkrzak, J.A. Dura, J. Fu, J.T. Conner, A.L. Plant, *Biophys. J.*, **1998**, *74*, 1388.

31 T.M. Bayerl, R.K. Thomas, J. Penfold, A. Rennie, E. Sackmann, *Biophys. J.*, **1990**, *57*, 1095.

32 S.J. Johnson, T.M. Bayerl, D.C. McDermott, G.W. Adam, A.R. Rennie, R.K. Thomas, E. Sackmann, *Biophys. J.*, **1991**, *59*, 289.

33 L.G. Parrat, *Phys. Rev. E*, **1954**, *95*, 359.

34 J.F. Nagle, R. Zhang, S. Tristram-Nagle, W.-J. Sun, H.I. Petrache, R.M. Suter, *Biophys. J.*, **1996**, *70*, 1419.

35 C.A. Helm, H. Möhwald, K. Kjaer, J. Als-Nielsen, *Europhys. Lett.*, **1987**, *4*, 697.

36 J. Seelin, *Biochim. Biophys. Acta*, **1978**, *515*, 105.

37 K. Tajima, N.L. Gershfeld, *Biophys. J.*, **1985**, *47*, 203.

38 M.C. Wiener, S.H. White, *Biophys. J.*, **1992**, *61*, 434.

39 B.W. Derjaguin, L. Landau, *Acta Physicochim. (USSR)*, **1941**, *14*, 633.

40 K. Mysels, H. Huisman, R. Razouk, *J. Phys. Chem.*, **1966**, *70*, 1339.
41 D. Exerowa, T. Kolarov, K. Khristov, *Colloids Surf.*, **1987**, *22*, 171.
42 P. Musselwhite, I. Kitchener, *J. Colloid Interface Sci.*, **1967**, *24*, 80.
43 D. N. Platikanov, G. P. Yampolskaya, N. I. Rangelova, Z. K. Angarska, L. E. Bobrova, V. N. Izmailova, *Colloid J. (USSR)*, **1981**, *43*, 149.
44 D. Clark, M. Coke, A. Mackie, A. Pinder, D. Wilson, *J. Colloid Interface Sci.*, **1990**, *138*, 207.
45 K. Marinova, T. Gurkov, O. Velev, I. Ivanov, B. Campbell, R. Borwankar, *Colloids. Surf. A*, **1997**, *123/124*, 155.
46 T. Peters, *Adv. Protein Chem.*, **1985**, *37*, 161.
47 D. E. Graham, M. C. Phillips, *J. Colloid Interface Sci.*, **1979**, *70*, 415.
48 J.-J. Benattar, M. Nedyalkov, J. Prost, A. Tiss, R. Verger, C. Guilbert, *Phys. Rev. Lett.*, **1999**, *82*, 5297.
49 J.-J. Benattar, F. Millet, M. Nedyalkov, D. Sentenac, in *Emulsion, Foams and Thin Films*, K. Mittal, P. Kumar (eds.). Marcel Dekker, New York, **2000**, Ch. 14, p. 251.
50 J.-J. Benattar, M. Nedyalkov, J. Prost, A. Tiss, R. Verger, C. Guilbert, *Phys. Rev. Lett.*, **1999**, *82*, 5297.
51 V. Petkoval, M. Nedyalkov, J.-J. Benattar, *Colloids Surf.*, **2001**, *190*, 9.
52 Z. I. Lalchev, P. J. Wilde, A. R. Mackie, D. C. Clark, *J. Colloid Interface Sci.*, **1995**, *174*, 283.
53 Jobe, Ikegami, *Am. Rev. Resp. Dis.*, **1987**, *136*, 1256.
54 Z. I. Lalchev, P. J. Wilde, D. C. Clark, *J. Colloid Interface Sci.*, **1994**, *167*, 8.
55 C. S. Vassilieff, I. Panaiotov, E. D. Manev, J. E. Proust, Tz. Ivanova, *Biophys. Chem.*, **1996**, *58*, 97.
56 D. Cho, G. Narsimhan, E. I. Franses, *Langmuir*, **1997**, *13*, 4710.
57 S. Sundaram, K. J. Stebe, *Langmuir*, **1997**, *13*, 1729.
58 K. Nag, J. Perez-Gil, A. Cruz, N. H. Rich, K. M. W. Keough, *Biophys. J.*, **1996**, *71*, 1356.
59 C. Tanford, *J. Mol. Biol.*, **1972**, *67*, 59.
60 E. D. Goddard, K. P. Ananthapadmanabhan (eds.), *Interactions of Surfactants with Polymers and Proteins*. CRC Press, Boca Raton, FL, **1993**.
61 S. Sundaram, K. J. Stebe, Langmuir, **1997**, *13*, 1729.
62 S. Sundaram, J. K. Ferri, D. Vollhardt, K. J. Stebe, Langmuir, **1998**, *14*, 1208.
63 D. E. Graham, M. C. Phillips, *J. Colloid Interface Sci.*, **1979**, *70*, 415, 427.
64 F. Uraizee, G. Narsimhan, *J. Colloid Interface Sci.*, **1991**, *146*, 169.
65 W. Norde, J. P. Favier, *Colloids Surf.*, **1992**, *64*, 87.
66 K. Anand, S. Damodaran, *J. Colloid Interface Sci.*, **1995**, *176*, 63.
67 B. S. Murray, *Langmuir*, **1997**, *13*, 1850.
68 N. Nishikido, T. Takahara, H. Kobayashi, M. Tanaka, *Bull. Chem. Soc. Jpn.*, **1982**, *55*, 3085.
69 R. R. Netz, D. Andelman, H. Orland, *J. Phys. II Fr.*, **1996**, *6*, 1023.
70 H. Matzumura, M. Dimitrova, *Colloids Surf. B*, **1996**, *6*, 165.
71 M. Dimitrova, H. Matzumura, V. Z. Neitchev, *Langmuir*, **1997**, *13*, 6516.
72 M. Dimitrova, H. Matzumura, *Colloids Surf. B*, **1997**, *8*, 287.
73 M. Dimitrova, H. Matzumura, V. Z. Neitchev, K. Furusawa, *Langmuir*, **1998**, *14*, 5438.
74 V. Petkova, J.-J. Benattar, M. Nedyalkov, *Biophys. J.*, **2002**, *82*, 541.
75 I. Javierre, M. Nedyalkov, V. Petkova, J.-J. Benattar, S. Weisse, R. Auzely-Welty, F. Djedaini-Pilard, B. Perly, *J. Colloid Interface Sci.*, **2002**, *254*, 120.
76 R. Auzély-Velty, B. Perly, O. Taché, T. Zemb, P. Jéhan, P. Guenot, J. P. Dalbiez, F. Djedaïni-Pilard, *Carbohydr. Res.*, **1999**, *318*, 82.
77 R. Auzély-Velty, F. Djedaïni-Pilard, S. Désert, B. Perly, T. Zemb, *Langmuir*, **2000**, *16*, 3727.
78 R. Auzély-Velty, C. Péan, F. Djedaïni-Pilard, T. Zemb, B. Perly, *Langmuir*, **2001**, *17*, 504.
79 F. Djedaïni-Pilard, J. Désalos, B. Perly, *Tetrahedron Lett.*, **1993**, *34*, 2457.
80 M. Gosnat, F. Djedaïni-Pilard, B. Perly, *J. Chem. Phys.*, **1995**, *92*, 1777.
81 G. Deinum, H. van Langen, G. van Ginkel, Y. K. Levine, *Biochemistry*, **1988**, *27*, 852.

82 T. P. W. McMullen, A. H. Lewis, R. N. McElhaney, *Biochemistry,* **1993**, *32,* 516.
83 T. P. W. McMullen, R. N. Mc Elhaney, *Curr. Opin. Colloid Interface Sci.*, **1996**, *1,* 83.
84 P. L. Yeagle, Biochim. *Biochim. Biophys. Acta*, **1985**, *822,* 267.
85 T. P. Trouard, A. A. Nevzorov, T. M. Alam, C. Job, J. Zajicek, M. Brown, *J. Chem. Phys.*, **1999**, *110,* 8802.

15
Phospholipid Foam Films: Types, Properties and Applications

Zdravko I. Lalchev

Abstract

In this chapter, some of the results concerning the formation, types, properties and applications of phospholipid foam films (PFFs) during the last three decades are summarized. The probability of the formation of stable PFFs and the dependences of the threshold concentration for PFF formation on the temperature and lipid phase state, as the main factors which determine the surface forces and extent of binding energy of the molecule in the film, are considered. The published results on phospholipid lateral diffusion (*D*) in the plane of PFF and dependences of *D* on the type and thickness of PFFs, molecular chain length, unsaturation and phase state of the phospholipids building the films are presented. The molecular interactions of PFFs with different surface-active agents present in the film-forming solution are also considered. Some applications of the PFFs for diagnostic aims and as a model system for studying lipid–protein interactions at interfaces, surface forces and binding energy of molecule in the film, the structure of alveolar surface and the properties of life-saving surfactant preparations used in medical practice are also presented.

15.1
Introduction

Thin liquid films at the air/liquid interface (foam films) have been known for a long time, dating back to Newton in 1704 and Plateau in 1873, who published some qualitative observations on them. Since then, a variety of systematic theoretical and experimental investigations have been carried out on foam films stabilized by numerous surfactants of the detergent and other types.

Studies of foam films from phospholipids started about three decades ago. Many data were collected on phospholipid foam films (PFFs) as they appeared to be a simple and adequate model system for studying foams, emulsions, foods, various problems in biology, pharmacokinetics, medicine and so forth.

Colloids and Interface Science Series, Vol. 1
Colloid Stability: The Role of Surface Forces, Part I. Edited by Tharwat F. Tadros

ISBN: 978-3-527-31462-1

The capability of PFFs to form mixed phospholipid–protein foam films suggested the possibility of using the model of PFFs for the investigation of lipid–protein interactions in highly ordered bilayer and multilayer membrane systems, in molecular cell biology, biotechnology and nanotechnology. The opportunities to investigate the surface forces and interactions of phospholipids (as the main components of cell membranes) with proteins and with other membrane-active agents in the film plane could involve PFFs as a model system for studying membrane–membrane interactions, cell fusion, alveolar stability and some aspects of lung physiology. Studies of the properties of different types of foam films composed of both artificial and natural mixtures of lipids and proteins are of increasing interest since they exist in nature, both *in vitro* and *in vivo*. One example of the functional importance of lipid–protein foam films *in vivo* is the lung surfactant.

Different types of stable PFFs can be formed *in vitro* spontaneously, hence it is possible to study them easily. The most attractive aspects of studying the PFFs are their formation, types and properties. This review aims to summarize some of the important investigations during the last three decades on PFFs in relation to the probability of their formation and lifetime (stability), kinetics of thinning, molecular binding energy and the surface phase transitions and lateral diffusion of the phospholipids in the film plane and to demonstrate some applications.

15.2
Formation and Types of PFFs

An important finding is that different types of stable PFFs can be formed experimentally. Some are illustrated schematically in Fig. 15.1, which shows a thick equilibrium film (1), which appears yellow or gray when viewed in reflected light under a microscope (the thickness of the film liquid core is in the range ca. 30–100 nm); (2) is a thick film with a black spot, signifying local thinning within the film; (3) is a common black PFF (with film thickness ca. 12–20 nm); and (4) is a Newton PFF, which is real bilayer film with thickness less than 8 nm.

The kinetics of formation of PFF are characterized by spontaneous thinning of the thick film (1) to the state of appearance of black spot(s) (2), which is expanded to cover the whole film area as a common (3) or Newton (4) black film. Some kinetic parameters during thinning of foam films of liposome suspensions [1] and the effect of the lipid phase state on the disintegration and transformation of liposomes into foam films have been published [2]. Data on direct measurements of the disjoining pressure in dimyristoylphosphatidylcholine (DMPC) foam films [3], thickness transitions in lysolecithin foam films [4], electrical conductivity in black films of lysophospholipids [5] and the effects of ions on the equilibrium film thickness [6] have been reported. The observations also of black films with "white spots" in them (i.e. lenses) [7] and stratified (multi-

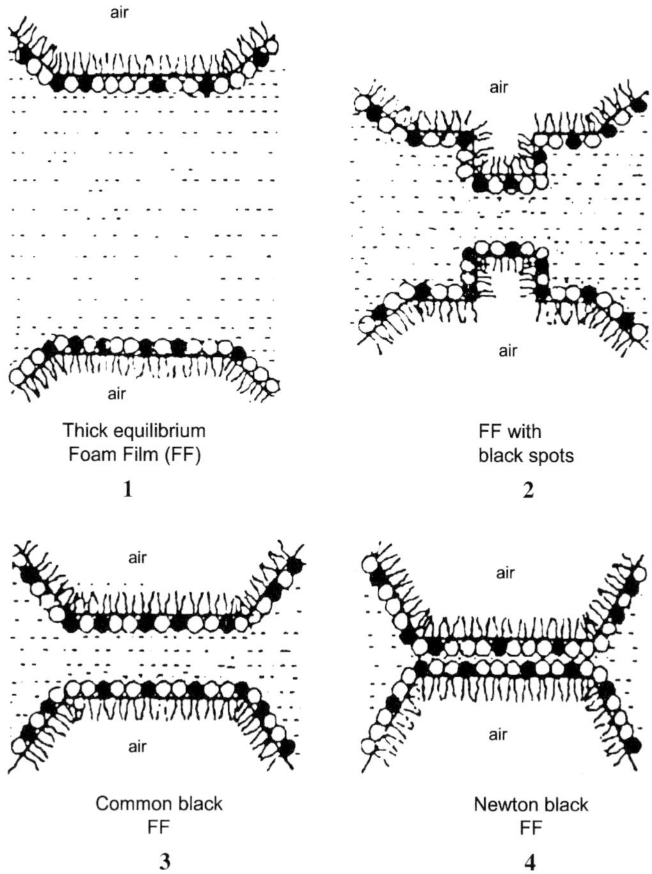

Fig. 15.1 Schematic diagram of phospholipid foam films (PFFs) of different types: (1) equilibrium thick foam film; (2) foam film with black spot; (3) common black PFF; (4) Newton black PFF. (From Lalchev [33], with permission).

layer) PFFs [8, 9] confirm the large variety of the types of PFFs that can be formed experimentally. Figure 15.2 shows the structure of the stratified foam film obtained from the phospholipid fraction of rat pulmonary lavages.

Figure 15.2 illustrates the applied external pressure (ΔP) versus film thickness (h) dependence, which shows a stepwise decrease in h from 32 to 12 nm with increase in ΔP, each step being about 5.5 nm. Four such steps are obtained until the film ruptures at $\Delta P \approx 2 \times 10^4$ N m^{-2}, which suggests that the stratified PFFs should be considered as multilayer structures consisting of several lamellae (bilayers) between the foam film monolayers (the structure represented on the upper right side of Fig. 15.2). It should be noted that systematic detailed investigations on the conditions and reproducibility of formation and properties of the stratified PFFs have still not been carried out.

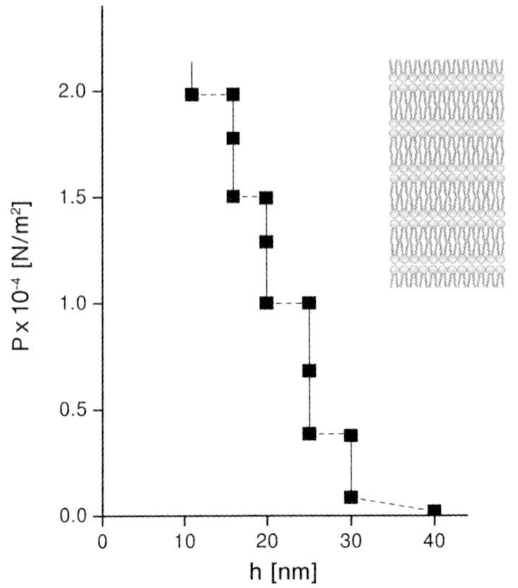

Fig. 15.2 Applied external pressure (ΔP) versus film thickness (h) dependence for foam films of the total phospholipid fraction in rat pulmonary lavages in the presence of 47.5% ethanol and 7×10^{-2} mol L^{-1} NaCl at 25 °C. (From Exerowa and Lalchev [8], with permission).

15.2.1
Probability of Formation of PFFs

The probability (W) of the formation of a stable black foam film and its lifetime (τ) depend strongly on the concentration (C) of the amphiphile molecules in the film=forming bulk solution, according to the hole-nucleation theory of Kashchiev and Exerowa [10, 11], as follows:

$$W = \exp(-t/\tau)$$

where t is the resolution time of the equipment, and

$$\tau = A(r)\exp[B/\ln(C_e/C)] \quad (1)$$

where $A(r)$ is connected with the hole-nucleation kinetics, B is proportional to the work for hole formation in the film and C_e is the concentration of molecules at which the diluted and the condensed phase of vacancies in the film are in thermodynamic equilibrium.

The $W(C)$ dependence was checked experimentally with a number of phospholipids and the results revealed that for the phospholipids W changed sharply from 0 to 1 in a very narrow concentration interval [12–14]. W can be calculated using the equation $W = \Delta N/N$, where N is the total number of trials and ΔN is the number of trials in which stable black films are formed, thus indicating that the films always rupture ($W = 0$) or always form stably ($W = 1$). The extremely steep character of the $W(C)$ dependence allows one to define a threshold concentration (C_t) as the minimum phospholipid concentration at which $W = 1$ and stable films always form.

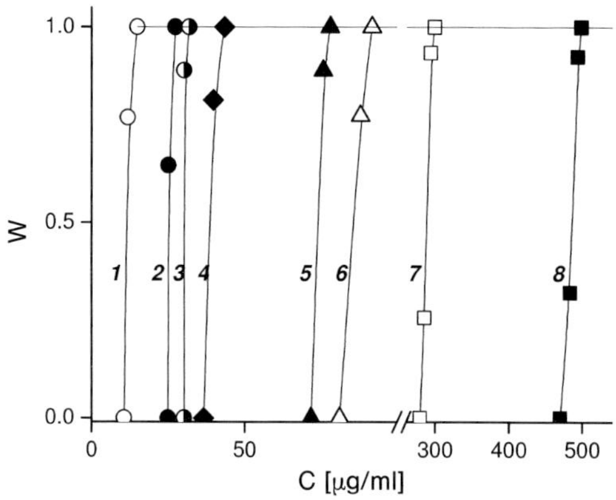

Fig. 15.3 Probability W of the formation of bilayer foam films as a function of the lipid concentration ($\mu g\ mL^{-1}$), in the presence of 47.5% ethanol, for a number of phospholipids: curve (1) phospholipids from amniotic fluid; (2) phosphatidylglycerol; (3) egg lecithin; (4) L-1,2-lecithin; (5) L-1,3-lecithin; (6) dilauroyllecithin; (7) phosphatidylinositol; (8) sphingomyelin. $T=25\,^{\circ}C$. (From Lalchev [14]).

It is important to note that C_t is strongly sensitive to the nature of the phospholipids (Fig. 15.3), composition of the film-forming solution (presence of ions and surface-active agents), temperature, phase state of the phospholipids and so forth [e.g. 14–18].

It can be seen in Fig. 15.3 that C_t differs significantly for different phospholipids, depending on their nature and structure.

15.2.2 Dependences of the Threshold Concentration (C_t) on Temperature and Lipid Phase State

The probability of film formation depends not only on concentration C (Eq. 1) but also on the phase state of the phospholipids. Figure 15.4 shows the $W(C)$ dependence for 1,2-dipalmitoleoyl-*sn-glycero*-3-phosphoethanolamine (DPoPE) in lamellar L_α, cubic Q_{II} (*Pn*3*m*) and inverse hexagonal H_{II} phases.

It can be seen in Fig. 15.4 that the respective C_t is $20\ \mu g\ mL^{-1}$ for the L_α phase, $30\ mg\ mL^{-1}$ for the *Pn*3*m* phase and $60\ mg\ mL^{-1}$ for the H_{II} phase. The order $L_\alpha > Pn3m$ at $22\,^{\circ}C$ and $Pn3m > H_{II}$ at $45\,^{\circ}C$ is observed with respect to the stability of the bilayer foam films formed from different DPoPE phases. These data reveal a greater ability of the L_α phase to disintegrate at the air/water interface in comparison with the *Pn*3*m*.

By studying the temperature dependence of C_t and film stabilization it is found that an increase in temperature causes increases in both C_t and phase transitions of the phospholipids in the film. Figure 15.5 shows the $C_t(1/T)$ dependence of foam bilayers of DMPC in gel and liquid-crystalline phase states

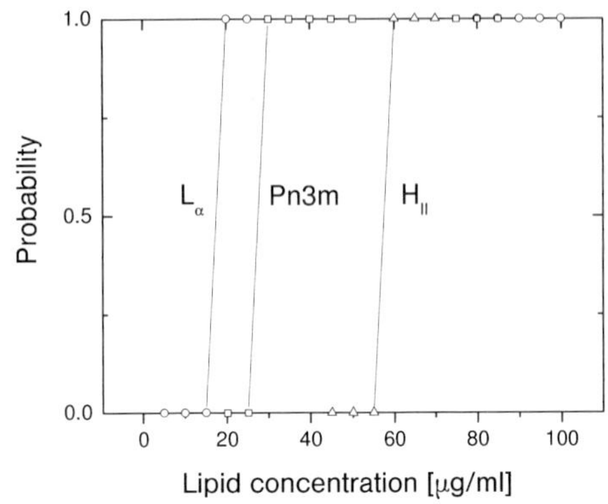

Fig. 15.4 Probability *W* of the formation of bilayer foam films as function of the lipid concentration (µg mL^{-1}) for DPoPE lamellar L_α (22 °C), cubic *Pn*3*m* (22 and 45 °C) and inverted hexagonal H_{II} (45 °C) DPoPE phases in 0.5 M NaCl solution. (From Jordanova et al. [19], with permission).

[18]. From this dependence, according to the theory [10, 11], the values of the binding energy of phospholipid molecule in the foam bilayers of different phase state are calculated for a number of lipids [18–21], according to the equation:

$$C_e = E_0 \exp(-Q/2kT) \qquad (2)$$

where C_e is the critical concentration, C_0 is a reference concentration and Q is the binding energy of the molecule in the bilayer. The values of Q for the DMPC molecule in the foam bilayer (Fig. 15.5), determined from the slopes of the curves [18], are 1.93×10^{-19} J in the gel state and 8.03×10^{-20} J in the liquid-crystalline state of the foam bilayer. The sharp change in the binding energy of a DMPC molecule in the foam bilayer is obviously due to the occurrence of a

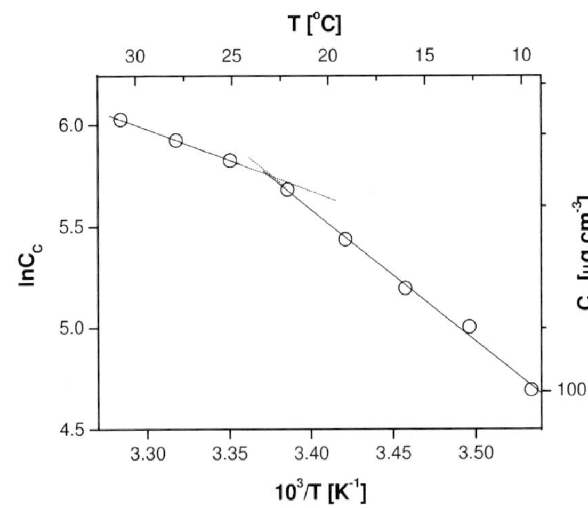

Fig. 15.5 Arrhenius plot of the dependence of the critical concentration for formation of DMPC foam bilayer on temperature: circles, experimental results; lines, theoretical dependence according to Eq. (2). (From Nikolova et al.[18], with permission).

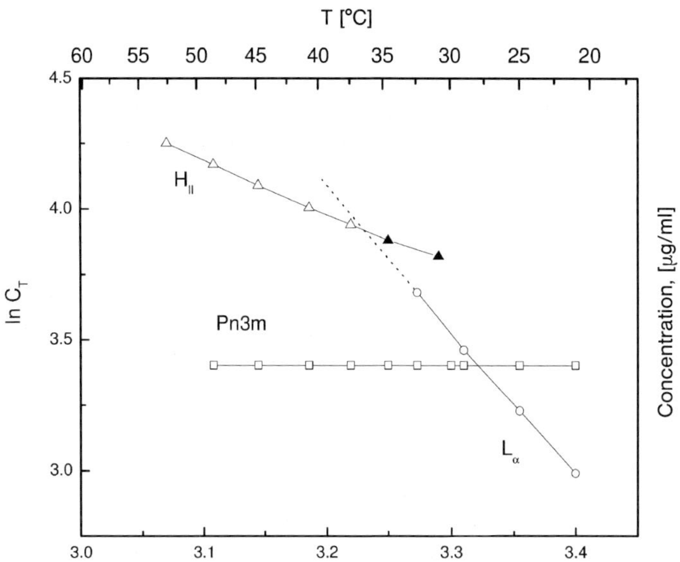

Fig. 15.6 Arrhenius plots of the threshold lipid concentrations C_t for formation of stable black foam bilayers from the *Pn3m*, L_α and H_{II} phases of DPoPE in 0.5 M NaCl solution. (From Jordanova et al. [19] with permission).

chain-melting phase transition in the bilayer. The higher value of *Q* for the gel state is natural, as it refers to a state with a higher degree of order. In Fig. 15.6 is shown also the Arrhenius plot of C_t versus temperature for foam films of DPoPE obtained from L_α, H_{II} and *Pn3m* cubic phases [19].

It can be seen from Fig. 15.6 that the slope of the curve from the L_α phase is higher than that from the H_{II} phase, indicating higher degree of order in the L_α than in the H_{II} phase. A sharp change in the slope takes place at about 39 °C, which corresponds to the temperature of the $L_\alpha \rightarrow H_{II}$ phase transition in the bulk for unsaturated DPoPE. Surprisingly no temperature dependence of C_t for the *Pn3m* phase is found (Fig. 15.6). Similar breaks in the Arrhenius dependence of C_t for saturated phospholipids DMPC [20] and DPPC [21], due to the gel $\rightarrow L_\alpha$ transition at 23 and 41 °C, respectively, have been reported.

15.3 Properties of PFFs

Black PFFs (Fig. 15.1) (3) and (4) are formed spontaneously as two-dimensional systems with a high degree of ordering. Newton black films are composed of two mutually adsorbed monolayers of phospholipid molecules with a "head-to-head" orientation, in contact with a gas phase and the adjacent solution (film-forming solution). The transition from thick PFF to common or Newton black films (depending on the experimental conditions) occurs spontaneously and is related to the film thickness. On the other hand, transitions in relation to the lipid phase state in the PFF also take place (as considered above). The phase state in PFFs is the main factor which determines the surface forces and extent of binding energy of the molecule in the film (as considered above) and the properties of PFFs, such as the molecular lateral diffusion in the film plane, the strength of the interactions of film molecules with the molecules (agents) present in the adjacent solution and so forth.

15.3.1 Molecular Lateral Diffusion in PFFs

The lateral mobility of lipids adsorbed at interfaces is of great importance in many instances, for example, in emulsion and foam stability, the food industry, lipid vesicles, pharmaceuticals, cell membrane functions and so forth. A number of methods, such as electron spin resonance, nuclear magnetic resonance, fluorescence correlation spectroscopy, time-resolved spatial photometry and fluorescence recovery after photobleaching (FRAP) [22] have been applied for measuring molecular diffusion in monolayers [23], liquid-crystal multilayers [24], black lipid membranes (BLMs) and the membranes of living cells [25]. FRAP is a non-invasive method used by most investigators and recently applied to PFFs. Using FRAP, the lateral diffusion coefficients (D, $cm^2\,s^{-1}$) are measured from the rate of recovery of fluorescence in the bleached local region in the film plane using the expression $D=\omega^2/4\tau_d$, where ω is the diameter of the bleached spot on the PFF and τ_d is the recovery time.

FRAP measurements of D in equilibrium PFFs were first reported in [26, 27] and later in [28–31]. These studies show that the values of D in foam films stabilized by phospholipid(s) depend on two main groups of factors. The first is related to the composition, type and thickness of the film. The second factor is related (within the limits of given film type and composition) to the molecular characteristics of the phospholipid(s) building the film, mainly the molecular charge and length, lipid phase state and so forth. Some examples are given below.

15.3.1.1 Dependence of the Diffusion Coefficient (*D*) on the Type and Thickness of PFFs

By using a specially constructed chamber [32] for FRAP experiments on PFFs, the dependence of D of the fluorescent lipid analog DiI-C18 in egg lecithin films on film thickness and type was measured [26, 28]. The dependence of D on the film type composed of pure dimyristoylphosphatidylglycerol (DMPG) and DMPC as a function of the temperature is shown in Figs. 15.7 and 15.8 [27, 29, 33].

It can seen in Figs. 15.7 and 15.8 that at any temperature the diffusion coefficients of the Newton black films were lower than those of the common films.

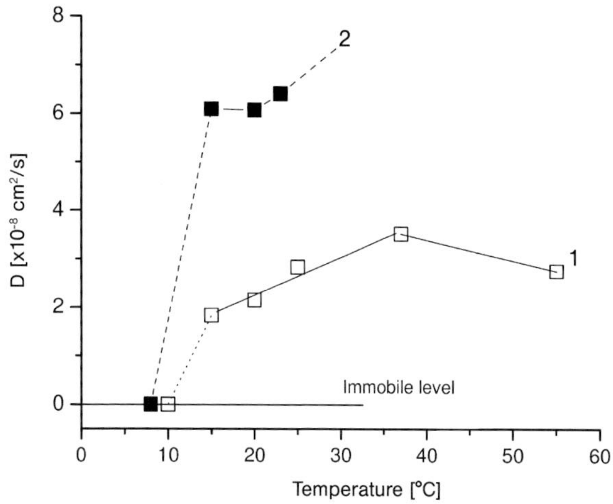

Fig. 15.7 Temperature dependence of the diffusion coefficient (*D*) of 5-*N*-(octadecanoyl)aminofluorescein (ODAF) in black foam films stabilized by DMPG; (1) common black foam films; (2) Newton black foam films. (From Lalchev et al. [29], with permission).

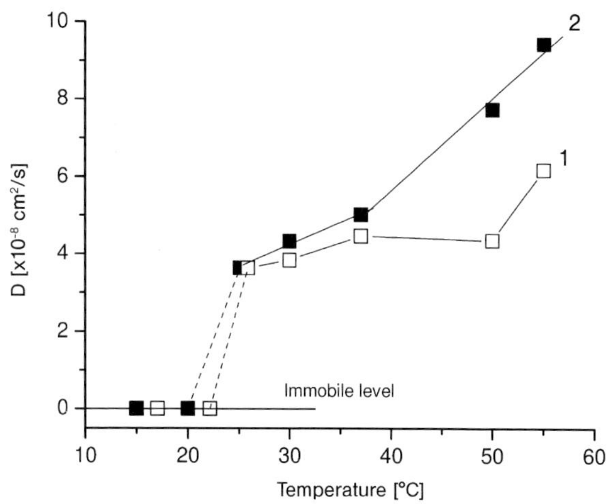

Fig. 15.8 Temperature dependence of the diffusion coefficient (*D*) of 5-*N*-(octadecanoyl)aminofluorescein (ODAF) in black foam films stabilized by DMPC; (1) common black foam films; (2) Newton black foam films. (From Lalchev et al [29], with permission).

During heating, the Newton PFFs of DMPG and DMPC show significantly slower diffusion in comparison with the common black films. It is detected that on going through the Newton and common black films to the thick equilibrium films of DMPC, the diffusion coefficients increase from approximately (3–5) $\times 10^{-8}$ to 1.2×10^{-7} cm^2 s^{-1} [27]. The increase in D with increase in film thick-

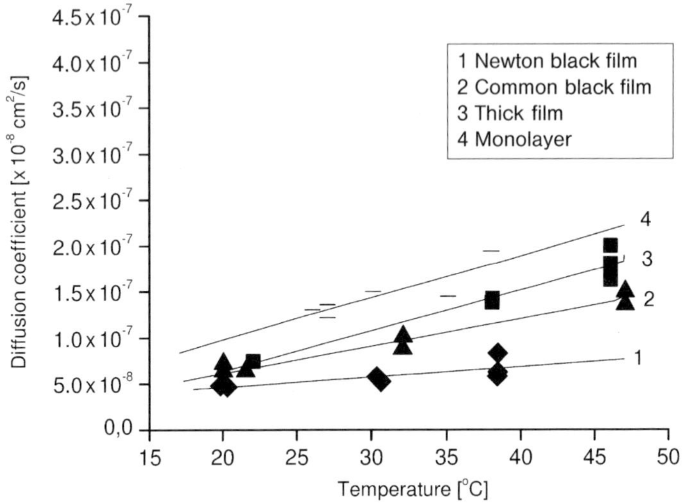

Fig. 15.9 Temperature dependence of the diffusion coefficient (*D*) of the probe NBD-PE in egg PC foam films of different types and in monolayer. (1) Newton black film; (2) common black film; (3) thick foam film; (4) monolayer ($\gamma = 37$ mN m^{-1}). NBD-PE probe: *N*-(7-nitrobenz-2-oxa-1,3-diazol-4-yl)-1,2-dihexadecanoyl-*sn-glycero*-3-phosphoethanolamine, triethylammonium salt. (From Lalchev and Mackie [31], with permission).

ness was studied in details for egg lecithin foam films and a comparison with the diffusion of the same lipid in a monolayer (Fig. 15.9) was reported [31].

It can be seen in Fig. 15.9 that the general trend of the curves for the thin Newton and common black films and the thick films reveals that the thicker films have faster diffusion, particularly at higher temperatures. One can see that the values of *D* corresponding to Newton black films and thick films are 5.0×10^{-8} and 13.3×10^{-8} cm^2 s^{-1}, respectively ($T = 37\,°C$). For the thick films it is found that on increasing the thicknesses between 30 and 90 nm, the values of *D* vary from approximately 1×10^{-7} to 2×10^{-7} cm^2 s^{-1}. Ultimately, the latter approaches the characteristic value for *D* in egg lecithin monolayers (Fig. 15.9, curve 4) at a macroscopic air/water interface above an infinitely thick subphase. Since the diffusion in monolayers depends on the molecular surface packing, *D* values of curve 4 are measured at monolayer compression to surface tension $\gamma = 37$ mN m^{-1}, as for thick foam films.

15.3.1.2 Dependence of Diffusion Coefficient (*D*) on the Phospholipid Phase State and Nature of Molecular Chains and Polar Headgroups

The first systematic data concerning the lateral diffusion in phospholipid foam films were published in 1994 [27]. The investigations were undertaken with foam films stabilized by DMPG, DMPC, DMPE, dilauroylphosphatidylethanolamine (DLPE), dipalmitoylphosphatidylethanolamine (DPPE) and dioleoylphosphatidylethanolamine (DOPE) using the lipid analog 5-*N*-(octadecanoyl)amino-fluorescein (ODAF) as a reporter group for the FRAP measurements. The latter were performed within the temperature range 8–85 °C, in order to measure *D* below and above the gel to L_α phase transition of the phospholipids.

A summary of the measured diffusion coefficients of ODAF in PFFs composed of phospholipids with three different polar headgroups within the temperature range 10–75 °C is shown in Fig. 15.10. The results reveal that each lipid shows either limited or irreproducible mobility below a certain critical temperature, referred to as the "immobile level", where the observed molecular mobility is due mainly to flow rather than to diffusion.

It can be seen in Fig. 15.10 that the temperature at which the onset of surface diffusion is detected follows the order DMPG (15 °C), DMPC (24 °C) and DMPE (50 °C), which are near to where the transition in bulk solution from the gel to the liquid-crystalline phase for the same lipids is observed.

The influence of the fatty acid chain length on *D* shows that in the case of saturated phospholipids, DLPE, DMPE and DPPE, the onset of diffusion takes place in the temperature range 45–50 °C (Fig. 15.11).

At any given temperature, the initial magnitude of the diffusion coefficient follows the order DLPE > DMPE > DPPE and increases in a similar way with increase in temperature. The first measurable diffusion of unsaturated phospholipid, DOPE, occurs at 22.7 °C and the magnitude of *D* increases with temperature, attaining values between those of DLPE and DMPE (Fig. 15.11). Hence one can conclude that the main determinant of *D* is chain unsaturation, which determines the lipid phase state at given temperature. Provided that the lipid state does not change across a given temperature interval, the chain length controls the magnitude of *D* in the manner shown for DLPE, DMPE and DPPE

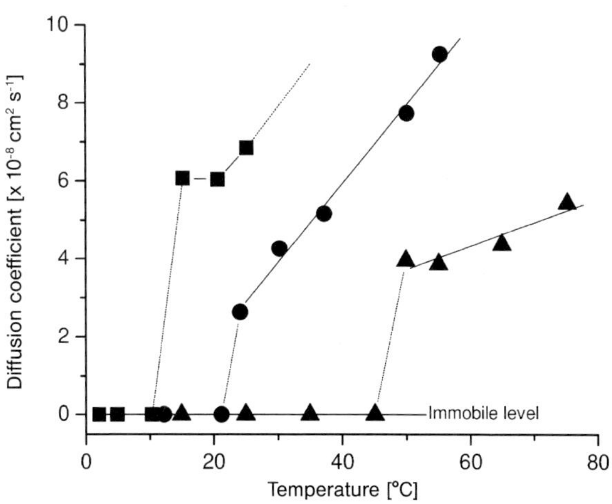

Fig. 15.10 Temperature dependence of the diffusion coefficient *D* of the lipid analog ODAF in foam films stabilized by DMPG (■), DMPC (▲) and DMPE (●). (From Lalchev et al. [27], with permission).

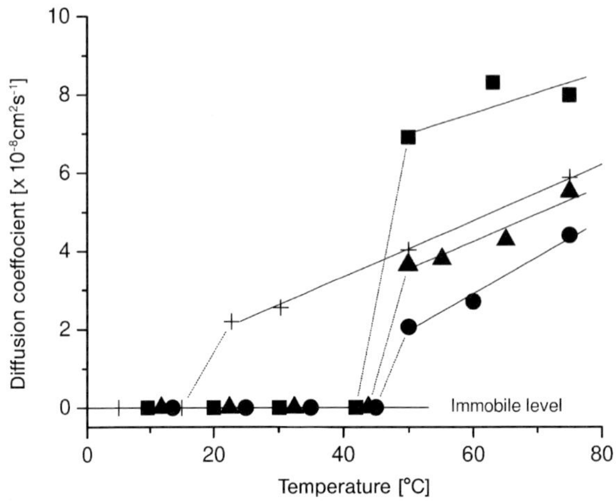

Fig. 15.11 Temperature dependence of the diffusion coefficient D of the lipid analog ODAF in foam films stabilized by DLPE (■), DMPE (●), DPPE (▲) and DOPE (+). (From Lalchev et al. [27], with permission).

(Fig. 15.11), that is, a shorter chain length determines higher diffusion. The lowest temperature during heating at which measurable diffusion occurs correlates with the gel to L_α phase transition of the phospholipid. This confirms the results on the lipid surface phase transition, measured by the temperature dependence of the $W(C)$ curve for DMPC [18, 34], which shows good agreement with DSC measurements of DMPC bulk phase transition.

A hysteresis phenomenon of D in PFFs during temperature cycling is observed. It is found [29] that $D(T)$ curves during heating lie under the curves of subsequent cooling. Probably this effect is complex and depends on the film composition, the thermal prehistory of the film, the lipid relaxation times during phase transitions and so forth. This hysteresis behavior of the films is considered to be normal since on going to a lower temperature after heating the system tends to retain its state at the previous higher temperature and transforms to the subsequent phase state according to its characteristic relaxation time. The same explanation applies for the effects of change in phase transition temperature during heating–cooling of several pure lipid samples observed by time-resolved X-ray diffraction and calorimetric studies [35–37].

15.3.2
Molecular Interactions of PFFs with Surface-active Agents

Interactions of the phospholipid molecules building the foam film with surface-active agents is studied by addition of agent(s) to the bulk solution which contacts the film (see Fig. 15.1). The addition of the agent(s) changes the stability of the film, which can be expressed by the probability of its formation (W) or its lifetime (τ) (see Eq. 1). The adsorption and penetration of the agent into the

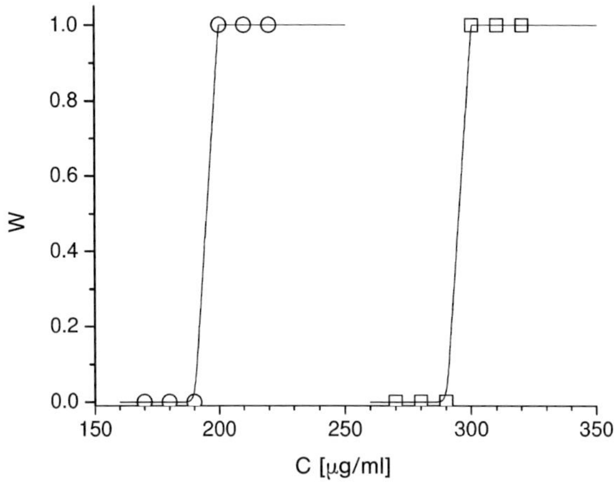

Fig. 15.12 Probability W of formation of common black foam films as a function of the lipid concentration (μg mL^{-1}) for DMPC. (□) DMPC films without DS-5000; (○) DMPC films with addition of DS-5000 (with concentration 8×10^{-7} M). $T=26\,°C$, $C_{el}=0.1$ M $CaCl_2$, pH=6.8–7.0. (From Lalchev [74]).

film result in a shift of the $W(C)$ dependence of pure phospholipid (see Fig. 15.3) to the left or right along the concentration axes, thus changing its C_t. Hence both $W(C)$ and the value of C_t and the $\tau(C)$ dependences of pure phospholipid are used in order to estimate the interactions of PFFs with the agents surrounding the film and in order to study the stabilizing (or destabilizing) effect of the agents on the film. The shift of the $W(C)$ dependence of pure phospholipid to higher C_t after addition of agent is interpreted as an effect of film destabilization and moving the $W(C)$ dependence to lower C_t means stabilization of the film (that is, an increase in its lifetime τ).

Figures 15.12 and 15.13 show the destabilizing effect of high molecular weight dextran sulfate (DS-5000) on the DMPC common black films and the stabilizing effect of lysolecithin on dilauroyllecithin (DLL) films, respectively.

It can be seen in Fig. 15.12 that the threshold concentration C_t of the DMPC film is 200 μg mL^{-1}, that is, at concentrations ≥200 μg mL^{-1} the films of DMPC are stable and never rupture ($W=1$). The addition of DS-5000 in the presence of Ca^{2+} ions causes rupturing of the films ($W=0$) in the concentration interval 210–290 μg mL^{-1}. The observed destabilizing effect of DS-5000 on the DMPC films agrees with literature data [38, 39] that the interactions of DS in the presence of Ca^{2+} with PC liposomes cause destabilization and induce liposome fusion. The results in Table 15.1 reveal that after further increases in DMPC concentration to 250 and to 300 μg mL^{-1}, it is possible to form stable films with increasing DS concentrations up to 9×10^{-8} and 8×10^{-7} M, respectively [40].

It is shown in Table 15.1 also that the addition of DS at concentrations up to 8×10^{-8} M does not change the film stability (at both DMPC concentrations) and stable black films are formed. The minimum concentration of 9×10^{-8} M DS is

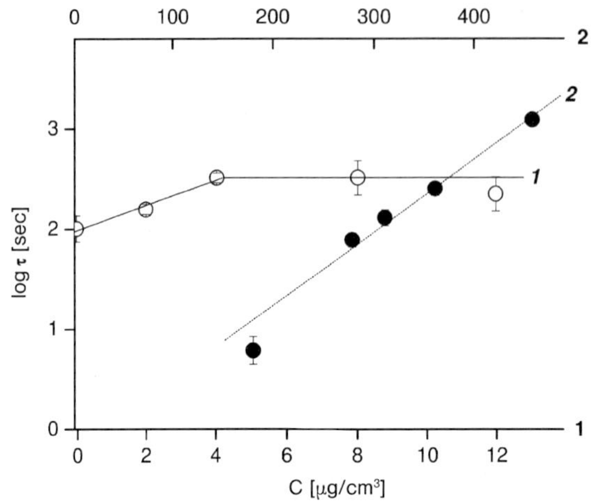

Fig. 15.13 Comparison of the stabilizing action of lysolecithin on the log (lifetime) of dilauroyllecithin (DLL) black foam films at a constant total concentration of lipids: DLL + lysolecithin = 44 μg mL^{-1} in 0.15 M salt solution (curve 1, abscissa 1) and on log(waiting time) for membrane fusion (curve 2, abscissa 2). (From Naydenova et al. [16], with permission).

determined at which DMPC films at 250 μg mL^{-1} rupture (after the appearance of black spots, BS), but a further increase in DMPC concentration to 300 μg mL^{-1} leads to the formation of a stable film again. A concentration of 8×10^{-7} M DS is found to be the minimum for rupturing a film formed at 300 μg mL^{-1} DMPC. In terms of the $W(C)$ dependence (Fig. 15.12), the above effect means that the $W(C)$ curve for pure DMPC is shifted to higher DMPC concentration after addition of DS, hence C_t is increased in the presence of DS.

Table 15.1 Effect of addition of dextran sulfate (DS-5000) on the stability of common black films (CBF) of DMPC and values of expansion time of black spot (BS) [a].

$C_{DS\text{-}5000}$ (M)	C_{DMPC} = 250 μg mL^{-1}		C_{DMPC} = 300 μg mL^{-1}	
	Thinning of the thick film	**Final state of CBF**	**Thinning of the thick film**	**Final state of CBF**
0 (pure DMPC)	Homogeneous	Stable (t_{1-2} = 30 s)	Homogeneous	Stable (t_{1-2} = 30 s)
2.5×10^{-8}	homogeneous	Stable (t_{1-2} = 30 s)	Homogeneous	Stable (t_{1-2} = 30 s)
9×10^{-8} [b]	Inhomogeneous	Rupture with BS	Homogeneous	Stable
2.5×10^{-7}	Inhomogeneous	Rupture with BS	Homogeneous	Stable
8×10^{-7} [c]	Inhomogeneous	Rupture before BS	Inhomogeneous	Rupture with BS
2.5×10^{-6}–10^{-2}	Inhomogeneous	Rupture before BS	Inhomogeneous	Rupture before BS

a) T = 26 °C; pH = 6.8–7.0; C_{el} = 0.1 M $CaCl_2$. The expansion time (t_{1-2}) is detected from the moment of formation of the black spot to its expansion to the whole film area.
b) The minimum $C_{DS\text{-}5000}$, at which the film ruptures at 250 μg mL^{-1} DMPC.
c) The minimum $C_{DS\text{-}5000}$, at which the film ruptures at 300 μg mL^{-1} DMPC.

It is assumed that interactions of DMPC with DS molecules result in dehydration of the DMPC polar heads in the presence of Ca^{2+} ions, which is in accordance with the observed dehydration effect of DS on phosphatidylcholine polar heads [41–43] in the presence of Ca^{2+} ions. Data have also been published on the interactions between phospholipid (PC, PC plus PE) polar heads in monolayers and other agents (PEG 8000–10 000) [44] in the presence of Ca^{2+} ions and also for a drastic change in the expansion time of the black spots in PFFs and for initiation of transitions from thick to Newton black films by PEG 4000 [45].

There are cases when interactions of PFFs with agents cause stabilization of the PFFs. The stabilizing effect of lysolecithin on the dilauroyllecithin (DLL) films was demonstrated by measuring the increase in the lifetime of DLL films after addition of lysolecithin, shown in Fig. 15.13 (curve 1, abscissa 1) [16]. The stabilizing effect of lysolecithin on the foam films corresponds with the increase of the waiting time for fusion of two contacting BLMs in the presence of lysolecithin in the bathing solution (curve 2, abscissa 2) [46].

The above data on the stabilization/destabilization effects of surface-active agents on PFFs confirms the adequacy of these films as a model to study molecular interactions in two-dimensional phospholipid systems, mechanisms of membrane-membrane interactions, fusion and so forth.

15.3.3
Recently Developed Techniques for Studying the Properties of PFFs

The microinterferometric technique of Scheludko and Exerowa was successfully combined with other techniques in studies of the PFF properties. Recently, in addition to FRAP, a number of methods have been applied that increased the measurable quantitative parameters of different PFFs. This extends the possibility of investigating some other properties of PFFs, such as electrical conductivity, gas permeation and so forth, and the variety of their interactions, for example with proteins, cyclodextrins and enhancers of gas permeation of the films, glycolipids and so forth. Such examples are described below.

A study of the interfacial behavior of a new type of amphiphilic cyclic oligosaccharides obtained by grafting a phospholipid on to a methylated cyclodextrin has been published [47]. These compounds are able to form stable black foam films, the structure of which can be determined using X-ray reflectivity. These films consist of a highly hydrated bilayer of modified cyclodextrins, which are remarkably thick owing to their abundant hydration core, which appears to be an efficient barrier against gas permeation. It was shown that the hydration of black films and the water core in them could be changed using IR irradiation [48]. The absorption of IR radiation by the water increases the film temperature and induces changing in the film thickness. In Newton black films, IR irradiation induces evaporation of the bonded water through the hydrophobic chain walls, whereas in common black films IR irradiation induces more complex hydrodynamic processes. It was shown also, using X-ray reflectivity, that the amount of bonded water depends on the nature of the lipids, in particular for zwitterionic lipids it was shown that the presence of water is controlled mainly by the chemical nature of the lipid headgroup and not by the electrostatic interactions [49].

The black foam films from DMPC alone with addition of the soluble DMPG have been studied in dynamic conditions and the dynamic contact angles for DMPC black films with and without DMPG were measured [50, 51] using the diminishing bubble method. It was shown also that the gas permeability coefficient is significantly reduced by the DMPG addition. It seems that the electrically charged DMPG anions, which determine a significant electrostatic disjoining pressure, play an important role in this specific behavior. The behavior of the dynamic contact angles is very different for DMPC Newton black films in the presence and absence of ethanol, showing larger gas permeability coefficients in the former case. The results could be connected with the change in the thickness and structure of the Newton black films in the presence of ethanol, taking into account both the solubility and hydration of the adsorption layers of the DMPC molecules.

The presence of ethanol in the film-forming suspensions influences the film tension of Newton black films obtained from DMPC [52]. In the case of DMPC in a water suspension, the film tension depends strongly on the film area, whereas in the case of DMPC films obtained from water–ethanol solution this dependence is less pronounced but still exists. The influence of ethanol on 1,2-distearoylphosphatidylcholine (DSPC) foam films [53] and DMPC films [54] was confirmed by parallel studies with grazing incidence X-ray diffraction and the Langmuir monolayer technique. The results on the thickness and free specific energy of formation of the films at different concentrations of ethanol show that both the thickness and tilt angle of the alkyl chains of the PC decrease with increasing concentration of ethanol. It is assumed that the ethanol causes a decreasing probability of the formation of hydrogen bonds of water molecules to the PC headgroups and a decrease in the surface excess energy per lipid molecule by increasing the van der Waals attraction between the film surfaces [53]. In addition, the contact angles between the film and the meniscus are detected as a function of the temperature in a range around the temperature of the main phase transition for the lipid [54]. Fluorescence microscopy was applied to investigate the distribution of a fluorescent lipid-like dye in the surface of the film and the meniscus and the results revealed that no structures are observed in the monolayers of the film [54].

Studies on black foam films from a mixture of phospholipids and a permeation enhancer (4-decyloxazolidine) were completed through the combination of three complementary techniques: surface tension measurements, X-ray reflectivity and the "diminishing bubble" method [55]. The thicknesses and the evolution of the coefficient of gas permeability in the presence of various 4-decyloxazolidine concentrations were examined.

An important topic concerning the mechanism of formation of PFFs is the kinetics of liposome disintegration at the film monolayer and the thinning processes of the PFFs. This has been addressed in several studies, usually by combination of the foam film technique with Langmuir monolayers [56, 57]. The kinetics of interfacial liposome breakdown of small unilamellar vesicles of DMPC were studied at temperatures above and below the temperature of the "melting" of the hydrophobic tails in the lipid aggregates. A change in the kinetic behavior was observed at both temperatures and, from the experimentally established time traces of the velocity of thinning of foam films, the rate constants of interfacial liposome disintegration were estimated. The foam film experiments confirm the existence of interfacial liposomal aggregates [56, 57].

It is reported that the method for insertion of soluble proteins within a Newton black film could also be used for phospholipid films and some properties of the mixed lipid–protein films obtained by this method have been described [58, 59]. For the first time foam films stabilized by glycolipids were studied and experimental and theoretical analyses of DLVO and non-DLVO forces in foam films obtained from a new class of glycolipids, rhamnolipids, were performed [60, 61].

15.4
Some Applications of PFFs

PFFs are self-assembled ordered systems, existing in unstable, metastable and stable states. The metastable and stable foam films have lifetimes ranging from several hours to days and can be formed and exist *in vitro* and *in vivo* in many systems in nature. For that reason, PFFs are widely used for studying foams stabilized by lipids [62], emulsions [63–66], coalescence phenomena in food emulsions and foams [67, 68], liposome suspensions [1, 2, 69, 70], lavages from animal and human lungs [12–14, 71–73] and as models for investigating the contact regions between membrane structures and cell membranes [16, 33, 74]. The type, structure and properties of PFFs are dominant factors for the stability of many lipid dispersions, pharmaceutical preparations, model membrane systems such as BLMs, multilamellar vesicles (MLVs) and Langmuir-Blodgett (LB) films on solid supports and so forth. Lately, Newton PFFs have acquired special interest. These structures exhibit some properties similar to those of BLMs, including thickness, refractive index and stability, but the molecular orientation in them is the reverse of that observed in BLMs. The comparison of the results collected for the bilayer structures, BLMs and Newton PFFs gave rise to the idea of the biological relevance of the latter structures.

The long-developed time theory and practice of foam films have been applied to PFFs, which have been used successfully during the last two decades as a model system for investigations of various problems in biology, pharmacokinetics, medicine and so forth. The interest in lipid and lipid–protein foam films is additionally amplified by the number of opportunities that the foam film model offers for studying lipid–protein interactions in highly ordered bilayer and multilayer membrane systems, in molecular cell biology, biotechnology and so forth. Many studies of pure PFFs [3–6, 13–16, 75, 76] and mixed lipid–protein foam films (LPFFs) obtained from human or animal lung surfactants and from amniotic fluids of pregnant woman [12–16, 77–79] demonstrate that they can be used as model systems for the investigation of inter-membrane interactions [4, 14, 15], cell fusion [16], fetal lung maturity and lung physiology [69–74, 79]. Reviews of all the above possibilities have been published [33, 74] and some examples of the potential applicability of PFFs are presented below.

15.4.1
Lipid–Protein Foam Film (LPFF) as a Model System for Studying Lipid–Protein Interactions at Interfaces

The model of LPFFs has been employed to study lipid–protein interactions at interfaces by estimation of the molecular lateral mobility in LPFFs and the binding energy of lipids existing in different phase states at the film interfaces. A study of the lateral diffusion coefficient (*D*) in foam films stabilized by lipids and lipid–protein mixtures isolated from porcine lung surfactant (LS) samples has been published [30].

Figure 15.14 shows the temperature dependence of *D* in foam films stabilized by the fraction of lung surfactant (LS) phospholipids before (curves 1 and 2) and after (curves 3 and 4) addition of the specific surfactant protein (SP-A). The *D* value in the pure DPPC (the main component of LS) films (curve 5) remains at the immobile level up to 45 °C and higher temperatures only induce small increases in *D*. In contrast, the *D* value in films stabilized by LS phospholipid fraction (curves 1 and 2) is much higher. An electrolyte dependence of *D* is

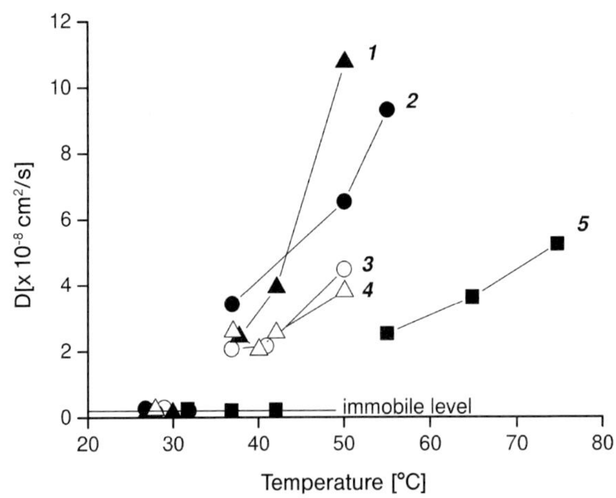

Fig. 15.14 Temperature dependence of the diffusion coefficient *D* of the lipid analog ODAF in foam films stabilized by fraction of lung surfactant phospholipids alone (curves 1 and 2) and in the presence of the specific surfactant protein SP-A (curves 3 and 4) in 0.125 M NaCl solution. Curve 5, the main component of lung surfactant phospholipids, DPPC. The type of the film is common black film. (From Lalchev [33], with permission).

detected, which could be due to small changes in the film thickness (the difference between curves 1 and 2). The higher *D* of the fraction compared with that of DPPC could be explained both by the presence of phospholipids other than DPPC in the fraction (some of which may be charged, probably have shorter acyl chain lengths and/or headgroup size, as shown in Figs. 15.10 and 15.11) and by a shift in the phase transition temperature of the fraction (consisting of unsaturated lipids) to a lower temperature comparable to DPPC, with resultant fluidization of the film surfaces (shown for DOPE in Fig. 15.11). The films from phospholipid–SP-A mixture are characterized by *D* values lower than those of the phospholipids alone but higher than that of DPPC (curves 3 and 4 in Fig. 15.14). This effect could be attributed to the surface interactions between the large, lipid-binding SP-A molecules with the phospholipids at the interface, resulting in reduced lateral diffusion.

In addition to the molecular surface mobility, the lipid–protein interactions influence the phase transition temperatures of the lipids at the film interfaces. Data for phase transitions of pure phospholipids in PFFs (e.g. [18, 20]) and in LPFFs [30] are consistent with literature data, where the transitions of lipids [35] and lung surfactant [80, 81] in the bulk were investigated by DSC. Whereas the individual phospholipids show sharp cooperative gel to liquid-crystalline transitions [27, 29], the presence of proteins (surfactant specific) in LS samples results in a broad temperature range transition, spanning from about 20 to 39 °C [80, 81]. Other data reveal that the increased content also of non-specific proteins correlates with a relatively large decrease in *D* and in a slight increase in the temperature where measurable diffusion is first observed due to the lipid–protein interactions [30, 33, 74]. Hence correlations are found between the increased content of proteins in PFFs both with the decreased lateral diffusion in the films and with change in the temperature where measurable diffusion in the film starts. The latter observations are connected with the effect of the lipid–protein interactions on the lipid phase state at the film interfaces.

15.4.2
LPFF as a Model System for Studying Alveolar Surface and Structure

Investigations of foam films composed of both artificial and natural mixtures of lipids and proteins are of increasing interest since they are widely present in nature, both *in vitro* and *in vivo*. An example of the functional importance of lipid–protein foam films *in vivo* is the lung surfactant (LS) that covers the alveolar surface. The surface activities of LS at the alveolar interface and its role in alveolar stability and lung integrity during the breathing process have been intensively discussed for a long time. Following Von Neergaard's work over 75 years ago, it has become generally accepted that the stabilization of the alveoli and anti-collapse phenomena in the lung are due to a combination of tissue and surface forces. The latter forces are exposed in the surface films of different types that exist *in vivo* in the lung. Initially, the existence of a monolayer lipid and lipid–protein film at the alveolar hypophase/air interface characterized by zero or near-zero surface tension and its physiological importance have long been the subject of considerable debate (e.g. [82, 83]). Hence model studies on "open" and "closed" monolayer films, such as "bubble" [84] or "captive bubble" [85] films, have been involved in studying the alveolar surface using LS samples. Later, model studies on bilayer and multilayer foam films of LS samples were introduced, based on the hypothesis of the existence *in vivo* of structures such as those of foam films [12, 14]. This hypothesis was based on the facts that bilayer and multilayer foam films (see Figs. 15.1 and 15.2) were formed experimentally *in vitro* from LS samples of rat, pig and rabbit lungs and that the conditions (electrolyte and phospholipid concentrations, pH capillary pressure, etc.) for their spontaneous formation *in vitro* were the same as in the lung [12–14, 70–72, 77–79]. The observations of foam in the lung [86, 87] and the possibility of spontaneous *in vitro* formation of stable lipid–protein foam films from different LS samples [12–14, 77–79] were the reasons for PFFs to be considered as realistic structural analogs of the surface films that exist *in vivo* in the alveoli. Subsequently, numerous papers have been published that prove experimentally the existence of structures such as bilayer and multilayer foam films in the alveoli [88, 89] and that consider the role of these films in the lung morphology and physiology [90–94]. The model of the PFFs opens up new possibilities for the investigation of lung surfactant and some respiratory problems (see the next section) and provides information additional to the monolayer model about the structure of alveolar films along with data on the film stability, thickness, kinetics of drainage and pressure–thickness isotherms, including the structure and forces in the stratified foam films [8, 9, 74]. The latter multilayer structures were first observed [95–97] from surfactants of detergent types and later from individual and mixed phospholipids [8, 9, 14]. Among the other models, the PFF model possesses several advantages for studying the properties and functional activity of the alveolar surface films and surfactant preparations. It was concluded [98], after comparative testing of the PFF model with four other models, that the black film method of Exerowa et al. "is the most discriminating of the tests studied. It provides a unique visual record of foam film formation and stability and clearly defines differences relative to both the nature and concentration of the preparations".

15.4.3
LPFF as a Model System for Studying Lung Maturity and Exogenous Surfactant Preparations

One practical application of the PFF model is to use the foam film stabilized by amniotic fluid (AF) as a diagnostic method for the assessment of fetal lung maturity [14, 77, 79]. The method is based on the possibility of the spontaneous formation of a stable, with respect to rupture, black foam film from mature AF, where the surfactant is sufficient. Whereas mature AF samples give stable black films, under conditions similar to those in the lung, the films from immature samples (with deficiency of phospholipids or surfactant proteins) are not stable and invariably rupture under the same conditions. Hence failure to obtain a stable foam film from AF samples indicates a high risk for the development of respiratory distress syndrome (RDS) in the newborn. The diagnostic "black foam film" method has been introduced in several hospitals and is easy to perform, fast, produces simple and unambiguous results and requires minimal sample volumes – all of which are advantages in comparison with the widely popular L/S ratio method [14, 78, 79].

Another application of the model of PFFs is in studying the behavior and surface characteristics of exogenous surfactant preparations (ESPs), which are widely used for the therapy of RDS and a variety of respiratory complications. The PFF model investigates the ESP *in vitro* at the same air/solution interface as in the alveoli, where the preparations spread and expose their functional activity after administration. The PFF model gives some new parameters for controlling ESP characteristics at the interface (some of them are briefly described below) and could be used to direct the production technologies of new ESPs.

Figure 15.15 shows the study of three commercial surfactant preparations, Infasurf (IN), Exosurf (EX) and Survanta (SU), using the PFF model. The probability (W_s) of black spot formation (see Fig. 15.1 (2)) is plotted against adsorption time as the preparation concentration is increased from 65 to 170 μg ml^{-1}. Black spot formation (W_s=1) by IN and SU requires adsorption times of up to about 10 min at each concentration, whereas EX requires about 40 min at the lowest concentration and about 12 min at higher concentrations. The black films (Fig. 15.1 (3)) are formed (W_f=1) by IN and EX only at the highest con-

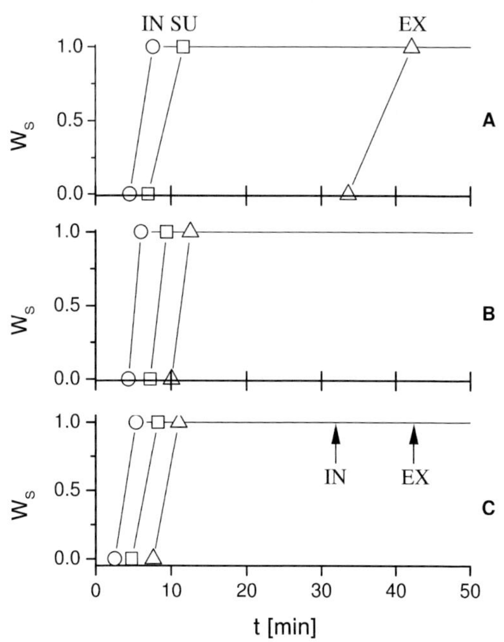

Fig. 15.15 Dependence of probability W_s of black spot formation in foam films on adsorption time at phospholipid concentrations of (a) 65, (b) 130 and (c) 170 μg mL^{-1} of the surfactant preparations Infasurf (IN), Exosurf (EX) and Survanta (SU). T=22 °C. (From Scarpelli et al. [70], with permission).

centration, when the adsorption time is increased to more than 30 min for IN and more than 40 min for EX (arrows in Fig. 15.15c). The films of SU always rupture ($W_f=0$) during the time of observation. In addition to the formation of black spots in the films, several hydrodynamic parameters of the foam films obtained from the preparations have also been studied [69, 70].

Photographs of the surface of IN, EX and SU foam films are shown in Fig. 15.16. The stages of black film formation are evident for IN and EX, namely thick film (line A in Fig. 15.16, comparable to Fig. 15.1 (1)); black spot formation and growth (line B in Fig. 15.16, comparable to Fig. 15.1 (2)) and black film formation (line C in Fig. 15.16, comparable to Fig. 15.1 (3)). In general, IN and EX films appear homogeneous in thickness; particles and aggregates are absent. Occasionally, white spots ("iceberg complexes") are seen in IN black films, probably representing lipid–protein complexes. Once formed, black films are very stable and persist for hours. In contrast, SU produces atypical transformations very slowly. Black spots are formed primarily at the periphery (line A in Fig. 15.16); their growth is very slow, taking up to several hours (line B); the film contains many particles and aggregates (line C) and the film thickness is

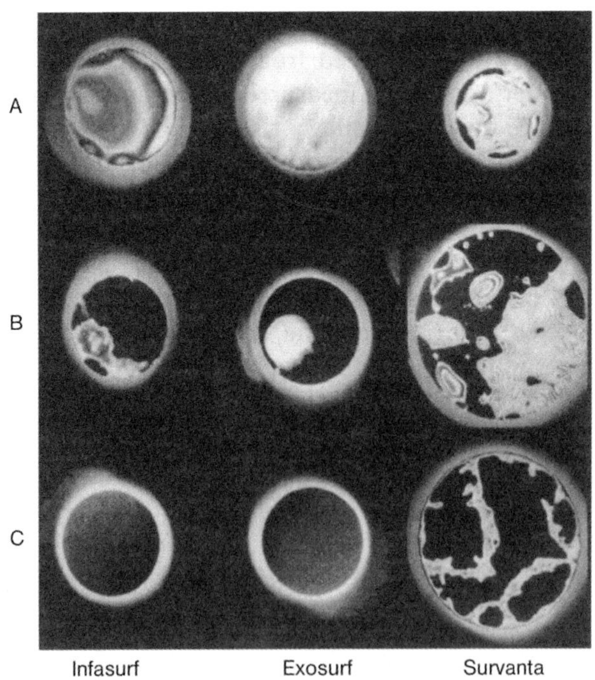

Fig. 15.16 Stages of foam film formation by Infasurf (IN), Exosurf (EX) and Survanta (SU). For IN and EX columns, row (A) shows thick films comparable to Fig. 15.1 (1); row (B) shows black spot formation and growth comparable to Fig. 15.1 (2); row (C) shows black film formation comparable to Fig. 15.1 (3). (From Scarpelli et al. [70], with permission).

heterogeneous. When films of IN, EX and SU are formed at higher concentration, which correspond to clinical levels, the different transformations of film architectures are analogous to those described [70].

A comparative study between foam films obtained from the preparations Alveofact, Curosurf, Survanta and Exosurf and from clinical samples of tracheal aspirates taken from newborns with RDS after surfactant therapy has been published [69]. A comparative analysis of the rheological behavior of the foam films obtained from surfactant preparations and from DPPC (as their main phospholipid component), under steady and transient flow conditions, is also available [99]. The differences observed are interpreted on the basis of the composition of the preparations, the presence of surfactant proteins and, in view of the bulk structures formed at different temperatures, the difference in the phase states between DPPC [100] and the other phospholipids present in the preparations [101].

In conclusion, it should be noted that the interest in the applications of phospholipid and lipid–protein foam films obtained from different artificial and natural mixtures (amniotic fluid, pulmonary lavages, tracheal aspirates liposomal surfactant preparations, etc.) is continuously increasing.

Acknowledgments

The author is grateful for financial support from grants from the BBSRC and The Royal Society (UK) and partially from grants (N BU-B-2/05 and N HT 1-04) from the Bulgarian Ministry of Education and Science.

References

1 Vassilieff C. S., Momtchilova I. G., Panaiotov I., Ivanova Tz., *Commun. Dept. Chem. Bulg. Acad. Sci.*, **1991**, *24*, 519–527.
2 Vassilieff C. S., Ivanova Tz., *Recent Res. Dev. Biophys. Chem.*, **2001**, *2*, 19–35.
3 Cohen R., Koynova R., Tenchov B., Exerowa D., *Eur. Biophys. J.*, **1991**, *20*, 203–208.
4 Cohen R., Exerowa D., Kolarov T., Yamanaka T., Muller V. M., *Colloids Surf.*, **1992**, *65*, 201–209.
5 Yamanaka T., Tano T., Tozaki K., Hayashi H., *Chem. Lett. Chem. Soc. Jpn.*, **1994**, 1143–1146.
6 Yamanaka T., Tano T., Kamegaya O., Exerowa D., Cohen R., *Langmuir*, **1994**, *10*, 1871–1876.
7 Vassilieff C. S., Manev E. D., *Colloid Polym. Sci.*, **1995**, *273*, 512–519.
8 Exerowa D., Lalchev Z., *Langmuir*, **1986**, *2*, 668–671.
9 Lalchev Z., Dimitrova L., Exerowa D., in *Proceedings of the 1987 International Congress on Membranes and Membrane Processes, ICOM, Tokyo*, Ref. N 13-OB 0918, pp. 727–728, **1987**.
10 Kashchiev D., Exerowa D., *J. Colloid Interface Sci.*, **1980**, *77*, 501–511.
11 Kashchiev D., Exerowa D., *Biochim. Biophys. Acta*, **1983**, *732*, 133–145.
12 Lalchev Z., Exerowa D., Koumanov K., *Ann. Univ. Sofia Fac. Chim.*, **1979**, *73*, 163–170.
13 Lalchev Z., *Ann. Univ. Sofia., Fac. Biol.*, **1982**, *4*, 38–46.

14 Lalchev, Z., Free liquid films of lipids, proteins and their mixtures. PhD Thesis, Bulgarian Academy of Science, Sofia, **1984**.
15 Exerowa D., Krugliakov P. M., in *Foam and Foam Films – Theory, Experiment, Application*, Mobius D., Miller R. (eds.). Elsevier, Amsterdam, **1998**, 741.
16 Naydenova S., Lalchev Z., Petrov A. G., Exerowa D., *Eur. Biophys. J.*, **1990**, *17*, 343–347.
17 Georgiev G., Lalchev Z., *Eur. Biophys. J.*, **2004**, *33*, 742–748.
18 Nikolova A., Exerowa D., Lalchev Z., Tsonev L., *Eur. Biophys. J.*, **1994**, *23*, 145–152.
19 Jordanova A., Lalchev Z., Tenchov B., *Eur. Biophys. J.*, **2003**, *31*, 626–632.
20 Exerowa D., Nikolova A., *Langmuir*, **1992**, *8*, 3102–3108.
21 Exerowa D., Todorov R., Nikolov L., *Colloids Surf. A*, **2004**, *250*, 195–201.
22 Schlessinger J., Axelrod D., Koppel D., Webb W., Elson E., *Science*, **1997**, *195*, 307–309.
23 Timbs M., Thompson N., *Biophys. J.*, **1990**, *58*, 413–428.
24 Timbs M., Poglitsch C., Pisarchick M., Sumner M., Thompson N., *Biochim. Biophys. Acta*, **1991**, *1064*, 219–228.
25 Schlessinger J., Koppel D., Axelrod D., Jacobson K., Webb W., Elson E., *Proc. Natl. Acad. Sci. USA*, **1976**, *73*, 4594–4598.
26 Lalchev Z., Ishida H., Nakazawa H., in *Proceedings of the VIth International Electro-Optics Symposium*, Jennings B. R., Stoilov S. P. (eds.). Institute of Physics Publishing, Bristol, pp. 239–244, **1991**.
27 Lalchev Z., Wilde P., Clark D., *J. Colloid Interface Sci.*, **1994**, *167*, 80–86.
28 Lalchev Z., Todorov R., Ishida H., Nakazawa H., *Eur. Biophys. J.*, **1995**, *23*, 433–438.
29 Lalchev Z., Wilde P., Mackie A., Clark D., *J. Colloid Interface Sci.*, **1995**, *174*, 283–288.
30 Lalchev Z., Todorov R., Christova Y., Wilde P., Mackie A., Clark D., *Biophys. J.*, **1996**, *71*, 2591–2601.
31 Lalchev Z., Mackie A., *Colloids Surf. B*, **1999**, *15*, 147–160.
32 Lalchev Z., Alexsandrov N., Todorov R., Exerowa D., Chamber for formation and investigation of thin liquid films. Bulgarian Patent No. 49729. Institute of Inventions, Sofia, **1990**.
33 Lalchev Z., Surface properties of lipids and proteins at bio-interfaces, in *Handbook of Surface and Colloid Chemistry*, Birdi K. S. (ed.). CRC Press, Boca Raton, FL, pp. 625–687, **1997**.
34 Exerowa D., Nikolova A., *Langmuir*, **1992**, *8*, 3102–3108.
35 Tenchov B., *Chem. Phys. Lipids*, **1991**, *57*, 165–177.
36 Koynova R., Tenchov B., Todinova S., Quinn P., *Biophys. J.*, **1995**, *68*, 2370–2375.
37 Koynova R., Tenchov B., Quinn P., Laggner P., *Chem. Phys. Lipids*, **1998**, *48*, 205–214.
38 Arnold K., Ohki S., Krumbiegel R., *Chem. Phys. Lipids*, **1990**, *55*, 301–307.
39 Budker V. G., Markushin Yu., Vahrusheva T. E., Kiseleva E. V., Mal'ceva T. V., Sidorov V. N., *Biol. Membr.*, **1990**, *7*, 419–427.
40 Georgiev G., Lalchev Z., *Eur. Biophys. J.*, **2004**, *33*, 742–748.
41 Huster D., Arnold K., *Biophys. J.*, **1998**, *75*, 909–916.
42 Huster D., Paasche G., Dietrich U., Zachömig O., Gutberlet T., Gawrisch K., Arnold K., *Biophys. J.*, **1999**, *77*, 879–887.
43 Steffan G., Wulff S., Galla H. J., *Chem. Phys. Lipids*, **1994**, *74*, 141–150.
44 Kuhl T., Guo Y., Alderfer J. L., Berman A. D., Leckland D., Israelashvili J., Hui S. W., *Langmuir*, **1996**, *12*, 3003–3014.
45 Todorov R., Alveolar surfactant-composition and properties studied by monolayer and thin liquid films. PhD Thesis, Sofia University, **1999**.
46 Chernomordik L. V., Koslov M. M., Melikyan G. B., Abidor I. G., Markin V. S., Chizmadzhev Yu. A., *Biochim. Biophys. Acta*, **1985**, *812*, 643–645.
47 Sultanem C., Moutard S., Benattar J.-J., Djedaini-Pilard F., Perly B., *Langmuir*, **2004**, *208*, 3311–3318.
48 Benattar J.-J., Shen Q., Bratskaya S., Petkova V., Krafft M. P., Pucci B., *Langmuir*, **2004**, *204*, 1047–1050.
49 Cuvillier N., Millet F., Petkova V., Nedyalkov M., Benattar J.-J., *Langmuir*, **2000**, *1611*, 5029–5035.

50 Petkova V., Platikanov D., Nedyalkov M., *Adv. Colloid Interface Sci.*, **2003**, *104*, 37–51.

51 Platikanov D., Nedyalkov M., Petkova V., *Adv. Colloid Interface Sci.*, **2003**, *100–102*, Suppl., 185–203.

52 Nedyalkov M., Petkova V., Platikanov D., *Colloids Surf. A*, **2003**, *220*, 35–43.

53 Brezesinski G., Muller H.-J., Toca-Herrera J. L., Krustev R., *Chem. Phys. Lipids*, **2001**, *110*, 183–194.

54 Toca-Herrera J. L., Krustev R., Muller H.-J., Mohwald H., *Colloid Polym. Sci.*, **2000**, *278*, 771–776.

55 Tranchant J.-F., Bonte F., Leroy S., Nedyalkov M., Platikanov D., Javierre I., Benattar J. J., *J. Colloid Interface Sci.*, **2002**, *249*, 398–404.

56 Vassilieff C. S., Panaiotov I., Manev E. D., Proust J. E., Ivanova Tz., *Biophys. Chem.*, **1996**, *58*, 97–107.

57 Horiuchi Y., Matsumura H., Furusawa K., *J. Colloid Interface Sci.*, **1998**, *207*, 41–45.

58 Cuvillier N., Petkova V., Nedyalkov M., Millet F., Benattar J. J., *Physica B*, **2000**, *283*, 1–5.

59 Petkova V., Benattar J. J., Nedyalkov M., *Biophys. J.*, **2002**, *82*, 541–548.

60 Cohen R., Ozdemir G., Exerowa D., *Colloids Surf. B*, **2003**, *29*, 197–204.

61 Cohen R., Exerowa D., Pigov I., Heckmann R., Lang S., *J. Adhesion*, **2004**, *80*, 875–894.

62 Lalchev Z., Wilde P., Clark D., *J. Colloid Interface Sci.*, **1997**, *190*, 278–285.

63 Sarker, D. K., Wilde P. J., *Colloids Surf. B*, **1999**, *15*, 203–213.

64 Rafai S., Sarker D., Bergeron V., Meunier J., Bonn D., *Langmuir*, **2002**, *18*, 10486–10488.

65 Manev E. D., Nguyen A. V., *Int. J. Miner, Proc.*, **2005**, *77*, 1–45.

66 Howbrook D., Sarker D., Lloyd A. W., Louwrier A., *Biotech. Lett.*, **2002**, *24*, 2071–2074.

67 Clark D., Mackie A., Smith L., Wilson D., in *Food Colloids*, Bee, R., Richmond, P., Mingins, J. (eds.). Royal Society of Chemistry, Cambridge, **1989**, 97.

68 Clark D., Wilde P., in *Gums and Stabilisers for the Food Industry*, Phillips G., Wedlock D., Williams P. (eds.). Oxford University Press, Oxford, **1992**, 343.

69 Lalchev Z., Georgiev G., Jordanova A., Todorov R., Christova E., Vassilieff C., *Colloids Surf. B*, **2004**, *33*, 227–234.

70 Scarpelli E., Mautone A., Lalchev Z., Exerowa E., *Colloids Surf. B*, **1997**, *8*, 133–145.

71 Lalchev Z., Christova Y., Todorov R., Alexandrov A., Stoichev P., Petkov R., *ACP Appl. Cardiopul. Pathophys.*, **1992**, *4*, 315–322.

72 Lalchev Z., Exerowa D., Koumanov K., Neycheva T., *Ann. Univ. Sofia Fac. Biol.*, **1983**, *77*, 35–44.

73 Todorov R., Lalchev Z., Vasilev D., Gerginova V., Gekova N., Dyankova D., Christova Y., *Acta Med. Bulg.*, **1995**, *1*, 10–19.

74 Lalchev Z., Properties and behavior of lipids and proteins in model membrane systems. DSc Thesis, Sofia University, **2004**.

75 Yamanaka T., *Membrane*, **1982**, *7*, 359–365.

76 Yamanaka T., Hayashi M., Matuura R., *J. Colloid Interface Sci.*, **1982**, *88*, 458–466.

77 Exerowa D., Lalchev Z., Marinov B., Ognianov K., Method for assessment of fetal lung maturity. Bulg. Authorship, Cert. No. 50291. Institute of Inventions and Rationalizations, Sofia, **1981**.

78 Exerowa D., Lalchev Z., Kashchiev D., *Colloids Surf.*, **1984**, *10*, 113–121.

79 Exerowa D., Lalchev Z., Marinov B., Ognyanov K., *Langmuir*, **1986**, *2*, 664–668.

80 Dluhy R., Reilly K., Hunt R., Mitchell M., Mautone A., Mendelsohn R., *Biophys. J.*, **1989**, *56*, 1173–1181.

81 Tchoreloff P., Gulik A., Denizot B., Proust J., Puisieux F., *Chem. Phys. Lipids*, **1991**, *59*, 151–165.

82 Pattle R., *Nature*, **1955**, *175*, 1120–1121.

83 Scarpelli E., Triangle, **1971**, *10*, 47–53.

84 Scarpelli E., *Pediatr. Res.*, **1978**, *12*, 1070–1076.

85 Schurch S., Bachofen H., Goerke J., Possmeyer F., *J. Appl. Physiol.*, **1989**, *67*, 2389–2396.

86 Scarpell E., Clutario B., Trave D., *Pediatr. Res.*, **1979**, *13*, 1285–1289.
87 Scarpelli E., *Surfactants and the Lining of the Lung*. Johns Hopkins University Press, Baltimore, **1988**.
88 Scarpelli E., *Anat. Rec.*, **1996**, *246*, 245–270.
89 Schurch S., *Biochim. Biophys. Acta*, **1998**, *1408*, 180–202.
90 Scarpelli E., *Colloids Surf. B*, **1997**, *8*, 133–145.
91 Scarpelli E., *Anat. Rec.*, **1998**, *251*, 491–498.
92 Scarpelli E. M., *Prenat. Neonat. Med.*, **2000**, *5*, 345–348.
93 Scarpelli E., Hills B., *J. Appl. Physiol.*, **2000**, *89*, 408–412.
94 Scarpelli E. M., *Prenat. Neonat. Med.*, **2001**, *6*, Suppl. 2, 15–20.
95 Perrin J., *Ann. Phys., New York*, **1918**, *10*, 160.
96 Bruil H., Specific ionic effects in free liquid films. PhD Thesis, Agricultural University, Wageningen, **1970**.
97 Larsson K., Lundquist M., Stallberg-Stenhagen S., Stenhagen E., *J. Colloid Interface Sci.*, **1969**, *19*, 268.
98 Cordova M., Mautone A., Scarpelli E., *Pediatr. Pulmon.*, **1996**, *21*, 373–382.
99 Antonova N., Todorov R., Exerowa D., *Biorheology*, **2003**, *405*, 531–543.
100 Exerowa D., *Adv. Colloid Interface Sci.*, **2002**, *96*, 75–100.
101 Exerowa D., *Prog. Colloid Polym. Sci.*, **2004**, *128*, 35–138.

Subject Index

Colloids and Interface Science Series, Vol. 1
Colloid Stability: The Role of Surface Forces, Part I. Edited by Tharwat F. Tadros

ISBN: 978-3-527-31462-1

d

e

Colloids and Interface Science Series

Colloid Stability
The Role of Surface Forces, Part I
Volume 1

2007
ISBN 13: 978-3-527-31462-1
ISBN 10: 3-527-31462-8

Colloid Stability
The Role of Surface Forces, Part II
Volume 2

2007
ISBN 13: 978-3-527-31503-1
ISBN 10: 3-527-31503-9

Colloid Stability and Applications in Pharmacy
Volume 3

2007
ISBN 13: 978-3-527-31463-8
ISBN 10: 3-527-31463-6

Colloids in Cosmetics and Personal Care
Volume 4

2007
ISBN 13: 978-3-527-31464-5
ISBN 10: 3-527-31464-4

Colloids in Agrochemicals
Volume 5

2007
ISBN 13: 978-3-527-31465-2
ISBN 10: 3-527-31465-2

Colloids in Paints
Volume 6

2008
ISBN 13: 978-3-527-31466-9
ISBN 10: 3-527-31466-0